Modern SECTIONAL ANATOMY

ALEXANDER LANE, Ph.D.
Coordinator of Anatomy and Physiology
Triton College, River Grove, Illinois
Visiting Associate Professor
University of Illinois at Chicago
Former Associate in Anatomy
Rush Medical College, Chicago, Illinois

HOSSEIN SHARFAEI, B.S.
Northwestern University Dental School
Chicago, Illinois

W.B. SAUNDERS COMPANY
Harcourt Brace Jovanovich, Inc.
Philadelphia ■ London ■ Toronto ■ Montreal ■ Sydney ■ Tokyo

W. B. SAUNDERS COMPANY
Harcourt Brace Jovanovich, Inc.

The Curtis Center
Independence Square West
Philadelphia, PA 19106

Library of Congress Cataloging-in-Publication Data

Lane, Alexander, Ph.D.
Modern sectional anatomy / Alexander Lane, Hossein Sharfaei.
p. cm.
Includes bibliographical references and index.
ISBN 0-7216-3217-3
1. Anatomy, Surgical and topographical. I. Sharfaei, Hossein.
II. Title.
[DNLM: 1. Anatomy, Regional. QS 4 L265m]
611′.9—dc20
QM531.L336 1992
DNLM/DLC 91-16738

Editor: Lisa Biello
Developmental Editor: Leslie E. Hoeltzel
Designer: Terri Siegel
Production Manager: Carolyn Naylor
Manuscript Editor: Wynette Kommer
Illustration Specialist: Brett MacNaughton
Page Layout Artist: Terri Siegel
Indexer: David Prout
Cover Design: Risa Clow

Modern Sectional Anatomy ISBN 0-7216-3217-3

Printed in the United States of America.

Last digit is the print number: 9 8 7 6 5 4 3 2 1

Preface

Anatomy as a science—a specific self-contained field of study—began in ancient Greece. Early Greek speculation about the nature of the structure of the body has been replaced today by modern empiric methods and research techniques. Current anatomic investigation is truly modern; however, questions about teaching modern anatomy remain. Has the teaching of anatomy kept pace with the science of anatomy? Can the teaching of anatomy be made modern? *Modern Sectional Anatomy* answers the first question negatively and the second one affirmatively.

During the long history of the study of anatomy, the anatomic scientist has established various approaches and methodologies, based on the state of knowledge at the time. The advent and the prevalence of today's diagnostic techniques in medicine warrant in-depth understanding of human anatomy from many different perspectives. No longer is anatomy presented in the systemic or regional approach sufficient to meet the challenge of the new expansion in the field of radiology. The advancement of diagnostic medicine rests on a different approach or view of the body.

Sectional anatomy, the new approach, has been deemed necessary because of the expansion of techniques for imaging the body, organs, and structures using a series of computer multiplane sections. This requires the radiologic technologist to master the identification of organs and structures as they appear in slices. Specifically, understanding relationships of anatomic structures in a three-dimensional array, from transverse, coronal, and sagittal segmental analyses, is essential to accurate interpretation by the radiologist of techniques such as ultrasound, magnetic resonance imaging (MRI), and computed tomography (CT) scans. The radiologic technologist works closely with the radiologist to produce relevant and clear images for the most accurate interpretation and diagnosis by the radiologist. The study of the three planes is needed because each has some distinct advantage. For example, the anterior/posterior comparison of each structure is visualized and a medial/lateral comparison is made in transverse sections. In other words, these dimensions can be viewed simultaneously. In coronal (frontal) sections, the comparative relationship of each structure is seen from superior/inferior as well as from medial/lateral perspectives. A sagittal section compares each structure from superior/inferior as well as from anterior/posterior dimensions. Thus, each plane presents different combinations of dimensional relationships.

Organs and tissues of the human body can be studied and observed in exquisite detail using computer techniques such as MRI, CT, and ultrasound. Modern anatomy, then, must include approaches for the understanding and interpretation of modern diagnostic techniques. The updated approach, the sectional approach, is not intended to replace all other established approaches but rather to make the teaching of anatomy relevant and modern. Some advantages of the sectional approach are (1) the observation of structures in situ; (2) the observation of structures in multiplane appearances, that is, each structure of each region is observed in transverse (axial), coronal (frontal), and sagittal planes; (3) the ability to correlate (pair) cadaver sections with computer imaging modalities of a living human body; (4) the provision of permanent reference material for mastery of the relationships of adjacent structures; and (5) more accurate interpretation and identification of structures from computer imagings.

The approach of *Modern Sectional Anatomy* is sectional, that is, multiplane cadaver sections and clinical images are presented for each body region. The anatomic structures of the transverse sections are viewed from the inferior surface. Modern computer imaging modalities view and display transverse images from the inferior surface. Therefore, both the cadaver transverse sections and the transverse images are viewed from the same surface. The coronal (frontal) sections are observed from the anterior surface. The sagittal sections are displayed from left lateral to right lateral and in some cases from right lateral to left lateral in a profile view. Some of the cadaver sections and clinical images are paired or correlated.

The cadaver specimens used in *Modern Sectional Anatomy* were derived from several cadavers. Some cadaver sections were prepared conventionally, whereas others are plastinated sections. (Plastination is a technique established and developed by Gunther von Hagen for the preparation of cadaver sections and other biologic specimens; it renders a high quality of permanent specimen for use in teaching and research.) The plastinated sections were obtained from the Carolina Biological Supply Company, Burlington, North Carolina. The contribution of Wilhelm Braune (1872) for some of the cadaver pictures is acknowledged with great respect for its historical significance. Coronal sections of the head and neck were adapted from *A Frontal Section: Anatomy of the Head and Neck,* written by Otto F. Kampmeier and colleagues and published by the University of Illinois Press, Urbana, Illinois, in 1957. To demonstrate small structures with more clarity, we magnified selected anatomic sections. Many of the transverse cadaver pictures were derived from sections of cadavers from the University of Illinois at Chicago and Triton College, River Grove, Illinois. Images used in *Modern Sectional Anatomy* were obtained from Westlake Community Hospital and Gottlieb Hospital, Melrose Park, Illinois; Silver Cross Hospital, Joliet, Illinois; and MRI Chicago, Chicago, Illinois.

Pathology is presented in *Modern Sectional Anatomy* from the clinical images, in some cases with the report of the radiologist. This allows the radiologic technologist and other imaging students and professionals to observe some disease processes demonstrated by various imaging modalities and enhances the ability to differentiate the normal structure from structural abnormalities. The radiologic technologist must be able to tailor the test images to aid the physician in determining the normal and abnormal structures.

The ultimate goal of *Modern Sectional Anatomy* is to significantly update and direct the radiologic technologist and other imaging personnel to recognize and identify structures in a multiplane vision. To accomplish this goal, a modern methodology has been incorporated into the writing of this book, including the classification of structures into four anatomic categories or units: musculoskeletal, neurovascular, visceral, and enclosing. A bibliography of current literature on the subject concludes the book. We hope that our book will enhance the learning of modern sectional anatomy for radiologic technology and imaging students and serve as a reference for practitioners. We wish to express our deep indebtedness to the W. B. Saunders Company for its help toward achieving this goal.

ALEXANDER LANE, PH.D.
HOSSEIN SHARFAEI, B.S.

Acknowledgments

Our appreciation and thanks go to so many it is difficult to name all of them. It occurs to us that the assistance rendered by these fine people cannot be repaid.

Some of the many whose assistance in the preparation of this text has been invaluable are listed below:

Staff Service Center, Triton College, Lorraine Wilson-Grace, Supervisor

Department of Biological Sciences–University of Illinois at Chicago, Sidney Simpson, Ph.D., Head, and Howard Buhse, Ph.D., Associate Head

Ulrich Danckers, M.D., Radiologist, Westlake Community Hospital, Melrose Park, Illinois

George Hallenbeck, M.D., Radiologist, Gottlieb Hospital, Melrose Park, Illinois

Joseph V. Stewart, M.D., for helpful suggestions

Jerry Banks, for preparation of cadaver sections and slides

Judith Solomon (Medical Student), Research Assistant

Jordy Sacksner (Medical Student), Research Assistant

Marzieh Sharfaei, for her patience, excellently prepared meals, and encouragement to the authors

Hossein Yousefnia, Ph.D., for his technical assistance

Zahra Yousefnia, for her kind assistance

W.B. Saunders Company, Lisa Biello, Editor-in-Chief, Health Related Professions, and Leslie E. Hoeltzel, Manager, Developmental Editorial

Finally, to each other—appreciation and gratitude for an excellent workable partnership.

ALEXANDER LANE
HOSSEIN SHARFAEI

A special note of recognition goes to Carolyn M. Lane-Wenner and Sondra R. Lane, my daughters; Mark D. Wenner, my son-in-law; and Alexander Lane Wenner, my grandson, for their encouragement and, most of all, their love.

ALEXANDER LANE

A special note of recognition goes to my parents, Sharfa and Rabeh Sharfaei, for their support, encouragement, understanding, patience, and, most of all, their love. A special note of gratitude and thanks go to all my sisters.

HOSSEIN SHARFAEI

Contents

PART VIII
The Lower Limb

PART I

ORIENTATION TO SECTIONAL ANATOMY

INTRODUCTION

Recent advances in the computer sectional scanning of body regions make the study of sectional anatomy mandatory. The interpretation of these body images requires a thorough knowledge of sectional human anatomy. The radiologic technologist's primary responsibility is to generate images of the human body needed by the imaging physician to make diagnoses. Because diagnosis and interpretation increasingly depend upon the study of multiplane body images in exquisite detail, layer by layer, the most logical approach to meet the educational need in anatomy of the radiologic technologist is sectional anatomy.

An image of the targeted body area is acquired by various imaging devices, such as x-ray and fluoroscope machines, and now more recently from computer body scanning system devices. The radiologic technologist's responsibility is to produce images using these machines. The technologist not only must recognize anatomic differences in images from x-ray and fluoroscope machines but also must appreciate clinical sectional anatomy as demonstrated in magnetic resonance imaging (MRI), computed tomography (CT), sonography (ultrasound), and positron emission tomography (PET).

The diagnosis of disorders and injuries is the primary concern of the imaging physician. An image of the body to be examined by the physician is performed by the radiologic technologist. This requires the technologist to be continually up-to-date in the field of anatomy in order to generate images with the clarity needed to aid the physician's diagnosis. There has been a significant expansion in the field of diagnostic medicine. For example, x-ray and fluoroscope machines were, until recently, the only instruments available in the field of diagnostic radiographic anatomy. Now several tomographic tools that provide three-dimensional images are in use, and these tools require a different perspective in the study of human anatomy. Sectional anatomy prepares the radiologic technologist for this expansion of the diagnostic medical field. The radiologic technologist and the radiologist (imaging physician) share a new vision in medicine.

The technologic advances in body imaging started in the 1970s. Although roentgenograms are still very useful in the process of diagnosis, they do not reveal the three-dimensional relationships of the body organs and structures. Another weakness of roentgenograms is their failure to demonstrate small differences in the densities of tissues. The advanced technology now in use by radiologic diagnosticians reveals body areas in multiplane sections and differentiates densities of tissues more thoroughly.

The study of sectional anatomy reveals three-dimensional relationships through multiplane slices of body organs. Because the computer body scanning systems are becoming the tools of choice for the technologist and radiologist, sectional anatomy must be included in the educational requirement of the radiologic technologist.

The three classic sectional planes have been included in *Modern Sectional Anatomy.* Therefore, typically each part of the text consists of a chapter on cross-sections, a chapter on coronal (frontal) sections, and a chapter on sagittal sections. Analyses of the sections are based on a written description and a series of labeled pictures isolated into four anatomic units.

To facilitate and enhance the methodology for the study of sectional anatomy (Lane, 1990), objectives have been formulated as guides to mastery. Students should be able to master these objectives, which apply to all cadaver sections and clinical images.

1. *Identify the section's origin.* Orientation lines (grids) are essential in both cadaver and *in vivo* sections regardless of plane. Another important part in identifying the origin of a section is to rely on the distance from Reid's baseline (infraorbital meatal line) in the head region for cross-section, the median for sagittal sections, and the midaxillary line for coronal sections. This guideline also can be duplicated in each region except for a different landmark for axial sections.
2. *Compare surface orientation of each plane.* The planes of cadaver sections must be correlated with the computed body scanning systems. For example, the inferior surface is emphasized in *in vivo* cross-sections. Emphasis is placed on the anterior surface in coronal sections. In sagittal sections the right surface seems to be preferred, but in some cases the left surface is also correlated with computed imaging scans. Thus, the emphasis in cadaver sectional surface should be the same as for computed body scans.
3. List and describe briefly the four units used in the analysis of sectional anatomy.
   a. *Musculoskeletal unit*—a group of bony features and muscular structures located in a section
   b. *Neurovascular unit*—a group of nerve structures and vessels, as well as dural sinuses, located in a section
   c. *Visceral unit*—a group of internal organs located in a section
   d. *Enclosing unit*—a group of connective tissue membranes and epithelial membranes, as well as spaces within the body cavities
4. Note the relationship of adjacent structures in each plane. Structures are discussed relative to their relationship with other nearby features, using anatomic directional terms.
5. Describe the structures as seen in different planes. The variation in the appearance of structures when viewed in different planes should be recognized.
6. Recognize the varying characteristics of each structure within the section. In structures of the musculoskeletal unit, the histologic class of bone (cancellous or compact) and the origin and insertion of muscles should be specified. In the neurovascular unit, the origin, distribution, and course of arteries should be noted. Veins and lymphatic vessels should be described in a similar manner, but their termination should be noted rather than their distribution, and course of each nerve seen in the section should be noted. For structures included in the visceral unit, size (as seen in the section), course, and major functional role should be discussed. However, for structures of the enclosing unit, it is adequate to discuss their description, location, and type of tissue.

The following objectives are applicable to all the *musculoskeletal* units, regardless of plane:

7. Identify and briefly describe each section of each structure in the skeletal part of the musculoskeletal unit.
8. Locate each skeletal feature of the musculoskeletal unit.
9. Identify and state the origin, insertion, innervation, vasculature, and action for each muscle in the muscular part of the musculoskeletal unit (see muscle tables presented in Parts II through VIII).
10. Locate each muscle of the muscular portion in the musculoskeletal unit.

The following objectives are applicable to all the *neurovascular* units, regardless of plane:

11. Identify and state origin/distribution and course of each artery in the neurovascular unit of each section.
12. Locate each artery identified in the neurovascular unit.
13. Identify and state origin/termination and course of each vein in the neurovascular unit, as well as dural sinuses.
14. Locate each vein identified in the neurovascular unit.
15. Identify and state origin/termination and course of lymphatic vessels in the neurovascular unit.
16. Locate each lymph vessel or node identified in the neurovascular unit.
17. Identify and state origin/distribution and course of nerves in the neurovascular unit.
18. Locate each nerve or nerve structure identified in the neurovascular unit.

The following objectives are applicable to all the *visceral* units, regardless of plane:

19. Identify each visceral structure of the visceral unit.
20. Describe the slice of each viscus.

21. Follow the course of each viscus in appropriate sections, including its relationships to its adjacent structures.

The following objectives are applicable to all the *enclosing* units, regardless of plane:

22. Identify connective tissue and membranes seen in each section of the enclosing unit.
23. Locate each enclosing unit structure.
24. Apply the relevant preceding objectives (1 to 23) to each anatomic unit, including the clinical images.
    a. Transverse (axial) sections of the head, including clinical images
        i. 8 cm above Reid's baseline
        ii. 6 cm above Reid's baseline
        iii. 4 cm above Reid's baseline (plastinated)
        iv. 1 cm above Reid's baseline (plastinated)
        v. Reid's baseline
        vi. 1 cm below Reid's baseline
    b. Coronal (frontal) sections of the head, including clinical images
        i. Section that passes through the crista galli of the ethmoid and the symphysis of the mandible
        ii. Section that passes through the posterior ethmoid air cells, posterior extent of frontal sinuses, maxillary sinuses, posterior portion of hard palate, and upper right second molar tooth
        iii. Section that passes through the anterior horns of the lateral ventricles of the brain, pituitary gland, body of the sphenoid, opening of the auditory (eustachian) tubes into the nasopharynx, and body of the hyoid bone
        iv. Section that passes through the posterior horn of the lateral ventricles, posterior margin of the foramen magnum, and tentorium cerebelli
    c. Sagittal section of the head, including clinical images
        i. Midsagittal cadaver section of the head, neck, and upper thorax, showing the frontal sinus, falx cerebri, cerebellum, corpus callosum, and pineal gland. (The head structures are described and discussed in this module.)
    d. Transverse (axial) sections of the neck, including clinical images
        i. Upper neck and lower face at cervical vertebra 2
        ii. Cervical vertebra 3 (plastinated)
        iii. Cervical vertebra 4
        iv. Cervical vertebra 5
        v. Cervical vertebra 6
25. Apply relevant objectives 1 to 23 to the following frontal/coronal section of the neck:
    a. Section that passes through the posterior atlanto-occipital membrane, spine of the cervical vertebrae, and the lower part of the cervical spinal cord
26. Apply relevant objectives 1 to 23 to the following sagittal sections:
    a. Head region—midsagittal
    b. Neck region—midsagittal (at median)
27. Apply relevant objectives 1 to 23 to the thoracic cross-sections and sagittal and thoracic frontal/coronal sections.
    a. Thoracic cross-sections:
        i. Thoracic vertebral level 1
        ii. Thoracic vertebral level 2
        iii. Thoracic vertebral level 4
        iv. Thoracic vertebral level 5
        v. Thoracic vertebral level 6
        vi. Thoracic vertebral level 7
        vii. Thoracic vertebral level 8
        viii. Thoracic vertebral level 9
        ix. Thoracic vertebral level 10
        x. Thoracic vertebral level 12
    b. Sagittal sections:
        i. Midclavicular line (right lateral line)
        ii. Median (midsagittal)
        iii. Midclavicular line (left lateral line)
    c. Thoracic frontal/coronal sections:
        i. 1/2 distal from midaxillary line to the anterior surface of the body
        ii. At midaxillary line
        iii. 1/2 distal from the midaxillary line to the posterior surface of the body
28. Apply relevant objectives 1 to 23 of abdominal cross-sections and abdominal sagittal and frontal/coronal sections.
    a. Abdominal cross-sections:
        i. Lumbar vertebral level 2
        ii. Lumbar vertebral level 4
    b. Abdominal sagittal sections:
        i. Midclavicular line (right lateral line)
        ii. Median (midsagittal)
        iii. Midclavicular line (left lateral line)
    c. Abdominal frontal/coronal sections:
        i. 1/2 distal from the midaxillary line to the anterior surface of the body
        ii. At midaxillary line
        iii. 1/2 distal from the midaxillary line to the posterior surface of the body
29. Apply relevant objectives 1 to 23 to the following cross-sections of the pelvis and pelvic sagittal and frontal/coronal sections.
    a. Pelvic cross-sections:
        i. Sacral vertebral level 1
        ii. Sacral vertebral level 2
    b. Pelvic sagittal sections:
        i. Midclavicular line (right lateral line)
        ii. Median (midsagittal)
        iii. Midclavicular line (left lateral line)
    c. Pelvic frontal/coronal sections:
        i. 1/2 distal from the midaxillary line to the anterior surface of the body
        ii. At midaxillary line
        iii. 1/2 distal from the midaxillary line to the

posterior surface of the body

30. Apply relevant objectives 1 to 23 to the following upper limb sagittal sections:
    a. Upper limb cross-sections:
        i. Brachial region (arm) midshaft of humerus
        ii. Antebrachial (forearm) region
        iii. Midshaft of radius and ulna
    b. Upper limb sagittal sections:
        i. Median (midsagittal)
31. Apply relevant objectives 1 to 23 to the following lower limb cross-sections and lower limb sagittal sections:
    a. Lower limb cross-sections:
        i. Thigh region (midshaft) of femur
        ii. Leg region (midshaft) of tibia and fibula

CHAPTER 1

Agenda for the Study of Sectional Anatomy

Modern Sectional Anatomy presents human sectional anatomy in a multiplane format; that is, sections of each region of the body are depicted in the three classic planes: transverse (axial), coronal (frontal), and sagittal. Most of the sections represent pictures from conventionally prepared cadaver sections along with some drawings, as well as sheet plastinated specimens. The process known as plastination was developed by Gunther von Hagen in 1979 (University of Heidelberg, Heidelberg, Germany). Plastinated sections of the human body are effective teaching materials because they are resistant to mechanical damage, they have a dry surface with minimal health hazard, storage and maintenance are simple, the structures are natural-appearing and vivid, and there is little or no smell (Lischka, 1987). Another major factor that makes the pictures of plastinated specimens in this book relative to images is that the sections are 5 mm in thickness, as are many of the magnetic resonance imaging (MRI) scans.

What should the methodology and approach for the study of each section of the body part include? It is suggested that *Modern Sectional Anatomy* be used in the following manner:

1. Identification of the section based on location is one of the first steps in organizing the approach. *Modern Sectional Anatomy* provides a grid for each section as well as a grid for each clinical image regardless of its plane. The grid should be consulted as the first step in determining the level and the area of the region from which the section is derived. Another important practice in identifying the level of the section is referred to in the statement at the beginning of the information for each anatomic section.

2. Each section is divided into four anatomic classes: musculoskeletal, neurovascular, visceral, and enclosing units. Once the identification by location has been established, the technologist should focus on each anatomic category separately.

Development of sectional anatomy based on four anatomic units provides a strategy and logical sequence to deal with any multiplane cadaver section as well as with any multiplane clinical image. Cadaver sections are needed to establish the normal location and size, dimensional morphology, and dimensional orientation of each structure within the section. Without study of the cadaver sections in a different array, the result of clinical images could be questioned. No one would know exactly what structures are to be seen in paired (correlated) clinical images. In other words, each structure varies in appearance, location, dimensional morphology, and dimensional orientation, depending upon the level or area of the body from which it is derived. Further, the *correlation* between cadaver sections and clinical images is enhanced by the four-anatomic-unit concept. For example, some imaging modalities depict one anatomic class in a far superior way to another imaging device. For instance, consider one region of the body, such as the head—experience teaches us that MRI depicts the brain (viscera) far better than does a CT scan. However, CT is better for the cranial vascular structures. Ultrasound is of no use for skull imaging except in the neonate or young child before the closure of the fontanelles.

The first anatomic category to be considered for the analysis of each image section is the **musculoskeletal unit.** Within this unit the bony framework and musculature establish the posture and foundation of the section. Of these two major structures within this unit, the bones are first identified and described and their location within the section determined, since dimensional orientation is included in each structure. The same is done for the musculature, except that in addition the origins, insertions, fiber direction, innervation, vasculature, and action of each skeletal muscle are to be mastered.

The study of the **neurovascular unit** follows the musculoskeletal unit in *Modern Sectional Anatomy.* There are four major structural categories in the neurovascular unit: arteries, veins, lymph vessels, and nerves. Therefore, when one is focusing on the neurovascular unit, it is recommended that the arteries, veins, lymphatic vessels, and nerves be identified and described in that sequence. The origin, termination, and distribution of these neurovascular structures are to be mastered. In addition, the course or route the structure traverses from origin to termination is presented for more thorough identification.

After study of the neurovascular unit, the **visceral unit** is presented. The internal organs are described based on characteristics, their location within the section that reinforces the course of the viscera, the level from which they start within the body, and their termination.

Finally, the **enclosing unit,** which includes membranes, depressions, fossae, spaces, body cavities, clefts, fasciae, and the visceral fats, is thoroughly examined. Each enclosing unit structure is described based on location, dimensional morphology, dimensional orientation, and relationship to adjacent structures.

3. In regard to the overall approach, *Modern Sectional Anatomy* has been organized into eight parts with 23 chapters as subheadings. Each part, except this first part, is divided into three chapters. The first chapter deals with the transverse (axial) plane of the cadaver sections or cadaver section drawings and clinical images. The second chapter in each part is concerned with the coronal (frontal) planes, and the third chapter of each part contains the sagittal cadaver sections, including similar clinical images. Each chapter is divided into modules based upon the level or the area from which the section has come. Some disease conditions are presented in several chapters and are demonstrated by various imaging modalities. The radiologic report is included in each pathologic finding.

The basic format for the presentation of skeletal musculature is a table, which is included in the information pertaining to each part. For example, the muscles of the head are listed in a tabular format in Part II. This allows quick reference for any level or area from which the section is derived and avoids redundancy with regard to one major aspect of the musculoskeletal unit. These tables include the name, origin, insertion, innervation, vasculature, and action of the muscles of each region.

Modern Sectional Anatomy includes one other major component: instructional objectives used to facilitate the learning process. The objectives are specific instructional guides for each section, to be followed when studying any section or clinical image.

The purposes of *Modern Sectional Anatomy* are to establish anatomic orientation and sectional anatomy information and to show how this knowledge correlates with clinical images. At least five computerized body scanning systems depend on the orientation of multiplane sectional anatomy. According to VanDevort, the information necessary in cross-sectional anatomy for radiography students should be at a very basic level (VanDevort, 1988), and Schneider suggests a strong base in cross-sectional anatomy for diagnostic ultrasound students prior to ultrasound training (Schneider, 1988). Others maintain the need for a comprehensive course in sectional anatomy. This book addresses these issues and several other concerns, such as the appropriateness of the text for the college reading level, the correlation between text and pictures, and the concise and programmed material.

Modern Sectional Anatomy is valuable for the entire imaging profession and particularly useful for the radiologic diagnostic technologist who will generate the various images directly from the patients. Essential in the modern approach to imaging is the technologists' knowledge of dimensional morphology, location, dimensional body orientation, and dimensional orientation of structures that are imaged, which will be gained from this book. Multiplane images are the most common type technologists generate with present-day imaging devices to produce a facsimile of structures for medicine's new frontiers.

References

Lane, Alexander. Teaching sectional anatomy with sheet plastinated sections. *Journal of the International Society of Plastination* 3(1): 38, 1989.

Lane, Alexander. Sectional anatomy: Standardized methodology. Paper from Fourth Annual Conference on Teaching and Technology: The Role of Computers in the Educational Process at University of Illinois at Chicago and City Colleges of Chicago Partnership Program. Chicago, May 4, 1990.

Lane, Alexander. Sectional anatomy: Standardized methodology. In: *Curriculum Resources for Teaching Sectional Anatomy Report* #2. The American Sectional Anatomy Consortium, 1990.

Lane, Alexander. Sectional anatomy: Standardized methodology. *Journal of the International Society of Plastination* 4: 16–22, 1990.

Lane, Alexander, and Hossein Sharfaei. *Modern Anatomy: The Sectional Approach.* River Grove, IL: Triton College Press and Triton College Staff Service Center, 1988.

Lischka, Martin. Plastination of whole-body slices with polymerizing emulsion. *Journal of the International Society for Plastination* 1 (1): 17–22, 1987.

Schneider, Jim. Identify Sectional Anatomy Needs of Ultrasonographers. Report of the Conference on Teaching Sectional Anatomy, Report No. 1, July 1988, pp. 17–20.

Stuart, C. Sectional Anatomy: Modern look at "flat world." *At Chicago* 8(20), March 7, 1990.

VanDevort, Lyn. Identify Sectional Anatomy Needs of Computed Tomography Technologists. Report of the Conference on Teaching Sectional Anatomy, Report No. 1, July 1988, pp. 12–13.

CHAPTER 2

Computers in Modern Clinical Research and Diagnostic Dentistry and Medicine

Some of the "brave new worlds" of science fiction are becoming reality. Star Trek's Dr. McCoy diagnosed his patients with hand-held, whole body scanners, which produced computed diagnoses identical to those of magnetic resonance imaging and computed tomography (Ragan, 1988). Systemic and regional approaches to the study of anatomy were developed by the ancient Greeks and were based primarily on human cadaver dissections. Recently, however, a new vision of the human body in a living sectional imaging state has become possible, based in part on the use of computers in modern medical research as well as diagnostic medicine.

None of the four imaging techniques described in this chapter requires surgery. This criterion alone thrusts computer imaging techniques into a brave new world. One example is the application of *magnetic resonance imaging* (MRI) to clinical medicine, which has expanded rapidly. "In 1981, only two magnetic imaging devices were in use in the United States. In 1987, 600 were being used there and another 100 in Europe" (Abelson, 1988). "More than a thousand magnetic resonance scanners are in use today worldwide, and more than 3000 are expected to be in use by 1991. Yet MRI is still in its infancy" (MacFall, 1988).

Another modality, *ultrasound,* a non-ionizing device, is becoming widely used in medicine, both therapeutically and diagnostically. Today, many infants in the United States have been subjected to ultrasound examination in utero (Nyborg, 1983).

> A schoolteacher in her 28th week of pregnancy came for an ultrasound scan. It was noticed that she was much larger than she should have been, with excessive fluid around the body. The scan revealed that the baby's mouth was open and its tongue was stretched far forward. . . . On further investigation by ultrasound in utero [the physician] found a large tumor growing in the neck under the jaw (Sochurek, 1987).

With this forewarning, throat surgeons were able to open the breathing passageway blocked by the tumor immediately after the infant's birth.

Another sectional scanning technique, *computed tomography* (CT), reveals anatomic and pathologic details in imaging the living human body. This diagnostic technique combines the use of a computer and x-rays in scanning the body. X-rays are passed through the body at different angles to produce cross-sectional images ("slices") of the tissue being examined. Clearer and more detailed information is obtained than with x-rays alone. (CT was formerly called computed axial tomography [CAT]).

CT was first attempted in 1914 by Karol Mayer, a Polish radiologist. The first CT scanner was put into operation in 1972; it was developed to study the brain. Since then, CT brain scanning has marked a major advance in the diagnosis and treatment of tumors, abscesses, and hemorrhages in the brain, as well as strokes and head injuries. Formerly these diseases required tests, such as angiography, pneumoencephalography, and ventriculography (an outmoded technique for imaging the ventricles of the brain), that not only were difficult to perform, lengthy, and not always clear-cut in their findings but also entailed risk to the patient. CT scanning, on the other hand, is simple, quick (about 1 hour), and efficient and carries a very modest exposure to radiation (Greenberg, 1983).

In addition to being essential for study of the brain, CT scanning is invaluable in investigating disease of any part of the trunk. It is a useful device in guiding the operator who is performing a needle biopsy. With a body CT scan, it is possible to detect and evaluate cancers and other diseases and abnormalities of the lung, liver, pancreas, kidney, spine, and abdomen (Greenberg, 1983). CT only recently has become the dominant imaging modality for heart, abdominal viscera, and associated structures (Morgan, 1983).

How it works: the computer-generated image varies with the density of the tissue. The density of the tissue causes variations in the CT number. The CT number indicates the attenuation coefficient of tissues—that is, some tissues are more easily penetrated by x-ray than are other tissues. CT numbers are higher for those tissues with higher densities (Phelps,

1973); for example, CT numbers are lower for fat and connective tissue than for parenchyma. In several fat-infiltrated tissues, such as liver (Mulhern, 1979; Stanley, 1976), adrenal tumors (Schaner, 1978), and atrophic muscle (Termote, 1980), the CT number is decreased compared with normal tissue of the same organ. A decrease in the CT number of the submandibular gland is well correlated with the adipose deposition occurring with age (Ida, 1989).

According to Norman (1977), "Computed tomography differs from other scanning techniques and from conventional radiology in that it does not primarily provide a sectional picture but provides numerical values of tissue density using an algorithm."

Positron emission tomography (PET) not only images an organ's structure but also reveals sectional images of the organ at work. For instance, PET allows one to visualize the brain in two-dimensional tomographic slices similar to those of a CT scan. Unlike the CT scan, which computes the transmission of externally generated x-rays, the positron emission tomograph computes the emission of positron-emitting isotopes injected into the patient's body. Such isotopes decay by ejecting a positron, which in turn collides with an electron, producing two gamma rays that leave the body at an angle of 180 degrees to each other. The detector sees the 180-degree simultaneous gamma rays and identifies the place where the positron was emitted. Because there are positron-emitting isotopes of carbon, oxygen, and nitrogen, one can make positron-emitting compounds of biologic significance and inject them into patients. Following the decay of these isotopes in specific areas of the brain allows one to calculate the metabolism of that area of brain. Thus, while a CT scan delineates the static anatomy of the area being examined, the positron emission tomograph holds out the hope of describing the dynamic physiology of the area.

Positron emission tomography has already proved clinically useful in identifying epileptic foci during interictal states and in measuring the metabolism of ischemic tissue. However, its full clinical role is still undefined. Because of the travel of the positron in tissue, resolution is likely to be no better than 1 cu cm, as opposed to 1 cu mm with CT scanning (McCullough, 1980). In other words, more detail is seen with the use of CT and other sectional computer imaging devices.

PET is most significant and useful in the measurement of specific physiologic processes, such as glucose and oxygen metabolism, blood flow and volume, amino acid metabolism and incorporation rates into protein, tissue acid-base balance, and membrane transport.

PET technology is used by some 40 centers in the United States for both research and diagnostic purposes (Montgomery, 1989). For example, tracing the activity of the human brain with a radioactive label appears to open a window into the mind. PET actually can look into one's brain and determine "whether one is smart." Begley (1988) reported the following study:

> First the subjects read or heard common nouns; in the next round they pronounced the words; finally they gave a use for each noun. In each round, the scientists used PET scans to see which new areas of the brain had kicked in to handle the new task. The results undercut the long-held theory that, when reading, the brain first speaks words silently before attaching meaning to them. This indicates that the brain can have parallel processing, that is, the brain can read a word and simultaneously imbue it with meaning.

This is true for highly talented readers, as determined by PET scanning. However, poor readers do need all three stages of reading, pronouncing, and giving a use for each noun. In poor readers, the brain appears to process one stage of this task independently instead of simultaneously, as occurs in "smart" readers.

SYSTEMIC AND REGIONAL APPROACHES

These new imaging modalities have required a fresh look and different approach to the study of the anatomy of the human body. No longer can anatomic scientists rely solely on the systemic and regional approaches. The *systemic approach* is usually followed when dissection is not an integral part of the course (Lane, 1990). In this approach, 11 organ systems are studied as separate units: (1) the skeleton, (2) the muscular system, (3) the nervous system, (4) the circulatory system, (5) the respiratory system, (6) the endocrine system, (7) the urinary system, (8) the reproductive system, (9) the digestive system, (10) the lymphatic system, and (11) the integumentary system.

The *regional approach* is followed when dissection *is* an integral part of the course. The regional approach emphasizes the description and identification of anatomic structures by (1) location, (2) dimensional morphology, (3) surface projection, (4) relationship of adjacent structures, and (5) major functional role (Lane, 1990). The location or position of a structure within the body is important in order to find the organ for treatment, including surgery. Dimensional morphology is a term used to describe the size, shape, and depth of an organ. Surface projection depicts the area of the outer body that covers or protects the organ. The relationship of adjacent structures enables the clinician to decide the effect of each adjacent organ on the other. It is well established that an organ has a profound effect on its neighbors. The major functional role of an organ is included because of the need to know what functions or activities would be missed without the organ.

Another aspect of the regional approach is that each organ has four structural characteristics: (1) Vasculature, which includes afferent and efferent blood vessels as well as lymph vessels. Afferent blood vessels are known as arteries and conduct blood to the organs of the body, whereas efferent blood vessels are known as veins and conduct blood from the organs. The lymph vessels remove the excess fluids not removed by blood vessels from organs. (2) Innervation, the nerve supply to the organ. Each organ is controlled by a nerve. (3) Parenchyma, the collection of functional elements of the organ. For example, there are 800,000 islets of Langerhans in the pancreas to secrete insulin, which will prevent diabetes mellitus. (4) Stroma, the embedding tissue of an organ, which supplies the foundation of all the other structures in the organ.

THE MODERN SECTIONAL APPROACH

The advent and prevalence of today's diagnostic techniques in modern dentistry and medicine warrant in-depth understanding of human anatomy from many different perspectives. Specifically, understanding relationships of anatomic structures in a three-dimensional array, from transverse, coronal, and sagittal segmental analyses, is essential for accurate interpretation of techniques such as ultrasound, magnetic resonance imaging, computed tomography, and positron emission tomography scans. Organs and tissues of the human body can be studied and observed in vivid detail with these computer techniques.

Sectional anatomy supports the study and interpretation of the imaging modalities. This updated approach is not intended to replace systemic and regional approaches but to make the teaching of anatomy relevant and modern. The advantages of the sectional approach are:

1. Observation of structures in situ as well as in a three-dimensional pattern (Fig. 2–1). The cross-sectional image shown in Figure 2–1*A* clearly shows the width and length of the cerebrum. The coronal (frontal) section (Fig. 2–1*B*) reveals both the height and width. Therefore, the cerebrum is viewed in three dimensions. In Figure 2–1*C*, the sagittal image shows the front-to-back length of the cerebrum as well as the height.
2. The ability to correlate (pair) cadaver sections with computer imaging modalities of a living human body.
3. Permanent reference material for mastery of relationships of adjacent structures.
4. More accurate interpretation and identification of structures from computer imagings (Lane, 1988).

MAGNETIC RESONANCE IMAGING

"The magnetic resonance imaging (MRI) phenomenon rests on the fact that nuclei with an odd mass number or an odd charge have a non-zero spin . . ." (MacFall, 1988). Nuclei having a non-zero spin interact with external magnetic fields to form different quantized energy states. Transitions between these energy states cause emission of electromagnetic radiation, which in turn can be detected and analyzed. Hydrogen nuclei have non-zero spin, and hydrogen is the most abundant element in the body, so magnetic resonance is particularly applicable to the human body. The local environment at each site in the body has a particular chemistry, and thus a particular density of hydrogen. The intensity of the pattern of imaging from the magnetic resonance thus will depend on the local density of hydrogen and therefore can be used to determine variations in tissues and the condition of the tissue being examined.

Normally, when the magnetic field is removed the hydrogen atoms will return to a random state by spinning into the original state. The quality or pattern of spin determines the condition of the tissue being examined. The computer translates the faint spinning signals of the hydrogen atoms into an image of the area scanned. Since hydrogen reflects water content, operators can use the image to make distinctions between tissues (Sochurek, 1987). The greater the water content of an organ, the clearer the MRI image. Bones are not clearly imaged by MRI because of the relatively small quantity of water they contain.

Clinical Applications. The application of MRI in dentistry and medicine requires a thorough knowledge of various organs and regions of the human body. This is especially true when surgical therapy of the organ is needed to rule out pathology. The first important aspect of the MRI survey is to be clear about the normal structure, including variations that are essentially normal. For example, if one is to do a survey of the forearm and hand, the examiner should know that 11 per cent of the human population has no long palmar (palmaris longus) muscle (Brash, 1951). MRI is most helpful for imaging soft tissue, such as skeletal musculature (Abelson, 1988), to rule out conditions such as myoma, myonecrosis, myoceles, and muscle atrophy.

The skeletal joints can be extensively studied and surveyed using MRI. Joints are of four classes: (1) fibrous joints, (2) primary cartilaginous joints, (3) secondary cartilaginous joints, and (4) synovial joints. Synovial joints are the most movable and most vulnerable to arthritis and injury. MRI can show the joints in exquisite detail and rule out pathology. This is especially true of the temporomandibular joint (TMJ). Gabler and colleagues used MRI to assess the disk-condyle relationship of the TMJ before and after arthroscopic surgery. In both open and closed positions of MRI images of the TMJ, the location of the disk was unchanged in 78 per cent of the group who had arthroscopic surgery. Most of the remaining postoperative patients showed greater translocation of the disk (Gabler, 1989). Thus, MRI has been successful in assessing pre- and postoperative arthroscopic studies of the TMJ. However, MRI has its limitation. For example, "teeth and bones, which contain little water, do not appear at all in MRI, enabling doctors to see tissue surrounded by bone." (Sochurek, 1987). The spinal cord in MRI is seen clearly because of the vertebrae surrounding the spinal cord.

MRI can help rule out pathology of the pelvis.

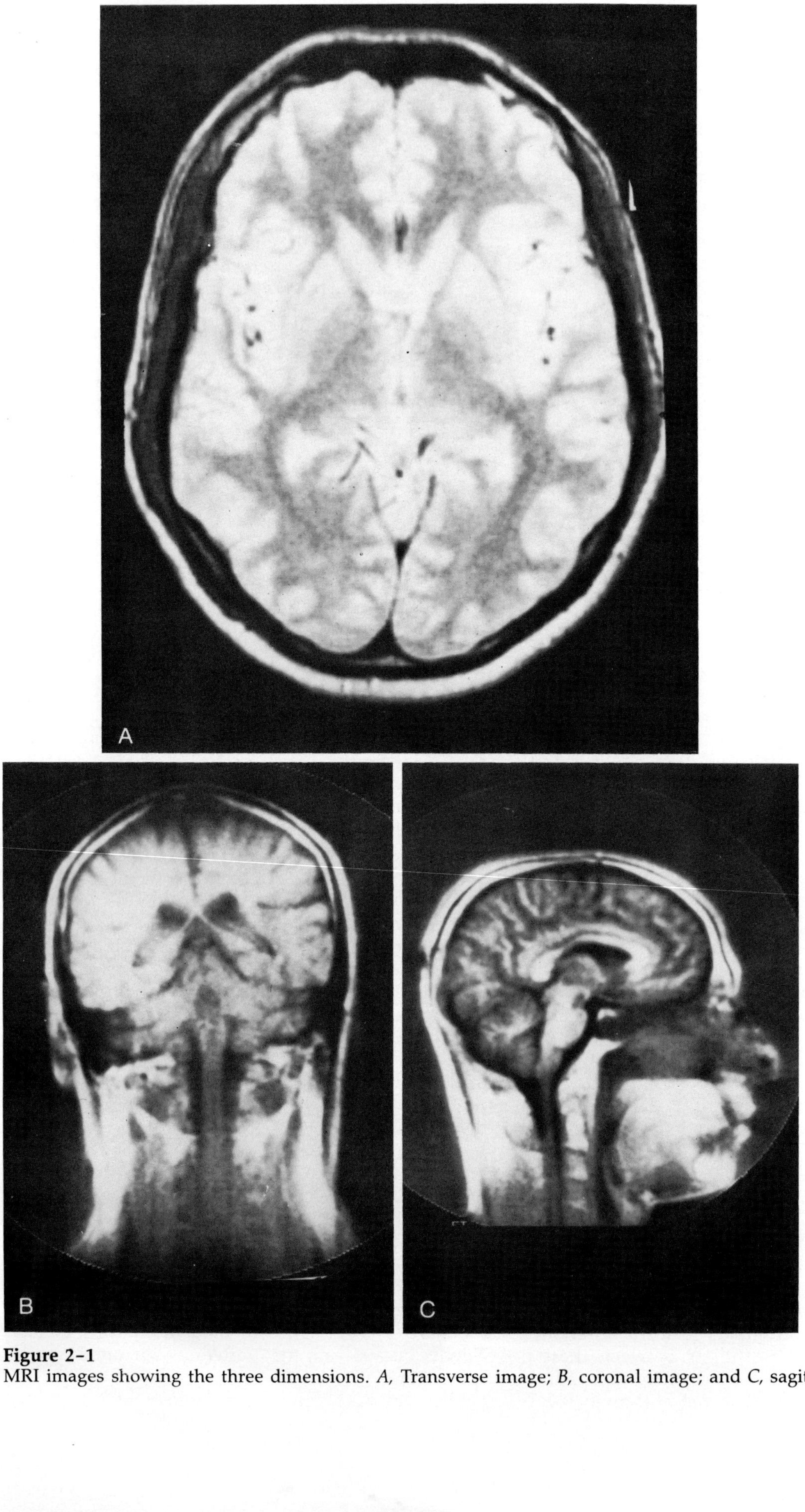

Figure 2–1
MRI images showing the three dimensions. *A,* Transverse image; *B,* coronal image; and *C,* sagittal image.

However, it must be noted that 40 per cent of the population have no psoas minor muscle (Brash, 1951). MRI can rule out prostate cancer in men or detect cancer of the prostate at a very early stage. In fact, MRI has been known to detect mere changes in the chemical makeup of tumor tissue. To delineate the chemical change that occurs before a tumor even starts to grow puts MRI diagnostics truly into the "brave new world." MRI is equally as helpful in the pelvic region of women. Uterine tumors and other pathology can be ruled out or detected at a very early stage. MRI surveys of the pelvic cavity in both women and men require no invasive procedures.

A survey of the brain by MRI to rule out pathology is highly recommended because of its ability to show exquisite detail as well as three-dimensional sectional images. Determination of the existence and location of tumors in the brain often can best be done by MRI. Further, one can follow the evolution of hematomas, such as subdural hematomas of the brain, over time (Abelson, 1988). In surveying the brain and head, caution should be observed because some structures in the brain vary with age. For example, 50 per cent of people over age 50 years have an ossified pineal gland (Brash, 1951). In one MRI study of the head, a very small percentage of the adult population showed a frontal suture. Thus, it is important to know what the common structures are as well as the variations in the human population.

It has been pointed out that MRI sectional images can detect chemical changes within soft tissue. MRI also can make observations on the distribution of phosphate compounds in normal and diseased states; an increase in phosphate within tissue indicates a tumor.

COMPUTED TOMOGRAPHY

The use of computers in medical imaging is almost synonymous with CT. "The fundamental concept underlying the technique of computed tomography is the capability of reconstructing or synthesizing a cross-section of the internal structure of an object from multiple projections . . . passing through the object" (Morgan, 1983). CT actually unites x-rays and computers and produces images by spatially mapping the local x-ray absorption within a slice (or section) through an object.

> One common utilization of the control minicomputer for clinical purposes is three-dimensional reconstruction. CT scanning has advanced to the point where it is possible to obtain a large number (100 or more) of serial sections, with 1 or 2 millimeters' thickness, within a single examination (Vannier, 1988).

Structures sectioned by this technique can be thoroughly searched for pathologic conditions.

> As in conventional radiology, structures are recognized by their size, shape, and position and only to a much lesser extent by tissue density. In the abdomen the presence of fat planes helps greatly in the recognition of the various organs, particularly the pancreas, aorta, and inferior vena cava (Norman, 1977).

Clinical Applications. Many regions of the body are best surveyed by CT scans. For example, a whole lung survey using the CT modality can rule out lung cysts.

> Blunt trauma to the chest may result in the formation of a traumatic lung cyst. The lesion itself is innocuous and requires no special treatment. Though it is important that traumatic lung cyst not be mistaken for a more serious complication of trauma requiring aggressive management, this distinction may be difficult to make on plain chest radiographs (Black, 1986).

A survey by CT of the mediastinum, which contains the heart and some other organs, can rule out mediastinal masses. On the CT sectional image, the tumor mass shows a greater density than normal mediastinal tissues. According to Feigin (1988), the new algorithm for mediastinal masses is thus based upon CT density, rather than location, as the primary criterion.

Carcinoma of the larynx and other organs has been demonstrated by CT scans. "Neoplasm is identifiable on CT because it contains tissue of greater density than the normal laryngeal soft tissues or because it distorts, displaces, or destroys normal anatomical structures" (Lee, 1983).

A pelvic survey by CT is another example of this modality's ability to rule out tumors in both men and women. Tissues can be viewed best using an intravenous contrast agent to define the bladder and the vascular characteristics of pelvic tumors. Lee (1983) recommends that direct coronal scans or reformatted images should be considered if suspicious densities are seen about the pelvic floor on the transaxial views.

For a survey of the kidneys by CT, an oral contrast agent is given 30 to 60 minutes before scanning (Lee, 1983). The cross-sections or slices are 10 millimeters thick. Nunez (1986) reported that a diagnosis of renal cell carcinoma was established by computed tomography–guided aspiration in two small solid lesions in the kidneys occurring in end-stage renal disease.

Oral contrast agent is needed 15 minutes prior to a CT scan image of a liver-spleen survey. Both pre- and postintravenous contrast scans may be necessary if one is searching for focal lesions such as hematomas and liver metastases (Nunez, 1986).

Other vital abdominal organs can be surveyed using CT including a liver-pancreas scan. Since the liver and pancreas share a common duct that enters the duodenum, diseases of one influence the other. A survey of this type may rule out common bile stones and pancreaticobiliary disease.

> Endoscopic retrograde cholangiopancreatography (ERCP) and computed tomography (CT) have made

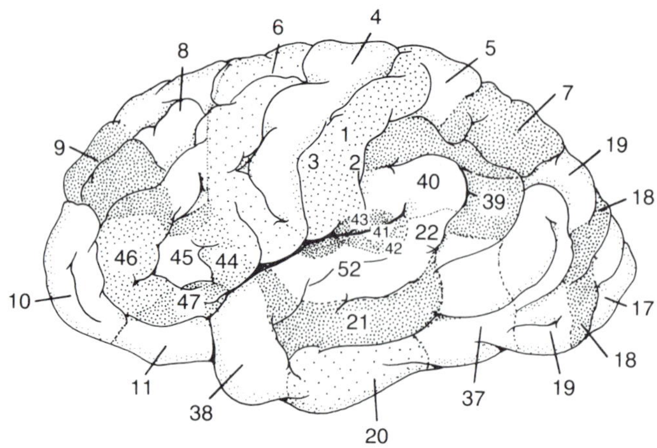

Figure 2–2
Brodmann's numbers on a cytoarchitectural map of the human brain.

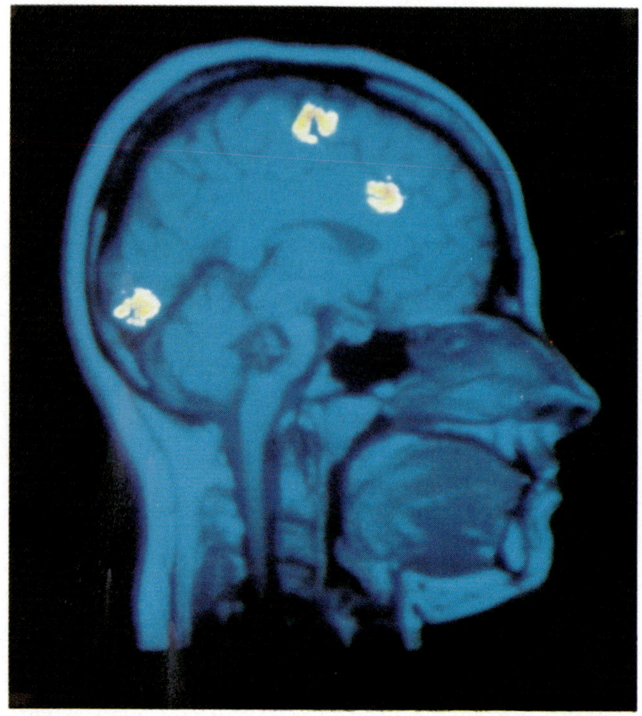

Figure 2–3
PET scan showing the three areas that are active during language tasks (see text).

significant contributions to the diagnosis and management of benign and malignant pancreaticobiliary disorders (Tobin, 1987).

ULTRASOUND

Ultrasound involves sound-wave vibrations reflected from the interface of tissues with different densities (Van De Graaff, 1989). For example, the densities of a fetus and its amniotic fluid are vastly different and can be revealed by ultrasound. The greater the density of the tissue, the greater the reflection of sound waves. These reflections are used by the computers to produce an image. The more reflections from the tissue, the clearer the image.

> These sound waves are reflected back to the crystal, which reconverts them into electric signals. A doctor places a transducer containing a crystal on the area to be scanned, such as the abdomen of a pregnant woman. Echoes from the fetus are translated into faint signals, which are processed by a computer into a video image (Sochurek, 1987).

Clinical Applications. Ultrasound has made a major contribution to the new vision of medicine. Certain regions and parts of the body are best surveyed by ultrasound. Scan of the pregnant uterus can reveal fetal sex and activity and abnormally sized and abnormally formed fetuses. Sochurek (1987) reports, "A picture made with ultrasound in the sixth month of pregnancy shows the face of a healthy fetus with mouth open in a yawn." Ultrasound gallbladder surveys can show cancer of the gallbladder, with or without gallstones. Liver ultrasound scanning can show soft tissue masses such as liver tumors.

POSITRON EMISSION TOMOGRAPHY (PET)

The PET scan requires radioactive tracer molecules to be within the tissue to be examined. The usual substance is glucose labeled with carbon-11, which is injected into the blood stream. The metabolic changes in glucose can be assessed via the PET scan. The PET scan photographs the gamma rays that are emitted from the patient's tissue through the skull (Van De Graaff, 1989).

PET scanning is gaining acceptance and strength every day. For example, a 2-year-old patient suffering from seizures was given little chance to survive until a PET scan using radioactive glucose revealed that the activity in the right hemisphere of his brain showed normal activity. Normal activity of an organ is shown by red, yellow, and green colors, based on the rate of glucose usage. Red indicates the fastest rate of glucose use, then yellow, and next green. The left hemisphere in this 2-year-old boy showed minimal activity, indicated by purple. A purple color in a PET scan indicates little or no activity of radioactive substances (Sochurek, 1987).

Clinical Applications. Various organs can be surveyed with PET; we will discuss its use for the brain and the female breast. PET can reveal a brain in its functional state; therefore, it is possible to delineate a functional map of the brain, which previously was possible only by autopsy or ablation techniques. In 1909, the German neurologist Korbinian Brodmann mapped a neocortical quiltwork of areas or fields of the brain (Nauta, 1986). Each area of control in the cytoarchitectural map of the human cortex has been numbered. For example, area 17 is for vision, 41 is the primary auditory (hearing) cortex, 44 is for speech, and writing areas are 22, 39, and 40 (Fig. 2–2). With PET scanning the scientist can map the human brain in a living human organism without the invasive techniques Brodmann used. Montgomery (1989) studied the language of the brain with PET:

> Bright spots generated by positron emission tomography (PET) show three areas in the left brain that are active during language tasks. The spot at the back lights up during reading. The middle area is active during speech. And the area of the front is involved in thinking about a word's meaning. These PET images of brain function have been laid over a magnetic resonance image of the brain's anatomy.

Figure 2–3 shows the three areas for language in a PET scan.

A survey of the mammary glands is another important area for the use of the PET scanning technique. "Now scientists have designed a traceable, 'hot' analogue of the female hormone estrogen that allows them to distinguish between the two major types of breast tumors—those with estrogen receptors and those without" (Weiss, 1988).

SUMMARY

Four computer imaging modalities have been described by which scientists are able to view the inner structures and functions of the living human body without surgery. Modern dental and medical research and diagnosis have profited enormously from these high-technology computers.

Some computer imaging techniques are most useful for the brain, some in surveying the pregnant uterus, some in identifying the heart and mediastinal structures, and some for abdominal visceral surveys. With all four imaging modalities, the entire living body can be monitored and described in clear detail. These devices have moved the study of the human body from the dissecting table to a computer screen. What a brave new world!

References

Abelson, H. Philip. New horizons in medicine. *Science* 25: 1109, 1988.

Begley, Sharon. How to tell if you're smart. *Newsweek* 29: 64, 1988.

Black, C. William, et al. Computed tomography of traumatic lung cyst: Case report. *Journal of Computed Tomography* 10: 33-35, 1986.

Brash, C. James. *Cunningham's Textbook of Anatomy.* New York: Oxford University Press, 1951.

Feigin, S. David, and Eric M. Padua. Mediastinal masses: A system for diagnosis based on computed tomography. *Journal of Computed Tomography* 10: 11–21, 1986.

Gabler, Michael J., et al. Effect of arthroscopic temporomandibular joint surgery on articular disk position. *Journal of Craniomandibular Disorders: Facial and Oral Pain* 3: 191–202, 1989.

Ida, M., and E. Honda. Age-dependent decrease in the computed tomographic numbers of parotid and submandibular salivary glands. *Dentomaxillofacial Radiology* 18: 31–35, 1989.

Lane, Alexander. *Functional Human Anatomy.* Edina, MN: Bellwether Press, 1990.

Lane, Alexander, and Hossein Sharfaei. *Modern Anatomy: The Sectional Approach.* River Grove, IL: Triton College Press, 1988.

Lee, K., Joseph Greenberg, Stuart S. Sagel, and Robert J. Stanley. *Computed Body Tomography.* New York: Raven Press Books, 1983.

MacFall, R. James, and Robert D. Jones. Computers in magnetic resonance imaging. *Computers in Physics* Sept./Oct. 1988, p. 102.

Montgomery, Geoffrey. The mind in motion. *Discover* March 1989, pp. 58–66.

Morgan, Lee Carlisle. *Basic Principles of Computed Tomography.* Baltimore: University Park Press, 1983.

Mulhern C.B., et al. Nonuniform attenuation in computed tomography study of the cirrhotic liver." *Radiology* 132: 399–402, 1979.

Nauta, H.J. Walle. *Fundamental Neuroanatomy.* New York: W. H. Freeman and Company, 1986.

Norman, David, Melvyn Korobkin, and Thomas H. Newton. *Computed Tomography.* St. Louis: C. V. Mosby Company, 1977.

Nunez, Diego, et al. Renal cell carcinoma complicating long-term dialysis: Computed tomography–guided aspiration cytology. *Journal of Computed Tomography* 10: 61–66, 1986.

Nyborg, L. Wesley. *Biological Effects of Ultrasound: Mechanisms and Clinical Implications.* Bethesda, MD: National Council on Radiation Protection and Measurements, 1983.

Phelps, M.E., M. H. Gado, and E. J. Hoffman. Correlation of effective atomic number and electron density with attenuation coefficients measured with polychromatic x-rays. *Radiology* 117: 585–588, 1973.

Ragan, Don. Medical physics: Turning science fiction into fact. *Computers in Physics* Sept./Oct. 1988, p. 102.

Schaner, E.G., et al. Adrenal cortical tumors with low attenuation coefficients: A pitfall in computed tomography diagnosis. *Journal of Computed Assisted Tomography* 2: 11–15, 1978.

Sochurek, Howard. Medicine's new vision. *National Geographic* Jan. 1987, pp. 2–41.

Stanley R.J., S.S. Sagel, and R.G. Levitt. Computed tomography of the body: Early trends in application and accuracy of the method. *AJR* 127: 53–67, 1976.

Termote, J.L., et al. Computed tomography of the normal and pathologic muscular system. *Radiology* 137: 439–444, 1980.

Tobin, S. Richard, et al. A comparative study of computed tomography and ERCP in pancreaticobiliary disease. *Journal of Computed Tomography* 11: 261–266, 1987.

Van De Graaff, M. Kent, and Stuart Ira Fox. *Concepts of Human Anatomy and Physiology.* Dubuque, IA: William C. Brown Publishers, 1989.

Vannier, W. Michael. Computers in computed axial tomography. *Computers in Physics* Sept./Oct. 1988, pp. 39–43.

Weiss, R. PET scan spots breast cancer receptors. *Science News* Oct. 1988, p. 276.

PART II

THE HEAD

INTRODUCTION

The pictures from cadaver transverse (axial) sections seen here in Part II of *Modern Sectional Anatomy* were derived from three different cadavers. The four drawings of the coronal (frontal) sections of the head region are adapted with permission from *A Frontal Section Anatomy of the Head and Neck* by Otto Frederic Kampmeier, Arthur R. Cooper, and Thomas S. Jones, published by the University of Illinois Press, 1957. The midsagittal cadaver pictures are adapted from *Topographisch-Anatomischer Atlas: nach Durchschnitten an gefrornen Cadavern,* published by Wilhelm Braune in 1872. The pictures of clinical images are from several patients. Some of the cadaver sections and drawings are correlated (paired) with appropriate clinical images.

The pictures from cadaver transverse sections and drawings of the coronal and sagittal planes are displayed in a designated sequence. That is, the most superior levels of transverse sections are discussed first, and the subsequent inferior levels are discussed in successive inferior sections of the head. In the coronal (frontal) plane, the sequence of the study of both cadaver sections and clinical images is from anterior to posterior. The sagittal section of the cadaver drawing is from the median. The sagittal clinical images are from left to right. In addition to MRI and CT scans, ultrasound scans of a newborn infant are depicted. Some of the scans reveal a pathologic condition with appropriate radiologic reports.

Part II contains Chapters 3, 4, and 5. Chapter 3 is concerned with the transverse sections of both cadaver and clinical images; Chapter 4 discusses the coronal slices of the head as well as coronal images. Chapter 5 is concerned with the sagittal plane. Included in this introduction is an illustration that shows the cranial afferent blood flow to the brain (Fig. II–1) to enhance the understanding of the vasculature of this region that is discussed in each chapter. For convenience, each route is named based on the main vessel involved: (1) the right carotid, (2) the left carotid, (3) the right vertebral, and (4) the left vertebral.

Also included here, as well as in the other Parts of *Modern Sectional Anatomy,* are tables of muscles, describing the origin, insertion, innervation, vasculature, and action of each muscle of the region.

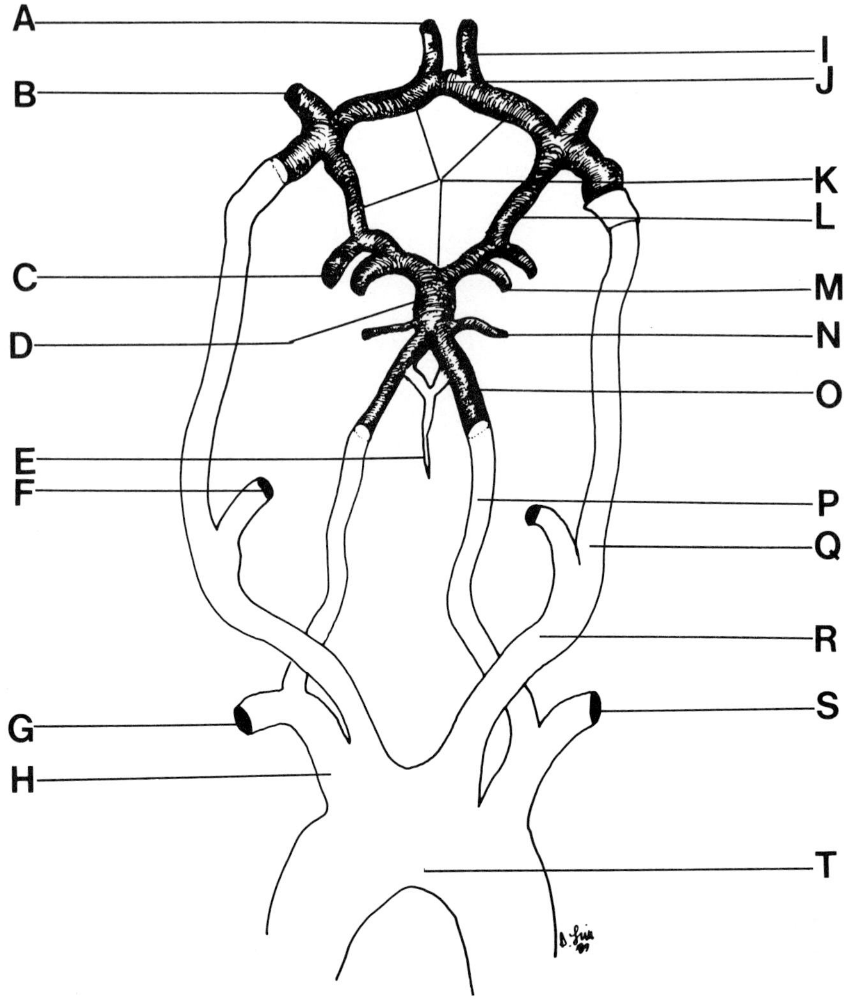

A. Right anterior cerebral artery
B. Right middle cerebral artery
C. Right posterior cerebral artery
D. Basilar artery
E. Anterior spinal artery
F. Right external carotid artery
G. Right subclavian artery
H. Brachiocephalic artery
I. Left anterior cerebral artery
J. Anterior communicating artery of the circle of Willis
K. Circle of Willis
L. Posterior communicating artery of the circle of Willis
M. Left superior cerebellar artery
N. Left anterior inferior cerebellar artery
O. Left cranial portion of vertebral artery
P. Left vertebral portion of the vertebral artery
Q. Left internal carotid artery
R. Left common carotid artery
S. Left subclavian artery
T. Arch of aorta

Figure II–1
Afferent cranial blood flow vessels, including the circle of Willis.

Table II–1 Muscles of Facial Expression

Muscle	Origin	Insertion	Innervation	Vasculature	Action
Occipitalis	Posterior inferior occipital bone and part of posterior temporal bone	Epicranial aponeurosis (galea aponeurotica)	Facial nerve	A posterior auricular branch known as the occipital branch, also as the descending branch of the occipital artery	Moves scalp posteriorly
Frontalis	Galea aponeurotica	Frontal bone superior to supraorbital orbital line	Facial nerve	A superficial temporal branch known as the frontal branch	Eyebrow elevation and frown
Zygomatic major	Outer surface of zygomatic bone	Superior angle of orbicular muscle of mouth	Facial nerve	A facial branch known as the superior labial branch	Laughing muscle (draws mouth angle posteriorly and superiorly)
Zygomatic minor	Outer surface of zygomatic bone	Lateral of orbicular muscle of superior lip	Facial nerve	A facial branch known as the labial branch	Pulls upper lip superiorly
Mentalis	Depression on anterior mandible known as incisive fossa	Skin over lower mandible (chin skin)	Facial nerve	A facial branch known as the interior labial branch	Lower lip elevation and protrusion, also wrinkles skin over chin
Superior levator muscle of lips	Lower orbital margin (infraorbital head)	Muscle of upper lip	Facial nerve	Angular and superior facial branches; labial	Raises upper lip
Levator of upper lip and alae of nose	Frontal process of maxilla adjacent to nose	Skin of superior lip (labial part), alae of nose (nasal part)	Facial nerve	Angular and superior facial branches; labial	Upper lip elevation and nostril dilation
Buccinator	Maxilla and mandible (alveolar process)	Angle of mouth on orbicular muscle of mouth	Facial nerve	Facial artery, muscular branches, and buccal branch of maxillary	Angle retraction and cheek compression
Orbicular muscle of mouth	Muscle fibers surround mouth opening	Angle of mouth	Facial nerve	Labial branches (superior and inferior) of facial	Closes lip
Orbicular muscle of eye	Frontal bone and maxilla	Eyelids	Facial nerve	Superficial temporal (zygomatic and frontal branches) and facial (angular branch)	Orbital part: closes the eyelids forcibly. Palpebral part: closes the eyelids gently

Table II–2 Muscles of Mastication

Muscle	Origin	Insertion	Innervation	Vasculature	Action
Temporalis	Temporal fossa of lateral skull, including deep surface of temporal fascia	Coronoid process of mandible	Mandibular branch of trigeminal	Anterior and posterior deep temporal branches of maxillary artery; also, median branch of the superficial temporal artery	Jaw elevation as in chewing; retracts mandible and clenches teeth
Internal (medial) pterygoid	Medial (internal) surface of pterygoid plate	Medial surface of ramus of mandible	Mandibular branch of trigeminal	Pterygoid branch of maxillary and muscular branches of facial arteries	Jaw elevation as in chewing. Jaw protraction; assists in rotary movement while chewing
Masseter	Deep part from medial surface of zygomatic arch; superficial part from anterior part of inferior edge of zygomatic arch	Superior half of ramus of mandible and mandibular angle	A branch of mandibular nerve of trigeminal	Branch of superficial temporal (transverse facial); branch of maxillary (masseteric); muscular branch of facial	Jaw elevation; clenching teeth
External (lateral) pterygoid	Greater wing of sphenoid and lateral surface of lateral pterygoid plate	Anterior of neck of mandible; inferior coronoid process	A branch of mandibular nerve of trigeminal	Pterygoid branch of maxillary artery	Mandibular protrusion; draws articular capsule anteriorly; assists in rotary movement of chewing

Table II–3 Muscles of the Soft Palate

Muscle	Origin	Insertion	Innervation	Vasculature	Action
Tensor velum palatini	Sphenoid bone (scaphoid fossa and wall of auditory tube)	Aponeurosis of soft palate and back part of palatine bone	Mandibular branch of trigeminal	Facial (ascending palatine branch) and maxillary (descending palatine branch)	Tightens soft palate during swallowing; opens auditory tube
Uvula	Back nasal spine and palatine of aponeurosis	Uvula (mucous membrane)	Branches of glossopharyngeal vagus and cranial portion of spinal accessory; also some sympathetic fibers from upper thoracic nerves	Facial (ascending palatine branch) and maxillary (descending palatine branch)	Elevates uvula

Table II-3 Muscles of the Soft Palate (*Continued*)

Muscle	Origin	Insertion	Innervation	Vasculature	Action
Palatoglossus (glossopalatinus)	Front surface of soft palate	Back and lateral parts of tongue	Branches of glossopharyngeal vagus and cranial portion of spinal accessory; also some sympathetic fibers from upper thoracic nerves	Facial (ascending palatine branch) and maxillary (descending palatine branch)	Reduces opening of fauces and raises posterior part of tongue and nasal pharynx
Palatopharyngeal (pharyngopalatinus)	Palatine aponeurosis	Back edge of thyroid cartilage and aponeurosis of throat	Branches of glossopharyngeal vagus and cranial portion of spinal accessory; also some sympathetic fibers from upper thoracic nerves	Facial (ascending palatine branch) and maxillary (descending palatine) as well as ascending pharyngeal (palatine branch)	Aids in deglutition
Levator velum palatini	Inferior surface of petrous portion of temporal bone and medial surface of cartilage of auditory tube	Palatine aponeurosis	Branches of glossopharyngeal, vagus, and cranial portions of spinal accessory; also some sympathetic fibers from upper thoracic nerves	Facial (ascending palatine branch) and maxillary (descending palatine branch)	Elevates soft palate during swallowing

Table II-4 Intrinsic Muscles of Tongue

Muscle	Origin	Insertion	Innervation	Vasculature	Action
Inferior longitudinal (superior longitudinalis lingua)	Root of tongue	Anterior part of tongue	Branches of cranial nerves 5, 7, 9, and 12	Lingual arteries (deep lingual branch) and branches from facial	Shortens and curls anterior part of tongue superiorly
Superior longitudinal (superior longitudinalis lingua)	Root of tongue	Anterior part of tongue	Branches of cranial nerves 5, 7, 9, and 12	Lingual arteries (deep lingual branch) and branches from facial	Shortens and curls anterior part of tongue inferorly
Transverse (transversus lingua)	Median sulcus of tongue	Back and side edges of tongue	Branches of cranial nerves 5, 7, 9, and 12	Lingual arteries (deep lingual branch) and branches from facial	Varies shape of tongue, reduces size and elongates tongue
Vertical (verticalis lingua)	Back of tongue	Inferior part of tongue	Branches of cranial nerves 5, 7, 9, and 12	Lingual arteries (deep lingual branch) and branches from facial	Flattens and widens tongue

Table II-5 Extrinsic Muscles of Eye

Muscle	Origin	Insertion	Innervation	Vasculature	Action
Superior rectus	Back part of orbital cavity near optic foramen	Top part of eyeball	Oculomotor	Ophthalmic (muscular branch). *Note:* This is a branch of cerebral part of internal carotid arteries	Moves eye superiorly and medially
Inferior rectus	Back part of orbital cavity near optic foramen	Bottom part of eyeball	Oculomotor	Maxillary (infraorbital branch) and inferior muscular branch of ophthalmic	Moves eye inferiorly and medially
Lateral rectus	Back part of orbital cavity near optic foramen	Side part of eyeball	Abducens	Ophthalmic (inferior muscular branch)	Moves eye laterally
Medial rectus	Back part of orbital cavity near optic foramen	Medial part of eyeball	Oculomotor	Ophthalmic (inferior muscular branch)	Moves eye medially
Superior oblique	Back part of orbital cavity near optic foramen	Top part of eyeball deep to superior rectus after passing through trochlea	Trochlear	Ophthalmic (superior muscular branch)	Moves eye inferiorly and laterally as well as medial rotation of eyeball
Inferior oblique	Floor of eye socket lateral to nasal lacrimal canal	Inferior and side parts of eyeball	Oculomotor	Maxillary (infraorbital branch) and ophthalmic (inferior muscular branch)	Rotates eyeball superiorly and laterally
Levator palpebrae superioris	Upper part of posterior orbital cavity	Torsal plate of superior eyelid (distal of eyelid)	Oculomotor	Ophthalmic (branch of internal carotid artery) and supraorbital artery (branch of ophthalmic)	Elevates upper lid

CHAPTER 3

Transverse (Axial) Cadaver Sections and Images of the Head

This chapter includes 6 different transverse (axial) cadaver sections and 14 clinical images. Two modules are based on sheet plastinated transverse (axial) sections. Others are conventional cadaver sections. Each module includes labeled images of musculoskeletal, neurovascular, visceral, and enclosing units. In the case of clinical images, two anatomic units are included on one picture as well as some cadaver sections.

Each module represents a transverse (axial) area of the head. The first module represents one area of the skull at 8 cm above Reid's baseline. The second module is 6 cm above Reid's baseline. The third module is 4 cm above Reid's baseline. The fourth module is 1 cm above Reid's baseline. The fifth module is at Reid's baseline, and the sixth module is 1 cm below Reid's baseline. Modules 3 and 4 are sheet plastinated specimens; all others are cadaver sections done in the conventional method.

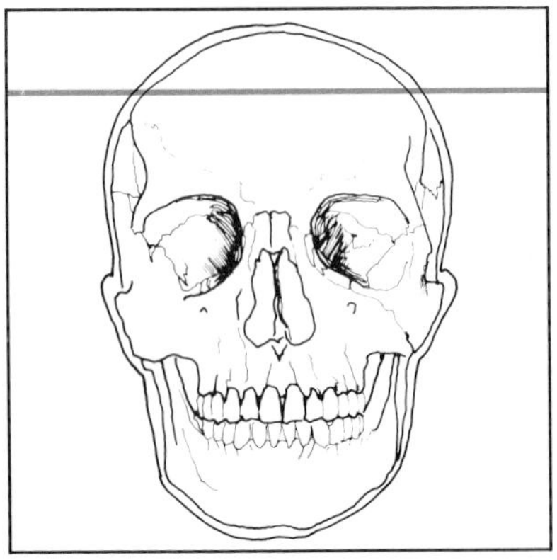

MODULE 1

This is a transverse (axial) section of the head at the level of 8 cm above Reid's baseline.

MUSCULOSKELETAL UNIT (Fig. 3–1)

I. Skeletal Portion

Frontal Bone (*A*). At this level this is seen as a curved bony structure anterior and lateral to the frontal lobes of the brain. This bone forms the skeletal part of the forehead in two portions, namely vertical (or frontal) and horizontal (or orbital-nasal). The vertical portion is located at the anterior part of the cranium, which forms the forehead. The horizontal portion forms the roof of the nasal fossae and the roof of the orbits. Squamous, orbital plates, and nasal and zygomatic processes are parts of the frontal bone.

Diploë (*B*). A spongy bone layer between the outer and inner tables of the flat, compact bones of the skull. This cancellous (spongy) bone contains red bone marrow.

Parietal Bone (*C*). At this level this is a slightly curved feature located bilaterally near the periphery of the section. These two bones form the superior and lateral parts of the skull and join at the sagittal sutures. They articulate with the frontal bone at the coronal suture, to the occipital bone at the lambdoidal suture (parieto-occipital suture), and to the temporal bones at the squamosal suture.

Occipital Bone (*F*). At this level this is a flat, curved bone posterior and lateral to the occipital lobes of the brain. This feature belongs to the posterior and inferior parts of the skull. The foramen magnum is a large hole that traverses this bone. It is joined to the parietal bones by the upper part at the lambdoidal suture.

Internal Occipital Protuberance (*H*). A marked inward projection in the median plane of the occipital bone.

Outer Table (*E*). An external compact bone layer of the skull bones. This layer is thicker than the inner table.

Inner Table (*D*). A internal compact bone layer of the flat skull bones. This layer is thinner than the outer table.

II. Muscular Portion

Galea Aponeurotica (*I*). A helmet-shaped aponeurosis (a broad sheet of tendon) that covers the upper skull. It serves as the insertion for both the frontalis and occipitalis muscles. This structure is also called the epicranial aponeurosis. It is joined to the skin by the fibrous superficial fascia. The association of this feature to the pericranium allows its movement.

NEUROVASCULAR UNIT (Fig. 3–2)

I. Arteries

Middle Meningeal Artery (*A*). At this level this is seen as a small, hollow blood vessel located on either side of the median directly medial to the anterior part of the parietal bones, and anterior to the midaxillary line.

Origin. Maxillary artery.

Distribution. Cranial bones and dura mater.

Course. This artery enters the cranium through the foramen spinosum and continues sideward toward the greater wing of the sphenoid bone. Then this structure divides into posterior and anterior branches. The posterior branch diverges posteriorly, producing a sulcus in the squamous portion of the temporal bone. Its course is located close to the temporal sulcus. The anterior branch proceeds superiorly, producing a sulcus in the superior part of the greater wing of the parietal bone. Its course corresponds to and lies medially anterior to the central sulcus.

II. Veins

No veins are seen in this conventionally prepared cadaver section at 8 cm above Reid's baseline owing to its preparation.

III. Vascular Venous Sinuses

Superior Sagittal Sinus (*B, C*). A triangular, hollow, vascular structure seen on both anterior and posterior median parts of this section.

Origin. Crista galli.

Termination. Confluence of sinuses.

Course. It arises from the crista galli and courses in a groove in the median of the internal skull (parietal and occipital bones). When it approaches the internal occipital protuberance, it usually splits into the right and left limbs, and each limb joins the right and left transverse sinuses, respectively.

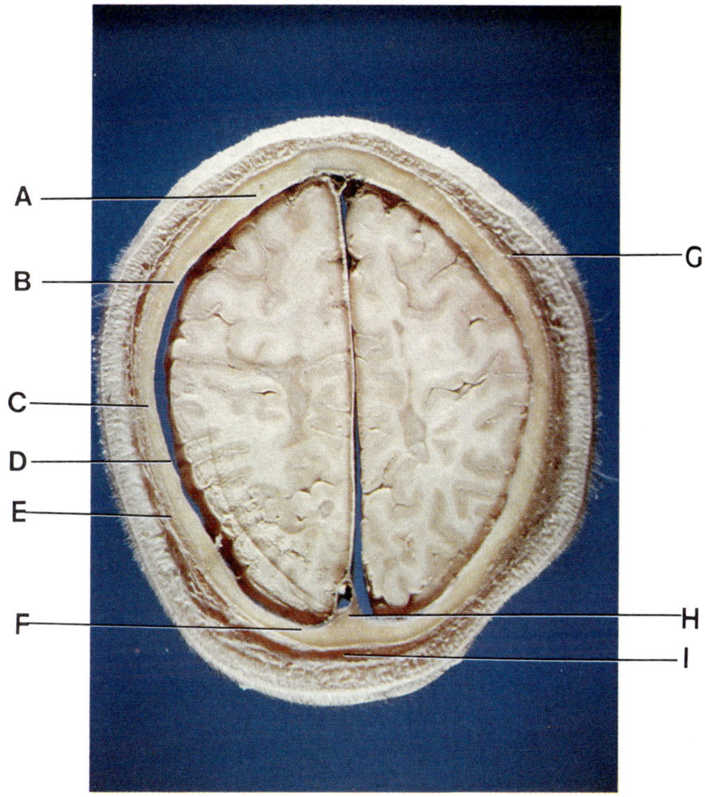

A. Frontal bone
B. Diploë
C. Parietal bone
D. Inner table
E. Outer table
F. Occipital bone
G. Temporalis muscle
H. Internal occipital protuberance
I. Occipitalis muscle

Figure 3-1
Transverse section of the head at 8 cm above Reid's baseline (Frankfort Horizontal Plane), highlighting the musculoskeletal unit structures.

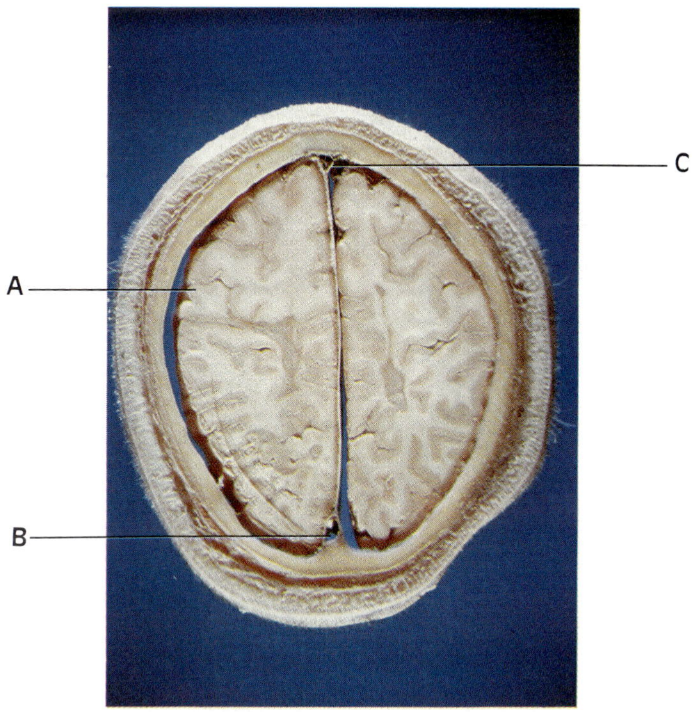

A. Middle meningeal artery
B. Posterior portion of superior sagittal sinus
C. Anterior portion of superior sagittal sinus

Figure 3-2
Transverse section of the head at 8 cm above Reid's baseline, highlighting the neurovascular unit structures.

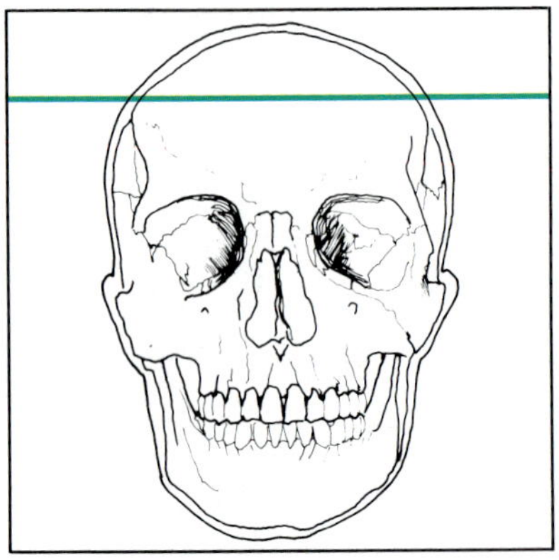

Figure 3–3 is a picture of a CT transverse image with contrast. This figure highlights several structures in both the musculoskeletal and neurovascular units.

VISCERAL UNIT (Fig. 3–4)

Occipital Lobe (*E*). At this level these are seen bilaterally as oval brain features located posterior to the midaxillary line and deep to the occipital bones.

Course. The occipital lobe is usually delimited by drawing a line inferior and slightly anterior from the parieto-occipital fissure to the bottom side edge of the hemisphere.

Parietal Lobe (*C*). In this cross-section, this is seen as a thick brain structure between the frontal and occipital lobes, anterior to the midaxillary line and bilateral to the median deep to the parietal bones.

Course. The anterior boundary of this lobe is the central sulcus, and the posterior boundary is the parieto-occipital sulcus. Its inferior part extends down to the occipital lobe, and its superior part is adjacent to the longitudinal cerebral fissure.

Frontal Lobe (*A*). In this section this is seen bilaterally as an oval anterior brain structure located deep to the frontal bones.

Course. This structure extends as far back as the central sulcus after having started at the anterior cranial cavity. Its lowest part rests on the anterior cranial fossa while its superior part extends to the longitudinal cerebral fissure. The medial part of the frontal lobe extends from the upper part of the longitudinal fissure to the corpus callosum.

Cerebral White Matter (*H*). A flat structure deep to the gray matter. It is the innermost layer of the entire cerebrum.

Cerebral Gray Matter (*G*). The peripheral layer of the brain superficial to the white matter. It appears darker than the white matter because of a greater number of nerve cell bodies. Each nerve cell body contains a dark nucleus.

Parieto-Occipital Fissure (*D*). This is seen here bilaterally as a thin slit separating the parietal and occipital lobes.

Cerebral Cortex (*F*). The outer layer of the cerebrum, which contains the cerebral gray matter.

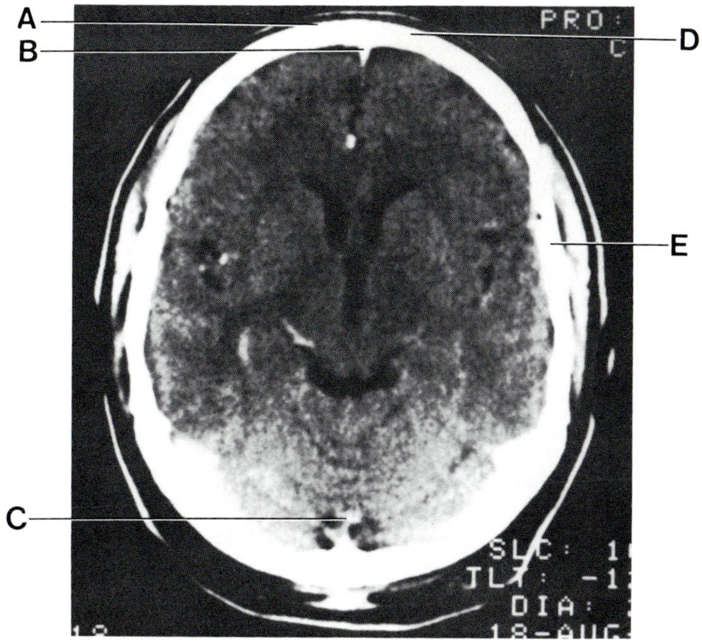

A. Frontal bone
B. Frontal crest
C. Superior sagittal sinus
D. Diploë of frontal bone
E. Diploë of parietal bone

Figure 3-3
CT (with contrast) transverse image of the head, highlighting both the musculoskeletal and neurovascular unit structures.

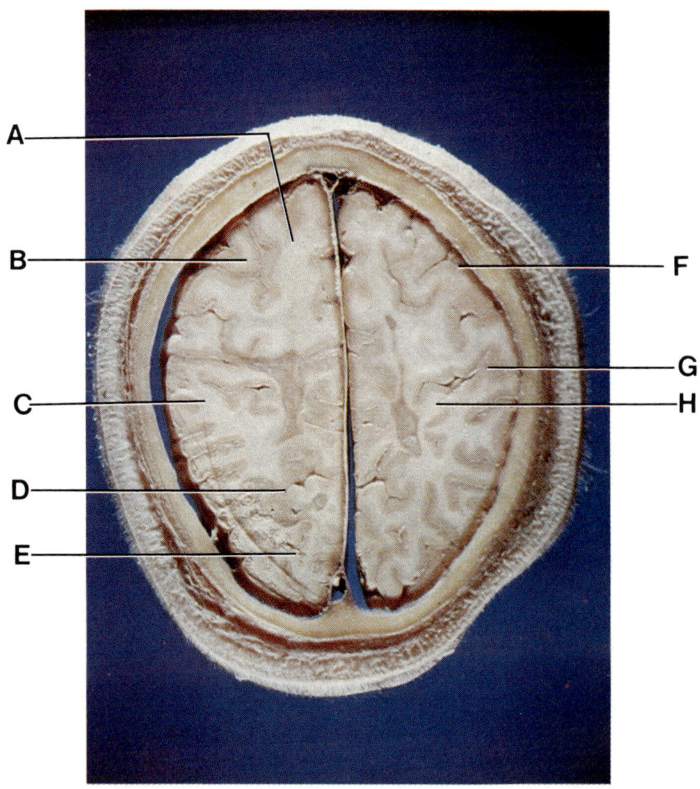

A. Frontal lobe
B. Middle frontal gyrus
C. Parietal lobe
D. Parieto-occipital fissure
E. Occipital lobe
F. Cerebral cortex
G. Cerebral gray matter
H. Cerebral white matter

Figure 3-4
Transverse section of the head at 8 cm above Reid's baseline, highlighting the visceral unit structures.

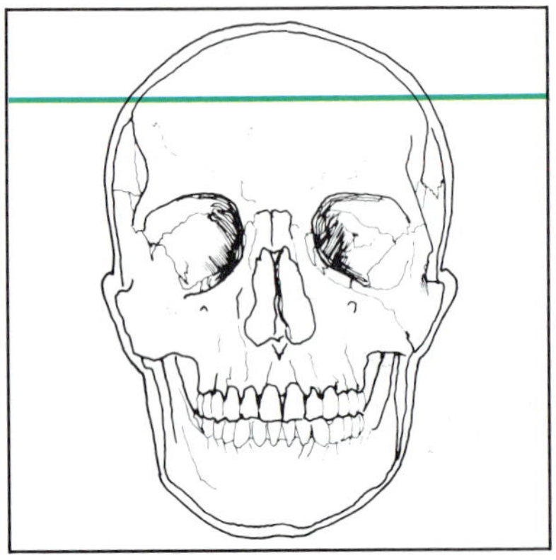

Enclosing Unit (Fig. 3–5)

Scalp (*A*). At this level this is seen as a thick layer surrounding the cranial bones. It consists of five layers: (1) the skin, a thick and protective outer layer; (2) connective tissue, dense and containing abundant blood vessels; (3) epicranial aponeurosis, the musculoaponeurotic sheath; (4) loose areolar tissue, containing emissary veins and permitting the superficial layers to move easily; and (5) periosteum, continuous with the outer layer of the dura mater through the foramina and sutures. This structure covers the bones of the cranium.

Falx Cerebri (*B*). The largest dural fold; it is seen at this level as a sheet of thin membrane between the two hemispheres of the brain. It is a double fold of the inner layer (meningeal) of the dura mater and lies within the longitudinal cerebral fissure. This structure is attached anteriorly to the crista galli and posteriorly to the tentorium cerebelli. The falx becomes wider as it extends from anterior to posterior parts of the skull. The upper border of the falx contains the superior sagittal sinus, and the lower border contains the inferior sagittal sinus.

Dura Mater (*D*). A tough membrane; it is seen at this level as a double membrane lining the medial aspect of the cranial bones. This structure consists of two layers: the outer periosteal and the inner meningeal. The latter forms the dural folds.

Subdural Space (*C*). An empty area seen here between the dura mater and arachnoid.

Figure 3–6 is a picture of a CT scan with contrast in the transverse plane. Visceral and enclosing unit structures are indicated with labels. This head slice is derived approximately 8 cm from Reid's baseline.

Visceral and enclosing unit structures are emphasized by labels in an MRI transverse head image in Figure 3–7.

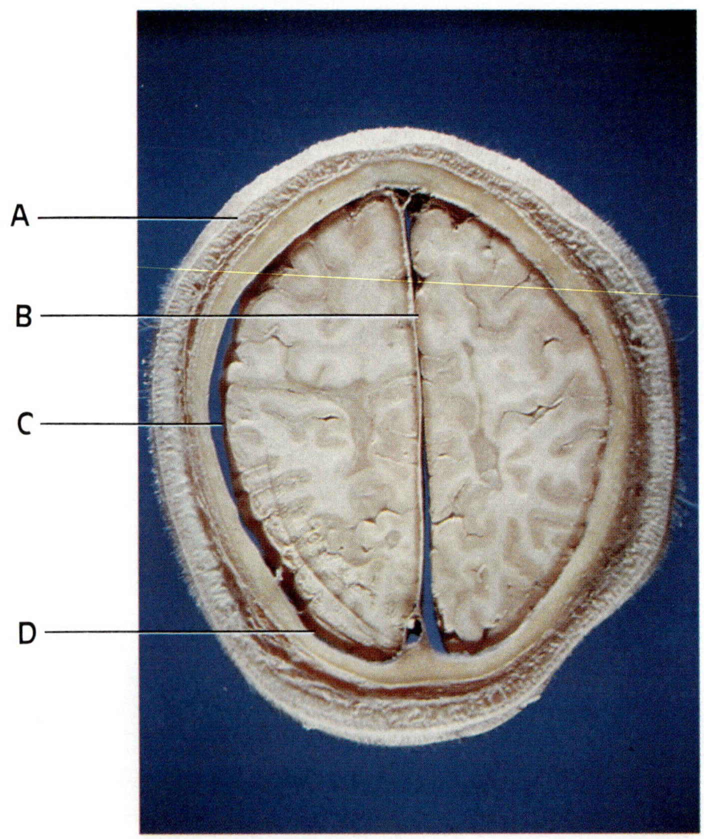

A. Scalp
B. Falx cerebri
C. Subdural space
D. Dura mater

Figure 3–5
Transverse section of the head at 8 cm above Reid's baseline, highlighting the enclosing unit structures.

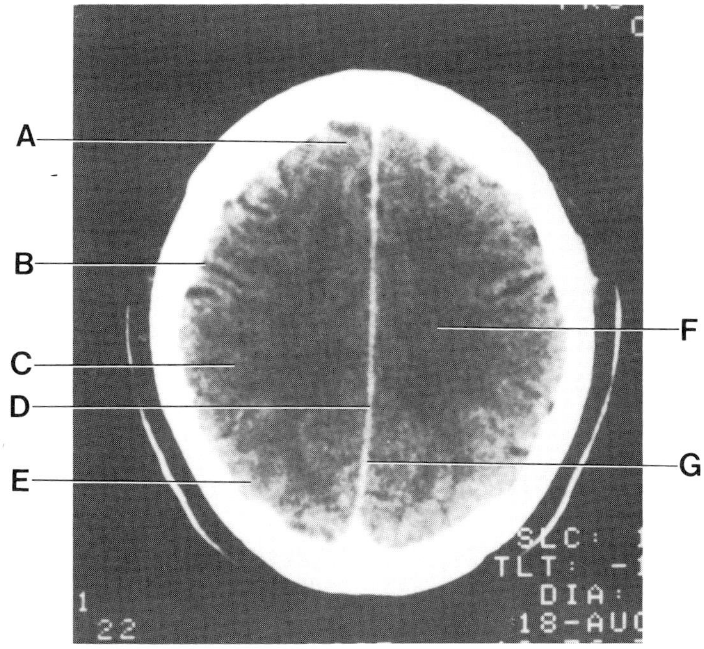

A. Frontal lobe
B. Central sulcus
C. Parietal lobe
D. White matter of cerebrum (centrum ovale)
E. Area of falx cerebri
F. Gray matter of cerebrum
G. Longitudinal cerebral fissure (interhemispheric fissure)

Figure 3-6
CT (with contrast) transverse image of the head, highlighting both the visceral and enclosing unit structures.

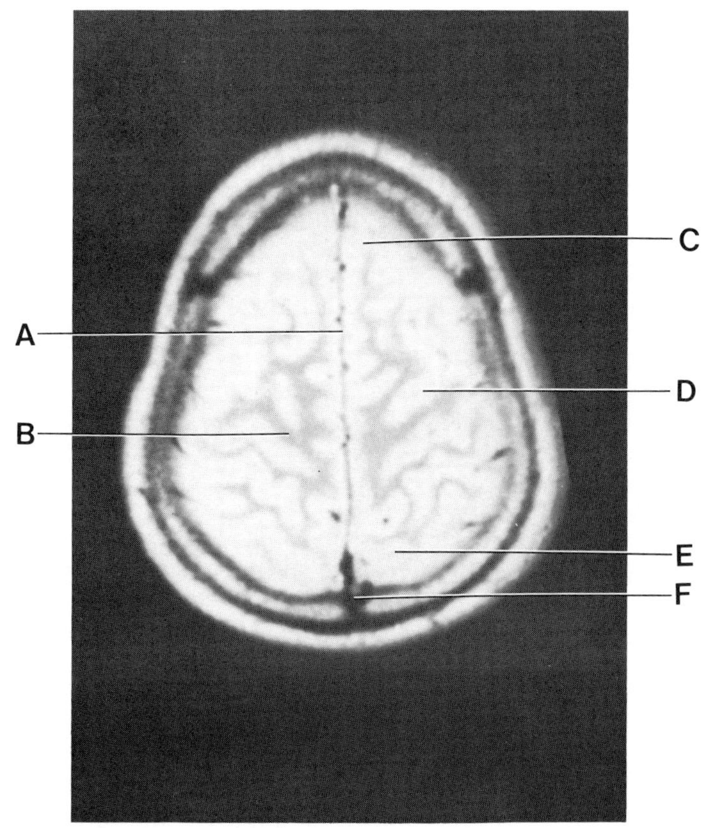

A. Longitudinal cerebral fissure
B. Centrum semiovale
C. Frontal lobe
D. Parietal lobe
E. Occipital lobe
F. Superior sagittal sinus

Figure 3-7
MRI transverse image of the head, highlighting neurovascular, visceral, and enclosing unit structures.

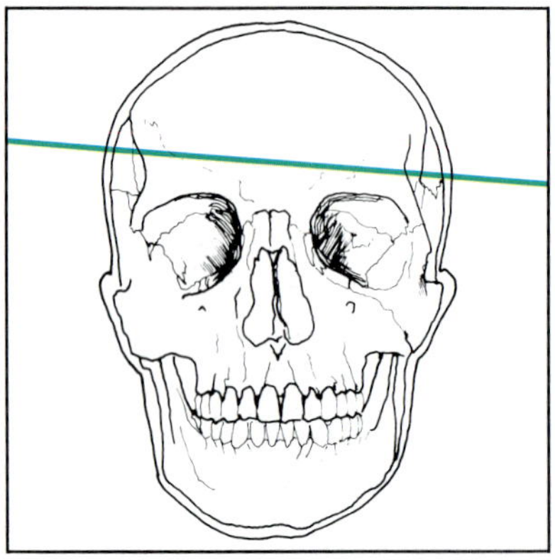

MODULE 2

This is a transverse (axial) section of the head at the level of 6 cm above Reid's baseline.

MUSCULOSKELETAL UNIT (Fig. 3–8)

I. Skeletal Portion

Frontal Bone (*A*). At this level this is seen as a narrow, elongated, circular bony structure located anterior to the midaxillary line and extending bilaterally from the median superficial to the frontal lobes of the brain.

Parietal Bone (*B*). This is seen at this level on the lateral parts of the skull. This flat bone is bilateral to the median and extends anteriorly and posteriorly from the midaxillary line. Its anterior limit is the coronal suture, and its posterior boundary is the parieto-occipital suture.

Occipital Bone (*I*). This is seen at this level on the posterior part of the skull. This flat bone is superficial to the occipital lobes.

Internal Occipital Protuberance (*C*). A marked inward projection of the occipital bone posterior to the cerebellar vermis and anterior to the occipitalis muscle.

Outer Table (*H*). The hard bone (compact bone) that is superficial to the diploë.

Diploë (*F*). The soft (cancellous or spongy) bone that is sandwiched between two layers of hard bone.

Inner Table (*G*). The hard bone (compact bone) that is deep to the diploë.

II. Muscular Portion

Temporalis Muscle (*E*). A thin band of muscle lateral to the parietal bone on both sides. This structure lies in the temporal fossa and is a muscle of mastication. The direction of the anterior portion of the fibers is vertical, and their function is to raise the mandible when they contract. The direction of the posterior portion of the fibers is horizontal, and their function is to lower the mandible. The superficiality of this muscle is related to the following: skin, the front and back portions of the auricular muscles, the temporal fascia, the superficial temporal vessels, the auriculotemporal nerve, the zygomaticotemporal nerve, the arch of the zygomatic bone, the superior part of the masseter, and the galea aponeurotica.

Occipitalis Muscle (*D*). This is seen here bilaterally as a transversely elongated muscle posterior to the occipital bones.

NEUROVASCULAR UNIT (Fig. 3–9)

I. Arteries

Choroid Plexus of Lateral Ventricle (*A*). This is seen here bilaterally as an oval feature on the posterior portion of each lateral ventricle. These choroid plexuses allow only certain materials (the normal content of the cerebrospinal fluid [CSF]) to enter the brain ventricles and therefore the brain tissue. The choroid plexuses, located in all four brain ventricles, secrete CSF.

Origin. Choroid artery (a branch of the cerebral part of the internal carotid artery).

Distribution. A choroid plexus is located in each ventricle and functions in the secretion of CSF.

II. Veins

Greater Cerebral Vein. A small, hollow structure anterior to the cerebellar vermis and posterior to the falx cerebri between the right and left lateral ventricles.

Origin. Paired internal cerebral veins.

Termination. Straight sinus.

Course. Turns superiorly and joins the inferior sagittal sinus above the corpus callosum.

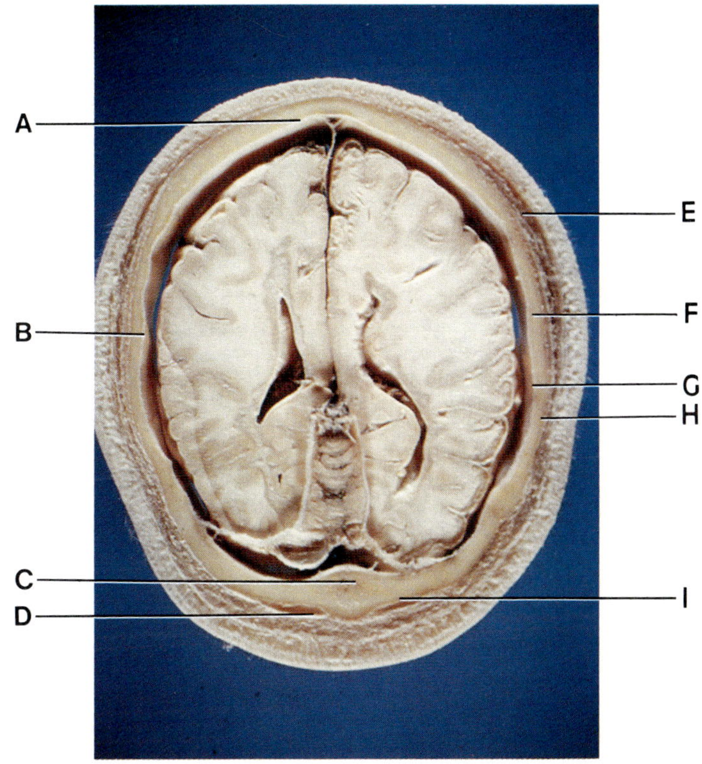

A. Frontal bone
B. Parietal bone
C. Internal occipital protuberance
D. Occipitalis muscle
E. Temporalis muscle
F. Diploë of parietal bone
G. Inner table of parietal bone
H. Outer table of parietal bone
I. Occipital bone

Figure 3–8
Transverse section of the head at 6 cm above Reid's baseline, highlighting the musculoskeletal unit structures.

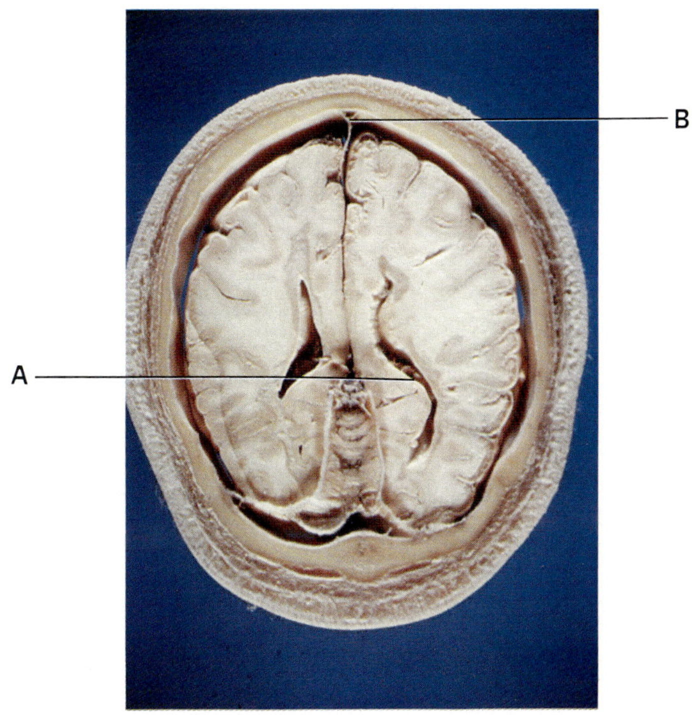

A. Choroid plexus of lateral ventricle
B. Superior sagittal sinus (anterior portion)

Figure 3–9
Transverse section of the head at 6 cm above Reid's baseline, highlighting the neurovascular unit structures.

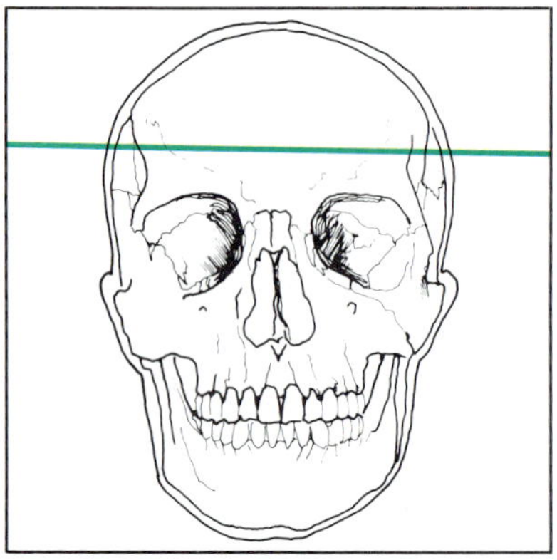

III. Vascular Venous Sinuses

Superior Sagittal Sinus (*B*). This is seen here in the median plane at both the anterior and posterior parts of the skull. The anterior portion of the superior sagittal sinus is located posterior to the frontal bone and anterior to the right frontal lobe. The posterior portion of the sagittal sinus is located posterior to the cerebellar vermis and anterior to the right occipital bone near the median.

Origin. Crista galli.

Termination. Confluence of sinuses.

Course. Refer to Module 1.

Figure 3–10 is a picture of a CT (with contrast) transverse image of the head, which is marked by labels of the musculoskeletal unit structures. Note the orientation line for the derivation of this slice.

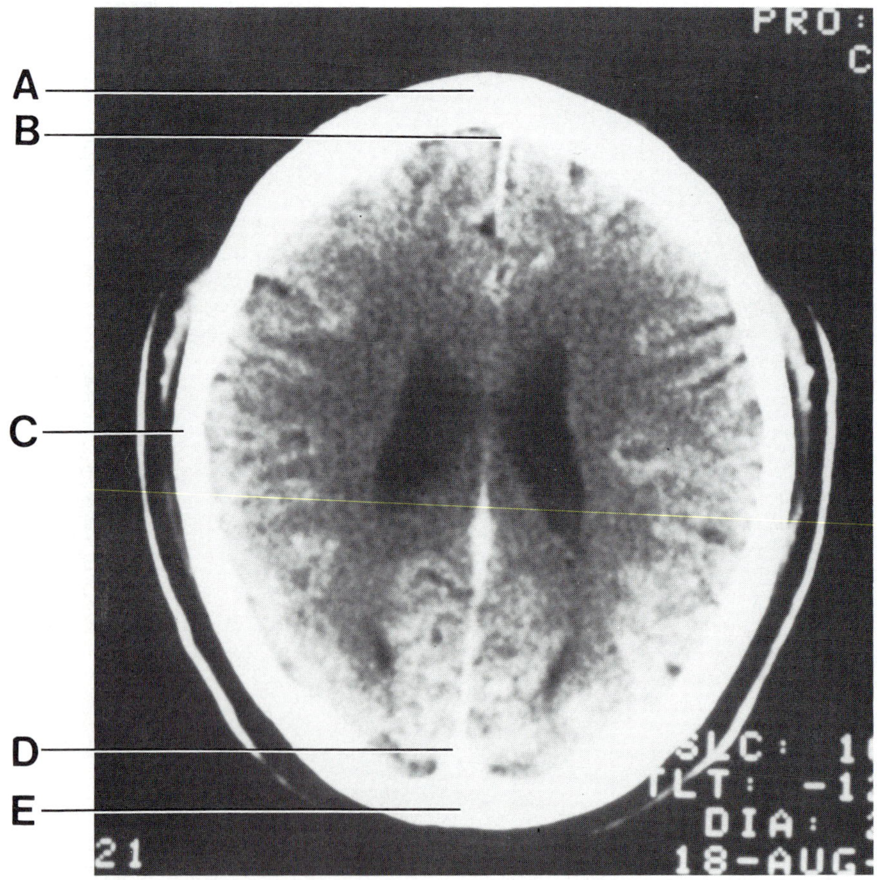

A. Frontal bone
B. Frontal crest
C. Parietal bone
D. Occipital crest
E. Occipital bone

Figure 3–10
CT (with contrast) transverse image of the head, highlighting the musculoskeletal unit structures.

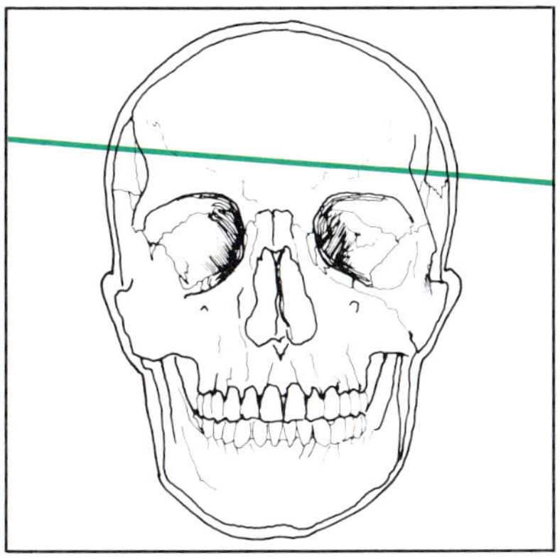

VISCERAL UNIT (Fig. 3–11)

Lateral Ventricle (*E*). At this level, this is seen bilaterally as a crescent-shaped feature lateral to the falx cerebri. The lateral ventricle is subdivided into an anterior horn (in the frontal lobe), a body (in the parietal lobe), a posterior horn (in the occipital lobe), and a descending horn (in the temporal lobe). Each lateral ventricle communicates with the third ventricle through the foramen of Monro.

Frontal Lobe (*B*). This is seen at this level as an irregularly shaped, convoluted brain structure on its peripheral parts, extending bilaterally from the median. The medial surface of each is nearly a straight surface. It is located posterior to the frontal bones and anterior to the parietal lobes.

Occipital Lobe (*K*). At this level, this is seen as an irregularly convoluted brain structure extending bilaterally from the median. It is located posterior to the parietal lobes and anterior to the occipital bones.

Parietal Lobe (*D*). This is seen at this level as an irregularly convoluted brain structure on its peripheral parts bilateral to the median, located between the frontal and occipital lobes. The medial surface of each lobe is adjacent to the longitudinal cerebral fissure.

Cerebral White Matter (*I*). This is seen medial to the cerebral gray matter. This is an area of high concentration of myelin, which surrounds axons of neurons.

Cerebral Gray Matter (*G*). This is a peripheral brain structure seen throughout the cerebrum. This area of the brain contains numerous nerve cell bodies.

Central Sulcus (Fissure of Rolando) (*C*). This is seen bilaterally as a thin slit between the frontal and parietal lobes.

Cerebellar Vermis (*F*). A worm-shaped brain feature on the median, anterior to the internal occipital protuberance and posterior to the falx cerebri between the two hemispheres of the cerebellum.

Posterior Horn of the Lateral Ventricle (*J*). This horn curves back and medially into the occipital lobe of the brain.

A. Superior frontal gyrus
B. Frontal lobe
C. Central sulcus (fissure of Rolando)
D. Parietal lobe
E. Lateral ventricle
F. Cerebellar vermis
G. Cerebral gray matter
H. Longitudinal cerebral fissure
I. Cerebral white matter
J. Posterior horn of lateral ventricle
K. Occipital lobe

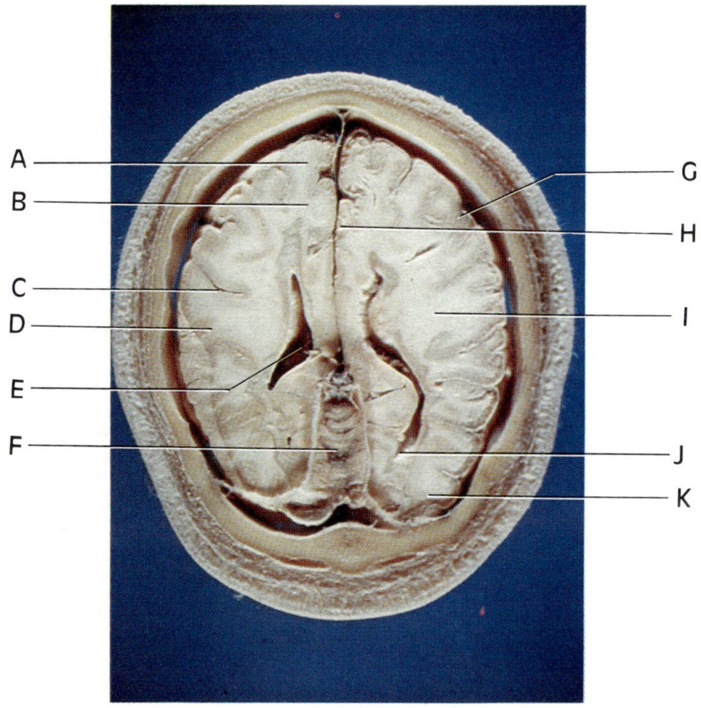

Figure 3–11
Transverse section of the head at 6 cm above Reid's baseline, highlighting the visceral unit structures.

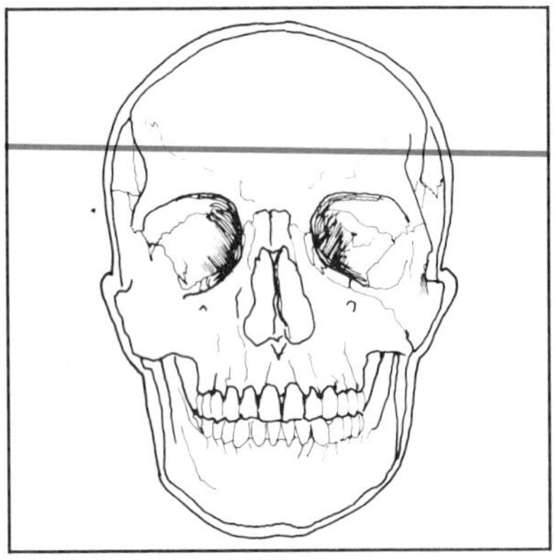

ENCLOSING UNIT

Falx Cerebri. A membrane between the right and left cerebral hemispheres of the brain. It is located anterior to the cerebellar vermis.

Dura Mater. The outer meninx (membrane) (*plural:* meninges) lining the inner wall of the cranium.

Subdural Space. Commonly described as the potential space between the dura mater and the arachnoid. It contains a thin film of fluid.

Visceral and enclosing unit structures are highlighted by labels in this CT (with contrast) transverse image (Fig. 3–12). This head section is approximately 4 cm above Reid's baseline and shows many internal structures of the brain, including part of the basal ganglia.

Figure 3–13 shows an MRI transverse scan that emphasizes by labels visceral and enclosing unit structures.

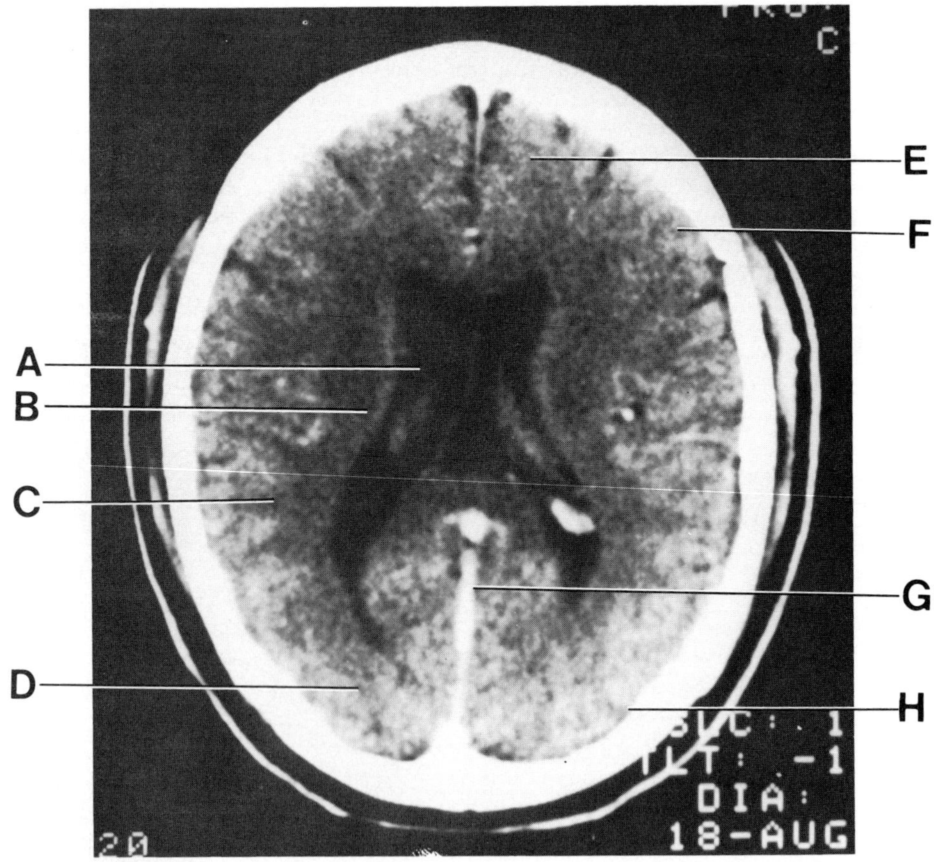

A. Lateral ventricle
B. Part of basal ganglia
C. Parietal lobe
D. Occipital lobe
E. Frontal lobe
F. Area of gray matter of cerebrum
G. Area of longitudinal cerebral fissure
H. Area of subarachnoid space

Figure 3–12
CT (with contrast) transverse image of the head, highlighting both the visceral and enclosing unit structures.

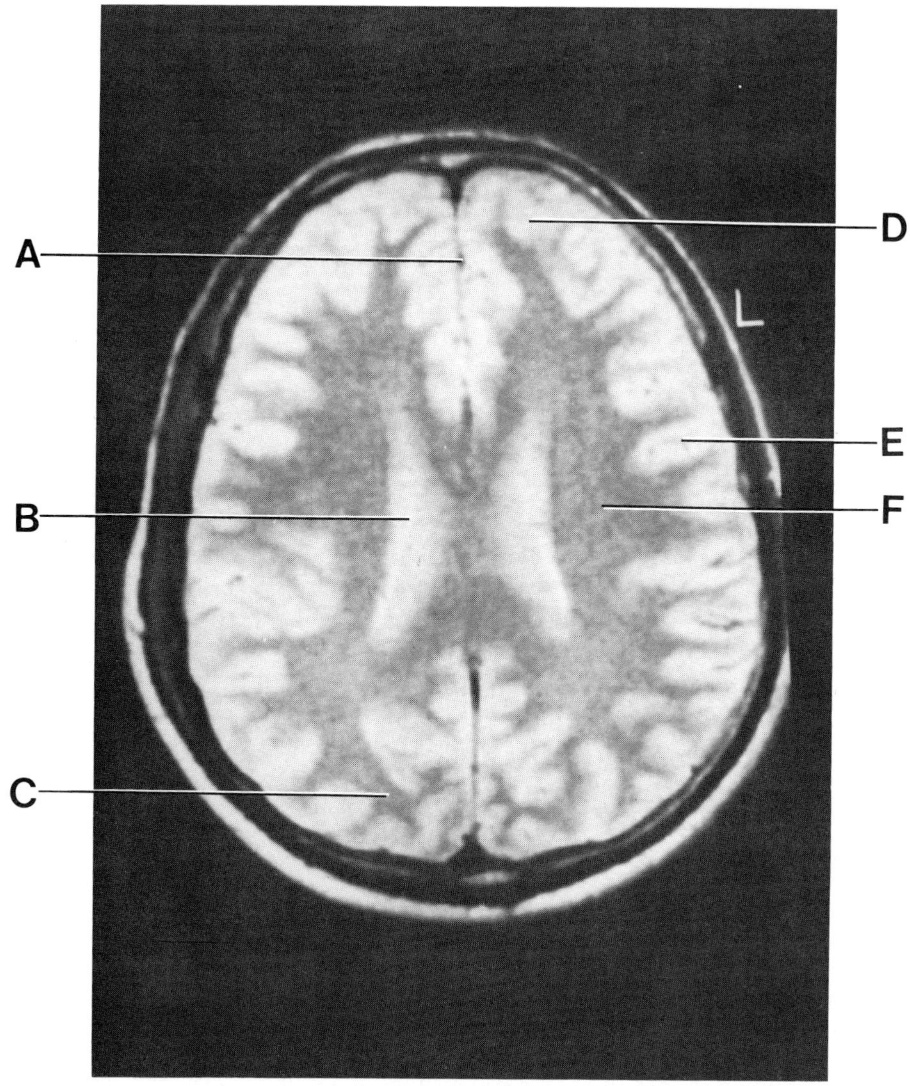

A. Longitudinal cerebral fissure
B. Lateral ventricle
C. Occipital lobe
D. Frontal lobe
E. Gray matter of cerebrum
F. White matter of cerebrum

Figure 3-13
MRI transverse image of the head, highlighting both visceral and enclosing unit structures.

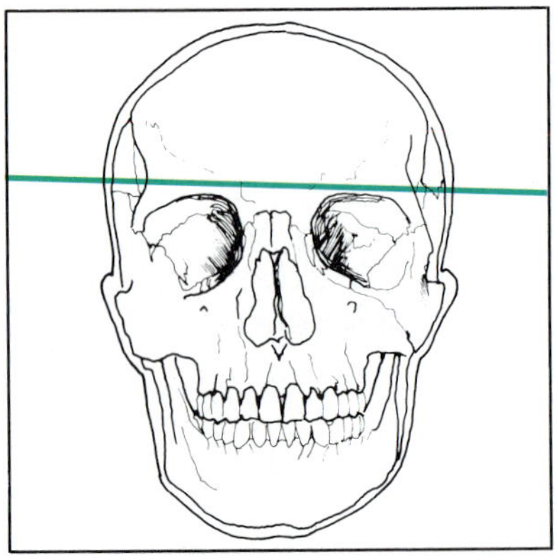

MODULE 3

This is a sheet plastinated (2.5 mm thick) transverse (axial) section of the head at the level of approximately 4 cm above Reid's baseline.

MUSCULOSKELETAL UNIT (Fig. 3–14)

I. Skeletal Portion

Frontal Sinus (*I*). Two rather large cavities between the inner and outer tables of the frontal bone.

Frontal Bone (*A*). A flat bone anterior and lateral to the frontal lobe of the brain.

Greater Wing of the Sphenoid (*B*). A lateral flat section of the sphenoid bone posterior and lateral to the eye socket. The greater wing of the sphenoid forms the posterior part of the orbital cavity and is lateral to the superior and inferior orbital fissures.

Parietal Bone (*C*). A flat, lateral bone located on the side of the cranial cavity. The parietal bone extends from the sagittal suture superiorly to the squamosal suture.

Inner Table (*F*). A thin sheet of compact bone seen in the frontal, parietal, and occipital bones in this section adjacent to the dura mater.

Diploë (*D*). A layer of spongy bone tissue, irregularly shaped, located between the inner and outer tables.

Outer Table (*E*). A thin sheet of compact bone seen in frontal, parietal, and occipital bones in this section adjacent to the scalp.

Occipital Bone (*G*). A thin sheet of flat bone on the posterior part of this section.

Internal Occipital Protuberance (*L*). Inward projection of the occipital bone immediately posterior to the superior sagittal sinus as seen in this section.

Upper Part of the Orbital Cavity (Orbital Plate of the Frontal Bone) (J). This oval feature is seen primarily on the anterior left part of the section. On the right side, the upper part of the orbital cavity is much smaller than on the left side in this section.

II. Muscular Portion

Frontalis Muscle (*H*). A thin slice of muscle tissue adjacent to the outer table of frontal bone.

Temporalis Muscle (*K*). A rather thick muscle slice that extends from the orbital area of this section to over half the distance to the posterior part in this section. The posterior part of this muscle is thinner than the anterior part.

NEUROVASCULAR UNIT (Fig. 3–15)

I. Arterial Structures

Choroid Plexus (*C*). A cluster of vascular tissue with a spongy appearance in this plastinated section of the head. A choroid plexus is found in each ventricle, although a choroid plexus is seen only in the lateral ventricles in this section.

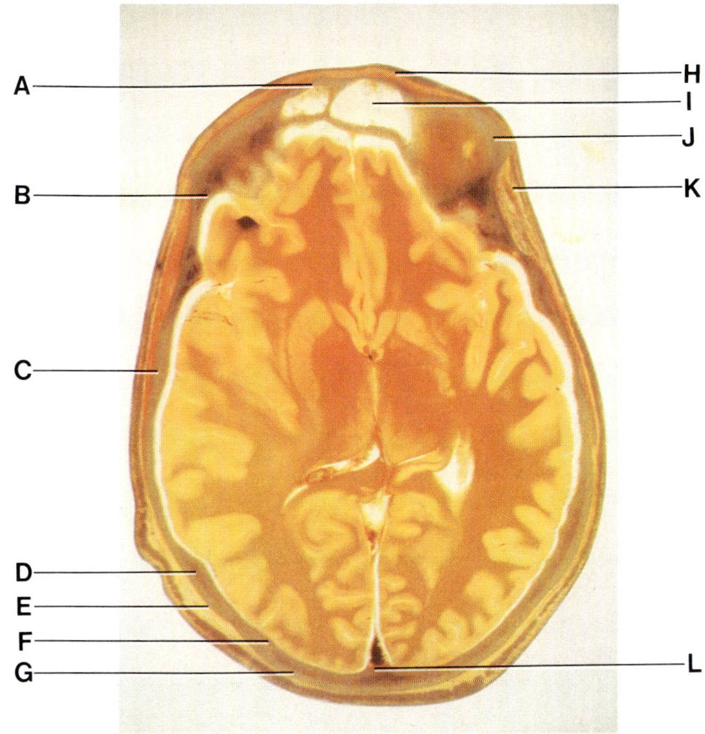

A. Frontal bone
B. Greater wing of sphenoid bone
C. Parietal bone
D. Diploë
E. Outer table
F. Inner table
G. Occipital bone
H. Frontalis muscle
I. Frontal sinus
J. Orbital plate of frontal bone
K. Temporalis muscle
L. Internal occipital protuberance

Figure 3–14
Plastinated transverse section of the head at the level of approximately 4 cm above Reid's baseline, highlighting the musculoskeletal unit structures.

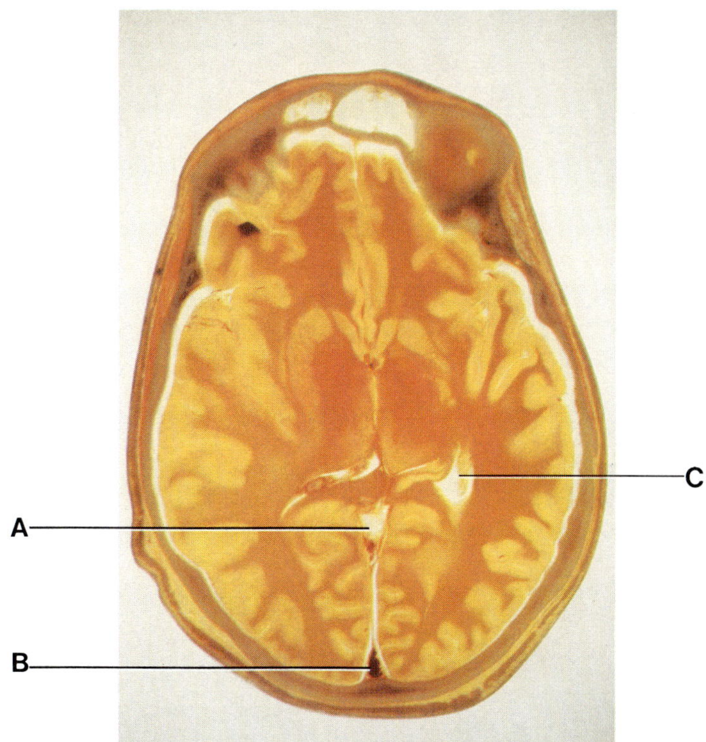

A. Straight sinus
B. Superior sagittal sinus (posterior portion)
C. Choroid plexus of lateral ventricle

Figure 3–15
Plastinated transverse section of the head at the level of approximately 4 cm above Reid's baseline, highlighting the neurovascular unit structures.

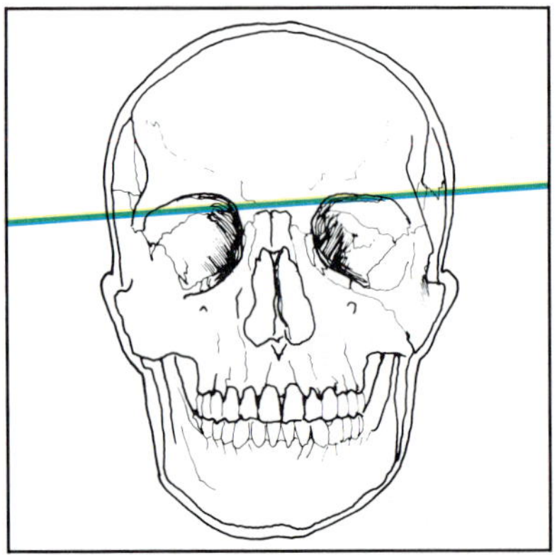

II. Vascular Venous Sinuses

Straight Sinus (*A*). A triangular vascular structure with its base on the anterior part.

Superior Sagittal Sinus (*B*). A triangular vascular structure formed by the meningeal layer of the dura mater immediately anterior to the internal occipital protuberance and posterior to the falx cerebri. The apex of this structure is adjacent to the falx cerebri.

Figure 3–16 is a CT scan (with contrast) in the transverse plane. This head slice emphasizes by labels both the musculoskeletal and neurovascular unit structures. Note the orientation line for derivation of this slice.

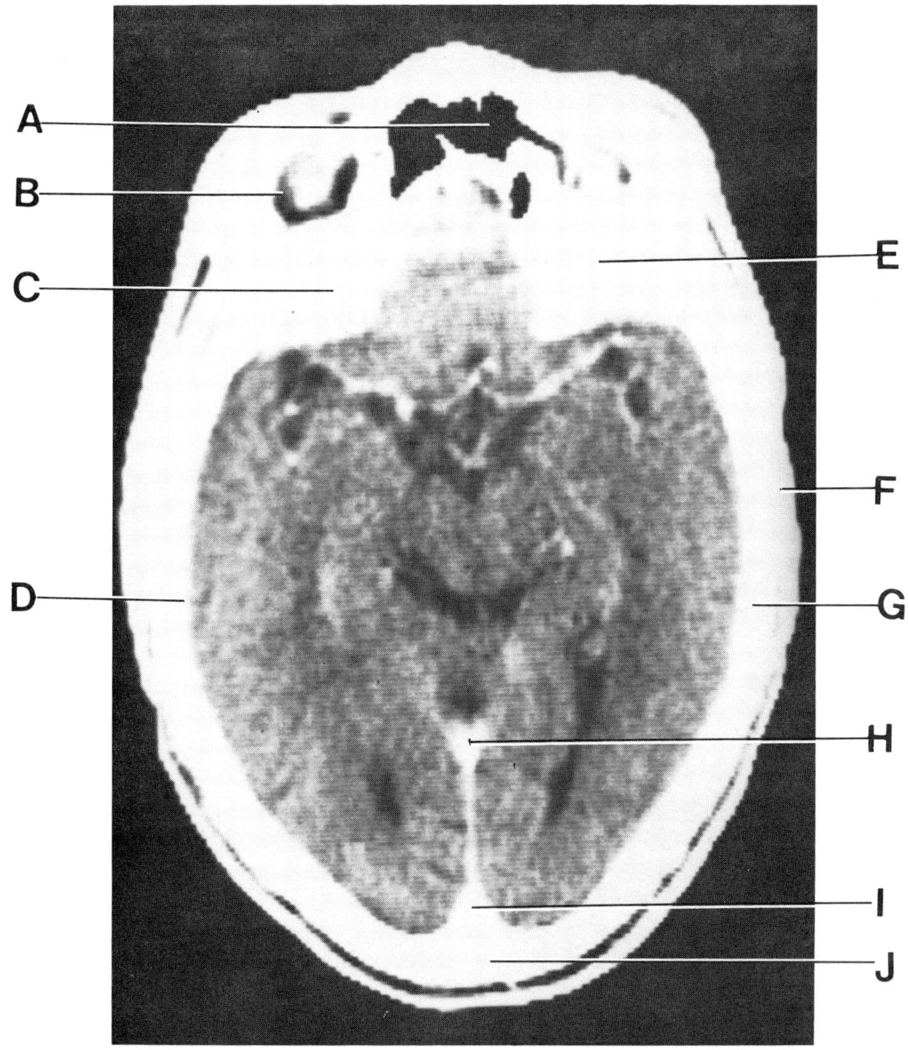

A. Frontal sinus
B. Orbit
C. Right orbital plate of frontal bone
D. Area of inner table
E. Left orbital plate of frontal bone
F. Area of outer table
G. Parietal bone
H. Straight sinus
I. Internal occipital protuberance
J. Occipital bone

Figure 3–16
CT (with contrast) transverse image of the head, highlighting both the musculoskeletal and neurovascular unit structures.

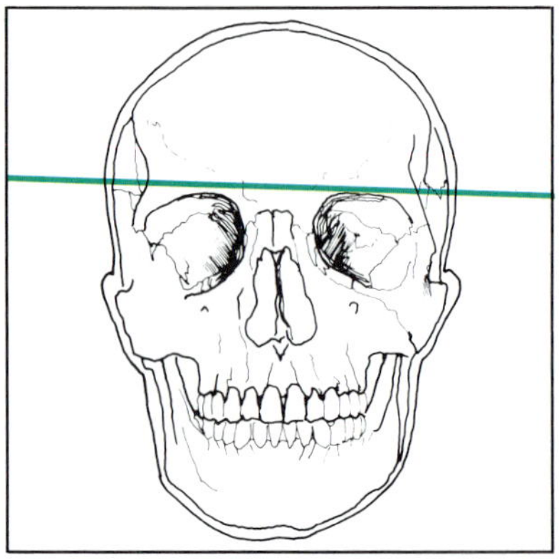

VISCERAL UNIT (Fig. 3–17)

Insular Lobe (*A*). Rather bilateral, large, deep to the lateral cerebral fissure and lateral to the extreme capsule. It is posterior to the frontal lobe as seen in this section.

Extreme Capsule (*B*). A crescent-shaped, rather thin sheet of white matter immediately medial to the peripheral edge of the insular lobe. The white matter in this picture of a plastinated specimen of the head appears brownish.

Caudate Nucleus (*F*). A gray matter area of the internal brain adjacent to the lateral ventricles.

Posterior Horn of the Lateral Ventricle (*G*). The back portion of the lateral ventricle located in the posterior cerebrum.

Claustrum (*C*). A thin strip of gray matter immediately medial to the extreme capsule and lateral to the external capsule. The gray matter in this picture of a plastinated specimen appears light in color.

Lentiform Nucleus (*E*). A mass of gray matter seen immediately medial to the external capsule and lateral to the internal capsule.

Internal Capsule (*D*). A rather thick layer of white matter immediately medial to the lentiform nucleus and lateral to the thalamus.

Third Ventricle (*L*). A median slit between the right and left thalamic brain parts. The third ventricle extends from the upper part of the thalamus to the lower part of the hypothalamus. This ventricle is between the structures (thalamus/hypothalamus) on the right and left sides.

Thalamus. (*L*). A rather large, oval part of the internal brain adjacent to the third ventricle, which forms its medial border. Its lateral border is adjacent to the internal capsule in this section. The thalamus is the most superior part of the diencephalon.

Splenium of Corpus Callosum (*H*). A rather thick, curved, white tissue with its concave surface on the posterior part. This part of the corpus callosum is immediately anterior to the straight sinus and posterior to the posterior horns of the lateral ventricles.

Longitudinal Cerebral Fissure (*I*). A slit seen between the frontal lobes immediately posterior to the frontal sinuses. Note: a part of the falx cerebri is seen within this section.

Frontal Lobe (*J*). A rather large, bilateral, triangular cerebral structure on the anterior part of this section. The apex of this lobe is toward the anterior part of the brain.

Lateral Cerebral Fissure (Fissure of Sylvius) (*K*). A slit on the side part of the brain. A section of the frontal lobe is seen anterior to this fissure, and the insular lobe is seen deep to it.

Cerebral Cortex (Gray Matter) (*M*). A lateral and peripheral folded part of the cerebrum. The folds are known as gyri, and the grooves are known as sulci, or fissures.

Posterior Horn of the Lateral Ventricle (*N*). A bilateral cavity of the posterior part of the lateral ventricle. The ventricle is immediately anterior to the splenium of the corpus callosum as seen in this section. These posterior horns are located in the anterior part of the occipital lobe.

Occipital Lobe (*O*). A large, bilateral, oval part of the cerebrum, with the posterior horn of the lateral ventricle lodged within its anterior part. Adjacent to its posterior medial part can be seen a section of the superior sagittal sinus.

ENCLOSING UNIT (Fig. 3–18)

Anterior Part of the Falx Cerebri (*A*). A thin sheet of meningeal dura mater between the frontal lobes lodged within the longitudinal cerebral fissure.

Subdural Space (*B*). A voided (empty) area deep to the dura mater and superficial to the arachnoid as seen in this section.

Arachnoid (*C*). A thin, irregular layer of tissue deep to the subdural space and superficial to the pia mater.

Pia Mater (*D*). A very thin sheet of tissue encasing the brain, as seen in this section, and deep to the arachnoid.

Posterior Part of the Falx Cerebri (*E*). A thin sheet of meningeal dura mater between the occipital lobes, lodged within the posterior part of the longitudinal cerebral fissure, as seen in this section.

Scalp (*F*). Consists of the skin and connective tissue structures that surround the entire section.

A. Insular lobe
B. Extreme capsule
C. Claustrum
D. Internal capsule
E. Lentiform nucleus
F. Caudate nucleus
G. Posterior horn of lateral ventricle (right side)
H. Splenium of corpus callosum
I. Longitudinal cerebral fissure
J. Frontal lobe
K. Lateral cerebral fissure
L. Third ventricle surrounded by the thalamic brain parts
M. Gray matter of cerebral cortex
N. Posterior horn of lateral ventricle (left side)
O. Occipital lobe

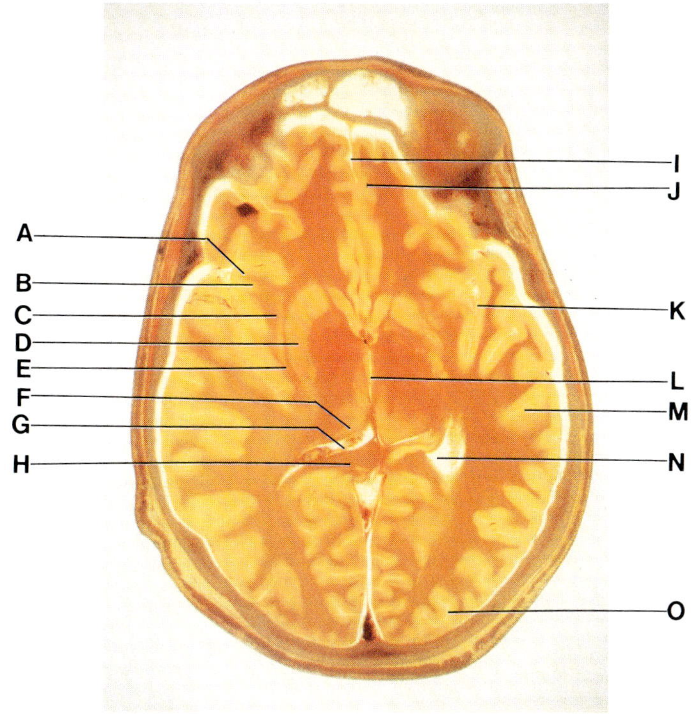

Figure 3–17
Plastinated transverse section of the head at the level of approximately 4 cm above Reid's baseline, highlighting the visceral unit structures.

A. Anterior part of falx cerebri
B. Subdural space
C. Area of arachnoid
D. Area of pia mater
E. Posterior part of falx cerebri
F. Scalp

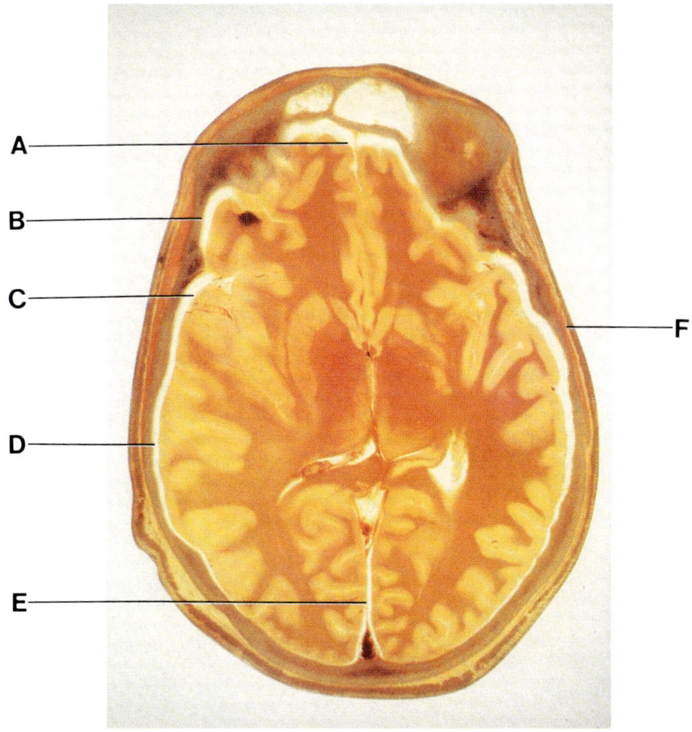

Figure 3–18
Plastinated transverse section of the head at the level of approximately 4 cm above Reid's baseline, highlighting the enclosing unit structures.

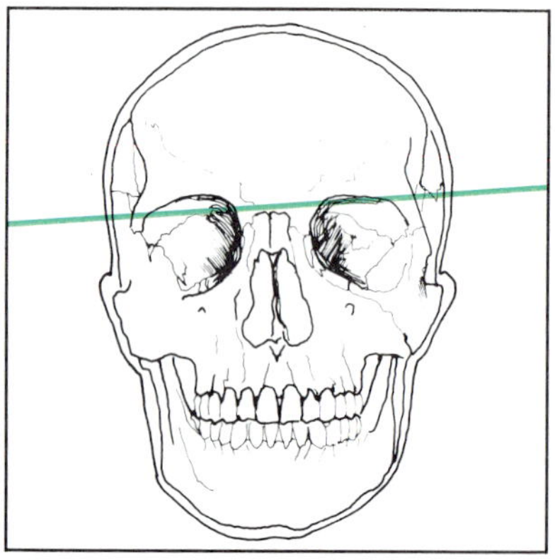

Figure 3–19 is a picture of a CT scan (with contrast) of the head region in the transverse plane. Visceral and enclosing unit structures are highlighted by labels. Note the orientation line for derivation of this head slice, that is, the body origin of the slice.

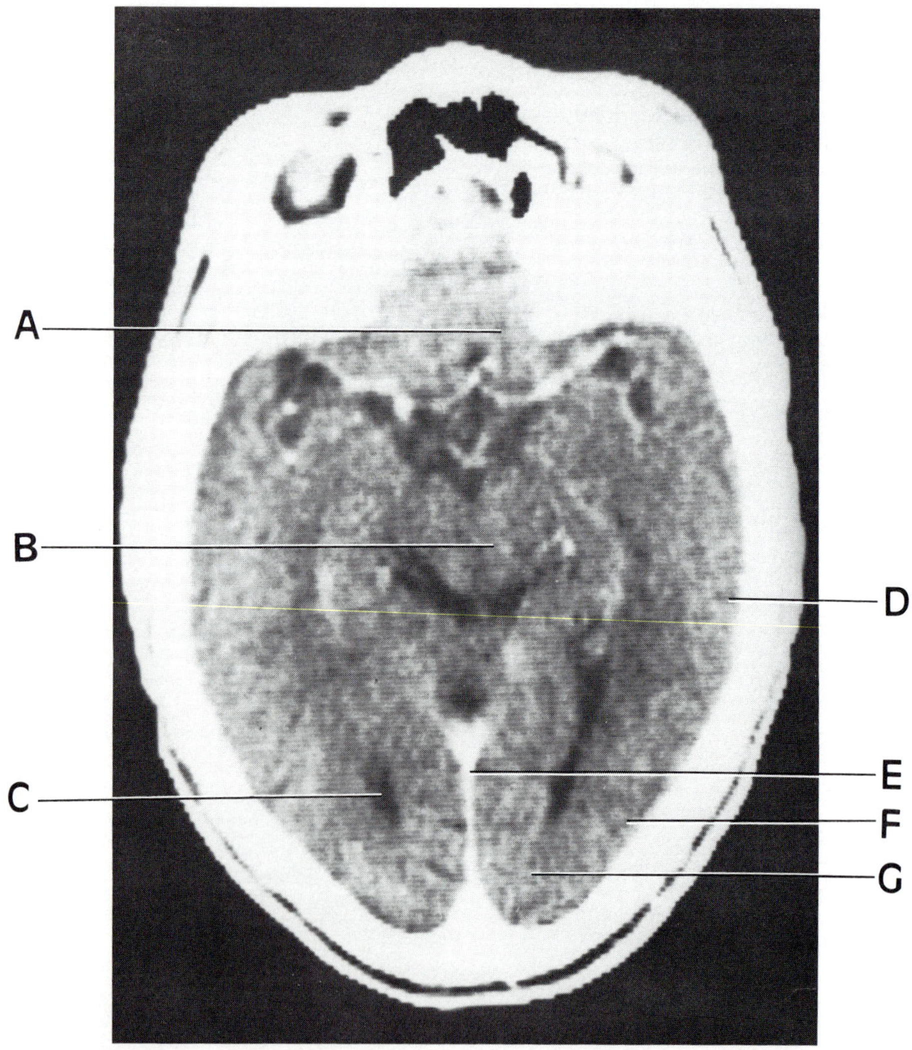

A. Rectus gyrus of frontal lobe
B. Midbrain
C. Occipital horn of lateral ventricle
D. Temporal lobe
E. Area of falx cerebri
F. Cortex
G. Occipital lobe

Figure 3–19
CT (with contrast) transverse image of the head, highlighting both the visceral and enclosing unit structures.

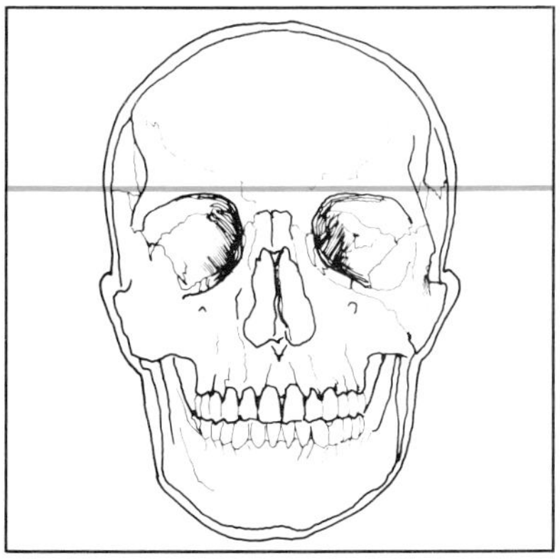

MRI transverse head image is shown in Figure 3–20. Visceral and enclosing unit structures, including many brain structures, are emphasized by labels.

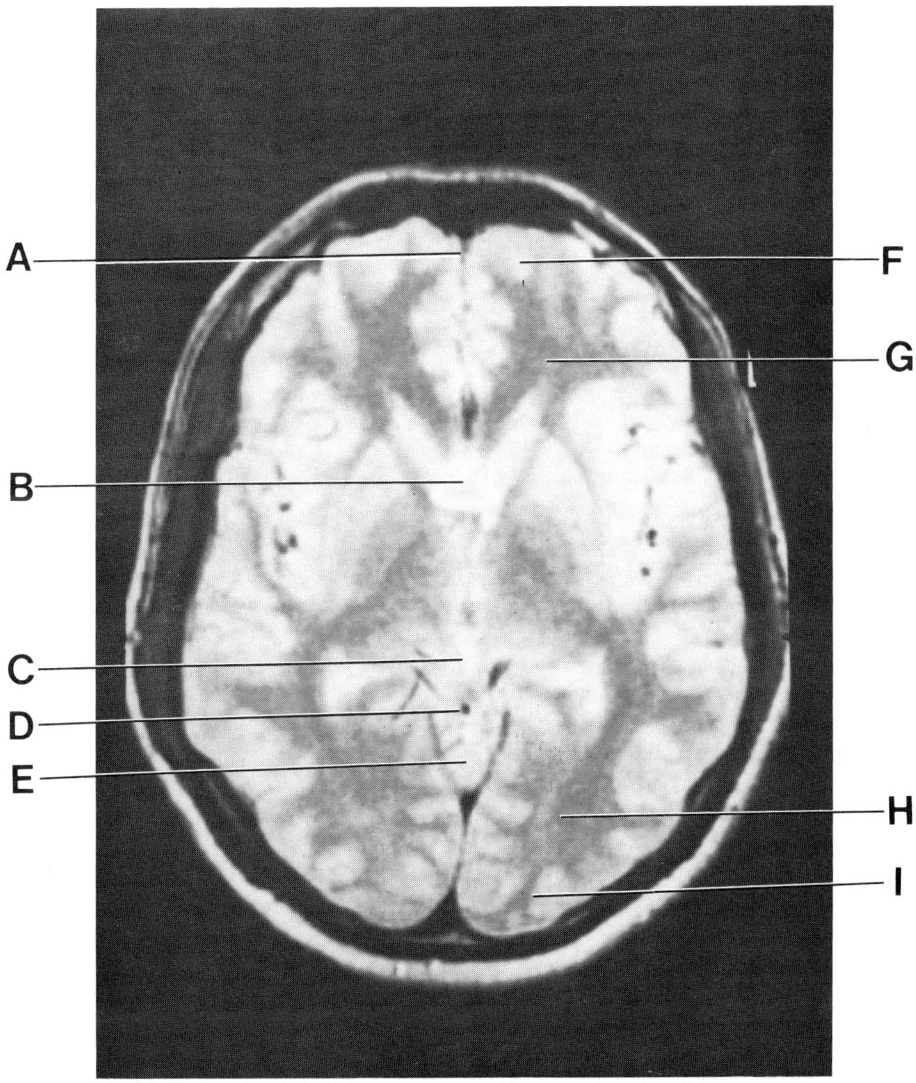

A. Anterior part of longitudinal cerebral fissure
B. Genu (a part of corpus callosum)
C. Splenium (a part of corpus callosum)
D. Aqueduct of Sylvius
E. Superior colliculus
F. Frontal lobe
G. White matter of frontal lobe
H. White matter
I. Occipital lobe

Figure 3–20
MRI transverse image of the head, highlighting both the visceral and enclosing unit structures.

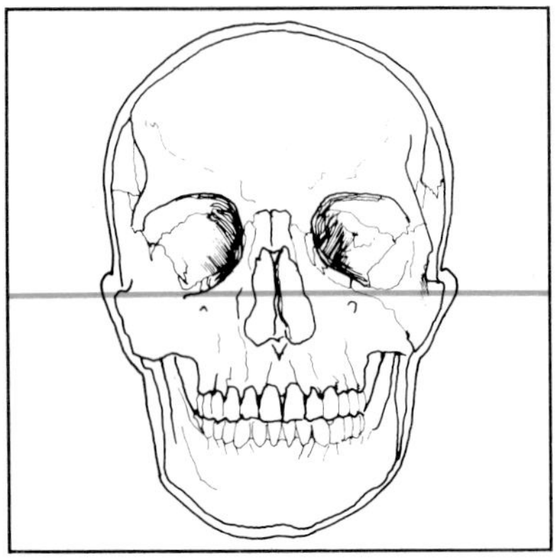

MODULE 4

This is a sheet plastinated (2.5 mm thick), transverse (axial) section of the head at the level of approximately 1 cm above Reid's baseline.

MUSCULOSKELETAL UNIT (Fig. 3–21)

I. Skeletal Portion

Septal Cartilage. A thin sheet of collagenous tissue that forms the anterior part of the nasal septum as seen in this section.

Nasal bone (*A, M*). An anterior strip of bone tissue immediately anterior to the septal cartilage. There are two of these bones. The bone slices in this plastinated section appear dark.

Nasal Portion of the Maxillary Bone (*N*). A laterally placed, rather thin section of bone tissue on the lateral wall of the nasal cavity. This bony structure runs posteriorly and obliquely. The posterior end is more lateral than the anterior end.

Nasal Septum (*B*). A thin sheet of tissue that separates the nasal cavity into two immediately lateral elongated spaces.

Ethmoid Sinus (*D*). Small, bilateral, irregularly shaped cavities immediately lateral to the nasal cavity and anterior to the maxillary sinus.

Maxillary Sinus (*F*). A rather large bilateral cavity lateral to the posterior part of the nasal cavity, as seen in this section, and medial to the zygomatic bone section.

Zygomatic Bone (*E*). An irregularly shaped bone slice lateral to the maxillary sinus and anterior to the temporalis muscle.

Sphenoid Sinus (*G*). A rather large bilateral cavity in the median posterior to the nasal septum and maxillary sinuses. This sinus is located within the body of the sphenoid bone.

Squamous Part of the Temporal Bone. A rather thick bone slice immediately lateral to the temporal lobe of the brain and medial to the temporalis muscle. The inner and outer layers of this bone slice appear compact, whereas the inner layer is spongy bone tissue.

Petrous Portion of the Temporal Bone (*H*). An inward projection of the temporal bone extending from the peripheral area to the posterior part of the sphenoid sinus. Spongy bone appears to be interspersed throughout the course of the hard bone.

Parietal bone (*I*). A rather thin sheet of flat bone seen bilaterally and posteriorly to the petrous portion of the temporal bone anterior to the occipital bone.

Parieto-Occipital Suture (Lambdoid Suture) (*J*). A bilateral seamlike feature that appears as a very narrow gap between the parietal and occipital bones.

Occipital Bone (*K*). A flat, curved bone that extends from the left-to-right parietal occipital sutures as seen in this section. The anterior part is the concave surface.

Internal Occipital Protuberance (*L*). An inward median projection of the occipital bone posterior to the median part of the cerebellum.

II. Muscular Portion

Orbicularis Oculi Muscle (*C*). A thin strip of bilateral muscle tissue anterior to the orbital area as seen in this section, which extends from the anterior medial orbit to the zygomatic bone slice.

Extrinsic Eye Muscles (*O*). Strips of muscle tissues located within the orbital area of this section.

Temporalis Muscle (*P*). An anterior-to-posterior elongated muscle slice lateral to the temporal bone and posterior to the orbital area, including the zygomatic bone section. The posterior end is thinner and extends as far back as the ear.

Occipitalis Muscle (*Q*). A thin, curved strip of muscle tissue immediately posterior to the occipital bone. The concave surface is adjacent to the occipital bone.

NEUROVASCULAR UNIT (Fig. 3–22)

I. Arteries

Internal Carotid Artery (*F*). An oval, bilateral, vascular structure located on the posterior part of the body of the sphenoid and medial to the temporal lobe.

II. Nerves

Optic Nerve (*E*). An elongated strip of nerve tissue seen entering the posterior orbital area in this section.

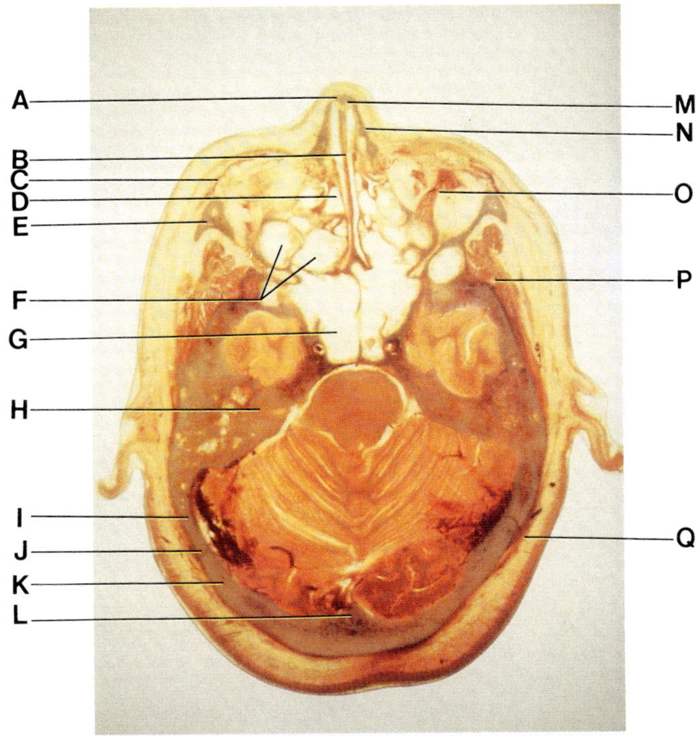

A. Nasal bone
B. Nasal septum
C. Orbicularis oculi muscle
D. Ethmoid sinus
E. Zygomatic bone
F. Maxillary sinus
G. Sphenoid bone
H. Petrous portion of temporal bone
I. Parietal bone
J. Parieto-occipital (lambdoid) suture
K. Occipital bone
L. Internal occipital protuberance
M. Nasal bone
N. Nasal portion of maxillary bone
O. Extrinsic eye muscles
P. Temporalis muscle
Q. Occipitalis muscle

Figure 3-21
Plastinated transverse section of the head at the level of approximately 1 cm above Reid's baseline, highlighting the musculoskeletal unit structures.

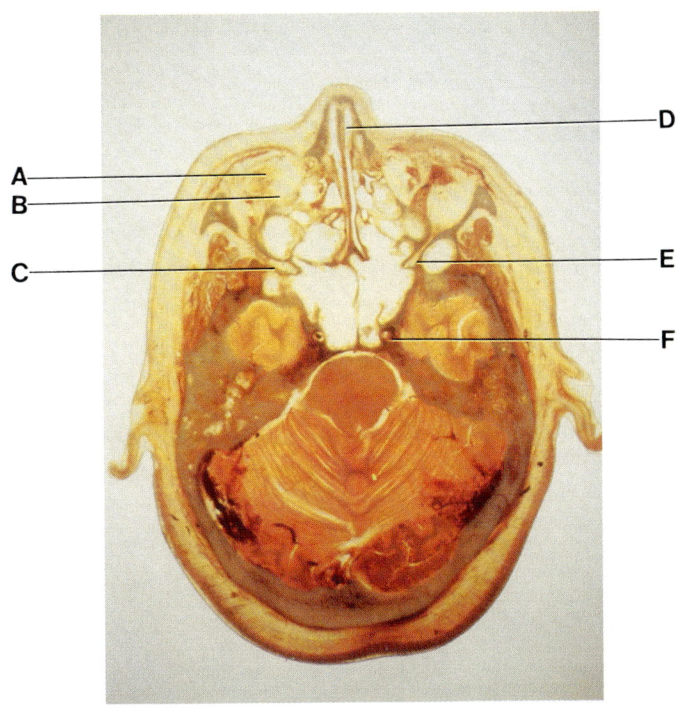

A. Orbital cavity
B. Orbital fat
C. Orbital foramen (a musculoskeletal structure)
D. Nasal cavity
E. Optic nerve
F. Internal carotid artery

Figure 3-22
Plastinated transverse section of the head at the level of approximately 1 cm above Reid's baseline, highlighting both the neurovascular and enclosing unit structures.

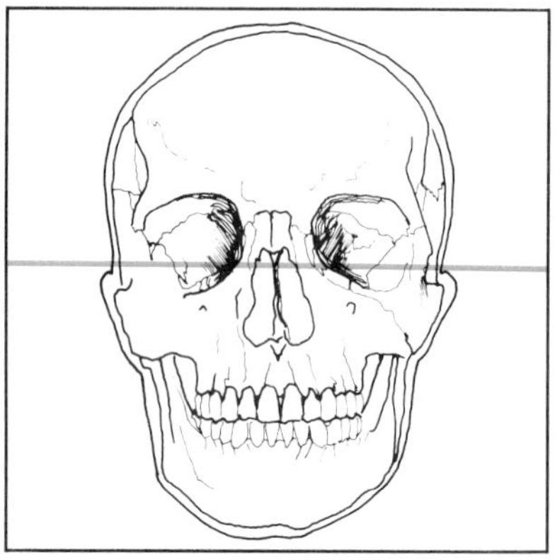

Figure 3–23 is a picture of a CT head scan (with contrast) in the transverse plane. Musculoskeletal unit structures are emphasized by labels.

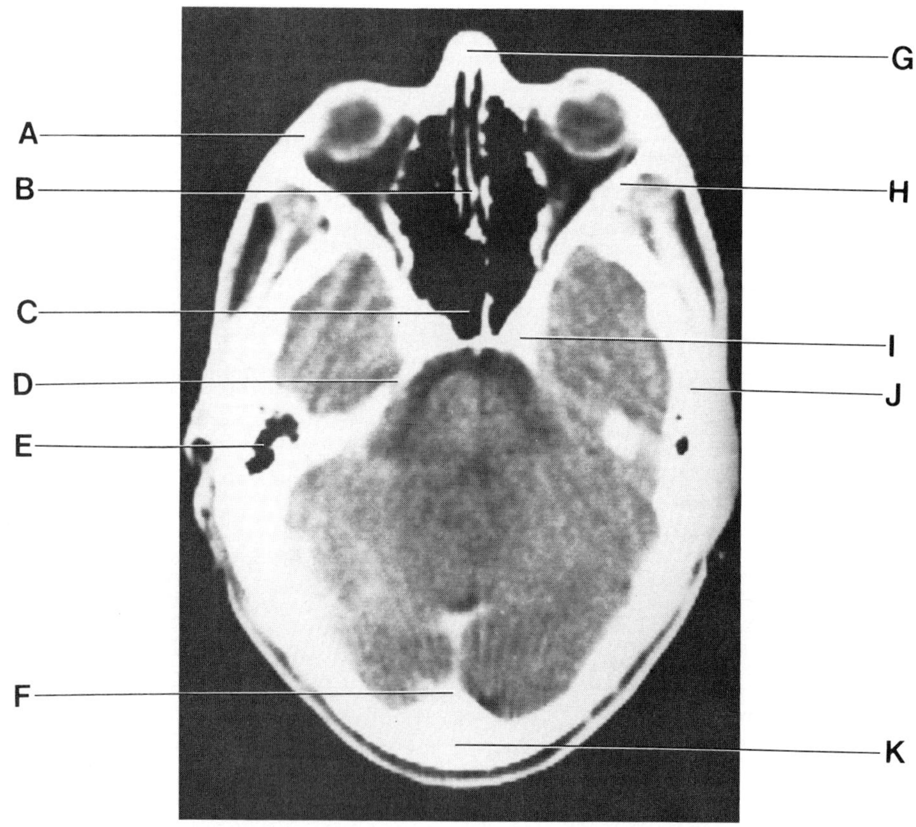

A. Zygomatic bone
B. Nasal septum
C. Sphenoid sinus
D. Apex of petrous part of temporal bone (right side)
E. Mastoid air cells of temporal bone
F. Internal occipital protuberance
G. Nasal bone
H. Greater wing of sphenoid bone
I. Apex of petrous part of temporal bone (left side)
J. Temporal bone
K. Occipital bone

Figure 3–23
CT (with contrast) transverse section of the head, highlighting musculoskeletal unit structures.

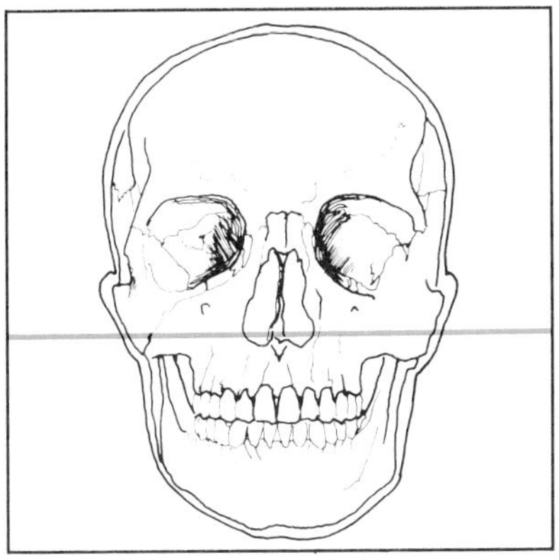

A CT head scan (with contrast) (Fig. 3–24) passes through the ethmoid sinuses as well as the clivus. This provides the opportunity to observe many sections of structures at this level in the head. Musculoskeletal structures are highlighted by labels in this transverse slice.

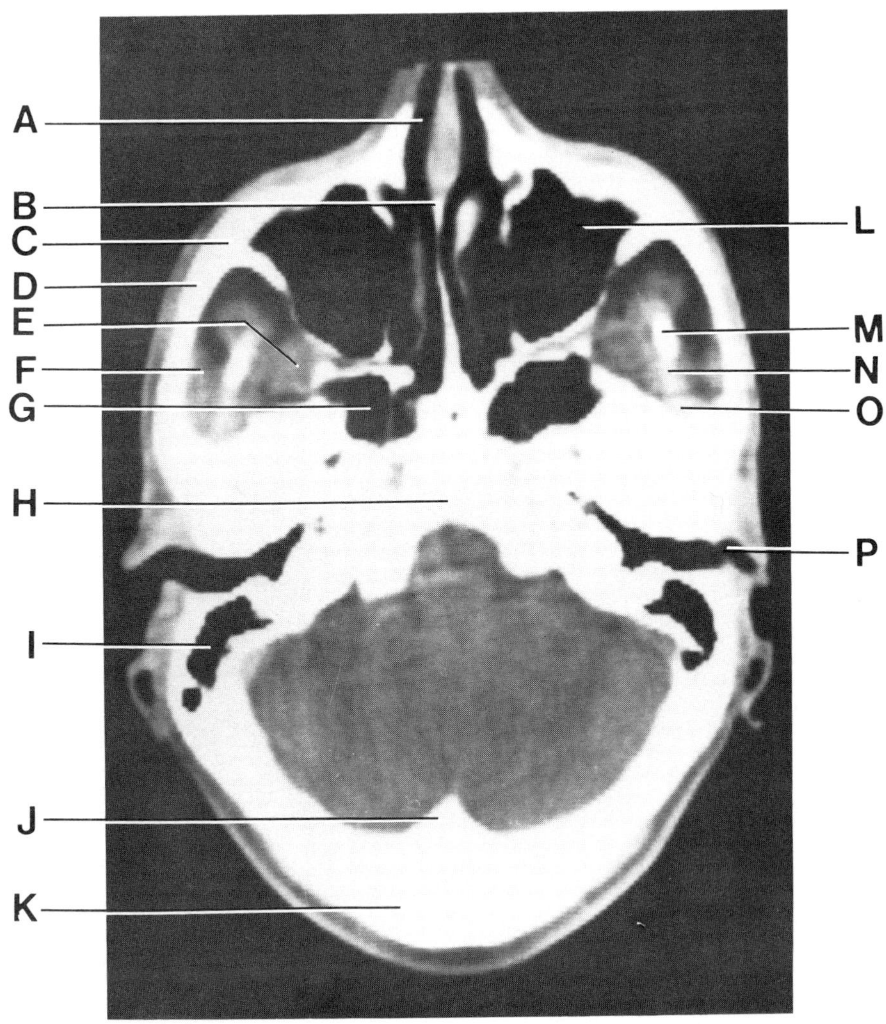

A. Nasal cavity
B. Nasal septum
C. Maxilla
D. Zygomatic arch
E. Temporalis muscle
F. Masseter muscle
G. Ethmoid sinus
H. Clivus of occipital bone
I. Mastoid air cells
J. Internal occipital protuberance
K. Occipital bone
L. Maxillary sinus
M. Coronoid process of mandible
N. Mandibular arch
O. Condyloid process of mandible
P. External acoustic (auditory) meatus

Figure 3–24
CT (with contrast) transverse section of the head, highlighting musculoskeletal unit structures.

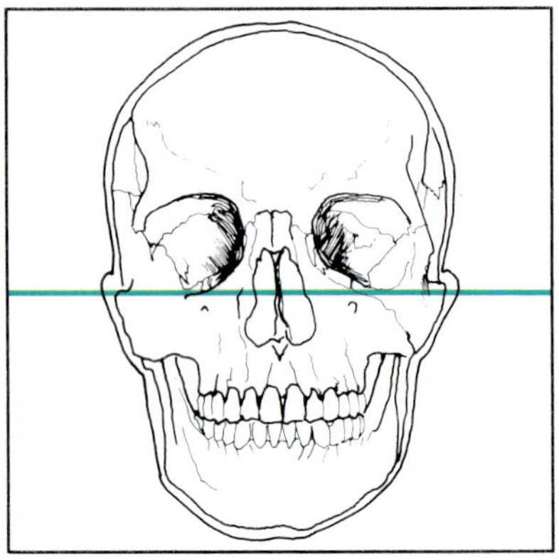

VISCERAL UNIT (Fig. 3–25)

Temporal Lobe (*A*). An oval, bilateral part of the cerebrum immediately lateral to the internal carotid artery, medial to the squamous portion of the temporal bone, and anterior to the petrous portion of the temporal bone.

Pons (*B*). A rather thick, curved median part of the brain stem anterior to the midbrain and posterior to the sphenoid sinus. The convex part of this structure is on its anterior surface.

Midbrain (*C*). An oval part of the brain stem immediately posterior to the pons in this section and anterior to the cerebellum. There is a hole in the posterior part of this midbrain slice.

Aqueduct of Sylvius (*E*). A hole in the posterior median of the midbrain. This feature conducts cerebrospinal fluid (CSF) from the third ventricle to the fourth ventricle.

Cerebellum (*F*). A large slice of brain tissue that is lodged in the posterior part of this section immediately posterior to the midbrain and petrous portion of the temporal bone.

Cerebellar Vermis (*D*). A rather large median part of the cerebellum is seen between the right and left hemispheres of the cerebellum. The three portions of the vermis seen in this section are rather oval.

Occipital Lobe (*G*). An oval bilateral brain structure located immediately lateral to the internal occipital protuberance and posterior to the cerebellum.

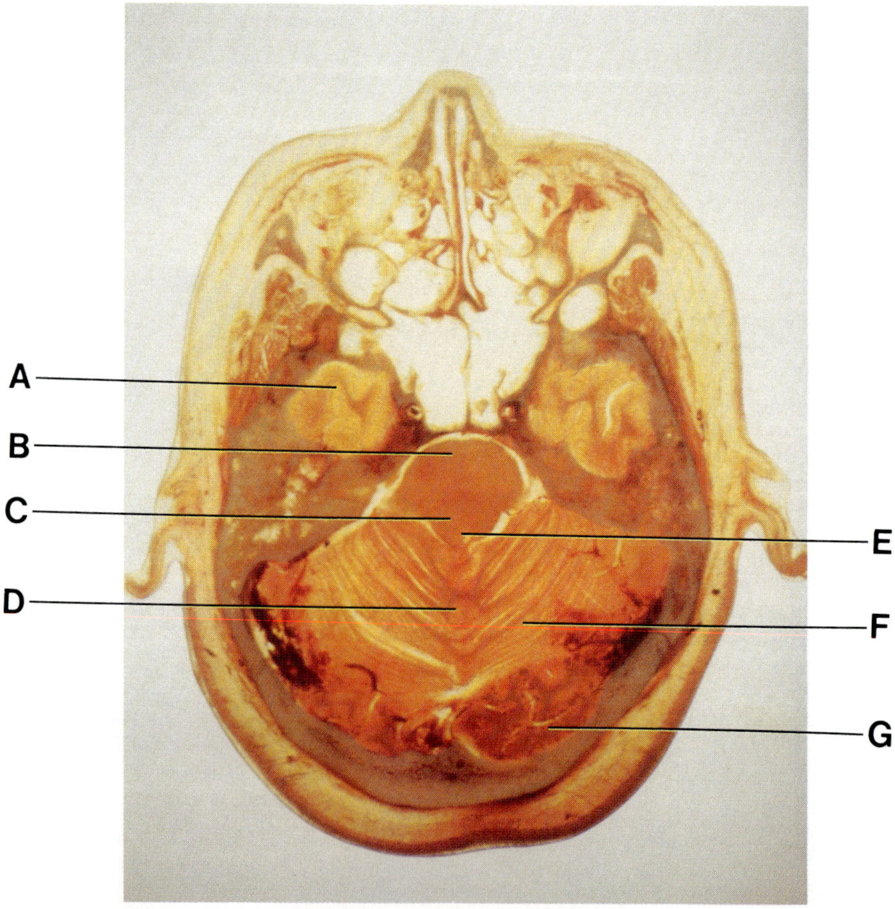

A. Temporal lobe
B. Pons
C. Midbrain
D. Cerebellar vermis
E. Aqueduct of Sylvius
F. Left cerebellar hemisphere
G. Occipital lobe

Figure 3–25
Plastinated transverse section of the head at the level of approximately 1 cm above Reid's baseline, highlighting the visceral unit structures.

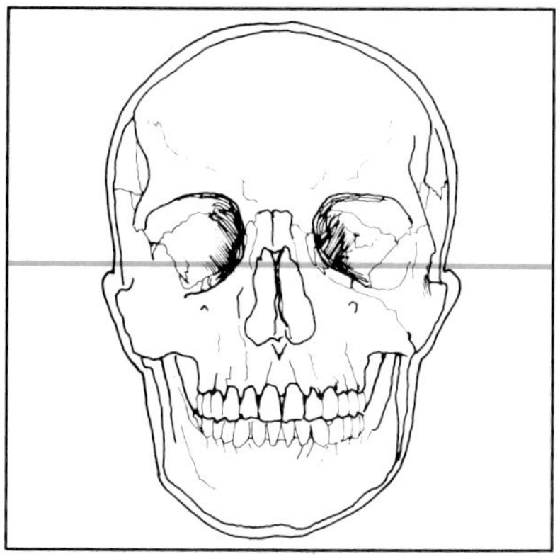

ENCLOSING UNIT (see Fig. 3–22)

Nasal Cavity (*D*). Two anterior-to-posterior elongated spaces seen on either side of the nasal septum.

Orbital Cavity (*A*). A thin, oval anterior lateral structure with fatty tissue lodged within the space.

Orbital Fat (*B*). Fatty tissue seen within the orbital area surrounding extrinsic muscles and the optic nerve.

Orbital Foramen (Orbital Canal) (*C*). An elongated hole posterior to the orbital area and anterior to the sphenoid sinuses. This canal has an oblique position, and the posterior end is more medial than the lateral end.

Figure 3–26 is a CT scan (with contrast) in the transverse plane as it passes through the pons.

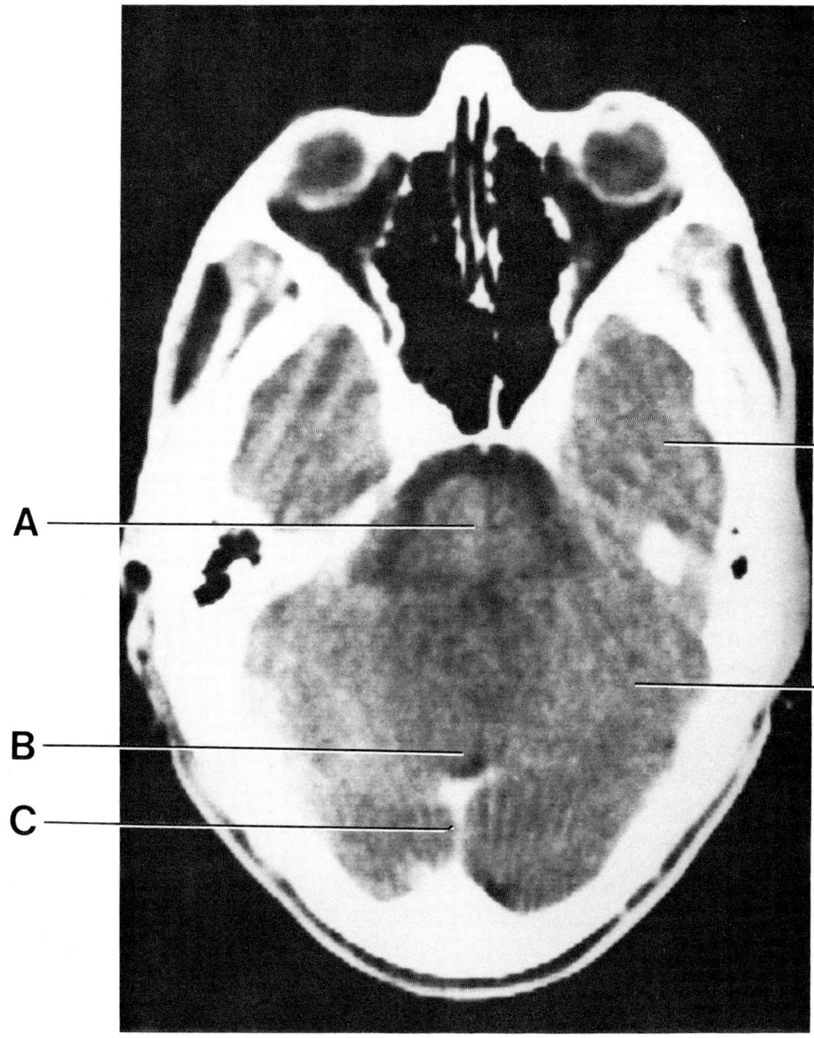

A. Pons
B. Cerebellar vermis
C. Area of falx cerebri
D. Temporal lobe
E. Left hemisphere of cerebellum

Figure 3–26
CT (with contrast) image of the head, highlighting both the visceral and enclosing unit structures.

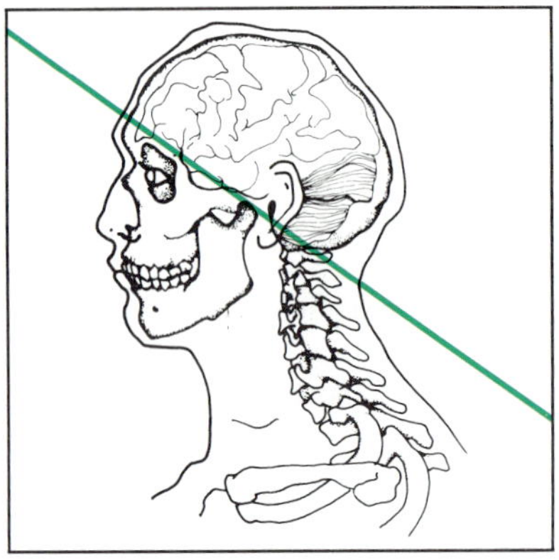

MODULE 5

This is a transverse (axial) section of the head at the level of Reid's baseline.

MUSCULOSKELETAL UNIT (Fig. 3–27)

I. Skeletal Portion

External Acoustic (Auditory) Meatus (*K*). An elongated, hollow feature anterior to the mastoid process of the temporal bone. The external auditory meatus is a canal of the outer ear.

Frontal Sinus (*F*). A hollow, oblong cavity on the median line anterior to the frontal lobe of the brain within the frontal bone.

Frontal Crest (*G*). An inner median ridge on the frontal bone.

Mastoid Process of the Temporal Bone (*L*). An oval bone structure of the temporal bone medial to the sternocleidomastoid muscle and posterior to the condyle of the mandible.

Sphenoid Sinus (*B*). A cavity inside the sphenoid bone. It is located in the median plane and medial to the temporal lobe of the brain.

Occipital Bone (*J*). This is seen on the median posterior to the sphenoid sinus.

Condyle of Mandible (*C*). This is seen here bilaterally in the median as a bony, oblong feature of the lower jaw anterior to the mastoid process of the temporal bone.

II. Muscular Portion

Trapezius (*Q*). A thin, elongated, transverse muscle posterior to the splenius capitis and the semispinalis capitis muscles.

Splenius Capitis (*O*). An elongated slice of muscle with a medial concave surface posterior and lateral to the semispinalis capitis muscle.

Semispinalis Capitis (*P*). This is seen here bilaterally in the median as a large, curved, sickle-shaped muscle posterior and lateral to the rectus capitis posterior major and minor muscles. It is deep to the splenius capitis muscle, and its concave surface is toward the median.

Obliquus Capitis Superior (*N*). This appears bilaterally as a spherical muscle anterior to the semispinalis capitis and medial to the splenius capitis muscles.

Rectus Capitis Posterior Minor (*E*). A round muscle slice between the right and left rectus capitis posterior major muscles and anterior to the semispinalis capitis muscle.

Rectus Capitis Posterior Major (*D*). This appears bilaterally as a thick, rectangular section of muscle lateral to the rectus capitis posterior minor muscle and deep to the semispinalis capitis muscle.

Sternocleidomastoid (*M*). This is seen bilaterally in the median as a transverse, elongated, thin, curved muscle. On the lateral neck its concave surface is toward the median. For more information, see the introduction to the neck.

Temporalis (*A*). This appears bilaterally as an elongated (anterior-to-posterior) muscle section lateral to the respective temporal bone. Each muscle slice is rather thick in its middle part and thinner on both ends.

Eye Muscles (*H, I*). These extrinsic eye muscles are seen embedded within the orbital fat. One is seen in the median of the orbital fat and the other is seen lateral to the orbital fat.

NEUROVASCULAR UNIT (Fig. 3–28)

I. Arteries

Internal Carotid Artery (*C*). This appears bilaterally as a hollow, round vessel anterior to the internal jugular vein and medial to the external auditory meatus.

Origin. Common carotid artery.

Distribution. This artery distributes blood to 10 branches. From these 10 branches, blood is conducted to the brain, eyes, pituitary gland, and forehead.

Course. The internal carotid artery arises from the common carotid artery in the carotid triangle of the neck. This vessel runs upward and enters the skull through the carotid canal of the temporal bone. It reaches the cavernous sinus and carotid groove on the side of the body of the sphenoid bone as it runs anteriorly. Then this artery terminates inferior to the perforated substance of the brain by splitting into middle and anterior cerebral arteries.

Superficial Temporal Artery (*B*). A small, round vessel superficial to the temporalis muscle.

Origin. External carotid artery.

Distribution. Masseter muscle, scalp, auricle, skin of face, and parotid gland.

Course. The superficial temporal artery starts at the level of the parotid gland posterior to the neck of the mandible. This vessel ascends on the superficial surface of the posterior part of the zygomatic process (temporal bone) and splits into anterior and posterior branches.

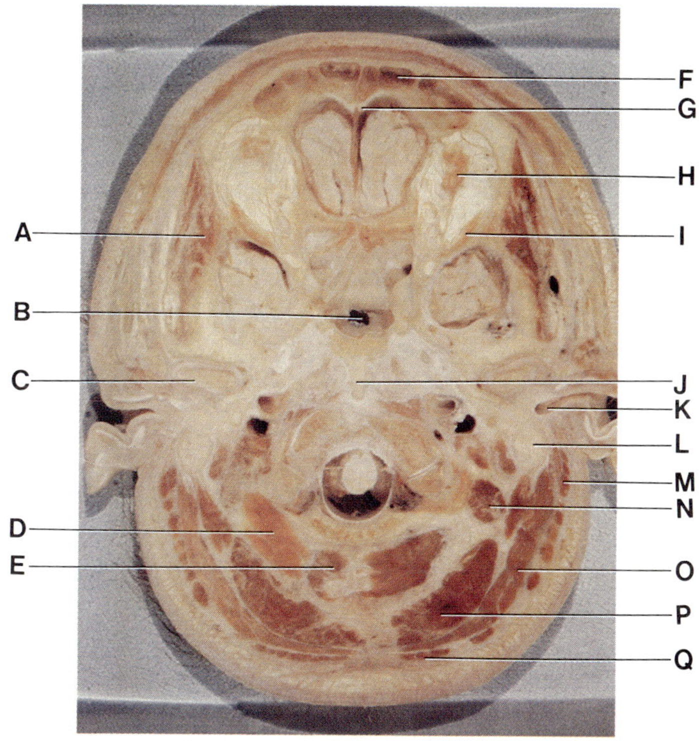

A. Temporalis muscle
B. Sphenoid sinus
C. Condyle of mandible
D. Rectus capitis posterior major muscle
E. Rectus capitis posterior minor muscle
F. Frontal sinus
G. Frontal crest
H. Extrinsic eye muscles
I. Extrinsic eye muscles
J. Occipital bone
K. External acoustic (auditory) meatus
L. Mastoid process of temporal bone
M. Sternocleidomastoid muscle
N. Obliquus capitis superior muscle
O. Splenius capitis muscle
P. Semispinalis capitis muscle
Q. Trapezius muscle

Figure 3–27
Transverse section of the head at the level of Reid's baseline, highlighting the musculoskeletal unit structures.

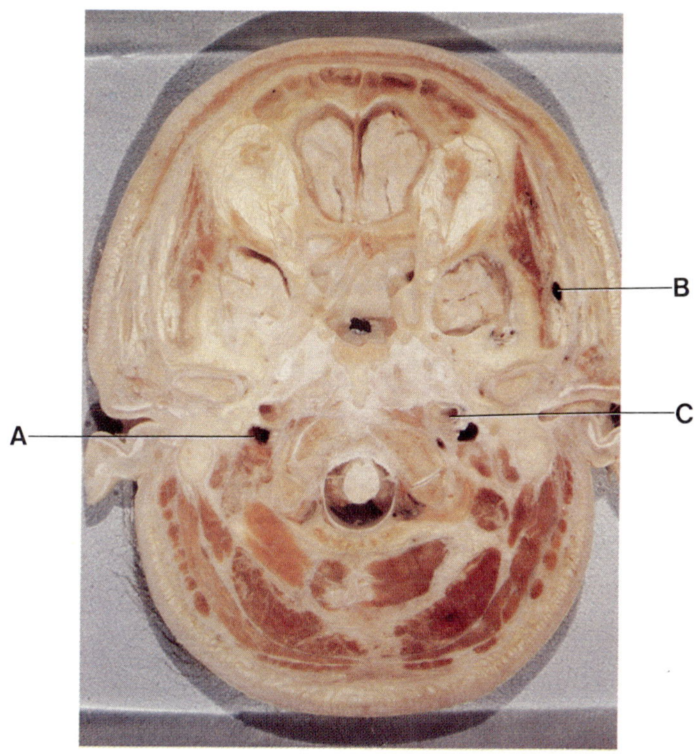

A. Internal jugular vein
B. Superficial temporal artery
C. Internal carotid artery

Figure 3–28
Transverse section of the head at the level of Reid's baseline, highlighting the neurovascular unit structures.

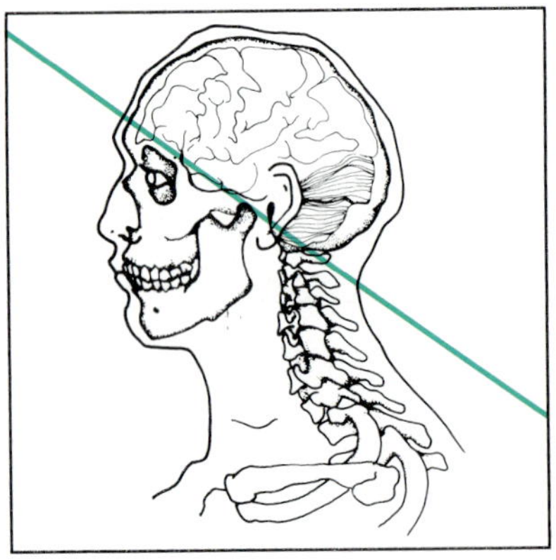

II. Veins

Internal Jugular Vein (*A*). This appears on both the right and left sides of the neck as a hollow, spherical vessel posterior to the internal carotid artery at this level.

Origin. Sigmoid sinus.

Termination. Joins with the subclavian vein to form the brachiocephalic vein.

Course. This vein is a direct continuation of the sigmoid sinus within the jugular foramen. This vein descends the neck within the carotid sheath lateral to the common carotid artery. It passes the posterior part of the sternal end of the clavicle and terminates in the brachiocephalic vein.

VISCERAL UNIT (Fig. 3–29)

Gyrus Rectus of the Frontal Lobe (*B*). This appears bilaterally in the median as a rather large, bulb-shaped brain part. These parts of the frontal lobe are between the right and left eye structures and lie on the medial part of the anterior cranial fossa.

Temporal Lobe (*D*). An oval shape on both right and left lateral sections of this part of the brain. Each slice of the temporal lobe is positioned in the lateral parts of the middle cranial fossa.

Medulla Oblongata (*E*). A full moon–shaped brain slice, the most inferior part of the brain stem. This brain part is located adjacent to the foramen magnum.

ENCLOSING UNIT

Dura Mater (*F*). A thick sheet of tissue that lines the inner wall of the cavity.

Orbital Fat (*C*). A white structure around the eye muscles within both eye sockets. Orbital fat is supportive and protective adipose tissue in the cone-shaped orbit.

Falx Cerebri (*A*). A thin, short membrane between the frontal lobe slices. In this view this structure is seen attached to the frontal crest anteriorly.

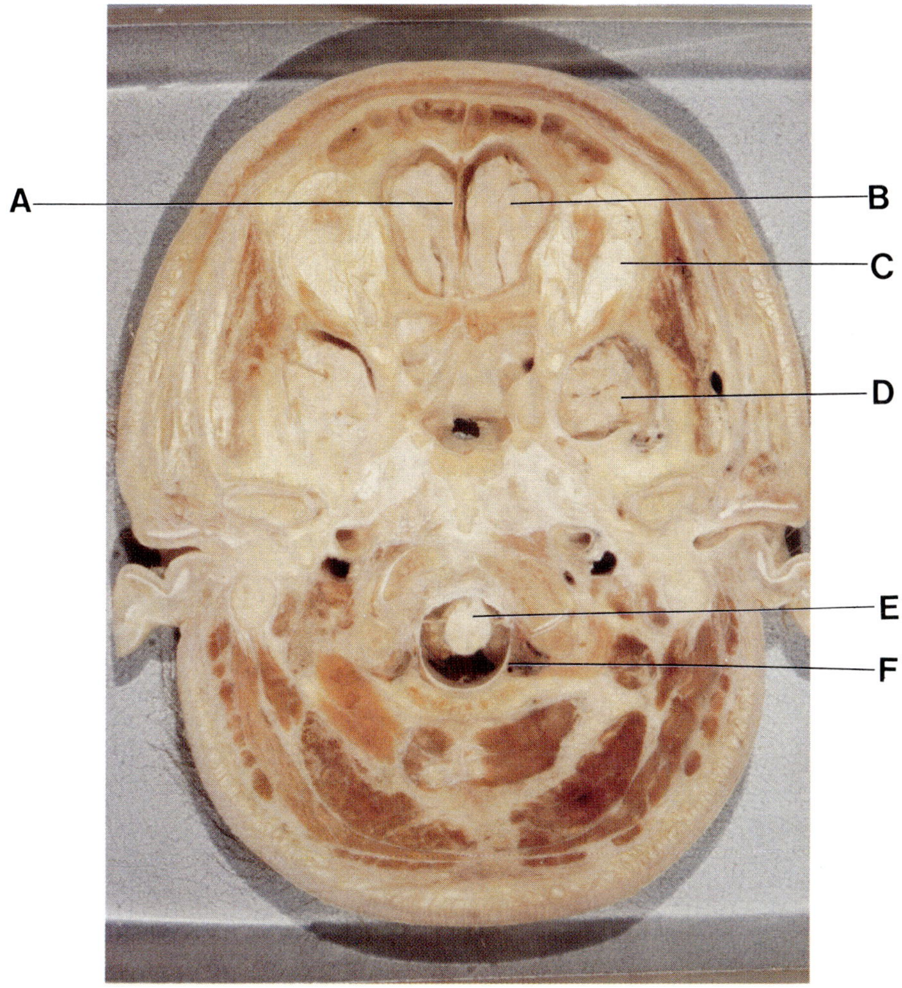

A. Falx cerebri
B. Gyrus rectus of frontal lobe
C. Orbital fat
D. Temporal lobe
E. Medulla oblongata
F. Dura mater

Figure 3–29
Transverse section of the head at the level of Reid's baseline, highlighting both the visceral and enclosing unit structures.

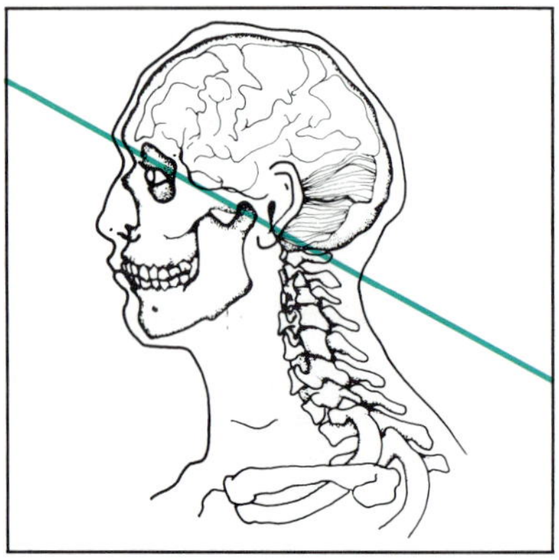

MODULE 6

This is a transverse (axial) section of the upper neck and head through the orbit approximately 1 cm below Reid's baseline.

MUSCULOSKELETAL UNIT (Fig. 3–30)

I. Skeletal Portion

Dens (Odontoid Process) (*W*). A bony, round feature anterior to the spinal cord.

Spinous Process of the Axis (*L*). A bifid bony structure adjacent to the median. This structure is the most posterior part of the axis.

Axis (*I*). A rather large, bony, round feature with a hole (spinal foramen) in the center. The axis is the second cervical vertebra.

Vomer Bone (*T*). A bony, elongated, midsagittal bone, the inferior part of the nasal septum.

Nasal Septum (*P*). This appears in the median as a bony sheet between the right and left nostrils.

Head of the Mandible (*U*). A flat bone appearing bilaterally in the median and posteriorly lateral to the lateral pterygoid muscle.

Frontal Process of the Zygomatic Bone (*C*). A triangular feature appearing bilaterally in the median and near the periphery of the head. Each process is located lateral to its respective eye.

Sphenoid Bone (*D*). A small, thin part of the sphenoid bone is seen in this section posterior to the orbital globe.

Infundibulum of the Frontal Bone (*O*). This appears as a small, hollow feature within the frontal bone. The infundibulum extends from the frontal sinus to the ethmoid sinuses.

Nasal Concha (Turbinate) (*A*). This appears on the lateral wall of the nasal cavity as an inward, elongated sagittal projection.

Spinal Foramen (*K*). A hole that passes through the vertebra.

Maxillary Sinus (*S*). Hollow spaces or cavities within the maxilla. Each maxillary sinus shows a similar shape.

II. Muscular Portion

Trapezius Muscle (*N*). A thin band of transverse muscle posterior to the splenius capitis muscle.

Sternocleidomastoid Muscle (*J*). This appears lateral to the neck as an elongated feature with a bulky anterior portion that is lateral to the axis and splenius capitis muscle.

Splenius Capitis Muscle (*M*). An elongated, curved muscle anterior to the trapezius and posterior and lateral to the semispinalis capitis. Each splenius capitis muscle shows a concave medial surface.

Semispinalis Capitis Muscle (*Y*). This appears bilaterally in the median as a bulky mass of muscle with a tapering medial end. Each is located anterior and medial to its respective splenius capitis muscle.

Temporalis Muscle (*E*). This appears bilaterally in the median as a mass of muscle medial to the masseter muscle and anterior to the lateral pterygoid muscle.

Lateral Pterygoid Muscle (*G*). This appears bilaterally in the median as an oval muscle slice lateral to the medial pterygoid and posterior to the temporalis muscles.

Longus Capitis Muscle (*V*). This appears as small, bilateral muscle slices anterior to the lateral part of the axis.

Obliquus Capitis Inferior Muscle (*X*). An elongated, slightly curved muscle slice anterior and lateral to the semispinalis capitis muscle.

Masseter Muscle (*F*). This appears bilaterally as a rather thick, elongated (sagittal), thin sheet of muscle on the lateral face.

Inferior Rectus Muscle (*R*). A small, bean-shaped muscle posteriorly within the orbit.

Lateral Rectus Muscle (*Q*). A small, sagittally oriented muscle slice lateral to the eyeball.

Medial Rectus Muscle (*B*). A small muscle section medial within the orbit and lateral to the nasal concha.

NEUROVASCULAR UNIT (Fig. 3–31)

I. Arteries

Internal Carotid Artery (*A*). This is seen here bilaterally in the median as a small, hollow structure anterior to the internal jugular vein and lateral to the longus capitis muscle.

Vertebral Artery (*B*). A small, dark, hollow structure located within the transverse foramen. This vessel is bilateral.

Origin. Subclavian artery.

Distribution. Lower portion of the cerebrum, spinal cord, cerebellum, neck muscles, and vertebrae.

Course. The vertebral artery arises from a subclavian artery, runs upward, and enters the transverse foramen of the sixth cervical vertebra and ascends the neck through each transverse foramen of successive cervical vertebrae. This artery enters the head through the foramen magnum and fuses with its paired vessel to form the basilar artery on the posterior surface of the clivus.

A. Nasal concha (turbinate)
B. Medial rectus muscle
C. Frontal process of zygomatic bone
D. Sphenoid bone
E. Temporalis muscle
F. Masseter muscle
G. Lateral pterygoid muscle
H. Medial pterygoid muscle
I. Axis
J. Sternocleidomastoid muscle
K. Spinal foramen
L. Spinous process of axis
M. Splenius capitis muscle
N. Trapezius muscle
O. Infundibulum of frontal bone
P. Nasal septum
Q. Lateral rectus muscle
R. Inferior rectus muscle
S. Maxillary sinus
T. Vomer
U. Head of mandible
V. Longus capitis muscle
W. Dens (odontoid process)
X. Obliquus capitis inferior muscle
Y. Semispinalis capitis muscle

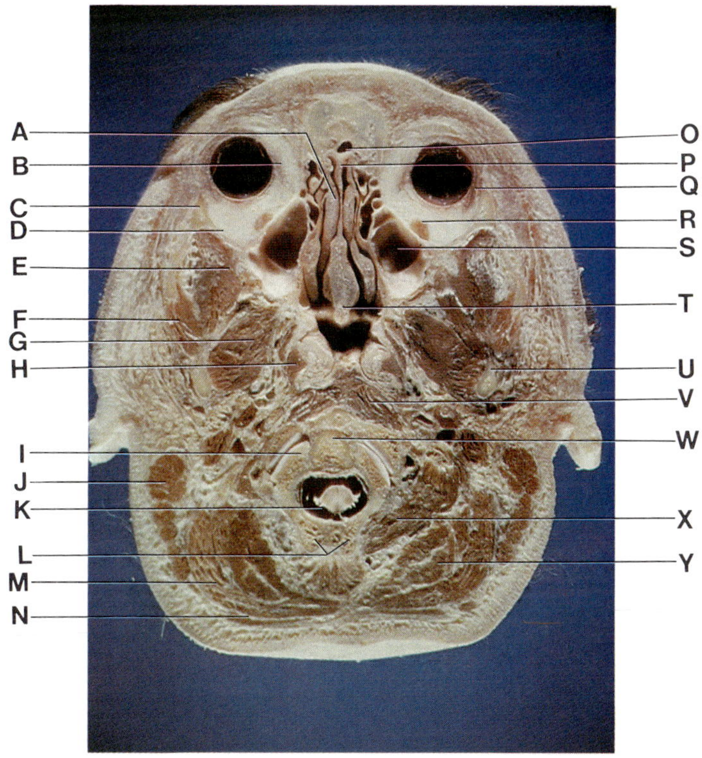

Figure 3-30
Transverse section of the head at the level of approximately 1 cm below Reid's baseline, highlighting the musculoskeletal unit structures.

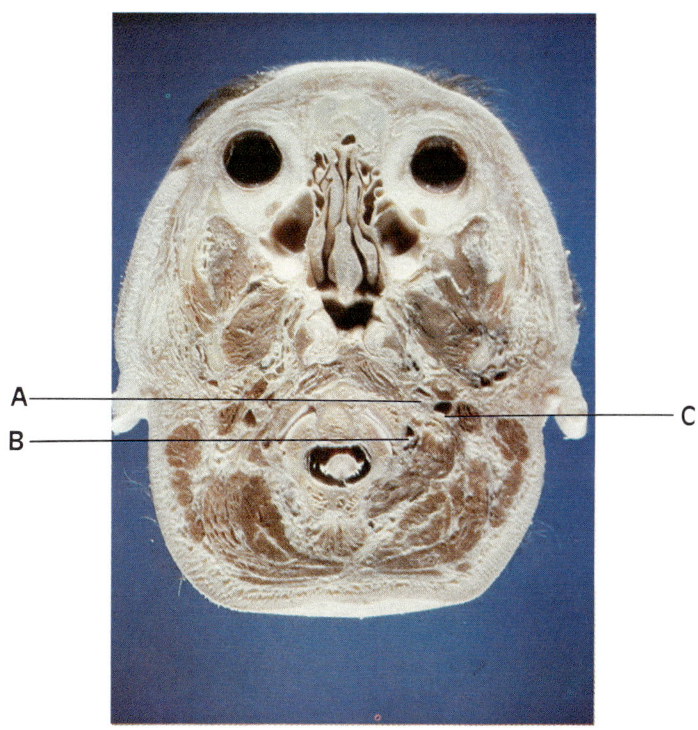

A. Internal carotid artery
B. Vertebral artery
C. Internal jugular vein

Figure 3-31
Transverse section of the head at the level of approximately 1 cm below Reid's baseline, highlighting the neurovascular unit structures.

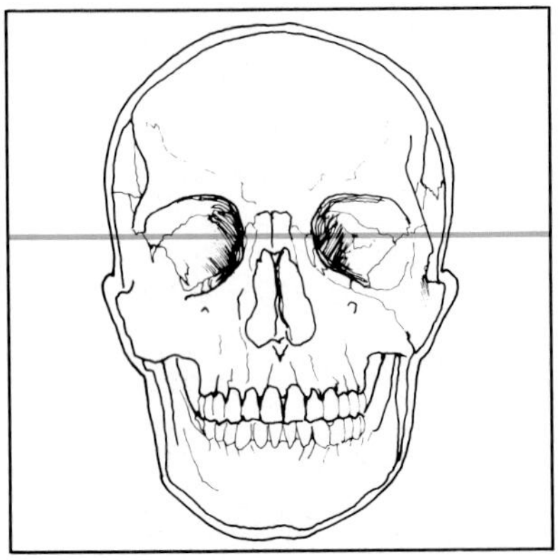

II. Veins

Internal Jugular Vein (C). This hollow, bilateral vessel is seen here posterior to the internal carotid artery and anterior to the obliquus capitis inferior muscle.

Figure 3–32 is a picture of a CT (with contrast) transverse scan of the head. This slice passes the ethmoid sinus, sphenoid sinus, and internal occipital protuberance. Both the musculoskeletal and neurovascular structures are highlighted by labels.

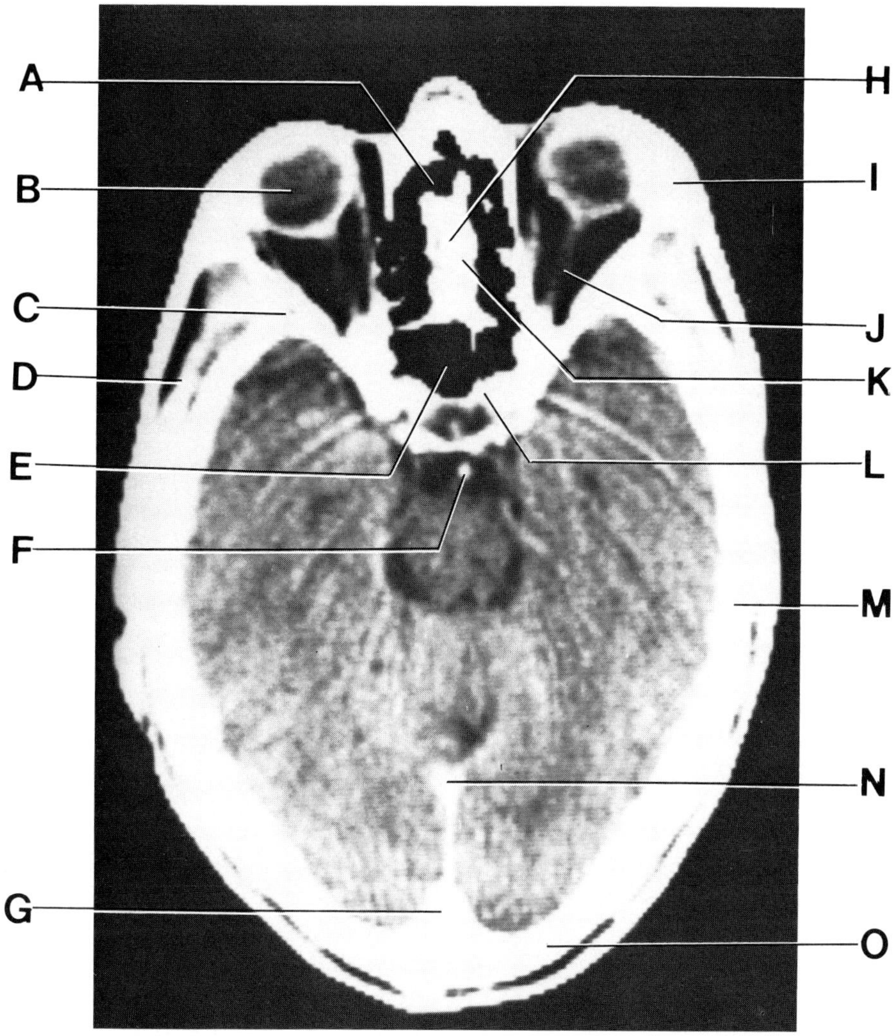

A. Ethmoid sinus
B. Orbital globe
C. Greater wing of sphenoid bone
D. Area of temporalis muscle
E. Sphenoid sinus
F. Basilar artery
G. Internal occipital protuberance
H. Area of crista galli of ethmoid bone
I. Zygomatic bone
J. Optic nerve
K. Cribriform plate of ethmoid bone
L. Dorsum sellae
M. Temporal bone
N. Straight sinus
O. Occipital bone

Figure 3-32
CT (with contrast) transverse image of the head, highlighting both the musculoskeletal and neurovascular unit structures.

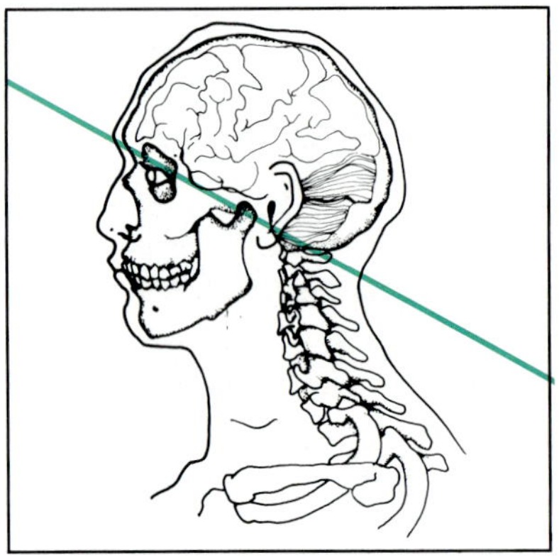

VISCERAL UNIT (Fig. 3–33)

Spinal Cord (*D*). A full moon–shaped nerve structure located within the spinal foramen (vertebral cavity).

Sclera of the Eye (*A*). The outer layer of the eye, which is tough, white, and hard, thereby furnishing support to the eyeball.

Vitreous Body of the Eye (*B*). A white structure in the orbital globe.

Nasopharynx (*C*). This is part of the throat and appears as a cup-shaped feature posterior to the vomer bone in this section.

ENCLOSING UNIT

Dura Mater. A thick membrane adjacent to the wall of the vertebral cavity.

Orbital Fat. A white adipose tissue around the eye structures.

Figure 3–34 is a transverse cadaver section of the head at approximately 1 cm below Reid's baseline. This is an enlarged picture at this level.

A picture of a transverse section at the level of the axis (Fig. 3–35) is enlarged to show structures in various anatomic units clearly.

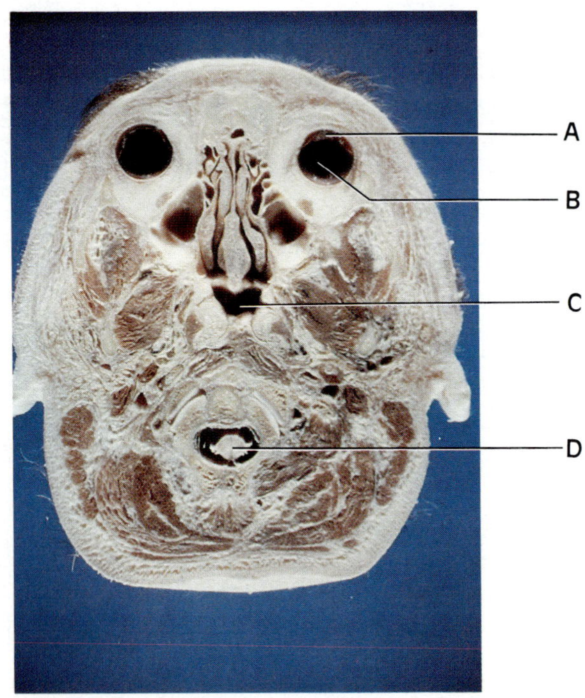

A. Sclera of eye
B. Vitreous body of eye
C. Nasopharynx
D. Spinal cord

Figure 3–33
Transverse section of the head at the level of approximately 1 cm below Reid's baseline, highlighting the visceral unit structures.

A. Nasal septum
B. Medial rectus muscle
C. Nasal concha (turbinate)
D. Orbital fat
E. Maxillary sinus
F. Sphenoid bone
G. Temporalis muscle
H. Lateral pterygoid muscle
I. Medial pterygoid muscle
J. Dens (odontoid process)
K. Axis
L. Spinal cord
M. Spinal foramen
N. Sclera of eye
O. Vitreous body of eye
P. Lateral rectus muscle
Q. Inferior rectus muscle
R. Vomer
S. Nasopharynx
T. Longus capitis muscle
U. Internal carotid artery
V. Internal jugular vein
W. Vertebral artery
X. Obliquus capitis inferior muscle
Y. Dura mater

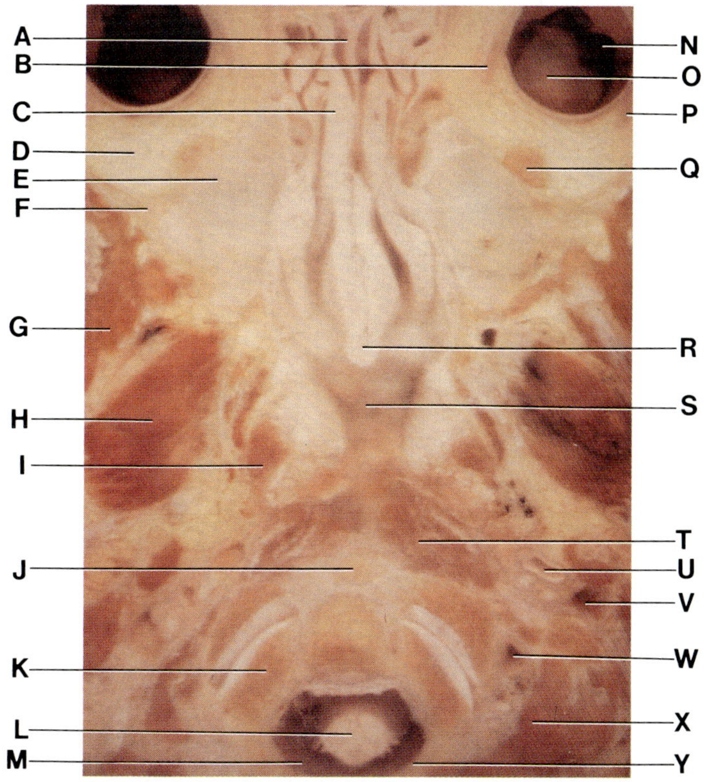

Figure 3-34
Enlargement of a transverse section of the head at approximately 1 cm below Reid's baseline, highlighting structures of some of the anatomic units.

A. Longus capitis muscle
B. Dens (odontoid process)
C. Spinal foramen
D. Sternocleidomastoid muscle
E. Spinous process of axis
F. Lateral pterygoid muscle
G. Internal carotid artery
H. Internal jugular vein
I. Vertebral artery
J. Axis
K. Spinal cord
L. Dura mater
M. Obliquus capitis inferior muscle
N. Semispinalis capitis muscle
O. Splenius capitis muscle
P. Trapezius muscle

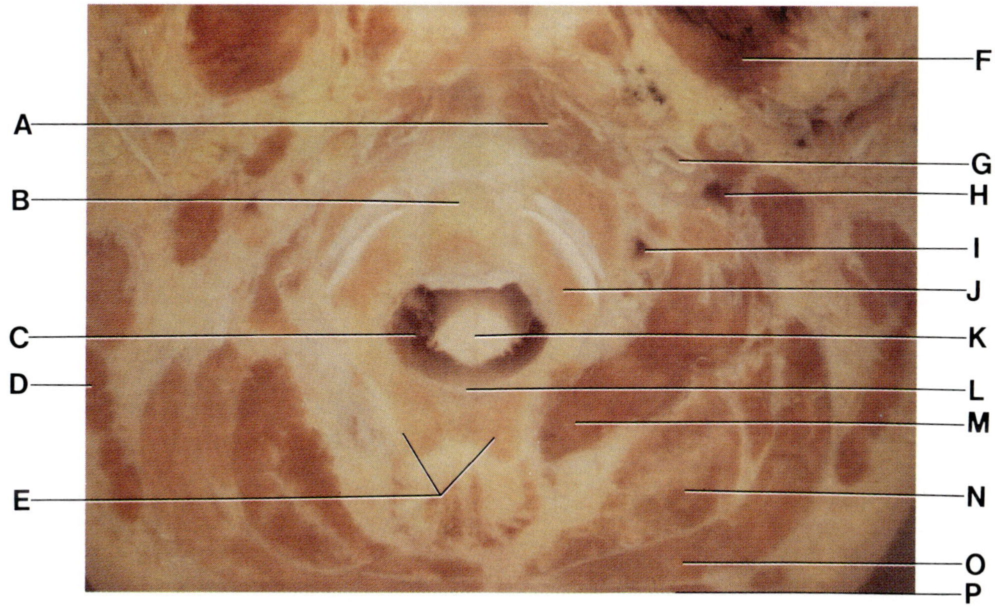

Figure 3-35
Enlargement of the transverse section of the head at the level of the axis, highlighting posterior structures of various anatomic units at this level.

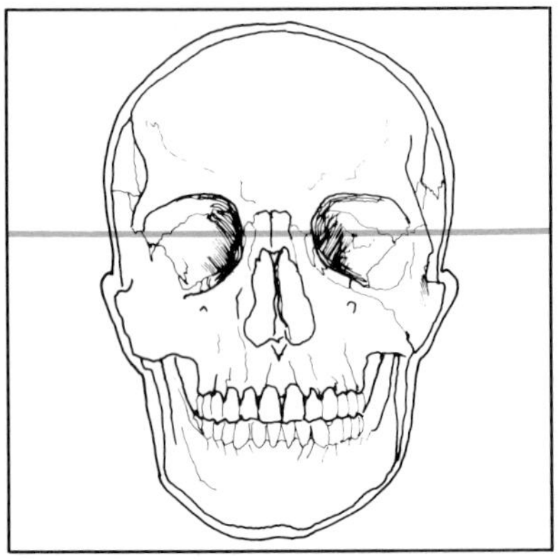

A transverse image of a CT scan (with contrast) (Fig. 3–36) passes through the midbrain and occipital lobe. Visceral and enclosing unit structures are emphasized by labels.

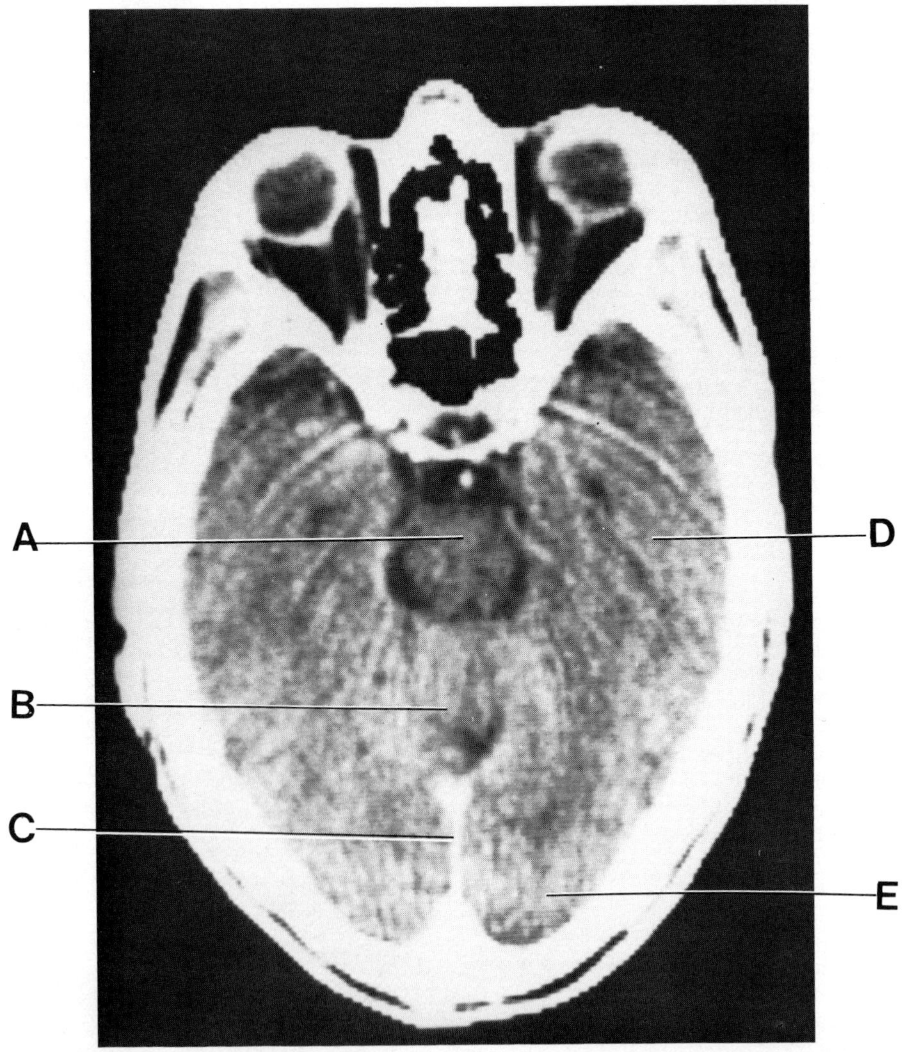

A. Midbrain
B. Cerebellar vermis
C. Area of falx cerebri
D. Temporal lobe
E. Occipital lobe

Figure 3–36
CT (with contrast) transverse image of the head, highlighting both the visceral and enclosing unit structures.

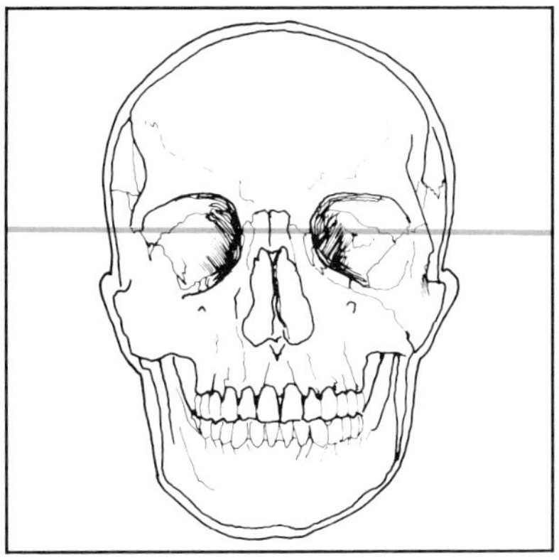

Figure 3–37 is a picture of an MRI transverse image of the head. Several structures that represent various anatomic units are labeled.

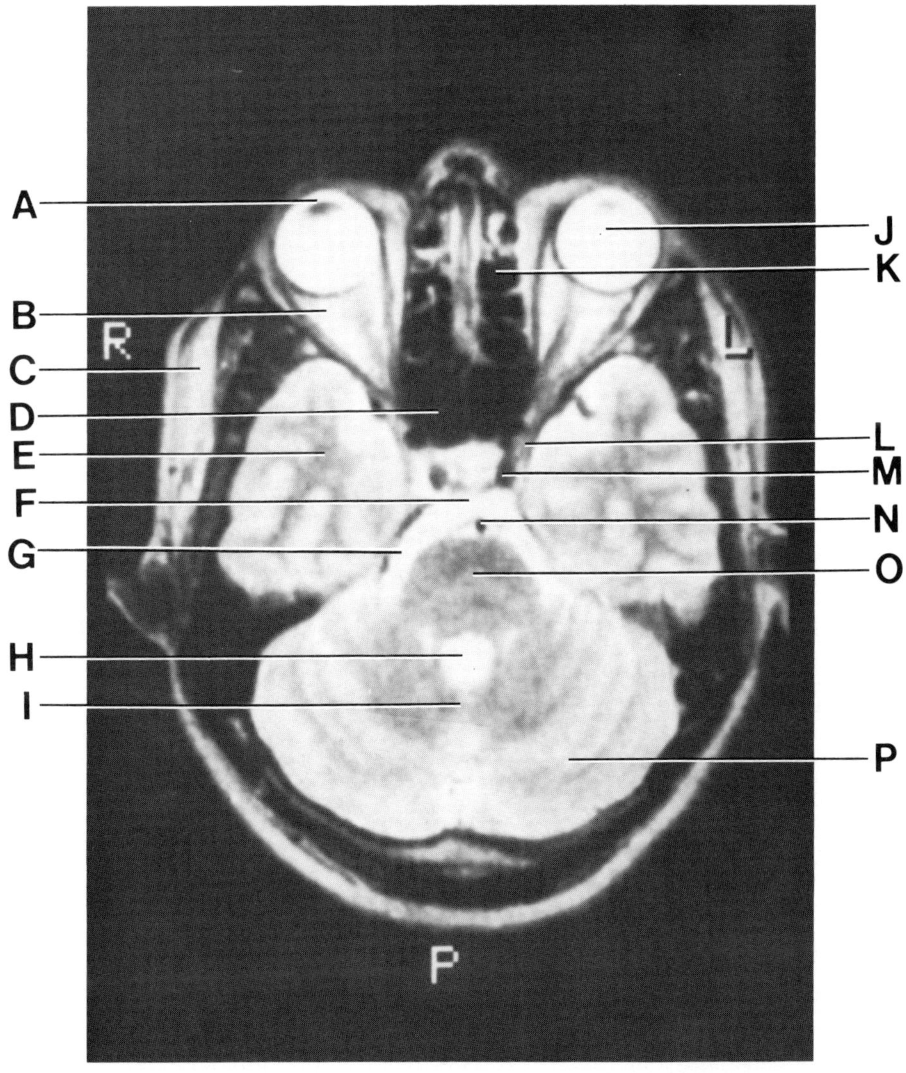

A. Lens of eye
B. Extrinsic muscles of eye
C. Temporalis muscle
D. Sphenoid sinus
E. Temporal lobe
F. Clivus
G. Cerebellopontine cistern
H. Fourth ventricle
I. Cerebellar vermis
J. Orbital globe
K. Ethmoid sinus
L. Optic nerve
M. Carotid canal
N. Basilar artery
O. Pons
P. Cerebellar hemisphere

Figure 3–37
MRI transverse image of the head, highlighting various anatomic unit structures.

CHAPTER 4

Coronal (Frontal) Cadaver Sections and Images of the Head

The four coronal (frontal) head illustration sections of this chapter are presented from anterior to posterior. Most of the sections reflect the labeling of four anatomic units. The anterior surface of each slice is viewed. The clinical images include three ultrasound and three magnetic resonance imaging (MRI) scans.

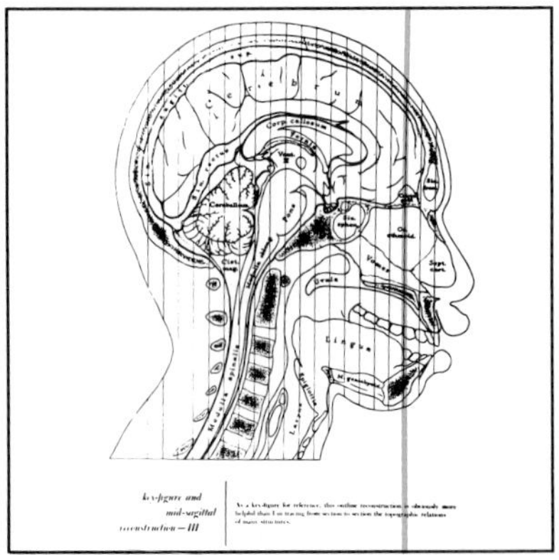

MODULE 1

We begin with the coronal (frontal) section, which passes through the crista galli of the ethmoid bone and the symphysis of the mandible. Noteworthy features are the extent of the frontal sinuses over the orbit and the relative straightness of the nasal septum but the asymmetry of the crista galli.

MUSCULOSKELETAL UNIT (Fig. 4–1)

I. Skeletal Portion

Frontal Bone (*A*). This is seen in this frontal (coronal) view as a dome-shaped bone with thickened bilateral edges. The inner and outer tables are clearly shown with the diploë sandwiched between them.

Crista Galli (*W*). A raised part of the ethmoid bone in the median of the anterior cranial cavity. It is clearly seen that part of the falx cerebri is attached to this feature of the ethmoid bone.

Maxillary Sinus (*P*). These are seen bilaterally as large cavities inside the maxillary bones immediately inferior to the orbits and lateral to the nasal cavity. These cavities extend from the inferior orbit to the roots of the teeth in the maxilla.

Perpendicular Plate. A thin, medial, bony feature of the ethmoid bone immediately inferior to the crista galli. Continuous inferiorly is another portion of the nasal septum known as the cartilaginous part.

Zygomatic Bone (*b*). Part of the inferior lateral orbit seen bilaterally next to the maxillary sinus.

Zygomaticomaxillary Suture (*J*). These joints are seen bilaterally as horizontal narrow junctions between the zygomatic and maxillary bones. They are lateral to the maxillary sinus and medial to the orbicularis oculi muscle.

Maxillary Bone (*S*). These are seen bilaterally next to the maxillary sinuses and superior to the oral cavity. This part of the oral cavity is dome-shaped and forms the anterior part of the hard palate. The part of the maxillary seen here separates the oral cavity from the nasal cavity. Teeth are seen bilaterally in the maxillary bone.

Mandible (*f*). A U-shaped, bony part at the inferior part of the oral cavity. Note teeth embedded in this bone.

Frontal Sinus (*X*). Bilateral cavities within the frontal bone superior to the eye socket, lateral to the crista galli.

Ethmoid Air Sinuses (*Y*). Several cavities are within the ethmoid bone medial to the eye socket, seen bilaterally next to the nasal septum.

Middle Nasal Meatus (*c*). A canal between the medial and inferior nasal conchae.

Inferior Nasal Concha (*d*). The most inferior medial projection from the lateral wall of the nasal cavity.

Middle Nasal Concha (*a*). A medial projection from the lateral wall of the nasal cavity. This structure is a part of the ethmoid bone and is located superiorly to the medial nasal meatus and inferiorly to the ethmoid sinuses.

Nasal Cavity (*e*). Bilateral spaces between the cribriform plates and maxilla. These cavities are separated by the nasal septum, which is made up of the perpendicular plate and cartilage.

Orifice of the Maxillary Sinus (*I*). This opening from the maxillary sinus extends superiorly and courses obliquely toward the middle to communicate with the ethmoid sinus.

Inferior Nasal Meatus (*K*). This is formed by the inferior nasal concha and the inferior part of the lateral wall of the nasal cavity. This conelike feature is bilateral.

Infundibulum (*G*). A funnel-shaped, canal-like cavity that communicates with both the frontal and ethmoid sinuses.

Cribriform Plate of the Ethmoid Bone. A bilateral, thin layer of bone immediately lateral to the bottom part of the crista galli. This feature contains many holes and transmits many fibers of the olfactory nerve.

II. Muscular Portion

Frontalis Muscle (*V*). A dome-shaped muscle superficial to the frontal bone. This muscle extends from the right temporal region to the left temporal region.

Superior Rectus Muscle (*D*). A transversely elongated, bilateral muscle in the posterior orbit inferior to the levator muscle of the upper eyelid and medial to the lacrimal gland.

Levator Muscle of the Upper Eyelid (Levator Palpebrae Superioris Muscle) (*C*). A transversely elongated, small part of muscle posterior to the superior orbit. This bilateral muscle is seen superior to the superior rectus and inferior to the supraorbital nerve.

Lateral Rectus Muscle (*F*). A vertical, elongated, bilateral muscle posterior to the lateral eye socket immediately inferior to the lacrimal gland and lateral to the eye bulb.

Inferior Oblique Muscle. A transversely oblong, bilateral muscle in the inferior orbit inferior to the inferior rectus.

Inferior Rectus Muscle (*H, Z*). A transversely

A. Frontal bone
B. Superior oblique muscle
C. Levator muscle of upper eyelid (levator palpebrae superioris muscle)
D. Superior rectus muscle
E. Medial rectus muscle
F. Lateral rectus muscle
G. Infundibulum
H. Inferior rectus (right) muscle
I. Orifice of maxillary sinus
J. Zygomaticomaxillary suture
K. Inferior nasal meatus
L. Orbicular muscle of eye (orbicularis oculi muscle)
M. Levator muscle of upper lips (levator labii superioris muscle)
N. Zygomatic minor muscle
O. Zygomatic major muscle
P. Maxillary sinus
Q. Buccinator muscle
R. Platysma
S. Maxillary bone
T. Depressor muscle of angle of mouth (depressor anguli oris [triangularis] muscle)
U. Depressor labii inferioris muscle
V. Frontalis muscle
W. Crista galli
X. Frontal sinus
Y. Ethmoid air sinuses
Z. Inferior rectus (left) muscle
a. Middle nasal concha
b. Zygomatic bone
c. Middle nasal meatus
d. Inferior nasal concha
e. Nasal cavity
f. Mandible
g. Mentalis muscle

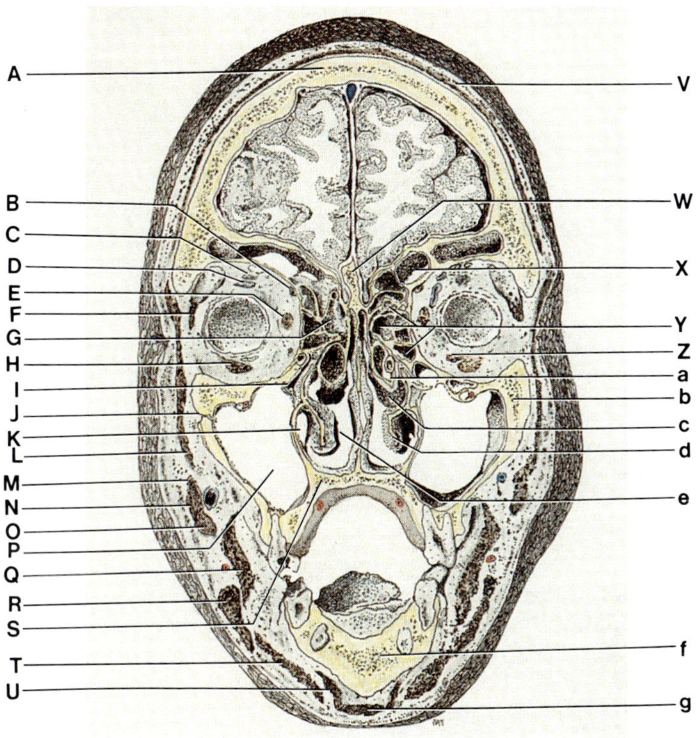

Figure 4–1
Coronal (frontal) section of the head, which passes through the crista galli of the ethmoid bone, highlighting the musculoskeletal unit structures.

elongated, small, bilateral muscle slice inferior to the eye bulb and superior to the medial part of the inferior oblique muscle.

Medial Rectus Muscle (*E*). An oval, bilateral muscle located in the medial part of the eye socket lateral to the ethmoid sinuses and medial to the eye bulb.

Superior Oblique Muscle (*B*). Small, oval, bilateral muscle tissue in the superomedial part of the eye socket.

Orbicular Muscle of the Eye (Orbicularis Oculi Muscle) (*L*). This bilateral, vertically elongated, lateral muscle extends from the lateral superior part of the orbit to the superficial fascia immediately above the levator muscle of the upper lip and the zygomatic minor muscles.

Levator Muscle of the Upper Lip (Levator Labii Superioris Muscle) (*M*). This bilateral muscle is seen in this figure as three muscles fused. The part immediately inferior to the upper part is the zygomatic minor, and the lower part is the zygomatic major. The three muscles fused together appear as a vertical, rather large muscle mass. This muscle mass is lateral to the facial vein and inferior to the orbicularis oculi muscle.

Zygomatic Minor Muscle (*N*). This is described in the preceding discussion of the levator muscle of the upper lip.

Zygomatic Major Muscle (*O*). This is described in the preceding discussion of the levator muscle of the upper lip.

Buccinator Muscle (*Q*). A vertically elongated muscle aggregate with two rather sharp, pointed ends. This bilateral muscle is seen medial to the external maxillary artery and lateral to the mouth cavity.

Platysma (*R*). A cone-shaped, superficial and bilateral muscle inferior and lateral to the buccinator muscle.

Depressor Muscle of Angle of Mouth (Depressor Anguli Oris Muscle) (Triangularis Muscle) (*T*). A

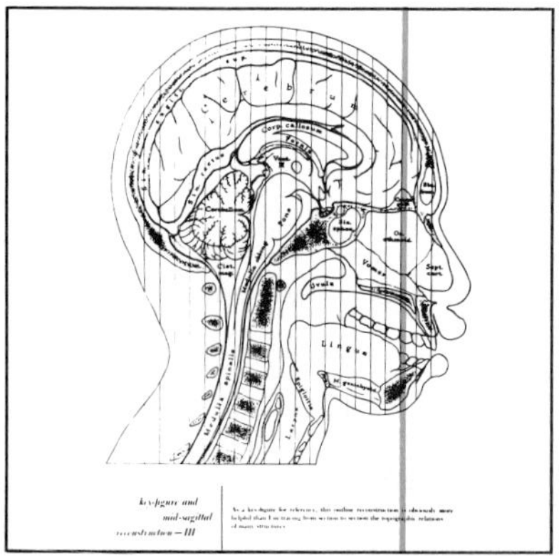

rather long, vertical, bilateral muscle in a slightly oblique position inferior to the lateral part of the mandible.

Mentalis Muscle (*g*). A rather thick, transversely elongated, bilateral muscle in the median and inferior to the middle of the mandible. Each end points superiorly toward the mandible in this area of the head.

NEUROVASCULAR UNIT (Fig. 4–2)

I. Arteries

Facial (External Maxillary) Artery (*D*). This is seen bilaterally, lateral to the buccinator muscle and superior to the triangularis muscle, as small, round vessels.

Origin. Arises from the external carotid artery.

Distribution. Face, tonsils, and palate.

Course. Deep to the submandibular gland as it passes superiorly, superficial to the inferior border of the mandible, 1 inch (2.54 cm) anterior to the angle of the mandible. This artery has an irregular course throughout the face and mouth, as well as the medial angle of the eye.

Greater Palatine Artery (*C*). Small, oval, bilateral vascular structures in the palatal soft tissue inferior to the lateral parts of the palatine process (maxillary bone).

Origin. Arises from the third part of its internal maxillary artery.

Distribution. Supplies the soft palate and palatine tonsil and anastomoses with the ascending palatine artery.

Course. This artery passes through the greater palatine canal and emerges on the oral surface of the palate through the greater palatine foramen and traverses forward in a groove near the alveolar edge of the hard palate to reach the incisive canal. It terminates after it passes through and anastomoses with a branch of the sphenopalatine artery.

Infraorbital artery (*G*). These are seen bilaterally as small rounded vessels inferior to the orbits and medial to the infraorbital nerve.

Origin. Arises from the maxillary artery.

Distribution. Maxillary sinuses and median face.

Course. Passes into the orbit by way of the inferior orbital fissure, runs anteriorly in the infraorbital groove, and then exits on the facial surface of the upper jaw at the infraorbital foramen.

II. Veins

Anterior Facial Vein (*H*). A bilateral, oval, vascular structure lateral to the maxillary bone and medial to the zygomatic major muscle.

Origin. By the union of the supraorbital and supratrochlear veins.

Termination. Common facial vein.

Course. This superficial vein drains the superficial tissue of the face. It starts in the medial angle of the orbit as the angular vein. It runs inferiorly and posteriorly as well as superficially to the facial muscles. This vein passes posterior to the facial artery as it courses to the lower border of the mandible and the front edge of the masseter muscle. It runs superficially to the submandibular gland, then joins the anterior branch of the retromandibular vein and forms the common facial vein.

Superior Ophthalmic Vein (*F*). This is seen here in the upper medial eye socket as a vertically elongated, vascular structure. This bilateral vein is located superiorly to the medial rectus muscle and laterally to the superior oblique muscle.

Origin. Combination of veins of the orbit.

Termination. Anterior end of cavernous sinus.

Course. It passes through the superior orbital fissure, then joins the central vein of the retina and courses directly to the cavernous sinus.

III. Vascular Venous Sinuses

Superior Sagittal Sinus (*A*). A triangular, hollow, vascular structure in the median immediately inferior to the frontal bone and superior to the falx cerebri.

IV. Nerves

Supraorbital Nerve (*E*). An oval structure bilaterally located in the superior part of each orbit.

Origin. Arises from the frontal nerve, which is the largest branch of the ophthalmic nerve.

Distribution. Upper eyelid and conjunctiva as well as skin of the scalp.

Course. This nerve courses between the levator muscle of the upper eyelid and the roof of the orbit and exits through the supraorbital notch (foramen). It runs upward in the superficial fascia of the forehead almost to the lambdoidal suture.

Infraorbital Nerve (*B*). A small, oval nerve struc-

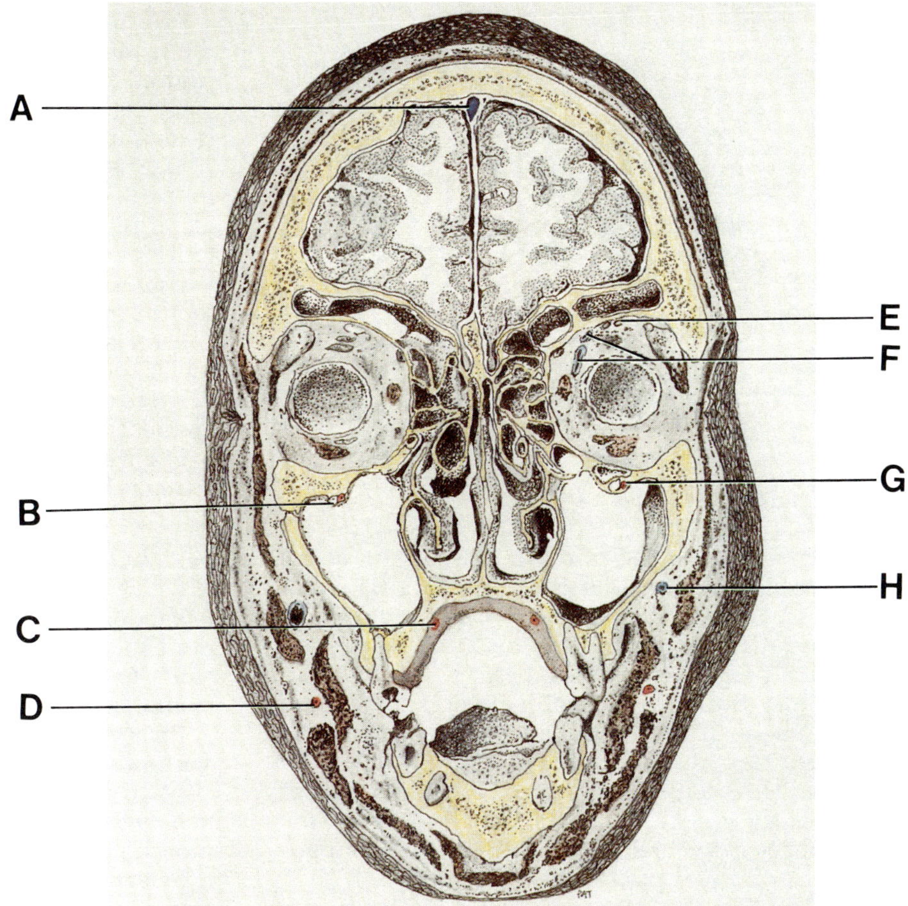

A. Superior sagittal sinus
B. Infraorbital nerve
C. Greater palatine artery
D. Facial (external maxillary) artery
E. Supraorbital nerve
F. Superior ophthalmic vein
G. Infraorbital artery
H. Anterior facial vein

Figure 4-2
Coronal section of the head, which passes through the crista galli of the ethmoid bone and the symphysis of the mandible, highlighting the neurovascular unit structures.

ture seen bilaterally, located medial to the superior part of the maxillary sinus and lateral to the upper part of the inferior nasal concha.

Origin. This nerve arises from the maxillary nerve (a branch of the trigeminal).

Distribution. Alae of the nose; mucous membranes of the cheeks, lower eyelids, and upper lip.

Course. This nerve exits from the skull through the infraorbital foramen and is continuous with filaments of the facial nerves.

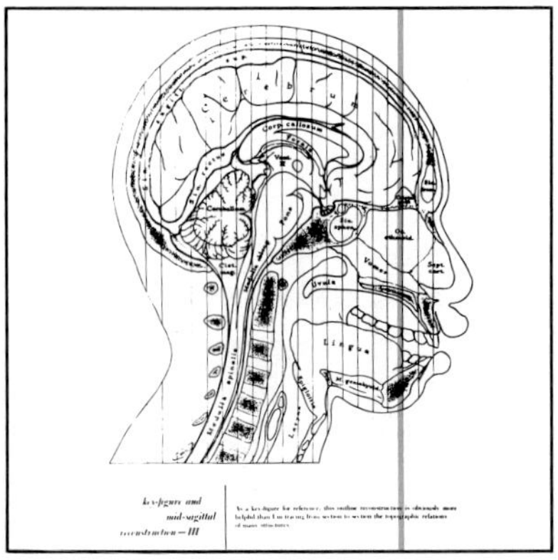

VISCERAL UNIT (Fig. 4–3)

Frontal Lobe of the Brain (*B*). This section of the head reveals the frontal lobes as bilateral, triangular nerve tissue. Folds such as the medial frontal gyrus (fold) and the inferior frontal gyrus are seen projecting on the peripheries of the lobes.

Olfactory Bulb (*I*). Bilateral hanging bulbs immediately lateral to the crista galli and medial to the frontal sinuses.

Course. They course inferior to the anterior part of the olfactory groove on the inferior surface of the frontal lobes of the brain. Further, they course superior to the cribriform plates.

Superior Frontal Gyrus (*A*). A fold of gray matter at the superior medial part of each frontal lobe.

Middle Frontal Gyrus (*F*). A lateral projection of the frontal lobe inferior to the superior frontal gyrus and superior to the inferior frontal gyrus.

Inferior Frontal Gyrus (*C*). A lateral inferior fold of the frontal lobe of the brain superior to the frontal sinuses in this view of the head.

Gyrus Rectus (*G*). A bilateral fold on the inferior medial frontal lobe superior to the olfactory bulb.

Lacrimal Gland (*H*). Oblong gland tissue located in the lateral and superior part of each orbit. This gland secretes tears that contain antibodies and bactericidal substances.

Upper Premolar Tooth (*J*). A typical premolar tooth in the upper jaw.

Inferior Premolar Tooth (*K*). A typical lower premolar tooth in the lower jaw (mandible).

Palatine Gland (*E*). Glandular tissue within the epithelial tissue.

Eye Bulb (*D*). A rounded, hollow eye feature. The vitreous humor is contained here in the living subject.

ENCLOSING UNIT (Fig. 4–4)

Falx Cerebri (*A*). A vertical, elongated sheet of dura mater that extends from the superior sagittal sinus to the crista galli between the right and left frontal lobes.

Fat Body (*F*). This is seen in the floor of the mouth posterior to the mandible.

Oral Cavity (*C*). This is surrounded by the teeth and both upper and lower jaws; it is known as the mouth proper.

Sclera (*E*). The thick outer circular layer of connective tissue surrounded by orbital fat.

Skin (*D*). This is seen covering the whole section.

Vestibule of the Mouth (*B*). A space between the lateral parts of the teeth and gums and the medial surface of the cheeks.

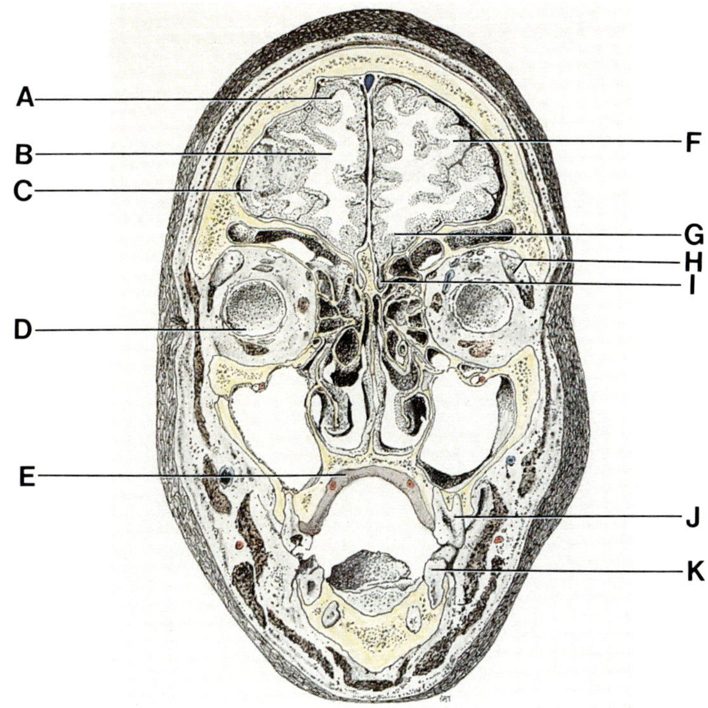

A. Superior frontal gyrus
B. Frontal lobe
C. Inferior frontal gyrus
D. Eye bulb
E. Palatine gland
F. Middle frontal gyrus
G. Gyrus rectus
H. Lacrimal gland
I. Olfactory bulb
J. Upper premolar tooth
K. Inferior premolar tooth

Figure 4–3
Coronal section of the head, which passes through the crista galli of the ethmoid bone and the symphysis of the mandible, highlighting the visceral unit structures.

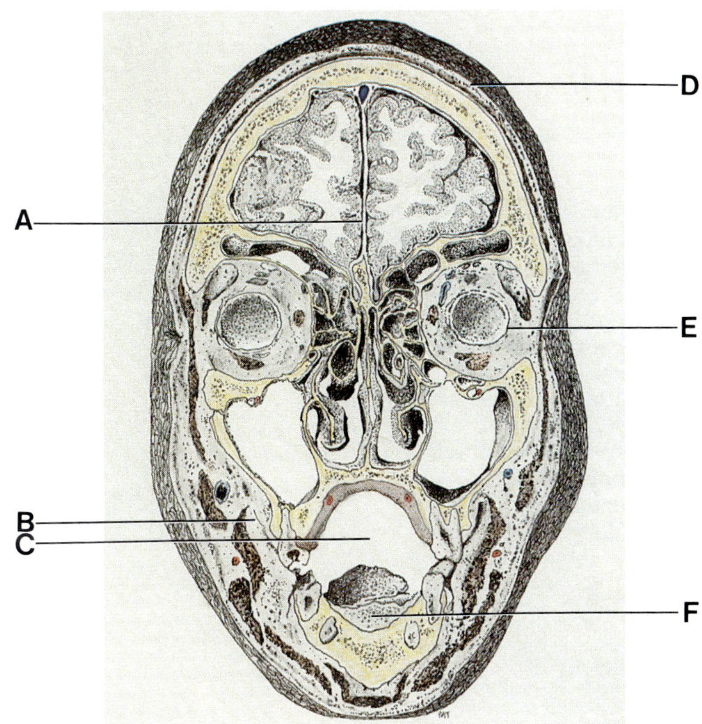

A. Falx cerebri
B. Vestibule of mouth
C. Oral cavity
D. Skin
E. Sclera
F. Fat body

Figure 4–4
Coronal section of the head, which passes through the crista galli of the ethmoid bone and the symphysis of the mandible, highlighting enclosing unit structures.

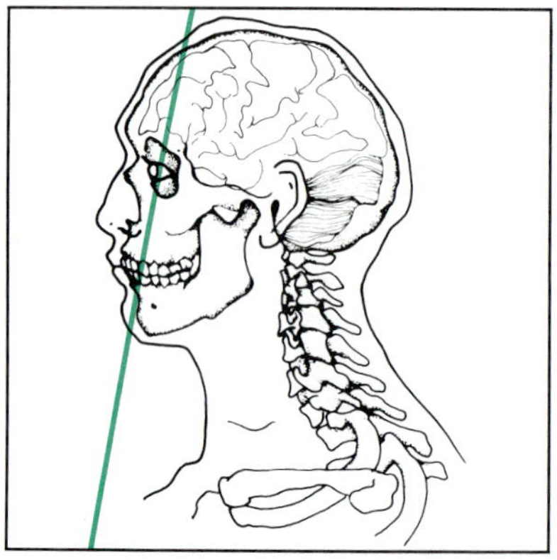

Figure 4–5 is a picture of an MRI coronal head image. This slice passes through such structures as the ethmoid sinus, the inferior nasal concha, and the frontal lobe of the brain. Note the orientation line that indicates its derivation. All four anatomic units are emphasized by labels.

Figure 4–6, an MRI coronal scan of the head, highlights all four anatomic unit structures. This slice passes through such structures as the temporalis muscle, inferior nasal concha, oral cavity, rectus gyrus of the brain, and maxillary sinus.

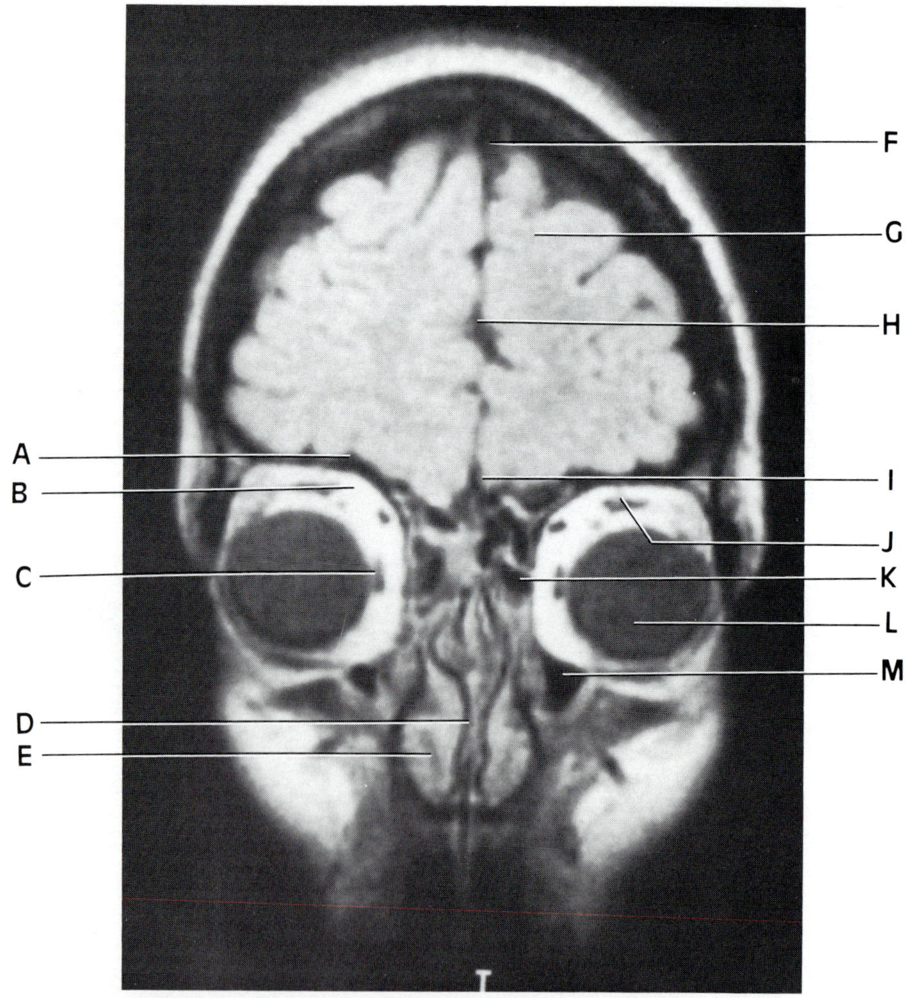

A. Orbital groove
B. Orbital fat
C. Medial rectus muscle
D. Nasal septum
E. Inferior nasal concha (turbinate)
F. Superior sagittal sinus
G. Frontal lobe
H. Longitudinal cerebral fissure
I. Rectus gyrus
J. Superior rectus muscle of eye
K. Ethmoid sinus
L. Orbital globe
M. Maxillary sinus

Figure 4–5
An MRI coronal image of the head, highlighting the musculoskeletal, neurovascular, visceral, and enclosing unit structures.

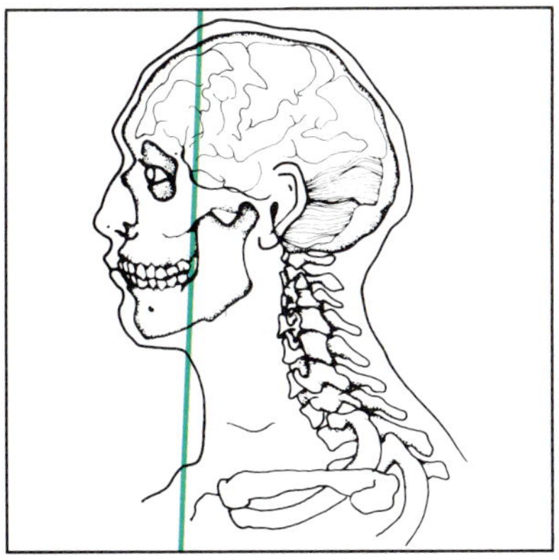

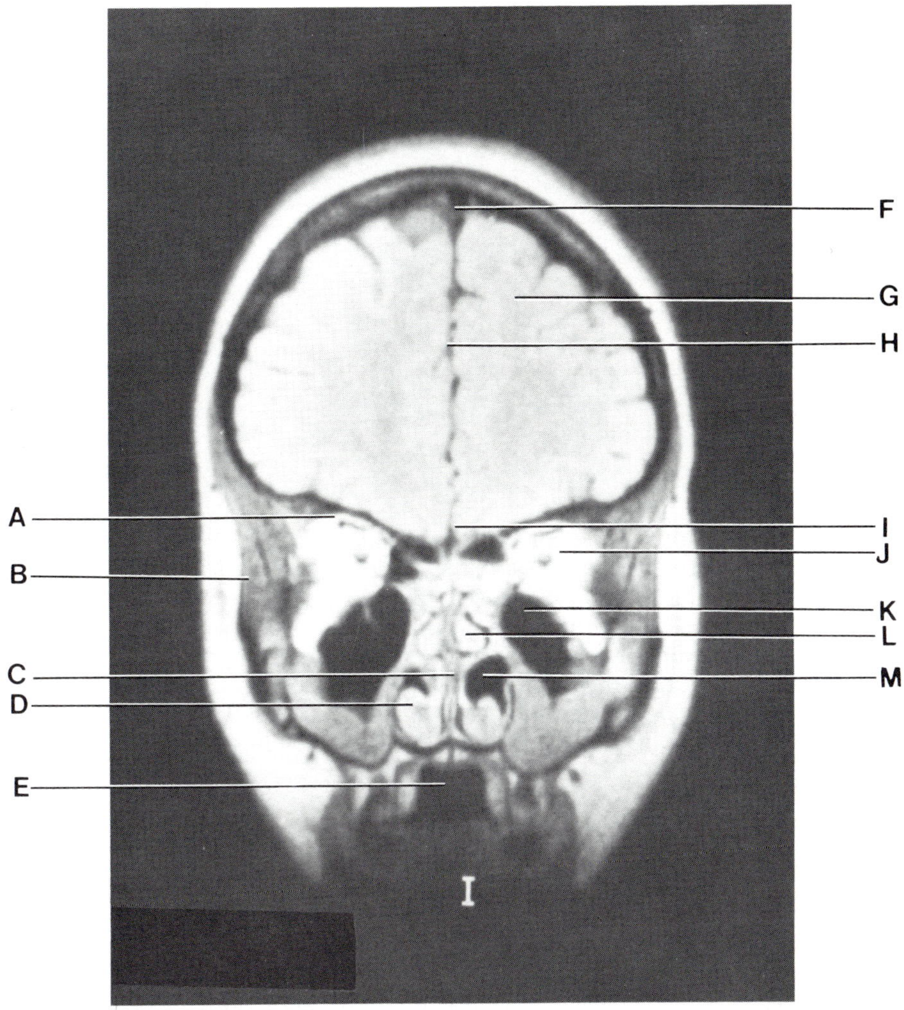

A. Superior rectus muscle
B. Temporalis muscle
C. Nasal septum
D. Inferior nasal concha (turbinate)
E. Oral cavity
F. Superior sagittal sinus
G. Frontal lobe
H. Longitudinal cerebral fissure
I. Rectus gyrus
J. Orbital fat
K. Maxillary sinus
L. Middle nasal concha (turbinate)
M. Nasal cavity

Figure 4–6
An MRI coronal image of the head, highlighting the musculoskeletal, neurovascular, visceral, and enclosing unit structures.

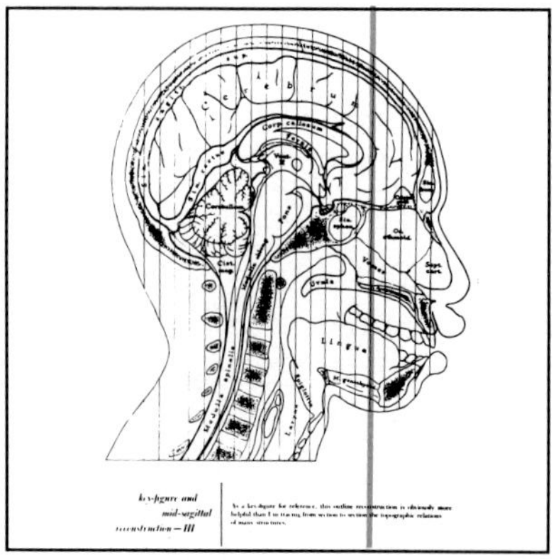

MODULE 2

The frontal (coronal) section passes through the posterior ethmoid air cells, the posterior extent of the frontal sinuses, the maxillary sinuses, the posterior portion of the hard palate (horizontal part of palatine bones), and the upper right second molar tooth. Attention is directed to the large size of the posterior ethmoid cells.

MUSCULOSKELETAL UNIT (Fig. 4–7)

I. Skeletal Portion

Frontal Bone (*A*). A dome-shaped bone covering the brain and extending as far as the eye socket. This bilateral bone is seen with two layers of compact bone. Between these two layers is less dense spongy bone.

Sphenoid Bone (*H*). A bilateral bone structure that extends from the inferior lateral part of the frontal bone to the lower part of the eye socket. The upper part is wider than the lower part in this frontal (coronal) slice.

Frontosphenoidal Suture (*G*). A fibrous joint that runs obliquely downward toward the median. This bilateral suture is between the frontal and sphenoid bones.

Superior Nasal Concha (*J*). An upper lateral feature of the nasal cavity. This bilateral concha is seen medial to the ethmoid sinuses.

Middle Nasal Concha (*M*). A curved, lateral projection on the lateral wall of the nasal cavity. This structure is bilateral inferior to the superior nasal concha and medial to the upper part of the maxillary sinus in this section.

Inferior Nasal Concha (*O*). A medial projection on the lateral inferior wall of the nasal cavity. This bilateral structure is located medial to the middle part of the maxillary sinus.

Vomer (*j*). A thin, lower bone part of the nasal septum. This bone is superior to the palatine bone and inferior to the perpendicular plate.

Perpendicular Plate of the Ethmoid Bone (*C*). A thin sheet of bone tissue that forms the upper part of the nasal septum.

Palatine Bone (*Q*). A transverse bone tissue across the roof of the mouth. In the median part of this bone is triangular bone tissue.

Maxilla (*P*). The maxilla appears bilaterally as a thin sheet of bone surrounding a rather large cavity. The inferior part of the maxilla shows an irregularly shaped part with the root of a tooth embedded.

Maxillary Sinus (*f*). A bilateral, relatively large cavity within the maxillary bone.

Mandible (*k*). A thick, oblong, bilateral slice of bone with compact bone on the periphery and cancellous bone forming the inner core. The upper part of this bone slice contains a tooth.

Zygomatic Bone (*e*). A bilateral, oblong bone in the vertical plane lateral and inferior to the eye socket and medial to the orbicularis oculi muscle in this section.

Cartilage of the Nasal Septum (*c*). A thin sheet of cartilage in the nasal septum between the perpendicular plate and the vomer.

Posterior Ethmoid Cell (*D*). Rounded bilateral openings medial to the eye socket and lateral to the upper part of the superior nasal concha.

Frontal Sinus (*E*). A cavity in the area of bone above the eye socket.

Superior Nasal Meatus (*K*). A canal on the lateral upper part of the nasal cavity between the superior nasal concha and the middle nasal concha.

Middle Nasal Meatus (*L*). A canal on the lateral wall of the nasal cavity immediately inferior to the medial nasal concha.

Inferior Nasal Meatus (*h*). A canal-like feature on the inferior lateral wall of the nasal cavity immediately inferior to the inferior nasal concha.

Inferior Orbital Fissure (*b*). As seen in this section, this structure is a relatively small slit at the lateral inferior part of the orbit.

II. Muscular Portion

Galea Aponeurotica (*W*). A thin sheet of tendinous tissue covering the upper and most of the lateral surface of the frontal bone.

Frontalis Muscle (*B*). A thin, bilateral sheet of muscle tissue adjacent to the lateral part of the frontal bone.

Medial Rectus Muscle (*F*). Moderately oblong, bilateral muscle tissue in the medial part of the eye socket.

Lateral Rectus Muscle (*I*). A rather large, oblong, bilateral muscle slice in the lateral part of the eye socket.

Zygomatic Major Muscle (*N*). A vertical, elongated muscle slice with tapering ends. This bilateral muscle is lateral to the inferior part of the zygomatic bone and lateral to the superior part of the masseter muscle.

Masseter Muscle (*g*). A rather large, oblong muscle slice inferior to the zygomatic bone slice. This bilateral muscle extends to the buccal fat in this section.

A. Frontal bone
B. Frontalis muscle
C. Perpendicular plate of ethmoid bone
D. Posterior ethmoid cell
E. Frontal sinus
F. Medial rectus muscle
G. Frontosphenoidal suture
H. Sphenoid bone
I. Lateral rectus muscle
J. Superior nasal concha
K. Superior nasal meatus
L. Middle nasal meatus
M. Middle nasal concha
N. Zygomatic major muscle
O. Inferior nasal concha
P. Maxilla
Q. Palatine bone
R. Superior longitudinal muscle of tongue
S. Inferior longitudinal muscle of tongue
T. Platysma
U. Mylohyoid muscle
V. Digastric muscle
W. Galea aponeurotica
X. Superior oblique muscle
Y. Levator muscle of eyelid
Z. Superior rectus muscle
a. Temporalis muscle
b. Inferior orbital fissure
c. Cartilage of nasal septum
d. Orbicularis muscle of eye (orbicularis oculi muscle)
e. Zygomatic bone
f. Maxillary sinus
g. Masseter muscle
h. Inferior nasal meatus
i. Buccinator muscle
j. Vomer
k. Mandible
l. Genioglossus muscle
m. Geniohyoid muscle

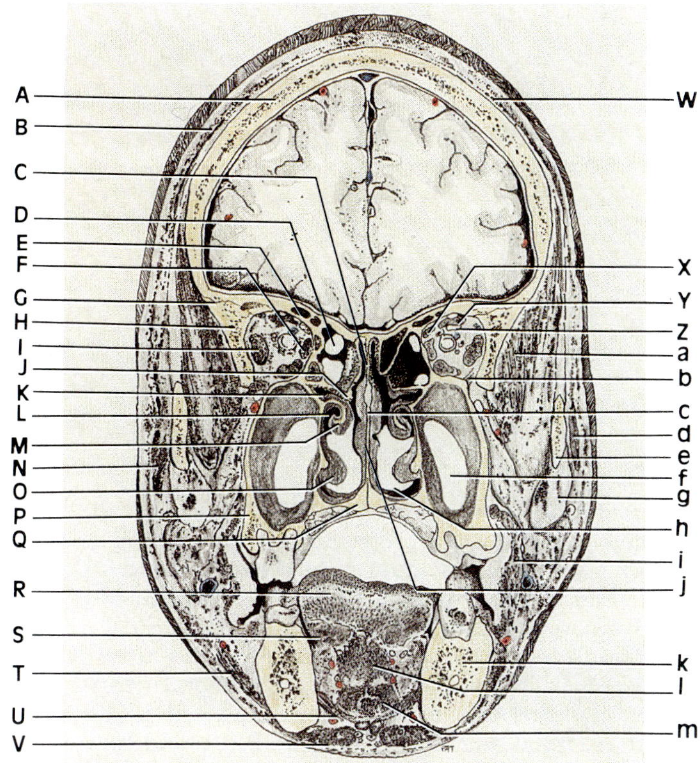

Figure 4–7
Coronal (frontal) section of the head, which passes through the posterior ethmoid air cells, the posterior extent of the frontal sinuses, the maxillary sinuses, the posterior portion of the hard palate, and the upper right second molar tooth, highlighting the musculoskeletal unit structures.

Inferior Longitudinal Muscle of the Tongue (*S*). A bilateral, oblong muscle of the inferior tongue located superiorly to the sublingual gland.

Superior Longitudinal Muscle of the Tongue (*R*). A continuous sheet of muscle tissue across the upper two thirds of the tongue.

Digastric Muscle (*V*). A transversely placed, oblong, bilateral muscle slice on the inferior part of the lower surface of the face.

Mylohyoid Muscle (*U*). A curved, rather thin, bilateral muscle sheet that extends from the right mandible to the left mandible. The upper part is concave.

Geniohyoid Muscle (*m*). A bilateral, oval muscle slice superior to the mylohyoid muscle and inferior to the genioglossus.

Genioglossus Muscle (*l*). A rather large, bilateral muscle section superior to the geniohyoid muscle and inferior to the intrinsic muscles of the tongue. This muscle is seen medial to the sublingual gland.

Platysma (*T*). A vertically bilateral muscle slice lateral to the mandible. The upper part of this muscle is thicker than the lower part in this section.

Buccinator Muscle (*i*). A vertically bilateral muscle section thicker in its medial part than in the superior and inferior parts. This muscle is located lateral to the vestibule of the mouth and medial to the buccal fat.

Orbicular Muscle of Eye (Orbicularis Oculi Muscle) (*d*). A rather thin, vertically placed bilateral muscle slice lateral to the zygomatic bone.

Temporalis Muscle (*a*). A vertically placed, rather thick bilateral muscle slice lateral to the frontal and sphenoid bones and medial to the zygomatic bone. This muscle's upper part is thinner than its lower part.

Superior Rectus Muscle (*Z*). A rather small, oval muscle slice in the upper part of the eye socket. This bilateral muscle is inferior to the levator palpebrae muscle.

Superior Oblique Muscle (*X*). A rather small, oval, bilateral muscle slice in the upper medial part of the eye socket. This muscle is superior to the medial rectus and medial to the superior levator palpebrae muscles.

Levator Muscle of the Eyelid (*Y*). A transversely placed muscle slice in the superior median eye socket. This bilateral muscle is superior to the superior rectus.

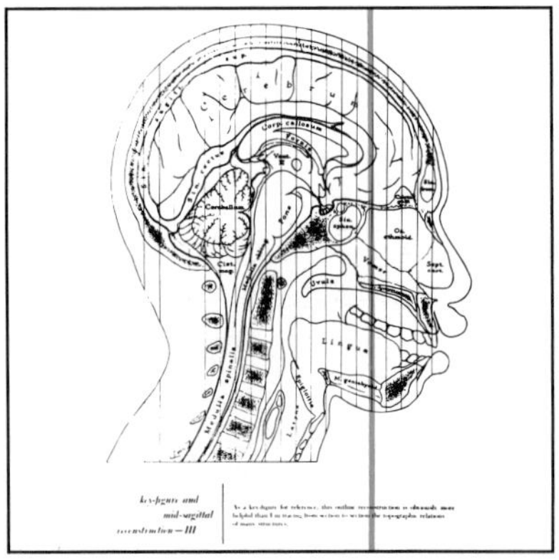

NEUROVASCULAR UNIT (Fig. 4–8)

I. Arteries

Anterior Cerebral Artery (*A*). A rather small, rounded, bilateral vascular structure seen at the upper part of the brain and immediately below the dura mater. This vascular structure is immediately lateral to the superior frontal gyrus.

Origin. Cerebral part of the internal carotid artery.

Distribution. Medial surface of the cerebral hemisphere, the part that extends from the front to the parieto-occipital sulcus. It also supplies blood to a narrow part of the cortex along the superior lateral parts.

Course. This artery runs anterior and medial to the optic nerve, comes into close contact with its opposite artery, and becomes connected to it by the anterior communicating artery.

Middle Cerebral Artery (*O*). A rounded, bilateral vascular structure seen in this section lateral to the inferior frontal gyrus on the left side. On the right side the vessel appears to be embedded in the lateral part of the frontal gyrus.

Origin. Cerebral part of the internal carotid artery.

Distribution. Most of the cerebral lateral surface except the occipital pole.

Course. The middle cerebral artery is lodged within the lateral cerebral sulcus, then passes posteriorly and superiorly on the surface of the insula. This artery divides into two branches: one is distributed to the insular lobe and the other to the lateral surface of the cerebrum except the occipital pole.

Infraorbital Artery (*F*). A small, rounded, bilateral vascular structure medial to the lower part of the temporalis muscle, lateral to the maxilla, and inferior to the infraorbital nerve.

Mylohyoid Artery (*L*). A small, rounded, bilateral vascular structure seen inferior to the lateral part of the mylohyoid muscle and superior to the digastric muscles.

Origin. Inferior alveolar artery.

Distribution. Mylohyoid muscle.

Course. It courses between the mylohyoid muscle and the anterior belly of the digastric muscle.

Sublingual Artery (*X*). A small, bilateral, vascular structure located between the inferior part of the genioglossus muscle and the inferior part of the sublingual salivary gland.

Origin. Lingual.

Distribution. Branches of the mylohyoid muscle and sublingual gland.

Course. It courses between the genioglossus muscle and the sublingual gland.

Deep Lingual Artery (Arteria Profunda Linguae) (Ranine Artery) (*V*). A small, rounded, bilateral vascular structure located between the genioglossus muscle and the sublingual gland immediately inferior to the lingual nerve.

Origin. Lingual artery.

Distribution. Mylohyoid and geniohyoid muscles.

Course. Ascends between the genioglossus and the inferior longitudinal muscle of the tongue.

Maxillary Artery (*W*). A rounded, bilateral vascular structure located inferior to the buccinator muscle and superior to the platysma in this section.

Origin. External carotid artery.

Distribution. Fifteen branches that distribute blood to the teeth, mouth, muscles of mastication, facial region, and meninges inside the skull through an artery known as the middle meningeal artery.

Course. This artery pierces the facial sheath of the parotid compartment, passes forward deep to the neck of the mandible, and passes the lateral pterygoid muscle. It continues on the lateral surface of the lateral pterygoid plate and terminates in the pterygopalatine fossa.

Greater Palatine Artery (*T*). A small, rounded, bilateral vascular structure lateral and slightly inferior to the anterior palatine nerve and medial to the lower part of the maxilla.

Origin. A branch of the descending palatine artery from the maxillary artery.

Distribution. Gums, palatine gland, and mucous membrane of the roof of the mouth.

Course. Passes through the greater palatine canal to supply the soft palate and tonsils. This artery enters the oral surface of the palate through the greater palatine foramen, passes anteriorly near the alveolar border of the hard palate to reach the incisive canal, enters the incisive canal, and terminates by anastomosing with a branch of the sphenopalatine artery.

Buccal Artery (*S*). A rounded, bilateral vascular structure medial to the inferior part of the temporalis muscle and lateral to the upper part of the sublingual gland in this section.

Origin. Maxillary artery.

Distribution. Buccinator muscle.

II. Veins

Superior Ophthalmic Vein (*D*). An oval, bilateral vascular structure immediately superior to the optic nerve in the orbital cavity.

Origin. Twigs unite from the inner angle of the orbit.

A. Anterior cerebral artery
B. Frontal nerve
C. Lacrimal vein
D. Superior ophthalmic vein
E. Infraorbital nerve
F. Infraorbital artery
G. Anterior palatine nerve
H. Anterior facial vein
I. Lingual nerve
J. Inferior alveolar nerve
K. Hypoglossal nerve (12th cranial nerve)
L. Mylohyoid artery
M. Superior sagittal sinus
N. Inferior sagittal sinus
O. Middle cerebral artery
P. Optic nerve (second cranial nerve)
Q. Ciliary nerve
R. Inferior branch of oculomotor nerve (third cranial nerve)
S. Buccal artery
T. Greater palatine artery
U. Buccal nerve (branch of seventh cranial nerve)
V. Deep lingual artery (arteria profunda linguae) (ranine artery)
W. Maxillary artery
X. Sublingual artery
Y. Submental lymph node

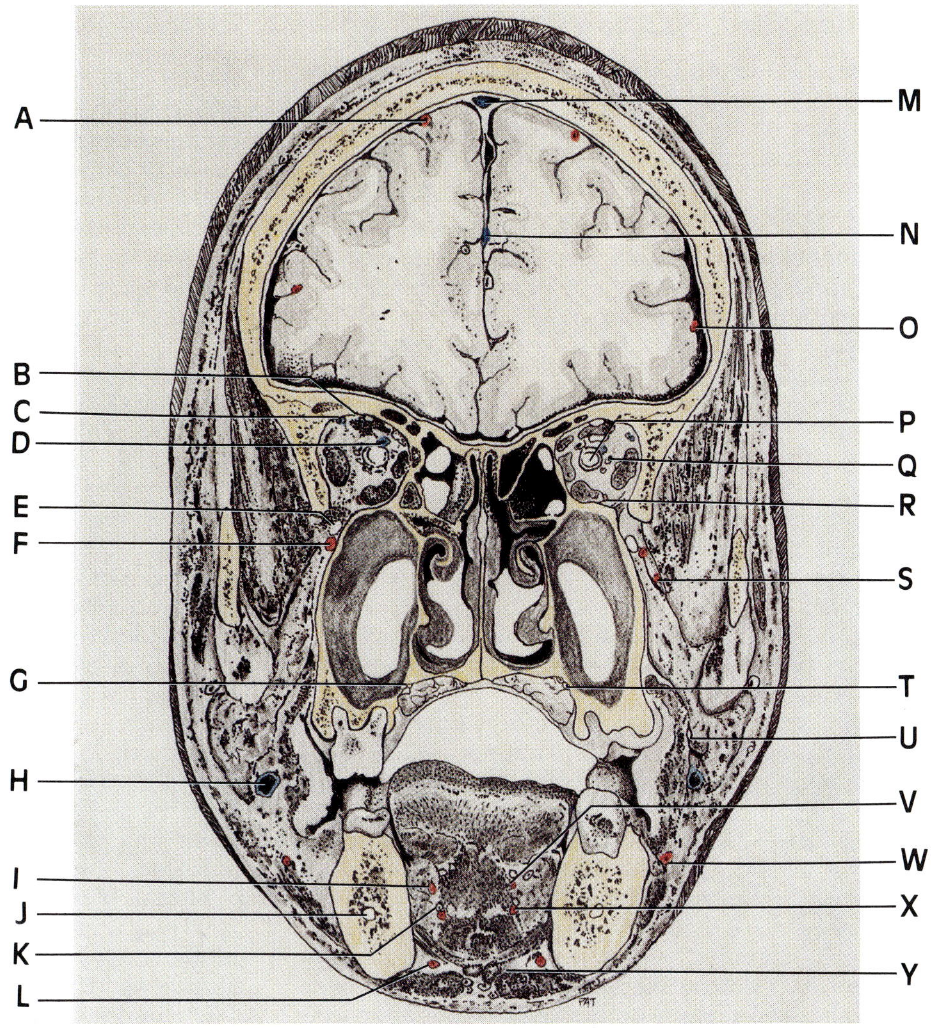

Figure 4–8
Coronal section of the head, which passes through the posterior ethmoid air cells, the posterior extent of the frontal sinuses, the maxillary sinuses, the posterior portion of the hard palate, and the upper right second molar tooth, highlighting the neurovascular unit structures.

Termination. Cavernous sinus.

Course. Follows the ophthalmic artery through the optic foramen on its superior surface.

Lacrimal Vein (*C*). A small, oval vascular structure located superiorly to the lateral part of orbital cavity and lateral to the superior oblique muscle and the levator palpebrae superioris muscle.

Anterior Facial Vein (*H*). A rather large, oval, bilateral vascular structure lateral to the buccinator and medial to the inferior part of the buccal fat pad.

Origin. Medial surface of the orbital edge as the angular vein.

Termination. Ophthalmic and supratrochlear veins.

Course. It passes inferiorly and posteriorly superficial to the facial muscles and behind the facial artery, to the inferior border of the mandible at the front edge of the masseter muscle, and penetrates the deep cervical fascia.

III. Vascular Venous Sinuses

Superior Sagittal Sinus (*M*). A median, triangular vascular structure immediately inferior to the frontal bone and superior to the falx cerebri.

Inferior Sagittal Sinus (*N*). An oval vascular structure between the right and left cingulate gyri and inferior to the falx cerebri.

Origin. Posterior two thirds of the free edge of the falx cerebri.

Termination. Straight sinus.

Course. Free inferior edge of the posterior falx cerebri.

IV. Lymphatic Structures

Submental Lymph Node (*Y*). These are two rather large, oval nodes inferior to the mylohyoid muscle in the median. These nodes are superior and medial to the digastric muscle.

V. Nerves

Frontal Nerve (*B*). A rounded sensory nerve tissue in the upper median of the orbit immediately superior to the levator palpebrae superioris.

Origin. Ophthalmic division of the trigeminal nerve.

Distribution. Supraorbital and supratrochlear branches.

Course. Passes anterior in the superior orbital cavity, splitting at about the middle of the cavity into larger supraorbital and smaller supratrochlear branches. These nerves supply tissues of the forehead and superior eyelid.

Infraorbital Nerve (*E*). An oval sensory nerve structure immediately superior to the infraorbital artery in this section and medial to the upper part of the maxilla.

Origin. Maxillary branch of the trigeminal.

Distribution. Alae of the nose, inferior eyelids, mucous membranes of the cheeks and skin, as well as the superior lip.

Course. It exits the skull through the infraorbital foramen to appear in the face and terminates as it divides into branches for its distribution. This division takes place under the levator labii superioris muscle.

Anterior Palatine Nerve (*G*). A small, rounded, bilateral nerve feature immediately superior to the lateral part of the palatine gland and medial to the inferior part of the maxillary sinus.

Origin. Maxillary branch of the trigeminal nerve.

Distribution. Gums, mucous membrane of the mouth, and palatine glands of the hard palate.

Course. Passes down through the greater palatine canal and comes to lie upon the hard palate through the greater palatine foramen. It terminates near the incisive teeth on the inferior part of the hard palate.

Inferior Alveolar Nerve (*J*). An oval, bilateral sensory nerve structure in the middle part of the mandible in this section.

Origin. Mandibular branch of the trigeminal nerve.

Distribution. Teeth of the lower jaw.

Course. This nerve is deep to the lateral pterygoid muscle and inferior to the anterior border of the inferior dental artery.

Lingual Nerve (*I*). A small, oval, bilateral sensory nerve structure immediately lateral to the upper part of the genioglossus muscle and immediately inferior to the inferior longitudinal muscle of the tongue.

Origin. Mandibular branch of the trigeminal nerve.

Distribution. Mucous membrane of the front two thirds of the tongue, the floor of the mouth, and the lateral wall of the mouth.

Course. The lingual nerve goes inferiorly and anteriorly between the ramus of the mandible and the medial pterygoid muscle. This nerve continues inferiorly to the inferior alveolar nerve, then passes inferiorly to the mandibular origin of the superior constrictor muscle of the pharynx. It leaves the gums to enter the side of the tongue.

Hypoglossal Nerve (12th Cranial Nerve) (*K*). A rounded, bilateral motor nerve structure immediately superior to the sublingual artery and lateral to the inferior part of the genioglossus muscle.

Origin. Hypoglossal nucleus in the medulla oblongata.

Distribution. Intrinsic muscle of tongue as well as hypoglossal and geniohyoid muscles.

Course. This nerve exits from the cranial cavity through the hypoglossal canal, enters the superior part of the carotid sheath, and curves around the vagus nerve. Branches from the first and second cervical nerves as well as the superior cervical sympathetic ganglion intertwine with this nerve. While in the carotid sheath it passes down between the internal jugular vein and the internal carotid artery near the border of the digastric muscle. Then this nerve runs anteriorly lateral to the origin of the occipital artery and superficial to the external carotid, lingual, and facial arteries in the carotid triangular. It exits from the carotid triangular as it passes deep to the digastric and stylohyoid muscles and enters the floor of the mouth between the mylohyoid and hyoglossus muscles.

Buccal Nerve (Branch of the Seventh Cranial Nerve) (*U*). A small, oval, bilateral nerve structure located between the upper part of the buccinator muscle and the buccal fat pad.

Origin. Facial nerve (seventh cranial nerve).

Distribution. Buccinator muscle, superior lip muscles, risorius muscle, lateral nose muscles, and inferior eyelid.

Course. It runs from the anterior ear deep to the buccal fat pad and superficial to the buccinator muscle to reach the muscles of the upper lip.

Optic Nerve (Second Cranial Nerve) (*P*). A rather large, rounded, bilateral nerve structure in the medial part of the eyeball, inferior to the superior rectus muscle and superior to the inferior rectus muscles. This structure is lateral to the medial rectus and medial to the lateral rectus muscles. This nerve is responsible for vision.

Origin. Retina.

Brain Termination. Lateral geniculate body of the thalamus.

Course. The optic nerve exits from the orbit through the optic canal, joins the optic chasm anterior to the pituitary gland, and they continue as the optic tract.

Ciliary Nerve (*Q*). A small, rounded, bilateral sensory nerve structure lateral to the optic nerve and medial to the lateral rectus muscle. More than one nerve structure represents the long ciliary nerve. *Note:* The ciliary nerve brevis (short ciliary nerve) is seen in the same location.

Origin. The origin of the long ciliary nerve is the

nasal ciliary nerve, which is a branch of the frontal nerve (the frontal nerve arises from the ophthalmic nerve of the trigeminal nerve). The origin of the short ciliary nerve is the ciliary ganglion.

Distribution. The distribution of both long and short ciliary nerves is to the ciliary body, cornea, and iris. *Note:* These nerves contains sympathetic fibers for the dilation of the pupil.

Course. Both long and short ciliary nerves penetrate the sclera adjacent to the attachment of the optic nerve and pass anteriorly between the sclera and choroid layers.

Inferior Branch of the Oculomotor Nerve (Third Cranial Nerve) (*R*). A small, rounded, bilateral motor nerve tissue immediately lateral to the inferior rectus muscle.

Origin. Midbrain.

Distribution. The oculomotor nerve distributes nerve branches to the extrinsic muscles of the eye, except the lateral rectus and superior oblique muscles.

Course. It exits from the skull through the superior orbital fissure to enter the orbit to distribute branches to each of the extrinsic muscles of the eye and the iris of the eye bulb.

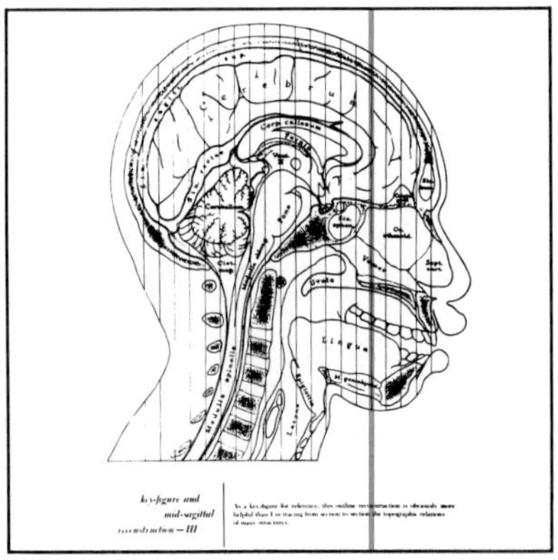

VISCERAL UNIT (Fig. 4–9)

Cingulate Sulcus (*B*). A bilateral slit at the superior borders of the cingulate gyrus. These slits are on the medial borders of the frontal lobes.

Medial Frontal Gyrus (*A*). A fold on the peripheral part of the frontal lobe deep to the dura mater, inferior to the superior frontal gyrus, and superior to the inferior frontal gyrus.

Superior Frontal Gyrus (*I*). A fold on the upper peripheral part of the frontal lobe adjacent to the upper part of the falx cerebri. This structure is bilateral.

Inferior Frontal Gyrus (*C*). A rather large fold on the peripheral lateral inferior part of the frontal lobe. It is lateral to the orbital gyri.

Lateral Orbital Gyrus (*D*). A bilateral fold at the inferior part of the frontal lobe medial to the inferior frontal gyrus and lateral to the medial orbital gyrus.

Medial Orbital Gyrus (*E*). An inferior bilateral fold of the cerebrum at its inferior part, located between the lateral orbital gyrus and the gyrus rectus.

Gyrus Rectus (*M*). A bilateral fold of the frontal lobe located between the longitudinal cerebral fissure and the medial orbital gyrus.

Parotid Duct (*F*). A rounded, tubelike structure inferior to the zygomatic major muscle and superior to the lateral part of the buccal fat pad.

Molar tooth (*Q*). This is seen embedded in both the upper and lower jaws in this section.

Sublingual Gland (*H*). A bilateral, vertically elongated glandular structure lateral to the genioglossus muscle, medial to the lower jaw, and inferior to the inferior longitudinal muscle of the tongue.

Tongue (*G*). A transversely elongated, rather thick muscular organ in the medial inferior part of the oral cavity.

Submandibular Duct (*R*). An oval, bilateral, tubelike structure that courses within the medial part of the sublingual gland superior and lateral to the lingual profundus artery.

Dorsum Linguae (*P*). The superior surface of the tongue.

Palatine Gland (*O*). A rather large, bilateral, glandular tissue deep to the mucosa and roof of the mouth and inferior to bone of the palate.

Olfactory Tract (*N*). An inferior, bilateral, oval nerve structure inferior to the lateral part of the gyrus rectus.

Frontal Lobe of the Cerebrum (*J*). A large, bilateral brain structure showing peripheral gray matter; the inner part shows white matter. The falx cerebri is seen between the two lobes at the upper half.

Corpus Callosum Sulcus (*L*). A bilateral groove on the medial part of the frontal lobe inferior to the inferior sagittal sinus and inferior to the cingulate gyrus.

Cingulate Gyrus (*K*). A medial, bilateral fold of the frontal lobe lateral to the inferior sagittal sinus and inferior to the cingulate sulcus. It is superior to the corpus callosum sulcus.

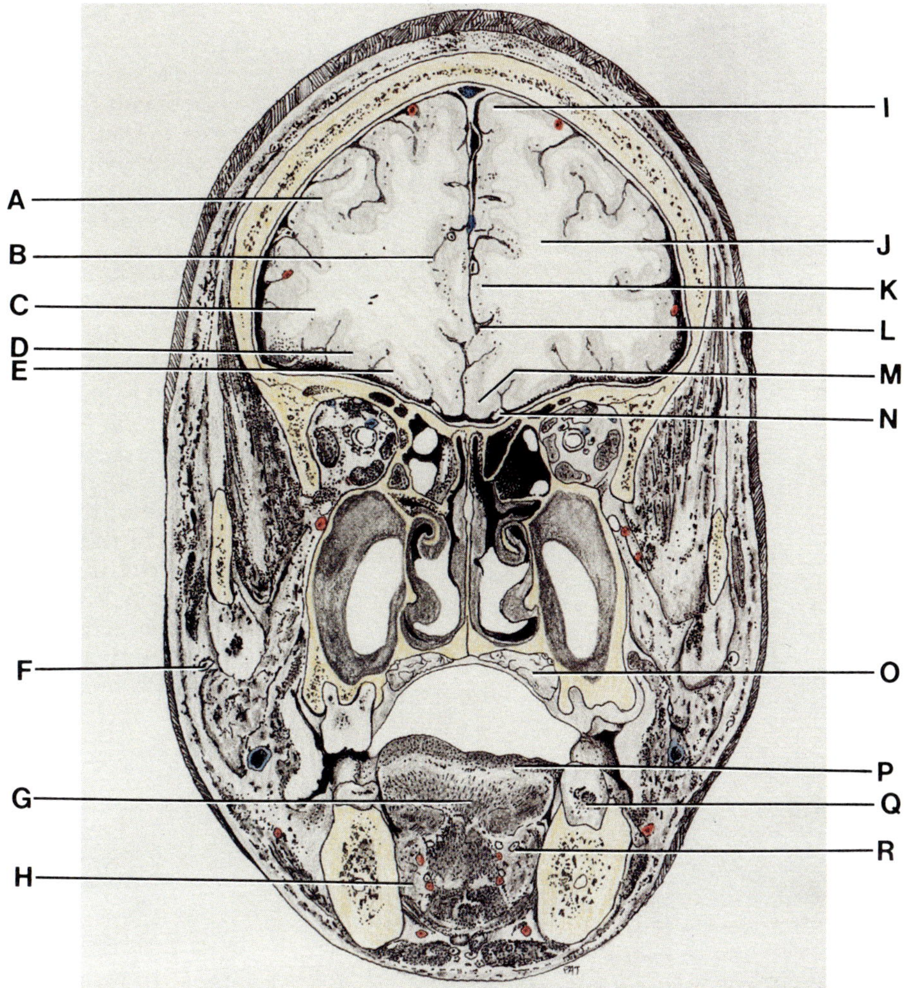

A. Medial frontal gyrus
B. Cingulate sulcus
C. Inferior frontal gyrus
D. Lateral orbital gyrus
E. Medial orbital gyrus
F. Parotid duct
G. Tongue
H. Sublingual gland
I. Superior frontal gyrus
J. Frontal lobe of cerebrum
K. Cingulate gyrus
L. Corpus callosum sulcus
M. Gyrus rectus
N. Olfactory tract
O. Palatine gland
P. Dorsum linguae
Q. Molar tooth
R. Submandibular duct

Figure 4-9
Coronal section of the head, which passes through the ethmoid air cells, the posterior extent of the frontal sinuses, the maxillary sinuses, the posterior portion of the hard palate, and the upper right second molar tooth, highlighting visceral unit structures.

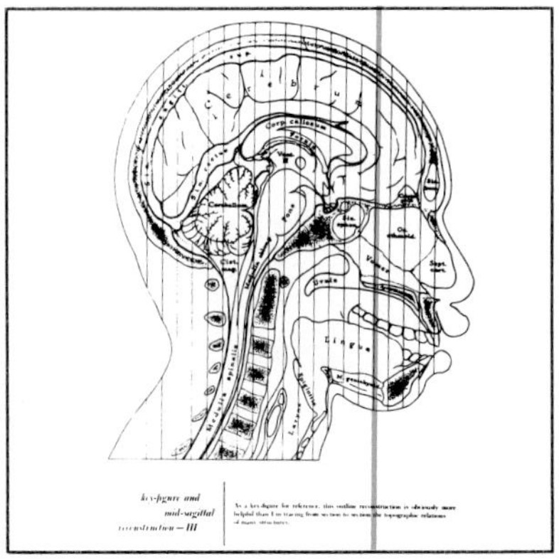

ENCLOSING UNIT (Fig. 4–10)

Falx Cerebri (*A*). A sheet of meningeal dura mater between the upper parts of the frontal lobes inferior to the superior sagittal and superior to the inferior sagittal sinus.

Dura Mater (*B*). A rather thick membrane lining the entire inner wall of the bone as seen in this section.

Vestibule of the Mouth (*D*). A vertically elongated space between the lateral surface of teeth/gums and the medial surface of the cheeks.

Nasal Cavity (*F*). Bilateral spaces lateral to the nasal septum and medial to the nasal concha.

Buccal Fat Pad (*G*). A large, irregular fat structure that extends from the lateral part of the maxillary sinus to the lateral part of the buccinator muscle. Some of this fat pad is seen surrounding the inferior part of the masseter muscle in this section.

Skin (*E*). A layer of tissue covering the entire section. The skin is the most superficial structure seen in this section.

Gingiva (*C*). A layer of mucous membrane on the lateral surface of the upper and lower jaws.

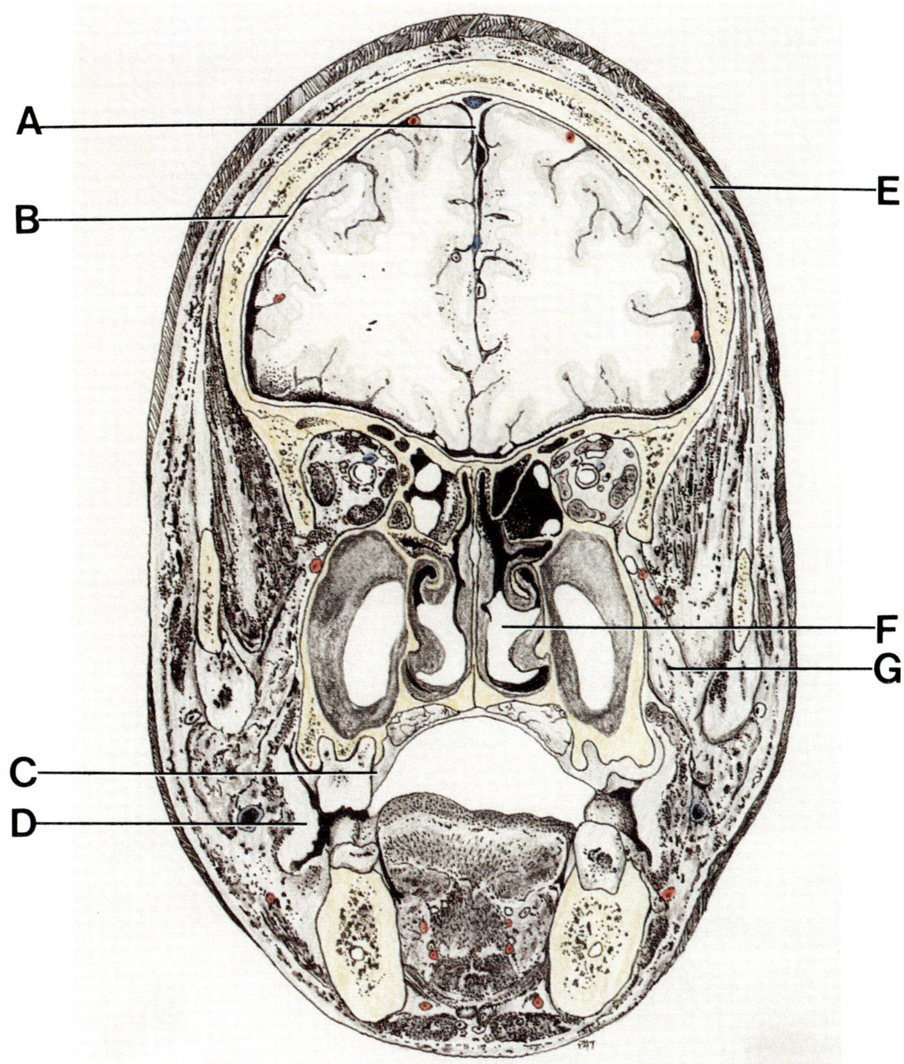

A. Falx cerebri
B. Dura mater
C. Gingiva
D. Vestibule of mouth
E. Skin
F. Nasal cavity
G. Buccal fat pad

Figure 4-10
Coronal section of the head, which passes through the posterior ethmoid air cells, the posterior extent of the frontal sinuses, the maxillary sinuses, the posterior portion of the hard palate, and the upper right second molar tooth, highlighting enclosing unit structures.

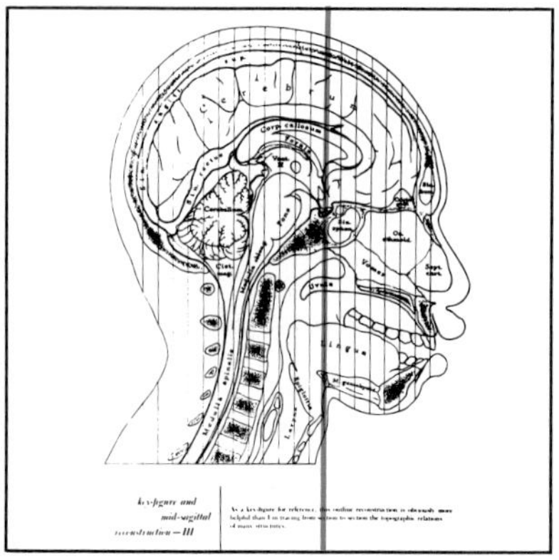

MODULE 3

This is a frontal (coronal) section that passes through the anterior horns of the lateral ventricles of the brain, the pituitary gland, the body of the sphenoid, the openings of the auditory (eustachian) tubes in the nasal pharynx, and the body of the hyoid bone. Note the size of the cavity within the pellucid septum, and also the close proximity of the pharyngeal tonsil to the mouths of the auditory tubes.

MUSCULOSKELETAL UNIT (Fig. 4–11)

I. Skeletal Portion

Frontal Bone (*A*). A bony, bilateral structure encasing the upper part of the brain in this section. As mentioned previously, there are inner and outer tables of this bone with a diploë sandwiched between them.

Parietal Bone (*Q*). A lateral part of the bone structure surrounding the brain as far down as the squamosal suture.

Temporal Bone (*S*). The inferior, lateral part of bone around the inferior part of the brain as seen in this section. A lateral projection extends from the inferior part of this bone that is known as the zygomatic process of the temporal bone.

Zygomatic Process of the Temporal Bone (*U*). A lateral projection from the inferior part of the frontal bone. It is seen inferior to the temporalis muscle and superior to the masseter.

Sphenoid Bone (*E*). This bone is seen at the floor of the cranial cavity with two large cavities in its medial part (sphenoid sinus).

Sphenoid Sinus (*T*). Bilateral, rather large cavities in the median of the cranial floor immediately inferior to the pituitary gland and immediately superior to the vomer.

Vomer (*R*). A median bone seen in this section as a transversely elongated bone immediately inferior to the sphenoid sinus and superior to the pharyngeal tonsil.

Anterior Clinoid Process (*C*). A triangular, bilateral bone section lateral to the internal carotid artery and superior to the oculomotor nerve. This process is a part of the sphenoid bone.

Coronoid Process of the Mandible (*F*). A rather sharp, superior bone structure of the mandible lateral to the internal maxillary artery and medial to the masseter muscle.

Mandible (*I*). A vertical, elongated, oblong bone section medial to the masseter muscle and lateral to the lingual nerve, as well as lateral to the internal pterygoid muscle. The upper part (coronoid process) is lateral to the external pterygoid muscle.

Hyoid Bone (*e*). A horseshoe-shaped bone with the free ends pointing downward, immediately inferior to the geniohyoid and medial to the mylohyoid muscles.

II. Muscular Portion

Galea Aponeurotica (*B*). A thin sheet of tendinous tissue immediately superficial to the frontal bone as seen in this coronal section.

Temporalis Muscle (*D*). A rather thin, bilateral muscle lateral to and adjacent to the parietal and temporal bones in this coronal (frontal) head section.

Masseter Muscle (*G*). Large, bilateral muscle tissue immediately lateral to the mandible and medial to the parotid gland. It is inferior to the zygomatic process of the temporal bone.

Pharyngopalatine (Palatopharyngeal) Muscle (*H*). A lateral strip of skeletal muscle tissue superior to the palatine tonsils as seen in this section. The lateral part of this muscle is immediately inferior to the internal pterygoid muscle. This vertical muscle tissue is in an oblique position. The superior end is medial to the lateral end.

Glossopalatine (Palatoglossus) Muscle (*J*). A narrow strip of muscle on the lateral part of the oral cavity immediately medial to the pterygoid muscle and the palatine tonsils. It is superior to the styloglossus muscle.

Styloglossus Muscle (*L*). An irregularly shaped, bilateral muscle tissue slice immediately lateral to the tongue.

Mylohyoid Muscle (*M*). Bilateral, rather elongated muscle tissue extending from the medial surface of the mandible to the hyoid bone as seen in this section.

Longitudinal Superior Muscle of the Tongue (*b*). An upper sheath of muscle tissue on the superior part of the tongue. This muscle extends the entire width of the tongue.

Hyoglossus Muscle (*K*). A bilateral muscle section, vertically elongated, immediately lateral to the styloglossus muscle and medial to the mylohyoid muscle.

Longitudinal Inferior Muscle of the Tongue (*N*). A bilateral, oval muscle section on the inferior part of the lateral part of the tongue. This muscle is immediately inferior to the lateral part of the transverse muscle of the tongue and medial to the mylohyoid muscle.

Digastric Muscle (*O*). An oval, bilateral muscle

A. Frontal bone
B. Galea aponeurotica
C. Anterior clinoid process
D. Temporalis muscle
E. Sphenoid bone
F. Coronoid process of mandible
G. Masseter muscle
H. Pharyngopalatine (palatopharyngeal) muscle
I. Mandible
J. Glossopalatine (palatoglossus) muscle
K. Hyoglossus muscle
L. Styloglossus muscle
M. Mylohyoid muscle
N. Longitudinal inferior muscle of tongue
O. Digastric muscle
P. Geniohyoid muscle
Q. Parietal bone
R. Vomer
S. Temporal bone
T. Sphenoid sinus
U. Zygomatic process of temporal bone
V. External (lateral) pterygoid muscle
W. Tensor muscle of velum palatini
X. Levator muscle of velum palatini
Y. Internal (medial) pterygoid muscle
Z. Muscle of uvula
a. Buccinator muscle
b. Longitudinal superior muscle of tongue
c. Genioglossus muscle
d. Sternohyoid muscle
e. Hyoid bone

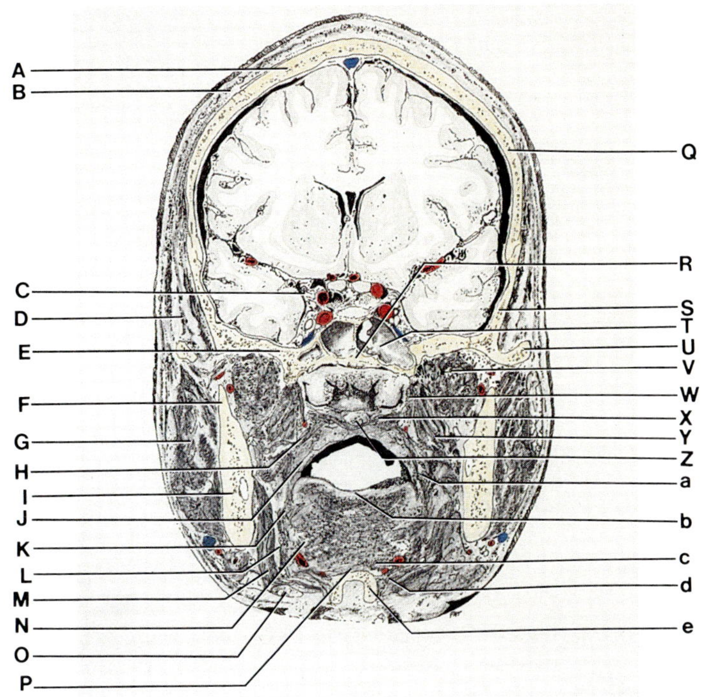

Figure 4-11
Coronal (frontal) section of the head, which passes through the anterior horns of the lateral ventricles of the brain, pituitary gland, body of the sphenoid, openings of the auditory tubes in the nasal pharynx, and body of the hyoid bone, highlighting musculoskeletal unit structures.

section medial to the submandibular gland and inferior to the mylohyoid muscle.

Genioglossus Muscle (*c*). A transverse strip of bilateral muscle tissue superior to the mylohyoid muscle and the hyoid bone as seen in this section. This muscle is medial to the deep lingual artery.

Sternohyoid Muscle (*d*). A bilateral, short, vertical muscle section immediately lateral to the hyoid bone, medial to the mylohyoid muscle, and inferior to the genioglossus muscle.

Geniohyoid Muscle (*P*). A thin strip of transversely placed, bilateral muscle tissue immediately superior to the hyoid bone as seen in this section.

Muscle of the Uvula (*Z*). Oval muscle tissue located in the median superior to the palatine gland and inferior to the pharyngeal tonsil.

Buccinator Muscle (*a*). A lateral, oblique strip of bilateral muscle tissue lateral to the palatine gland and medial to the internal pterygoid muscle.

Levator Muscle of the Velum Palatini (*X*). A bilateral muscle sheet, rather triangular in shape with the apex toward the median, located lateral to the muscle of the uvula and to the inferior lateral part of the pharyngeal tonsil. It is medial to the inferior part of the tensor muscle of the velum palatini.

Internal (Medial) Pterygoid Muscle (*Y*). Several strips of bilateral muscle tissue that extend from the lateral part of the pharyngeal tonsil to the medial part of the mandible as seen in this section.

External (Lateral) Pterygoid Muscle (*V*). A rather large, rounded, bilateral muscle slice lateral to the internal pterygoid muscle, inferior to the lateral part of the sphenoid bone, and medial to the internal maxillary artery.

Tensor Muscle of the Velum Palatini (*W*). A vertically placed, narrow strip of bilateral muscle slice immediately lateral to the pharyngeal tonsil.

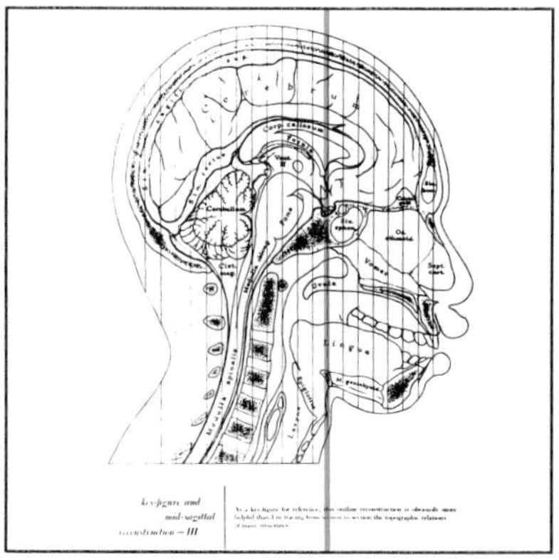

NEUROVASCULAR UNIT (Fig. 4–12)

I. Arteries

Internal Carotid Artery (*P*). Two oval, bilateral, vascular structures. The inferior part of the internal carotid artery is larger than the superior part. Two parts are seen on each side of median owing to the course of the internal carotid artery. The inferior part of the internal carotid artery in this section is immediately lateral to the pituitary gland. The upper part of this artery is seen medial to the anterior clinoid process and lateral to the optic nerve.

Masseteric Artery (*S*). An obliquely placed, hollow, rather small, bilateral, vascular structure superior to the coronoid process of the mandible, lateral to the external pterygoid muscle, and medial to the upper part of the masseter muscle.

Origin. Maxillary artery.

Distribution. Masseter muscle.

Course. Passes posterior to the tendon of temporalis and enters the mandibular notch, courses to the medial surface of the masseter muscle, and terminates by joining the facial arteries, including the transverse facial artery.

Internal Maxillary Artery (*H*). A rounded, rather large, bilateral, vascular structure located medial to the upper part of the coronoid process (mandible) and lateral to the external pterygoid muscle.

Origin. External carotid artery.

Distribution. Fifteen branches distribute blood to the maxilla and mandible, the teeth of both jaws, the muscles of chewing (temporalis, internal pterygoid, masseter, and external pterygoid), the palate, the nose, and the cranial dura mater.

Course. There are three parts of the maxillary artery. The first (mandibular portion) lies between the neck of the mandible and the sphenomandibular ligament. The second part (pterygoid portion) is superficial to the inferior head of the lateral pterygoid muscle. The third (pterygopalatine portion) runs between the superior and inferior heads of the lateral pterygoid muscle and ends in the pterygopalatine fossa.

External Maxillary (Facial) Artery (*K*). A rounded, vascular, bilateral structure immediately lateral to the submandibular gland and inferior to the masseter muscle.

Deep Lingual Artery (Arteria Profunda Linguae) (*Y*). An oval, bilateral structure immediately inferior to the genioglossus muscle and immediately superior to the sternohyoid muscle as seen in this section. *Note:* This is the third part of the lingual artery.

Submental Artery (*W*). A rounded, bilateral, vascular structure immediately inferior to the mandible as seen in this section and superior to the submandibular gland.

Origin. Facial artery.

Distribution. Inferior lip, chin, mylohyoid muscle, and two salivary glands (sublingual and submandibular).

Course. It passes anteriorly on the superior surface of the mylohyoid muscle below the body of the mandible.

Lesser Palatine Artery (*T*). A small, rounded, bilateral, vascular structure immediately lateral to the upper part of the buccinator muscle as seen in this section and inferior to the tensor muscle of the velum palatini.

Origin. Greater palatine artery (a branch of the maxillary artery).

Distribution. Supplies the tonsils and soft palate.

Course. Passes through the lesser palatine canal and joins with the ascending palatine artery.

Middle Cerebral Artery (*O*). A bilateral, rather elongated, oblique artery placed deep within the lateral cerebral fissure immediately inferior to the insular lobe and immediately superior to the temporal lobe.

Anterior Cerebral Artery (*N*). A rather large, bilateral, oval, vascular structure seen immediately above the optic chiasm at the point where the optic nerves join and inferior to the medial part of the insular lobe.

Sublingual Artery (*Z*). A rounded bilateral vascular structure immediately lateral to the genioglossus muscle and superior to the mylohyoid muscle.

II. Veins

Facial Vein (*J*). A large, oval, bilateral, vascular structure immediately inferior to the masseter muscle, superior to the external maxillary artery, and lateral to the inferior part of the mandible as seen in this section.

Origin. The union of the supratrochlear and supraorbital veins. Sometimes the supratrochlear veins unite into a single trunk, and this single trunk is the origin of the two anterior facial veins.

Termination. Blood from the facial vein flows into the cavernous sinus through the ophthalmic and deep facial veins.

Course. It starts in the medial angle of the eye and runs in an oblique position across the lateral face. It runs deep to the zygomatic major muscle and superficial to the anterior part of the masseter muscle. This

A. Optic chiasm
B. Optic (second cranial) nerve
C. Abducens (sixth cranial) nerve
D. Ophthalmic nerve
E. Maxillary nerve
F. Cavernous sinus
G. Masseteric nerve
H. Internal maxillary artery
I. Lingual nerve
J. Facial vein
K. External maxillary (facial) nerve
L. Submandibular lymph node
M. Superior sagittal sinus
N. Anterior cerebral artery
O. Middle cerebral artery
P. Internal carotid artery
Q. Occulomotor (third cranial) nerve
R. Trochlear (fourth cranial) nerve
S. Masseteric artery
T. Lesser palatine artery
U. Inferior alveolar (dental) nerve
V. Mylohyoid nerve
W. Submental artery
X. Hypoglossal (12th cranial) nerve
Y. Deep lingual artery (arteria profunda linguae)
Z. Sublingual artery

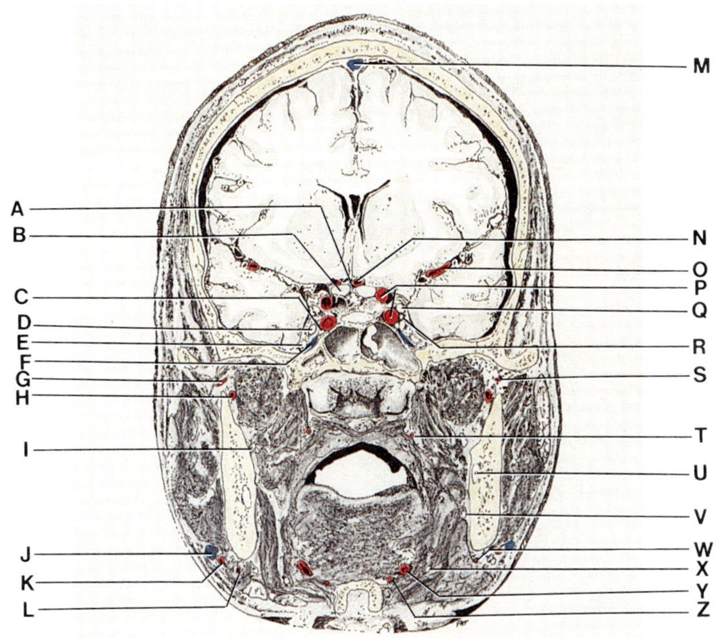

Figure 4-12
Coronal section of the head, which passes through the anterior horns of the lateral ventricles of the brain, pituitary gland, body of the sphenoid, openings of the auditory tubes in the nasal pharynx, and body of the hyoid bone, highlighting neurovascular unit structures.

vein connects with the internal jugular vein near the carotid triangular, also with a communicating branch to the retromandibular vein. *Note:* The facial vein has no valves, and blood may flow in both directions.

III. Vascular Venous Sinuses

Superior Sagittal Sinus (*M*). A rather large, oval, vascular structure in the superior medial inner cranium immediately superior to the falx cerebri.

Cavernous Sinus (*F*). An irregularly shaped, bilateral, vascular structure between the ophthalmic and the maxillary nerves.

Origin. Superior ophthalmic vein.

Termination. Inferior and superior petrosal sinuses.

Course. A cavernous sinus is found on the right and left sides of the body of the sphenoid bone on its cranial surface.

IV. Lymphatic Structures

Submandibular Lymph Node (*L*). A rounded, bilateral lymph tissue inferior to the mandible, immediately lateral to the submental artery in this section. *Note:* It is seen only on the right side in this section. This lymph node is located in the submandibular triangle, which is a subdivision of the anterior triangle of the neck.

V. Nerves

Ophthalmic Nerve (*D*). An oval, rather large, bilateral sensory nerve superior to the cavernous sinus, as seen in this section, and lateral to the abducens nerve.

Origin. The sensory branch of the trigeminal nerve (first division).

Distribution. Skin of the forehead, skin of the nose, mucous membrane of the nose, lacrimal gland, eyeball, and conjunctiva.

Course. This nerve exits from the skull through the superior orbital fissure and is associated with three main branches: lacrimal, frontal, and nasal ciliary nerves.

Maxillary Nerve (*E*). An oval, bilateral sensory nerve structure immediately inferior to the cavernous sinus and superior to the lateral part of the body of the sphenoid bone.

Origin. The sensory branch of trigeminal nerve (second division).

Distribution. Nasopharynx, roof of the mouth, teeth of the superior jaw, and skin of the cheek.

Course. Passes transversely anterior on the inferior part of the lateral wall of the cavernous sinus and exits from the skull through the foramen rotundum. It makes its way across the superior part of the pterygopalatine fossa, enters the orbit through the inferior orbital fissure, and becomes the infraorbital nerve.

Lingual Nerve (*I*). An oval, bilateral sensory nerve structure immediately lateral to the internal pterygoid muscle and the medial upper part of the maxilla.

Hypoglossal (12th Cranial) Nerve (*X*). An oval, bilateral motor nerve structure superior to the mylohyoid muscle and inferior to the hyoglossus muscle.

Mylohyoid Nerve (*V*). An oblong, bilateral motor nerve feature lateral to the lateral end of the mylohyoid muscle and medial to the mandible.

Origin. Inferior alveolar nerve.

Distribution. Mylohyoid muscle and the anterior belly of the digastric muscles.

Course. It penetrates the sphenomandibular ligament and goes down to the medial surface of the mandibular ramus within a groove below the mylohyoid line to reach the bottom surface of the mylohyoid muscle and the anterior belly of the digastric muscles.

Inferior Alveolar (Dental) Nerve (*U*). A rather large sensory and motor nerve found within the medial part of the mandible.

Origin. Mandibular branch (third branch) of the trigeminal nerve.

Distribution. Teeth of the mandible, skin of the chin, and the mylohyoid muscle.

Course. It enters the mandibular foramen after it passes between the sphenomandibular ligament and the mandibular ramus. It courses inferior to the teeth of the mandible to reach the mental foramen.

Abducens (Sixth Cranial) Nerve (*C*). A rather small, rounded motor nerve immediately medial to the ophthalmic nerve and inferior to the internal carotid artery. This sixth cranial nerve is superior to the sphenoid sinus.

Origin. Pons.

Distribution. Lateral rectus muscle of the eye.

Course. It exits from the skull through the superior orbital fissure to enter the orbit and reach the lateral rectus muscle.

Trochlear (Fourth Cranial) Nerve (*R*). A small, rounded motor nerve immediately inferior to the oculomotor nerve, superior to the ophthalmic nerve, and lateral to the internal carotid artery.

Origin. Midbrain.

Distribution. Superior oblique muscle of the eye.

Course. It exits from the skull through the superior orbital fissure to enter the orbit and reach the superior oblique muscle of the eye.

Oculomotor (Third Cranial) Nerve (*Q*). A rather large, oval, bilateral cranial (third) nerve immediately superior to the trochlear nerve, as seen in this section, and lateral to the superior part of the internal carotid artery.

Optic (Second Cranial) Nerve (*B*). A rather large, oval, bilateral sensory nerve tissue immediately medial to the internal carotid artery and lateral to the optic chiasm inferior to the anterior cerebral artery.

Optic Chiasm (*A*). A thin strip of nerve tissue between the right and left optic nerves, superior to the pituitary gland. This structure represents the union of right and left optic nerves.

Masseteric Nerve (*G*). A narrow, elongated, bilateral motor nerve superior to the mandible and masseteric artery. It is seen in this section extending to the superior surface of the masseter muscle.

Origin. Mandibular branch (third division) of the trigeminal nerve.

Termination. Masseter muscle, which it innervates.

Course. It proceeds laterally, superior to the lateral pterygoid muscle, anterior to the temporomandibular joint, and posterior to the tendon of the temporalis muscle.

VISCERAL UNIT (Fig. 4–13)

Corpus Callosum (*B*). A transversely curved, connecting brain structure that connects the right and left brain hemispheres. The concave portion of this structure is the superior edge. It is superior to the lateral ventricle and inferior to the longitudinal cerebral fissure.

Lateral Ventricle (*C*). A crescent-shaped, hollow, bilateral cavity immediately lateral to the septum pellucidum and medial to the head of the caudate nucleus. The concave portion of this crescent-shaped space is on its lateral surface.

Head of the Caudate Nucleus (*M*). An internal, bilateral gray mass immediately lateral to the lateral ventricle and immediately medial to the internal capsule.

Internal Capsule (*N*). A bilateral, curved strip of white matter between the head of the caudate nucleus and the lentiform nucleus.

Lentiform Nucleus (*D*). An internal, bilateral gray mass between the internal and external capsules. The lentiform nucleus is a part of the basal ganglia and is divided into a medial part known as the globus pallidus and a lateral part known as the putamen. The putamen is immediately medial to the external capsule. The globus pallidus is lighter in color and is located immediately lateral to the internal capsule.

External Capsule (*F*). A crescent-shaped, bilateral strip of white matter immediately lateral to the lentiform nucleus and medial to the claustrum. The concave surface of the external capsule is adjacent to the lentiform nucleus.

Claustrum (*G*). A rather small, bilateral strip of curved gray matter immediately lateral to the external capsule and medial to the extreme capsule. The concave surface is located adjacent to the external capsule.

Extreme Capsule (*P*). A rather large, bilateral strip of curved white matter immediately lateral to the claustrum and medial to the insular lobe. The concave surface is adjacent to the claustrum.

Insular Lobe (*O*). An outward, bilateral, folded mass of brain tissue with a peripheral edge of gray matter deep to the lateral fissure of Sylvius. This lobe is seen inferior to the frontal lobe and superior to the temporal lobe in this section.

Lateral Fissure of Sylvius (*E*). A bilateral slit between the frontal and temporal lobes, as seen in this section. This feature is located immediately lateral to the insular lobe.

Frontal Lobe (*A*). A large, bilateral brain structure showing peripheral gray matter; the inner part shows white matter immediately superior to the lateral fissure of Sylvius and lateral to the falx cerebri. The lower medial edge contains the lateral ventricle.

Temporal Lobe (*H*). A rather large, oval, bilateral brain structure showing peripheral gray matter, with an inner core of white matter immediately inferior to the lateral fissure of Sylvius and the insular lobe.

Pharyngeal Tonsil (*I*). A rather large, irregularly shaped lymphoid tissue in the median of this section, immediately inferior to the vomer and sphenoid sinus. It is superior to the muscle of the uvula.

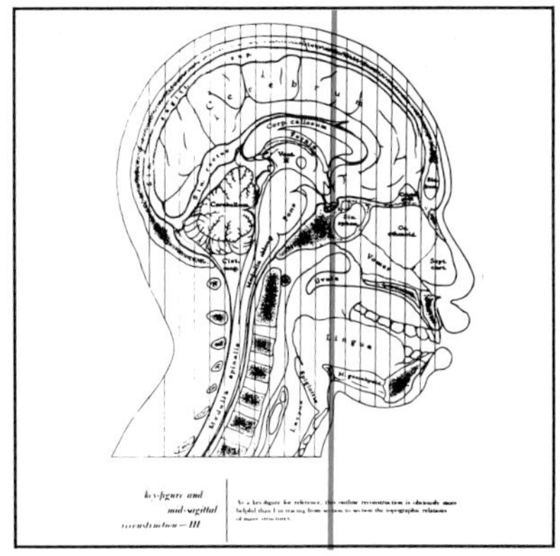

Submandibular (Wharton's) Duct (*V*). A rather small, oval tube located immediately medial to the mylohyoid muscle and lateral to the hyoglossus muscle.

Origin. Submandibular gland in the submandibular triangle of the neck.

Termination. Inferior to the tongue on the floor of the mouth at the most superior part of the small sublingual papilla, immediately posterior to the inferior incisive teeth.

Course. Passes anteriorly and superiorly to the deep surface of the mylohyoid muscle and lateral to the hyoglossus, genioglossus, and styloglossus muscles. It reaches the sublingual papilla after passing on the medial surface of the sublingual salivary gland.

Palatine Tonsil (*J*). Oval, bilateral lymphoid tissue immediately medial to the buccinator muscle as seen in this section. It is located inferior and lateral to the palatine gland.

Palatine Gland (*U*). Irregular, bilateral glandular tissue embedded within the mucosa in the roof of the mouth.

Submandibular Gland. A rather large, irregular, bilateral glandular tissue immediately inferior to the mandible and lateral to the mylohyoid muscle. This is a salivary gland that secretes both mucus and salivary amylase.

Parotid Duct (*T*). A rather small, bilateral tube located in the inferior part of the parotid gland immediately lateral to the masseter muscle.

Origin. Parotid gland.

Termination. In the mouth cavity medial to the crown of the superior second molar tooth.

Course. Sialography reveals a pattern of branchings of the parotid duct. It extends from the parotid gland along an imaginary line toward the alae of the nose. In its course it passes lateral to the masseter muscle and pierces the buccal fat pad and the buccinator muscle. It runs for a short distance between the buccinator muscle and the mucous membrane of the mouth.

Parotid Gland (*S*). A vertical, oblong, bilateral, glandular structure immediately lateral to the masseter muscle and inferior to the lateral part of the zygomatic process (temporal bone). It is the largest of the three salivary glands.

Pituitary Gland (Hypophysis) (*Q*). Transversely oblong, glandular tissue located between the internal carotid arteries as seen in this section. This gland is located immediately superior to the sphenoid sinus.

Pharyngotympanic Tube (Auditory or Eustachian Tube) (*R*). A bilateral, vertical slit immediately lateral to the pharyngeal tonsil and medial to the tensor muscle of the velum palatini.

Origin. Nasal pharynx lateral to the pharyngeal tonsil.

Termination. Ear drum (tympanic membrane). *Note:* Air proceeds between the nasopharynx and the tympanic cavity to equalize the air pressure on the medial and lateral surfaces of the tympanic membrane.

ENCLOSING UNIT (see Fig. 4–13)

Falx Cerebri (*K*). A sheath of meningeal dura mater between the right and left frontal lobes, as seen in this section, immediately inferior to the superior sagittal sinus.

Cavity Within Septum Pellucidum (*L*). A triangular space between the upper parts of the lateral ventricles and inferior to the corpus callosum. The base of the space is adjacent to the corpus callosum.

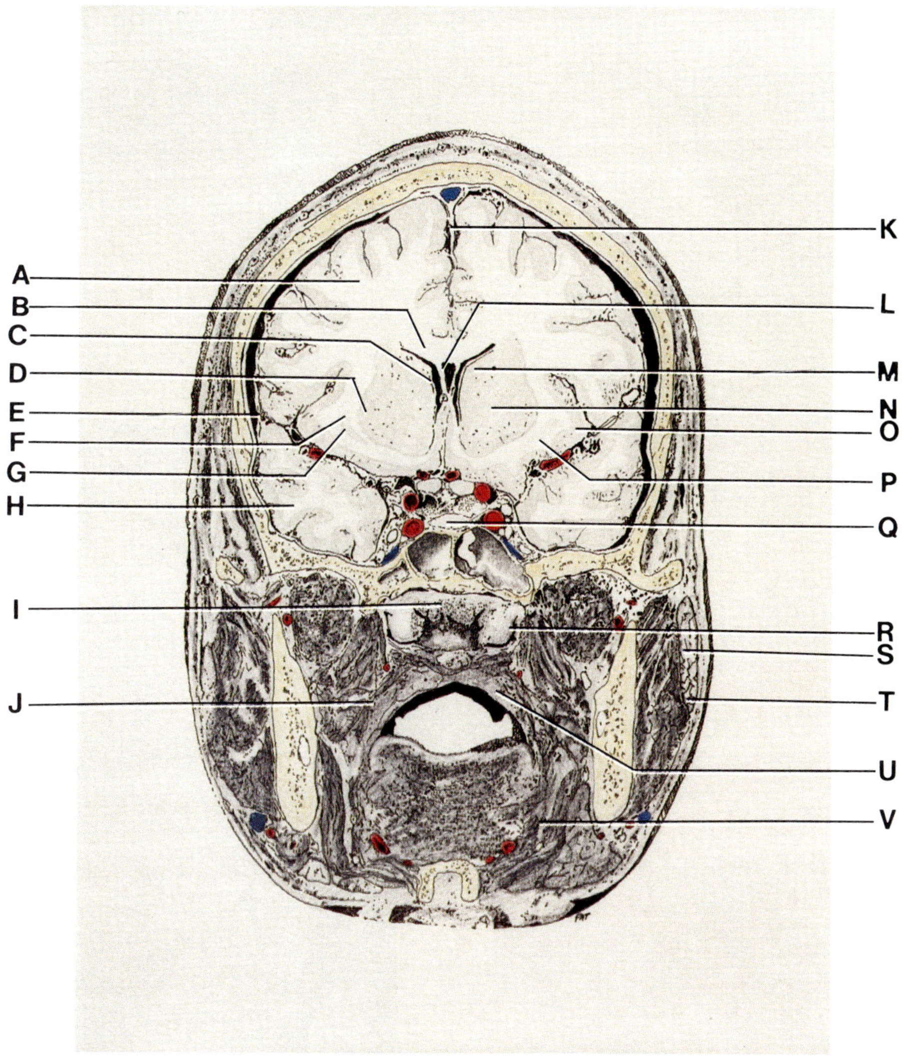

A. Frontal lobe
B. Corpus callosum
C. Lateral ventricle
D. Lentiform nucleus
E. Lateral fissure of Sylvius
F. External capsule
G. Claustrum
H. Temporal lobe
I. Pharyngeal tonsil
J. Palatine tonsil
K. Falx cerebri
L. Cavity within septum pellucidum
M. Head of caudate nucleus
N. Internal capsule
O. Insular lobe
P. Extreme capsule
Q. Pituitary gland (hypophysis)
R. Pharyngotympanic tube (auditory or eustachian tube)
S. Parotid gland
T. Parotid duct
U. Palatine gland
V. Submandibular (Wharton's) duct

Figure 4–13
Coronal section of the head, which passes through the anterior horns of the lateral ventricles of the brain, pituitary gland, body of the sphenoid, openings of the auditory tubes in the nasal pharynx, and body of the hyoid bone, highlighting both the visceral and enclosing unit structures.

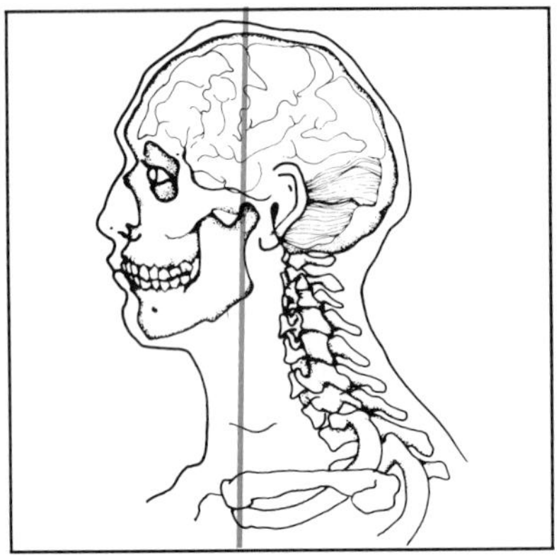

Figure 4–14 is a picture of an MRI coronal image that emphasizes by labels all four anatomic units. This slice passes through the sphenoid sinus, the lateral fissure of Sylvius, and the corpus callosum, as well as many other structures.

All four anatomic unit structures are highlighted on this MRI coronal image (Fig. 4–15). This slice passes through the superior sagittal sinus, the frontal lobe of the brain, the oropharynx, and many other structures.

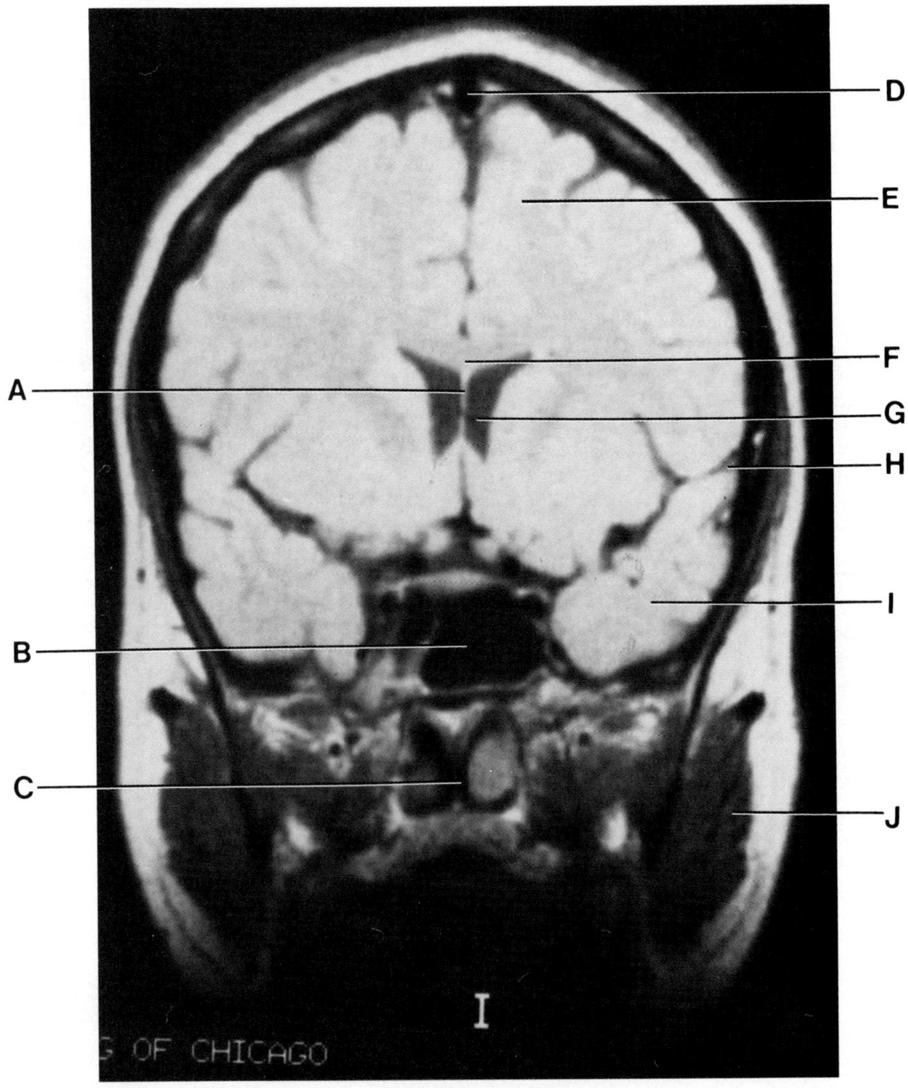

A. Septum pellucidum
B. Sphenoid sinus
C. Nasopharynx
D. Superior sagittal sinus
E. Frontal lobe
F. Corpus callosum
G. Lateral ventricle
H. Lateral fissure of Sylvius
I. Temporal bone
J. Masseter muscle

Figure 4–14
MRI coronal image of the head, highlighting the musculoskeletal, neurovascular, visceral, and enclosing unit structures.

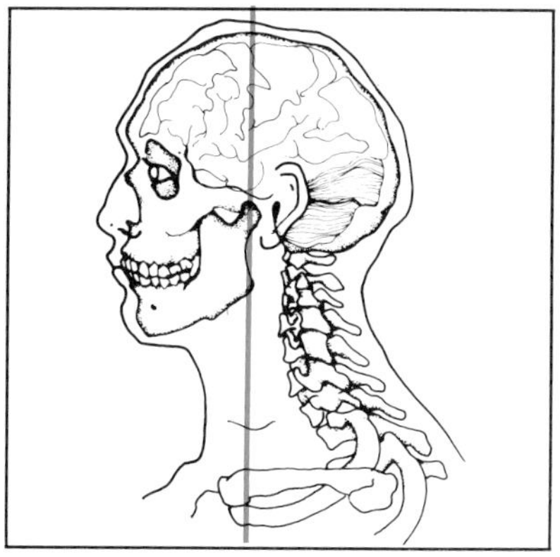

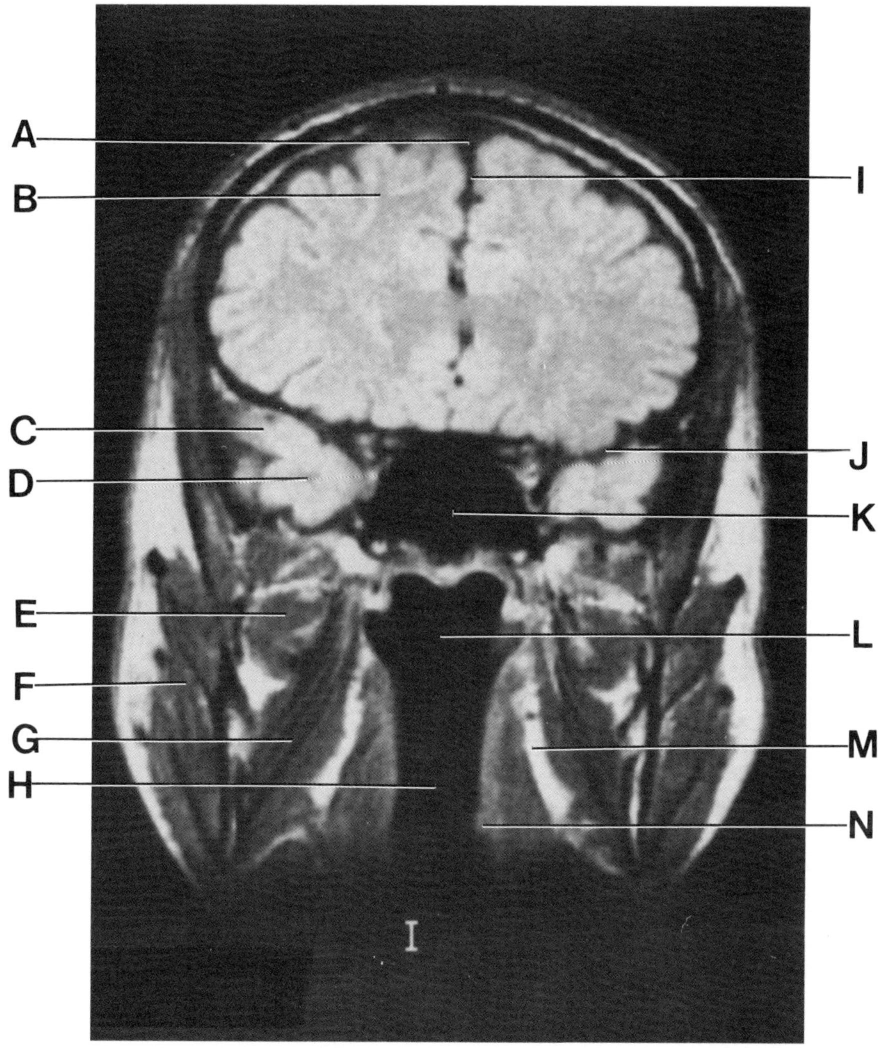

A. Superior sagittal sinus
B. Frontal lobe
C. Superior temporal gyrus
D. Temporal lobe
E. Lateral pterygoid muscle
F. Masseter muscle
G. Medial pterygoid muscle
H. Oropharynx
I. Longitudinal cerebral fissure
J. Lateral fissure (sylvian)
K. Sphenoid sinus
L. Nasopharynx
M. Parapharyngeal fat plane
N. Pharyngeal constrictors

Figure 4–15
MRI coronal image of the head, highlighting the musculoskeletal, neurovascular, visceral, and enclosing unit structures.

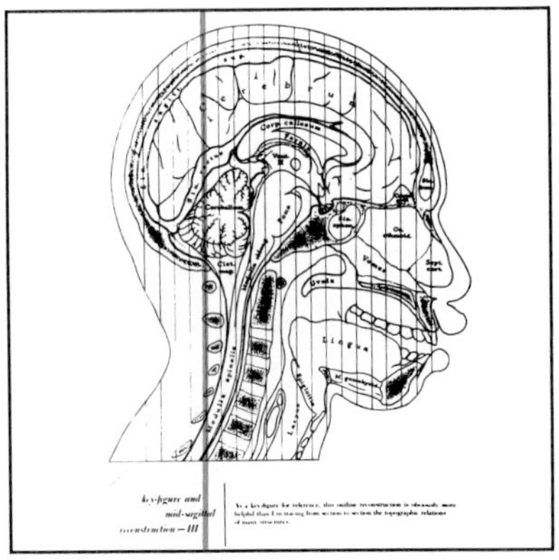

MODULE 4

Here is a frontal (coronal) section that passes through the posterior horn of the lateral ventricles, the posterior margin of the foramen magnum, and the straight sinus. Note the large size of the superior sagittal sinus, the smaller size of the right transverse sinus compared with the left transverse sinus, and the inclination of the straight sinus to the left.

MUSCULOSKELETAL UNIT (Fig. 4–16)

I. Skeletal Portion

Parietal Bone (*H*). A bilateral flat bone section encasing the brain. Each bone extends from the sagittal suture to the parietomastoid suture on its respective side. Each bone has an outer and an inner table with a diploë sandwiched between.

Sagittal Suture (*F*). An irregular, seamlike feature in the median superior skull between the right and left parietal bones.

Parietomastoid Suture (*J*). An irregular, seamlike structure between the inferior part of the parietal bone section and the mastoid process of the temporal bone. It is immediately lateral to the transverse sinus.

Mastoid Process of the Temporal Bone (*K*). An oval, bilateral bone feature between the parietomastoid and occipitomastoid sutures. Small cavities are seen within this bone structure.

Occipitomastoid Suture (*L*). An irregular, seamlike, bilateral feature between the mastoid process of the temporal bone and the occipital bone.

Occipital Bone (*M*). A bone section at the inferior part of the brain immediately inferior to the cerebellum. This bone shows compact bone on its peripheral surfaces and spongy bone in parts of its inner layer.

II. Muscular Portion

Galea Aponeurotica (*A*). A sheet of tendinous tissue covering the parietal bones, as seen in this section. Its lower edge is attached to the occipitalis muscle on the lateral inferior wall of the cranial cavity.

Occipitalis Muscle (*D*). A narrow, bilateral strip of muscle tissue immediately lateral to the inferior part of the parietal bone and the superior part of the mastoid process of the temporal bone.

Posterior Auricular Muscle (*E*). A small, irregular, bilateral muscle tissue lateral to the inferior part of the mastoid process and inferior to the occipitalis muscle.

Origin. Lateral superior surface of the mastoid process.

Insertion. Inferior division of the cranial surface of the ear.

Innervation. Posterior auricular nerve and temporal branches of the facial nerve.

Vasculature. Parietal branch of the superficial temporal artery and the auricular branch of the posterior auricular artery.

Action. Usually functionless.

NEUROVASCULAR UNIT (see Fig. 4–16)

I. Veins

Posterior Auricular Vein (*B*). A rather small, rounded, bilateral vasculature structure located in the superficial fascia of the scalp on the lateral parts of the head approximately midway between the sagittal and parietomastoid sutures.

Origin. This superficial vein arises from the parietal and posterior parts of the temporal head region.

Termination. External jugular vein.

Course. It proceeds posterior to the ear and connects with the retromandibular vein inferior to the parotid gland to form the external jugular vein.

II. Vascular Venous Sinuses

Superior Sagittal Sinus (*G*). A large, triangular vascular structure in the median superior cranial cavity immediately inferior to the sagittal suture and immediately superior to the falx cerebri.

Straight Sinus (*C*). An elongated, triangular vascular structure with its apex adjacent to the falx cerebri and its base adjacent to the tentorium cerebelli.

Origin. Great cerebral vein and inferior sagittal sinus.

Termination. Left transverse sinus.

Course. It passes posteriorly between the falx cerebri and tentorium cerebelli to reach the left transverse sinus.

Transverse Sinus (*I*). A large, oval, bilateral vascular structure seen immediately medial to the parietomastoid suture and lateral to the tentorium cerebelli.

Origin. The right transverse sinus begins as a continuation of the superior sagittal sinus, and the straight sinus is the beginning of the left transverse sinus.

Termination. Sigmoid sinus.

Course. It proceeds laterally and anteriorly to the back and side parts of the petrous feature of the temporal bone. It curves inferiorly and becomes the sigmoid sinus.

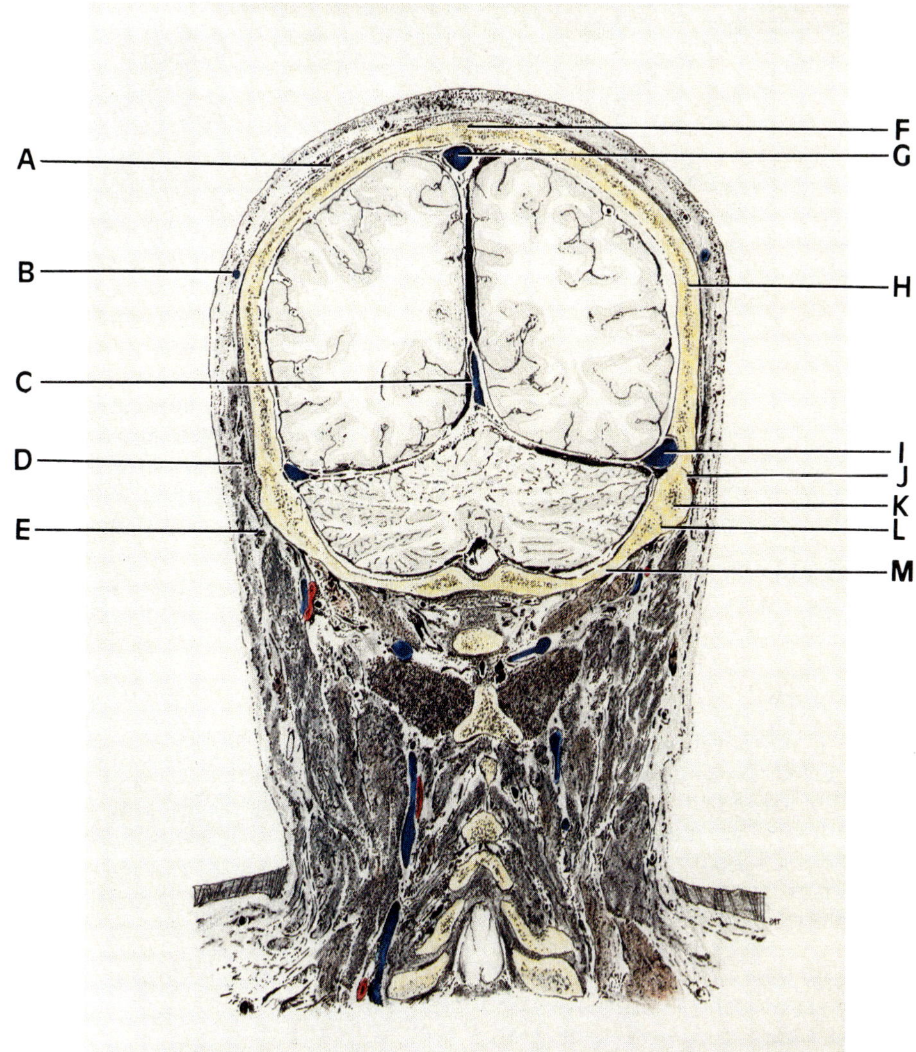

A. Galea aponeurotica
B. Posterior auricular vein
C. Straight sinus
D. Occipitalis muscle
E. Posterior auricular muscle
F. Sagittal suture
G. Superior sagittal sinus
H. Parietal bone
I. Transverse sinus
J. Parietomastoid suture
K. Mastoid process of temporal bone
L. Occipitomastoid suture
M. Occipital bone

Figure 4-16
Coronal (frontal) section of the head, which passes through the posterior horn of the lateral ventricles, the posterior margin of the foramen magnum, and the straight sinus, highlighting both musculoskeletal and neurovascular unit structures.

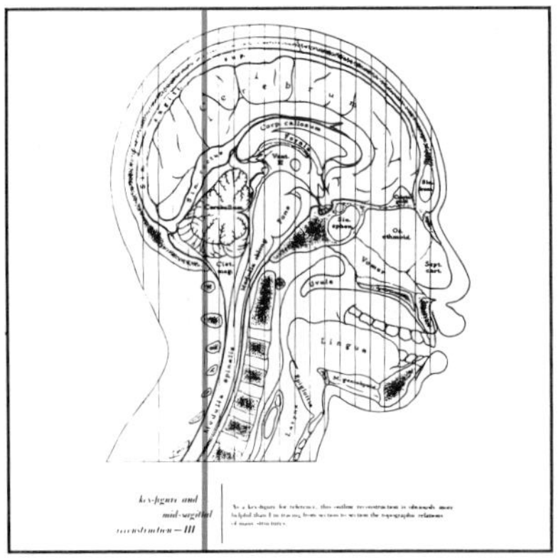

VISCERAL UNIT (Fig. 4–17)

Parietal Lobe (*O*). A large, bilateral brain structure showing the peripheral gray matter; the inner part shows white matter. The peripheral part of this lobe, as in all the lobes of the brain, shows many folds known as gyri (singular: gyrus). Each lobe is in the upper part of the cranial cavity in this section, superior to the temporal lobe.

Superior Gyrus of the Parietal Lobe (*L*). A bilateral, rather large fold on the upper peripheral part of the parietal lobe. It is superior to the inferior gyrus of the parietal lobe.

Inferior Gyrus of the Parietal Lobe (*N*). A bilateral, large fold on the inferior part of the parietal lobe, as seen in this section, immediately superior to the temporal lobe.

Temporal Lobe (*R*). A large, bilateral brain structure showing peripheral gray matter with its inner part of white matter and some dispersed gray matter. It is inferior to the parietal lobe and immediately superior to the occipital lobe as seen in this section.

Medial Temporal Gyrus (*P*). A large, bilateral, peripheral fold of the temporal lobe inferior to the parietal lobe and superior to the inferior temporal gyrus.

Inferior Temporal Gyrus (*Q*). A bilateral, peripheral fold superior to the occipital lobe as seen in this section and inferior to the medial temporal gyrus.

Occipital Lobe (*G*). A rather large, bilateral brain structure showing peripheral gray matter with white matter in its inner core and with some dispersion of gray matter. It is immediately superior to the tentorium cerebelli as seen in this frontal (coronal) section.

Calcarine Fissure (Sulcus) (*J*). A medial, bilateral slit of the peripheral part of the occipital lobe. It is immediately lateral to the straight sinus, inferior to the cuneus gyrus of the occipital lobe, and superior to the lingual gyrus.

Parieto-Occipital Fissure (*E*). A bilateral, slanted slit between the parietal and occipital lobes. The lateral end of the slit is superior to the medial end. This feature is immediately lateral to the straight sinus.

Cuneus Gyrus (*D*). A rather large, bilateral, medial fold of the superior occipital lobe. This feature is lateral to the straight sinus.

Lingual Gyrus of the Occipital Lobe (*K*). A bilateral, rather large, medial fold immediately inferior to the calcarine fissure and lateral to the medial part of the tentorium cerebelli.

Fusiform Gyrus (*I*). A bilateral, inferior, peripheral fold of the occipital lobe immediately medial to the lateral gyrus of the occipital lobe and superior to the tentorium cerebelli.

Lateral Occipital Gyrus (*T*). A bilateral fold located in the inferior lateral occipital lobe superior to the transverse sinus and medial to the inferior part of the parietal bone. It is lateral to the fusiform gyrus.

Precuneus (*C*). A bilateral, medial, gray matter fold of the parietal lobe immediately lateral to the falx cerebri and superior to the parietal occipital fissure.

Cerebellar Vermis (*V*). A median, rounded, worm-shaped structure between two hemispheres of the cerebellum, superior to the basilar plexus.

Posterior Horn of the Lateral Ventricle (*F*). A narrow, bilateral, elongated, slightly curved space located within the occipital lobe lateral to the cuneus gyrus and medial to the stratum sagittale.

Stratum Sagittale (*S*). Bilateral strips of gray matter within the inner core of the occipital lobe and lateral to the posterior horn of the lateral ventricle.

Calcarine Gyrus (*H*). A fold of gray matter near the medial surface of the occipital lobe, immediately lateral to the cuneus gyrus and medial to the upper part of the posterior horn of the lateral ventricle.

Cingulate Sulcus (*M*). A bilateral groove on the medial surface on the parietal lobe. The lateral end of this sulcus slants toward the temporal lobe.

ENCLOSING UNIT (see Fig. 4–17)

Falx Cerebri (*B*). A median sheet of meningeal dura mater between the superior sagittal sinus and the straight sinus in this section.

Tentorium Cerebelli (*U*). A sheet of tissue formed by the meningeal layer of the dura mater, located between the occipital lobe and the hemispheres of the cerebellum. In this section it is seen extending from the right transverse sinus to the left transverse sinus.

Dura Mater (*A*). A rather thick membrane seen lining the entire cranial cavity in this section, with its inner layer forming two major sheets of tissue. One sheet of tissue formed by the inner layer of the dura mater is known as the falx cerebri. Another sheet formed by the inner layer of the dura mater and seen in this section is the tentorium cerebelli. Within the edges of these folds, four dura mater sinuses are produced: the superior sagittal sinus, the straight sinus, and the right and left transverse sinuses.

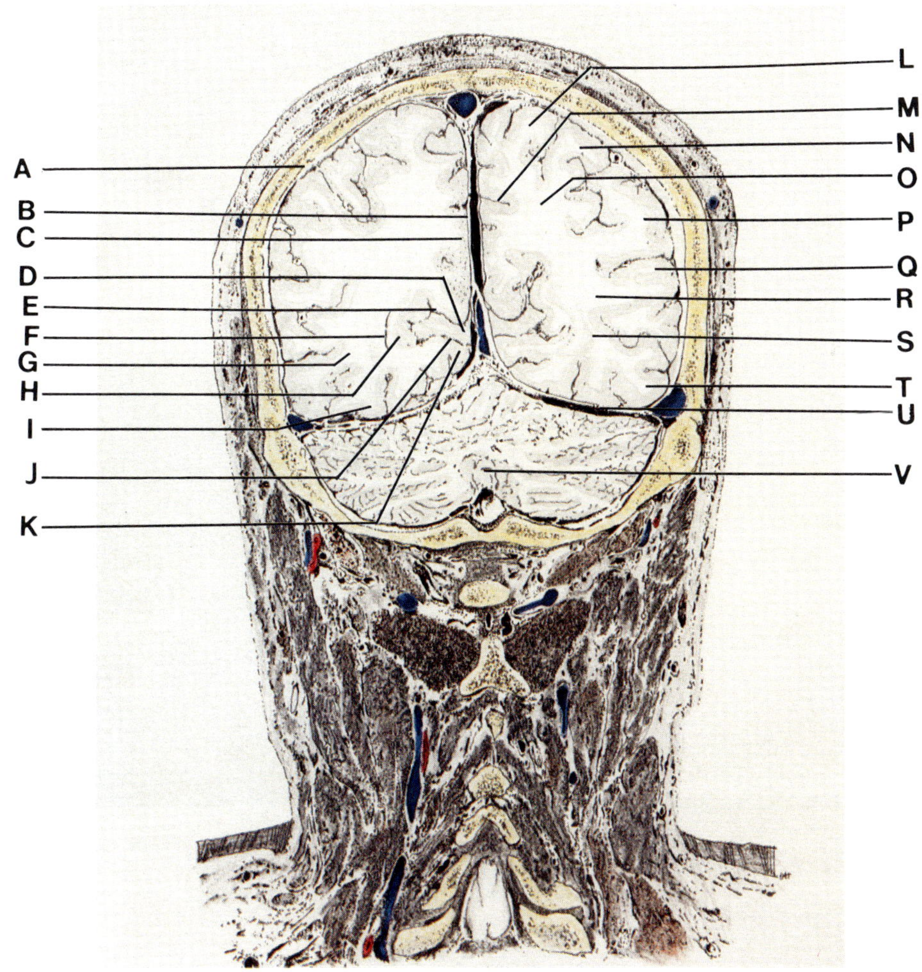

A. Dura mater
B. Falx cerebri
C. Precuneus
D. Cuneus gyrus
E. Parieto-occipital fissure
F. Posterior horn of lateral ventricle
G. Occipital lobe
H. Calcarine gyrus
I. Fusiform gyrus
J. Calcarine fissure (sulcus)
K. Lingual gyrus of occipital lobe
L. Superior gyrus of parietal lobe
M. Cingulate sulcus
N. Inferior gyrus of parietal lobe
O. Parietal lobe
P. Medial temporal gyrus
Q. Inferior temporal gyrus
R. Temporal lobe
S. Stratum sagittale
T. Lateral occipital gyrus
U. Tentorium cerebelli
V. Cerebellar vermis

Figure 4–17
Coronal section of the head, which passes through the posterior horn of the lateral ventricles, the posterior margin of the foramen magnum, and the straight sinus, highlighting both visceral and enclosing unit structures.

Figure 4–18 is a picture of an ultrasound coronal view of a fetal brain. Hydrocephalus is evident in the left lateral ventricle.

The foramen of Monro, the right frontal horn, and the thalamic area are seen in an ultrasound coronal picture (Fig. 4–19).

Figure 4–20 is an ultrasound coronal image of a fetal brain showing blood within the brain tissue. Several other structures are seen in this ultrasound picture.

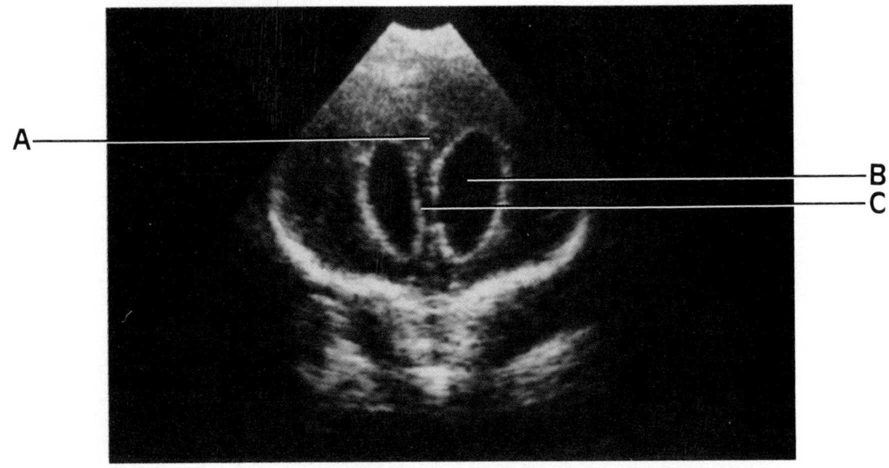

A. Corpus callosum
B. Left lateral ventricle
C. Septum pellucidum

Figure 4–18
An ultrasound coronal view of a fetal brain, showing the lateral ventricles. Note the enlargement of the left lateral ventricle (hydrocephalus).

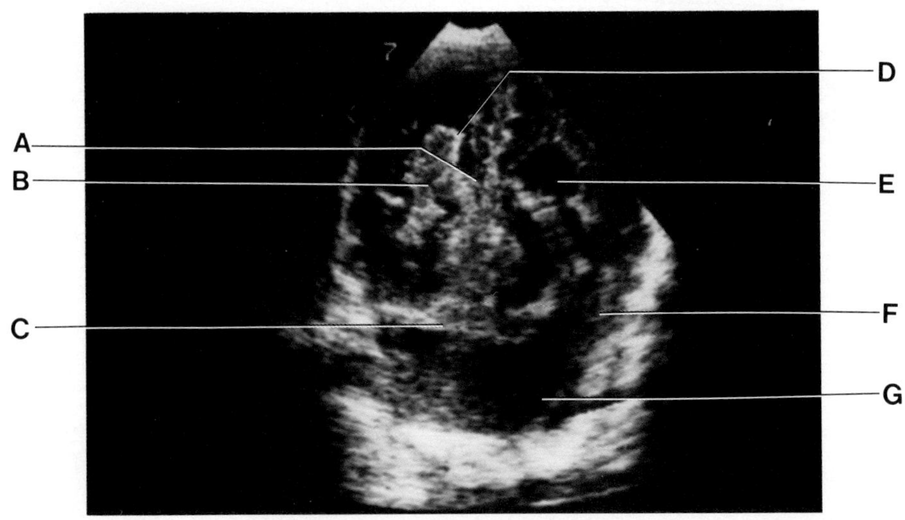

A. Foramen of Monro (interventricular foramen)
B. Blood in central nervous system
C. Blood
D. Anterior horn of right lateral ventricle
E. Anterior horn of left lateral ventricle
F. Blood in intraparenchymal brain tissue
G. Occipital horn of lateral ventricle

Figure 4–19
An ultrasound coronal view of a fetal brain, showing the foramen of Monro, right frontal horn, and thalamic area.

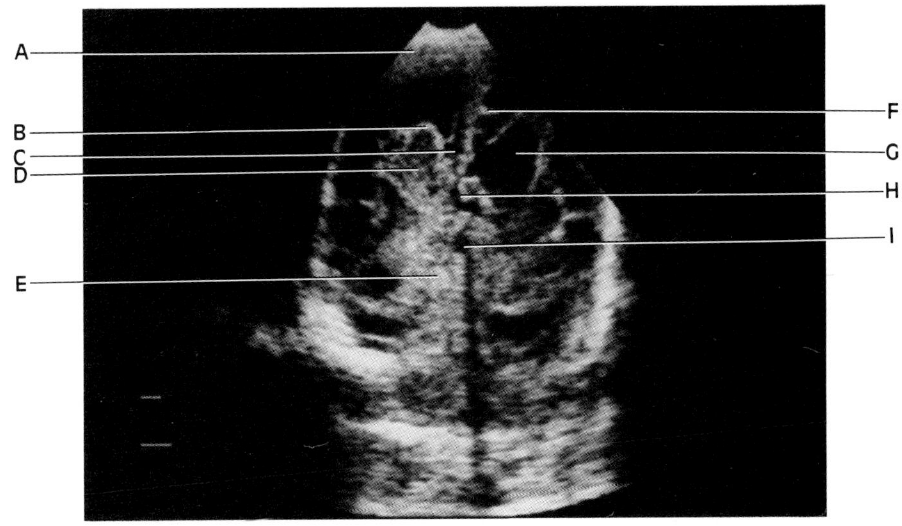

A. Blood within brain tissue
B. Anterior horn of right lateral ventricle
C. Choroid plexus of lateral ventricle
D. Blood within brain tissue
E. Blood within brain tissue
F. Brain sulcus
G. Anterior horn of left lateral ventricle
H. Foramen of Monro
I. Third ventricle

Figure 4–20
An ultrasound coronal view of a fetal brain, showing blood within the brain tissue.

CHAPTER 5

Sagittal Cadaver Sections and Images of the Head

In this chapter, sagittal sections of the head are studied. Features such as the intervertebral disks, spinal cord, and anterior and posterior longitudinal ligaments are seen clearly in a longitudinal plane. The attachment of the spinal cord to the brain is clearly shown. The most superior attachment of the anterior and posterior longitudinal ligaments is visible in this plane. These sagittal sections are seen from the left profile surface.

Ten MRI scans and three ultrasound scans are included in this chapter. These scans show many of the brain structures. MRI differentiates brain features with high resolution. The ultrasound scans of fetal brain are especially clear in visualizing the ventricles and hemorrhages.

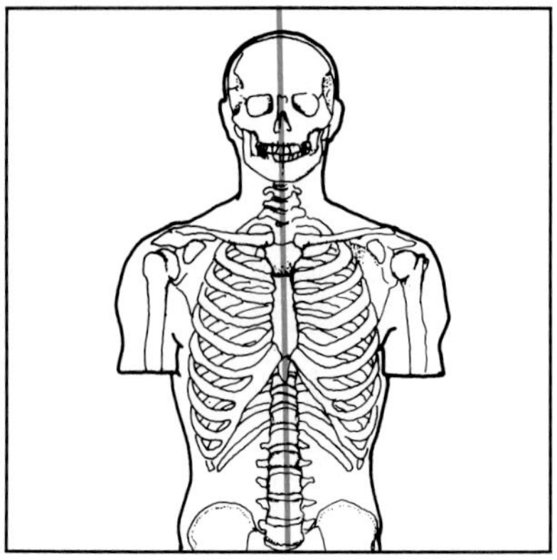

MODULE 1

This is a midsagittal cadaver section of the head, neck, and upper thorax, showing the frontal sinus, falx cerebri, cerebellum, corpus callosum, and pineal gland. The head structures are described and discussed in this module.

MUSCULOSKELETAL UNIT (Fig. 5–1)

I. Skeletal Portion

Frontal Sinus (*C*). Two vertically oblong cavities within the inferior part of the frontal bone, superior to the nasal cavity and anterior to the front part of the falx cerebri.

Diploë of the Frontal Bone (*B*). A sagittal thin slice of spongy bone between the inner and outer tables of the frontal bone. In this midsagittal section, this bone is seen between the outer and inner tables of the other cranial bones.

Outer Table of the Frontal Bone (*L*). A sagittal, thin slice of compact bone superficial to the diploë. This bone feature is seen superficial to the diploë of the other cranial bones, as seen in this midsagittal section.

Inner Table of the Frontal Bone (*A*). A sagittal, thin slice of compact bone deep to the diploë.

Palatine Bone (*G*). An elongated (anterior-to-posterior) bone section superior to the oral cavity and inferior to the nasal septum. The posterior part of this bone attaches to the anterior part of the soft palate.

Mandible (*K*). A sagittal slice of the lower jaw, which shows an oblong inferior part and a narrow superior part with a tooth projecting upward. This bone slice is posterior to the inferior lip and skin of the chin.

Body of the Sphenoid Bone (*E*). A curved slice of bone tissue immediately inferior to the pituitary gland and anterior to the midbrain. The superior part of this bone section shows a concave surface.

Nasal Septum (*D*). An irregular sheet of bone and cartilaginous tissue. The bone tissue is posterior to the cartilaginous tissue.

Nasopharynx (*O*). A hollow area immediately above the soft palate and uvula. This cavity is seen in this midsagittal section inferior to the posterior part of the nasal septum.

Oropharynx (*J*). A vertical hollow area immediately inferior to the uvula and superior to the epiglottis. This feature is also posterior to the base of the tongue.

Soft Palate (*H*). A section of curved soft tissue posterior to the palatine bone, inferior to the nasopharynx, and superior to the oropharynx. The posterior part of this structure is positioned inferiorly to the anterior part.

Uvula (*I*). A slice of soft tissue located at the most posterior and inferior parts of the soft palate immediately above the oropharynx.

NEUROVASCULAR UNIT (see Fig. 5–1)

I. Vascular Venous Sinuses

Transverse Sinus (*P*). A triangular, dural venous structure between the cerebellum and posterior inferior part of the occipital lobe. The base of the triangle is adjacent to a part of the occipital bone.

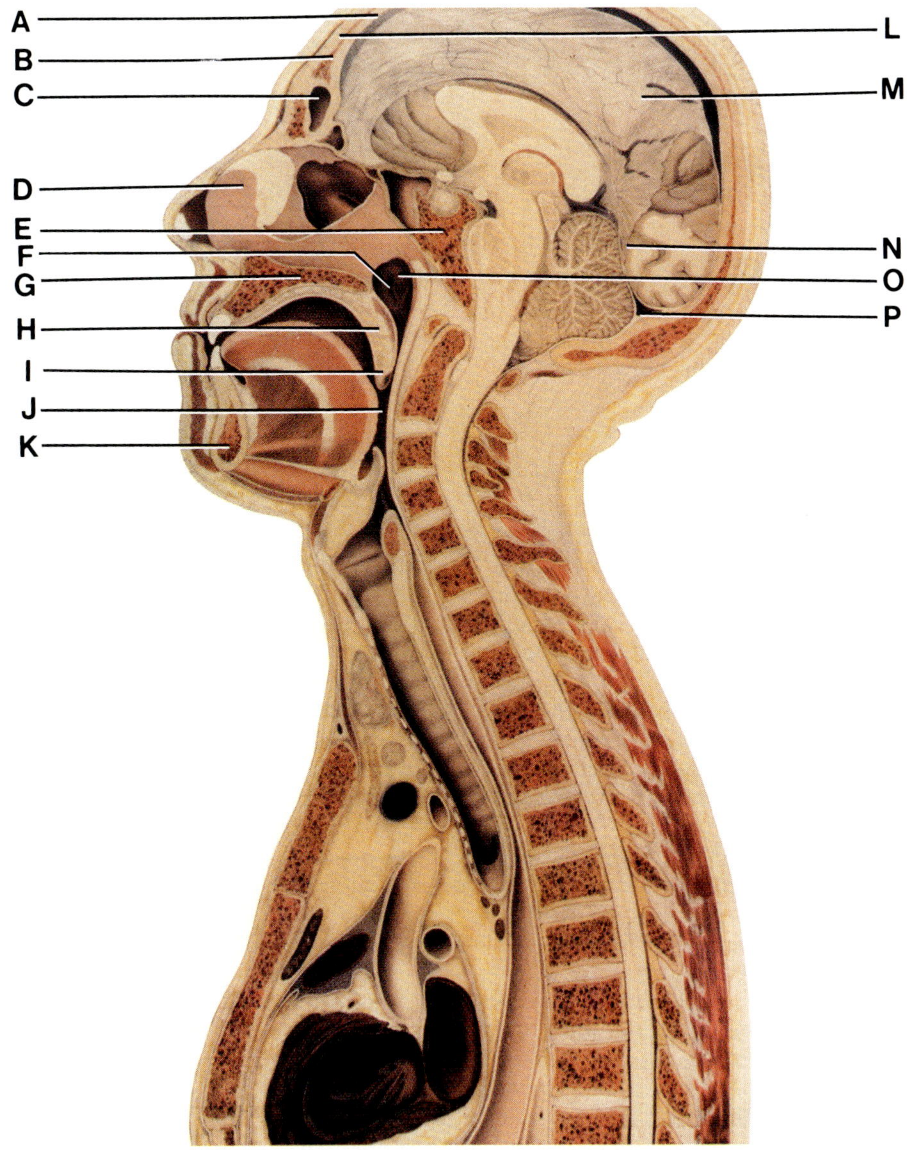

A. Inner table of frontal bone
B. Diploë of frontal bone
C. Frontal sinus
D. Nasal septum
E. Body of sphenoid bone
F. Eustachian (auditory) tube
G. Palatine bone (hard palate)
H. Soft palate
I. Uvula
J. Oropharynx
K. Mandible
L. Outer table of frontal bone
M. Falx cerebri
N. Tentorium cerebelli
O. Nasopharynx
P. Transverse sinus

Figure 5-1
Midsagittal cadaver section of the head, neck, and upper thorax, highlighting the head musculoskeletal, neurovascular, and enclosing unit structures.

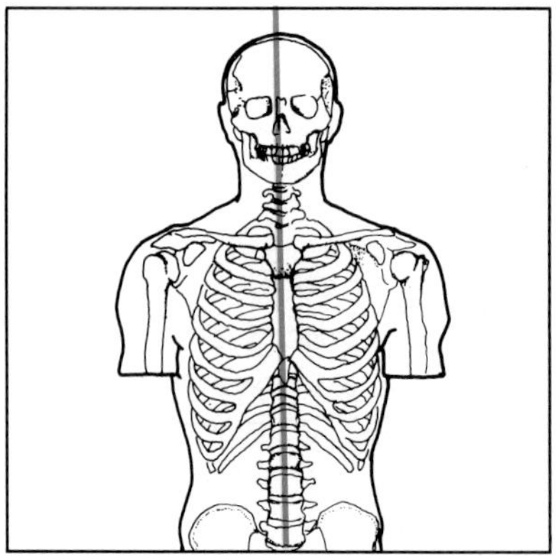

VISCERAL UNIT (Fig. 5–2)

Body of the Corpus Callosum (*A*). An elongated (anterior-to-posterior) slice of brain tissue inferior to the falx cerebri and superior to the septum pellucidum. This structure extends from the area adjacent to the superior part of the cerebellum to the medial surface of the frontal lobe.

Genu of the Corpus Callosum (*D*). An anterior, oval, projected extension of the corpus callosum medial to the posterior part of the frontal lobe. This structure is adjacent to the medial posterior part of the frontal lobe.

Splenium of the Corpus Callosum (*P*). A posterior, oval, projected extension of the corpus callosum adjacent to the medial part of the occipital lobe and superior to the cerebellum.

Septum Pellucidum (*C*). An irregularly triangular feature between the corpus callosum and the fornix. The base of the triangular structure is adjacent to the anterior part of the fornix.

Fornix (*B*). A vertical, oblique structure inferior to the corpus callosum. The inferior end of the anterior portion is more posterior than the superior end. The inferior end of the posterior portion is more anterior than the superior end. These inferior ends of the fornix join and form a triangular figure with the inferior surface of the corpus callosum.

Anterior Cerebral Commissure (*F*). An oval brain structure superior to the infundibulum and anterior to the mamillary body.

Posterior Cerebral Commissure (*I*). A small, rounded brain structure superior to the corpora quadrigemina of the midbrain and anterior to the pineal gland.

Corpora Quadrigemina (*R*). A vertical slice of midbrain posterior to the aqueduct of Sylvius and inferior to the posterior cerebral commissure. This midbrain structure is superior to the cerebellum.

Tegmentum. A vertical, oblique, broad slice of midbrain tissue superior to the pons and anterior to the aqueduct of Sylvius. This structure is regarded as the superior continuation of the dorsal portion of the pons. The superior end is more anterior than the inferior end.

Aqueduct of Sylvius (*S*). A vertical groove, as seen in this midsagittal section, between the corpora quadrigemina and the tegmentum of the midbrain.

Pons Varolii (Bridge of Varolius) (*J*). A rather large, oval brain structure located anterior to the tegmentum of the midbrain and superior to the medulla oblongata.

Medulla Oblongata (*K*). A vertical slice of brain tissue immediately inferior to the pons and tegmentum of the midbrain. This structure is wider at its superior end and attaches to the spinal cord by its inferior end.

Infundibulum (*G*). A vertical, short structure between the hypothalamus and the pituitary gland. This structure is attached to both the hypothalamus and the pituitary gland.

Pituitary Gland (*H*). A rounded, globular gland located within the sella turcica and inferior to the infundibulum. The sella turcica (Turk's saddle) is a feature of the superior part of the body of the sphenoid bone.

Mamillary Body (*E*). A rounded body between the hypothalamus and the tegmentum of the midbrain. This structure is at the inferior posterior part of the hypothalamus.

Foramen of Monro (*M*). A small opening located at the inferior part of the fornix and anterior to the mass intermedia of the thalamus. Cerebrospinal fluid is conducted from the lateral ventricle to the third ventricle through this structure.

Mass Intermedia (*N*). A rounded structure within the center of the thalamus posterior to the opening of the foramen of Monro and superior to the superior part of the tegmentum of the midbrain.

Thalamus (*O*). A circular portion of brain tissue, as seen in this midsagittal section, superior to the midbrain and posterior to the fornix. This structure is inferior to the posterior part of the corpus callosum. The anterior part of this structure is superior to the hypothalamus.

Hypothalamus. A triangular brain structure inferior to the anterior part of the thalamus. The base of the triangular feature is adjacent to the anterior part of the thalamus. The anterior part of this brain tissue is connected to the infundibulum. The mamillary body is the most posterior part of the hypothalamus.

Pineal Gland (*Q*). A rather small, oval gland superior to the corpora quadrigemina and posterior to the superior part of the aqueduct of Sylvius. It is inferior to the splenium of the corpus callosum.

Occipital Lobe of the Cerebrum (*T*). Irregularly oval, posterior cerebral brain tissue superior to the cerebellum and inferior to the parieto-occipital fissure.

Cerebellum (*U*). A rounded sagittal slice of brain tissue posterior to the fourth ventricle and inferior to the occipital lobe of the cerebrum.

Fourth Ventricle. A rather elongated space between the cerebellum and tegmentum of the midbrain. This space widens between the midpoint (superior-to-inferior) of the cerebellum.

Tongue (*L*). An elongated (anterior-to-posterior) slice of the tongue muscles.

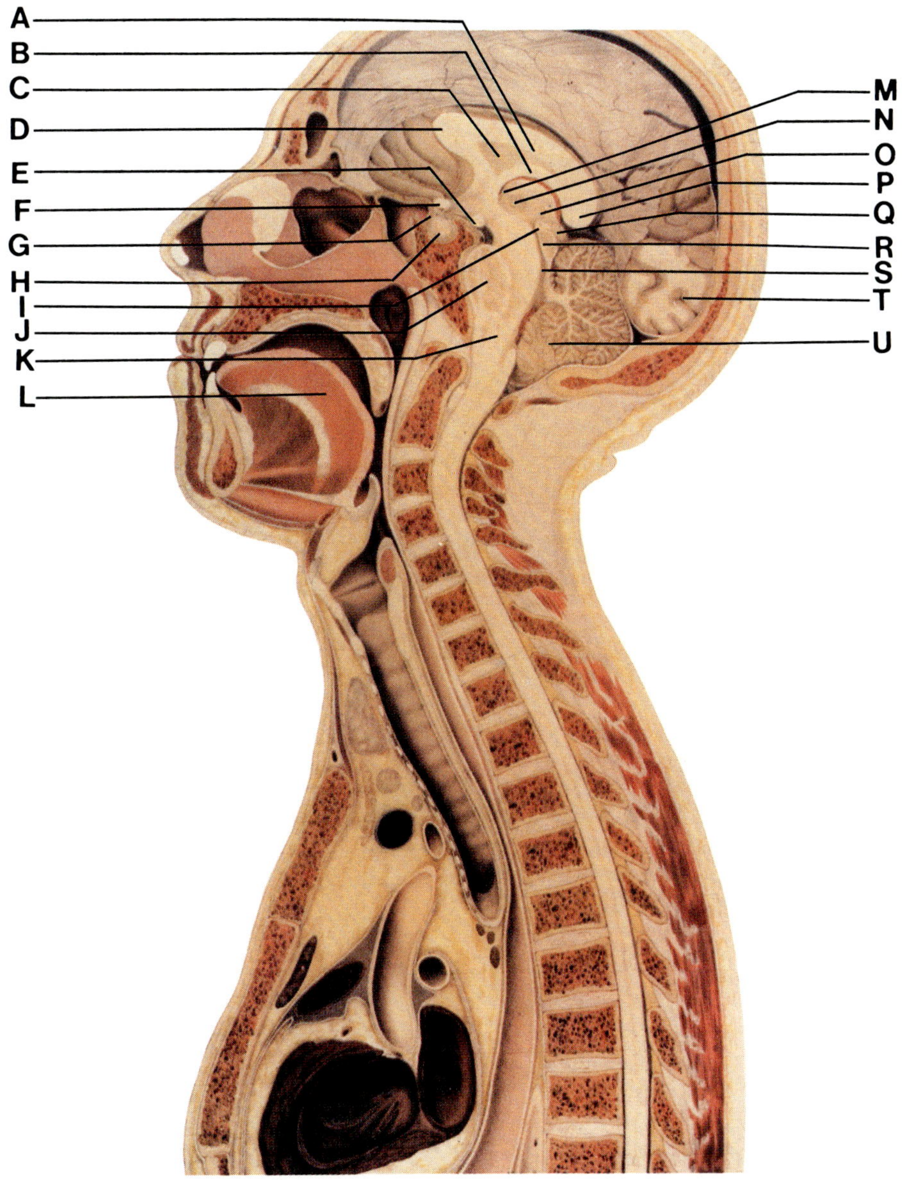

A. Body of corpus callosum
B. Fornix
C. Septum pellucidum
D. Genu of corpus callosum
E. Mamillary body
F. Anterior cerebral commissure
G. Infundibulum
H. Pituitary gland
I. Posterior cerebral commissure
J. Pons varolii (bridge of Varolius)
K. Medulla oblongata
L. Tongue
M. Foramen of Monro
N. Mass intermedia
O. Thalamus
P. Splenium of corpus callosum
Q. Pineal gland
R. Corpora quadrigemina
S. Aqueduct of Sylvius
T. Occipital lobe
U. Cerebellum

Figure 5-2
Midsagittal cadaver section of the head, neck, and upper thorax, highlighting the head visceral unit structures.

ENCLOSING UNIT (see Fig. 5–1)

Falx Cerebri (*M*). A sickle-shaped sheet of tissue adjacent to the medial portion of the cerebrum, as seen in this sagittal section. This structure extends from the occipital lobe of the cerebrum to the frontal crest of the frontal bone.

Eustachian (Auditory) Tube (*F*). An opening on the lateral wall of the nasopharynx. In this sagittal section, only the proximal part of this feature is seen.

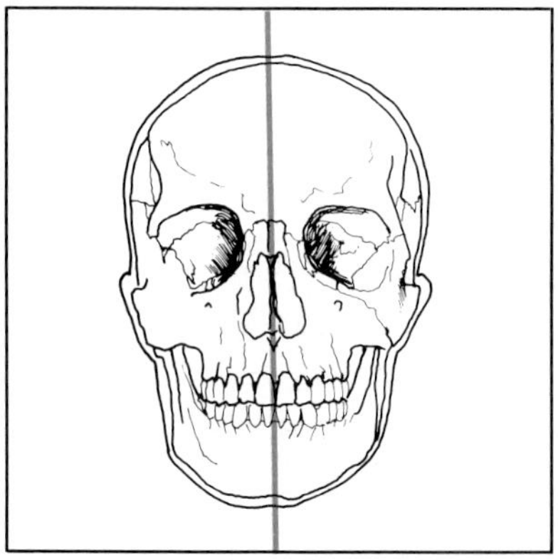

Figure 5–3 is an illustration of a molar tooth in the sagittal plane. Figure 5–4 is an MRI sagittal slice of the head. Both the musculoskeletal and neurovascular unit structures are highlighted by labels. This section passes through the frontal bone, sphenoid sinus, and several other structures. The visceral and enclosing unit structures are depicted in Figure 5–5. This MRI sagittal section passes through structures such as the pons, fourth ventricle, and anterior horn of the lateral ventricle, as well as several other structures. Note that this picture is the same as Figure 5–4, but different structures are highlighted on each.

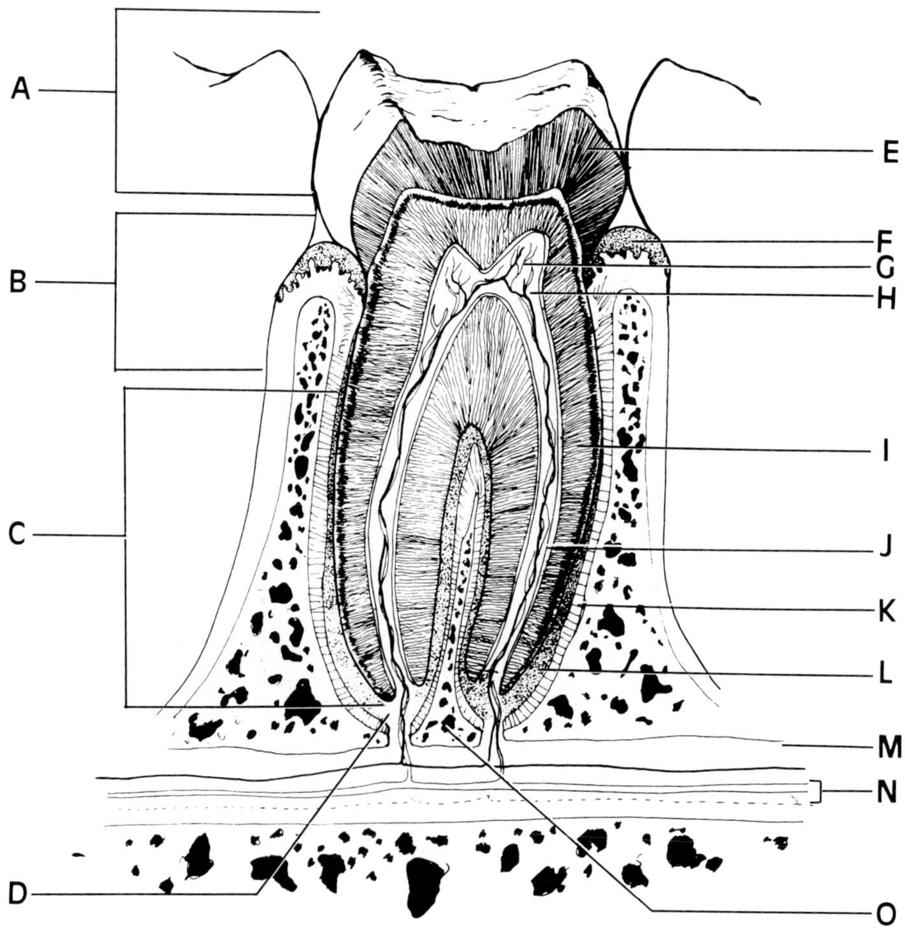

A. Crown
B. Neck
C. Root
D. Apical foramen
E. Enamel
F. Gingiva (gums)
G. Pulp
H. Pulp cavity
I. Dentin
J. Root (pulp) canal
K. Periodontal membrane
L. Cementum
M. Alveolar (dental) nerve
N. Alveolar vessels
O. Spongy bone of alveolar process

Figure 5–3
Sagittal section of a molar tooth and its related structures.

A. Bone marrow
B. Frontal bone
C. Inner table of frontal bone
D. Area of foramen of Monro
E. Optic chiasm
F. Sphenoid sinus
G. Inferior nasal turbinate (concha) of nasal cavity
H. Marrow within clivus of occipital bone
I. Parietal bone
J. Superior sagittal sinus
K. Internal cerebral vein
L. Interpeduncular cistern
M. Area of straight sinus
N. Occipital bone
O. Transverse sinus
P. Cisterna magna
Q. Basilar artery

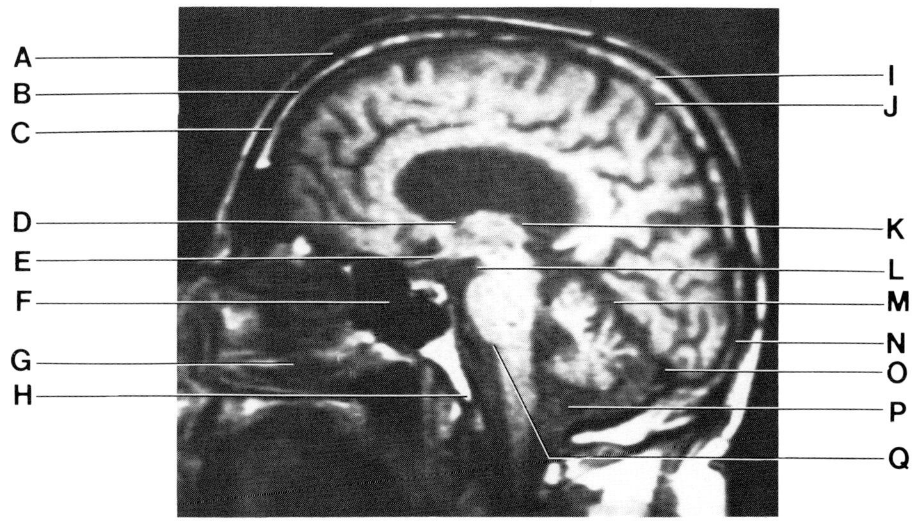

Figure 5-4
MRI sagittal section of the head, highlighting both musculoskeletal and neurovascular unit structures.

A. Corpus callosum
B. Cingulate gyrus
C. Anterior corpus callosum
D. Frontal lobe
E. Anterior horn of lateral ventricle
F. Anterior third ventricle
G. Area of infundibulum
H. Pituitary gland
I. Pons
J. Fourth ventricle
K. Soft palate
L. Area of tongue
M. Scalp
N. Thalamus
O. Parietal lobe
P. Lateral ventricle
Q. Splenium of corpus callosum
R. Parieto-occipital fissure
S. Occipital lobe
T. Cerebral peduncle of midbrain
U. Cerebellum
V. Medulla oblongata

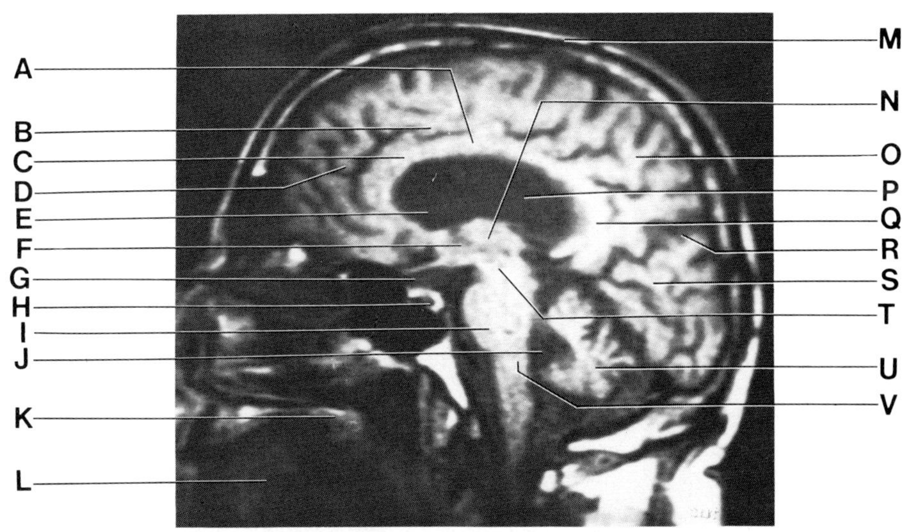

Figure 5-5
MRI sagittal section of the head, highlighting both visceral and enclosing unit structures.

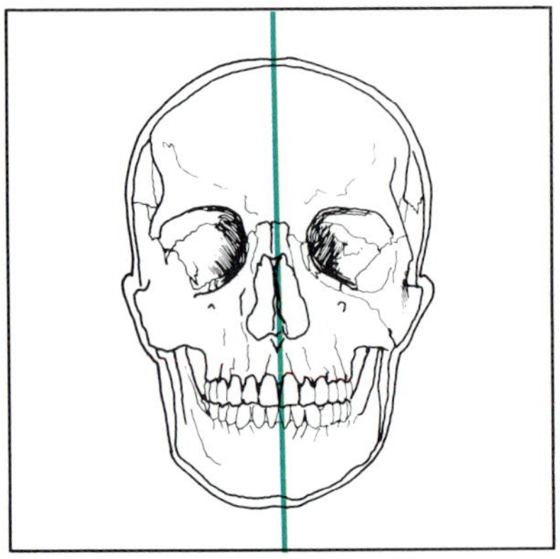

The MRI sagittal head scans in Figures 5–6 and 5–7 are the same pictures. Figure 5–6 emphasizes the musculoskeletal and neurovascular unit structures, whereas Figure 5–7 highlights by labels the visceral unit structures. Note the orientational pictures for the derivation of both.

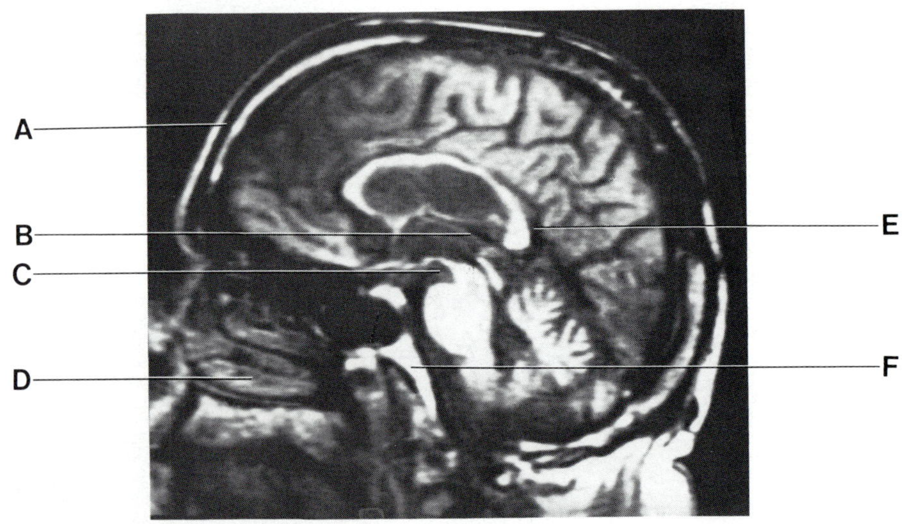

A. Area of bone marrow (diploë)
B. Internal cerebral vessels
C. Interpeduncular cistern
D. Inferior nasal turbinate (concha)
E. Area of vein of Galen
F. Marrow within clivus of occipital bone

Figure 5–6
MRI sagittal section of the head, highlighting both musculoskeletal and neurovascular unit structures.

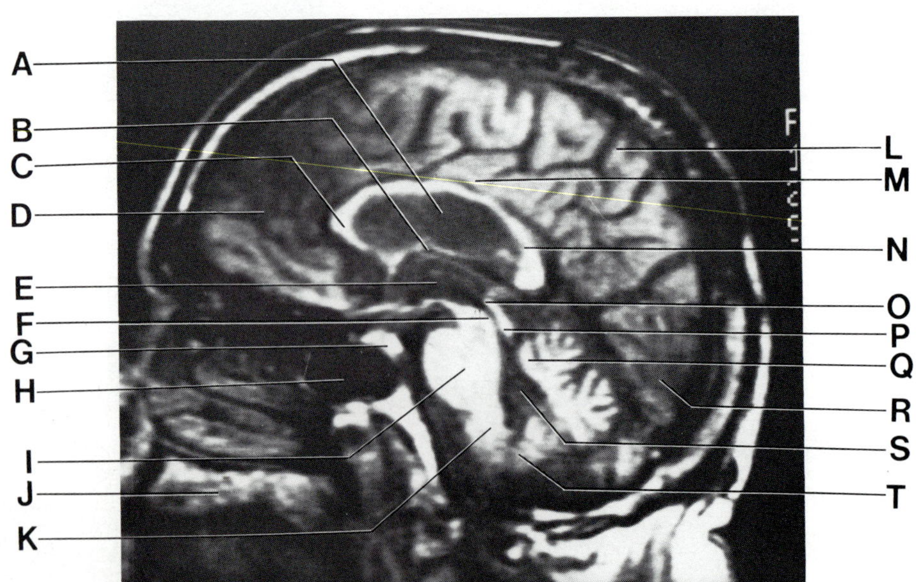

A. Lateral ventricle
B. Fornix
C. Anterior corpus callosum
D. Frontal lobe
E. Third ventricle
F. Aqueduct of Sylvius
G. Pituitary gland
H. Sphenoid sinus (a musculoskeletal structure)
I. Pons
J. Soft palate
K. Medulla oblongata
L. Parietal lobe
M. Cingulate gyrus
N. Splenium of corpus callosum
O. Superior colliculus of midbrain
P. Inferior colliculus of midbrain
Q. Central lobule of cerebellum
R. Occipital lobe
S. Fourth ventricle
T. Cerebellar tonsil

Figure 5–7
MRI sagittal section of the head, highlighting visceral unit structures.

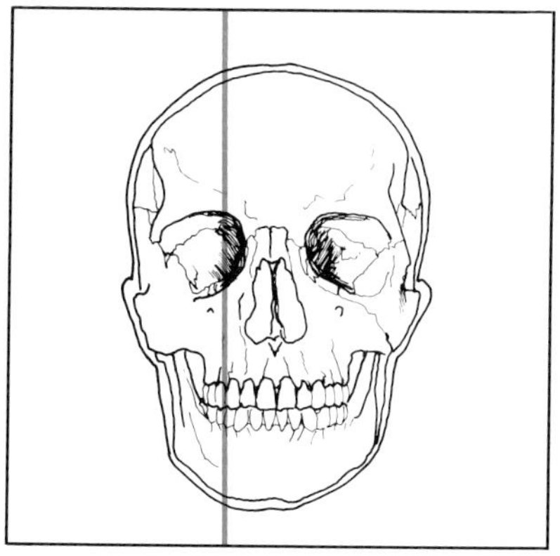

Figure 5–8 is an MRI sagittal section of the head. The musculoskeletal and neurovascular unit structures are emphasized. In Figure 5–9, the same type of picture as Figure 5–8 highlights the visceral and enclosing unit structures.

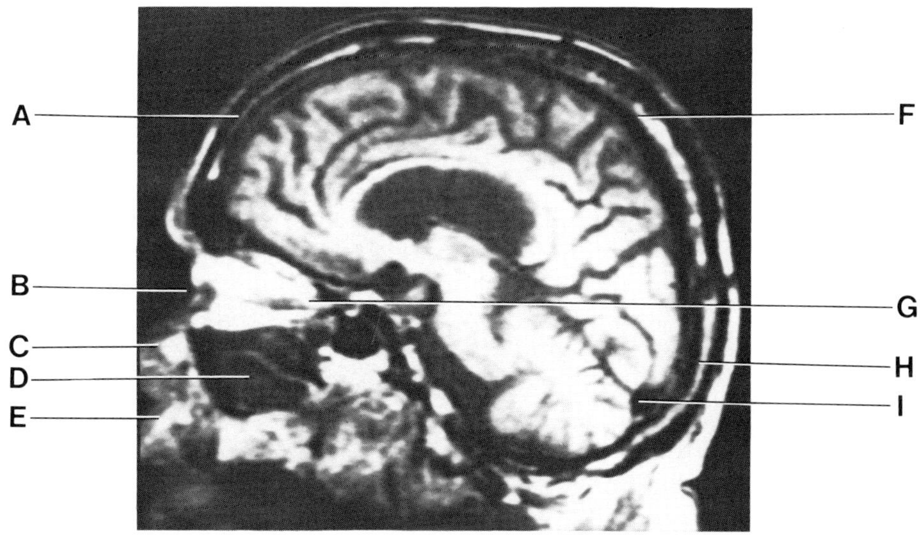

A. Frontal bone
B. Orbit
C. Nasal bone
D. Maxillary sinus
E. Maxilla
F. Parietal bone
G. Optic nerve
H. Occipital bone
I. Transverse sinus

Figure 5–8
MRI sagittal section of the head, highlighting both musculoskeletal and neurovascular unit structures.

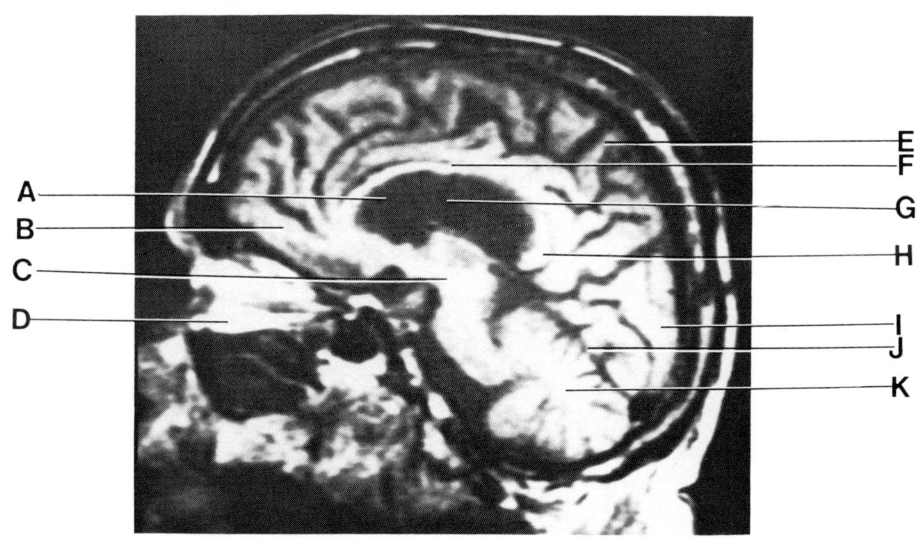

A. Anterior horn of lateral ventricle
B. Frontal lobe
C. Thalamus
D. Orbital fat
E. Parietal lobe
F. Corpus callosum
G. Lateral ventricle
H. Splenium of corpus callosum
I. Occipital lobe
J. Area of tentorium cerebelli
K. Cerebellum

Figure 5–9
MRI sagittal section of the head, highlighting both visceral and enclosing unit structures.

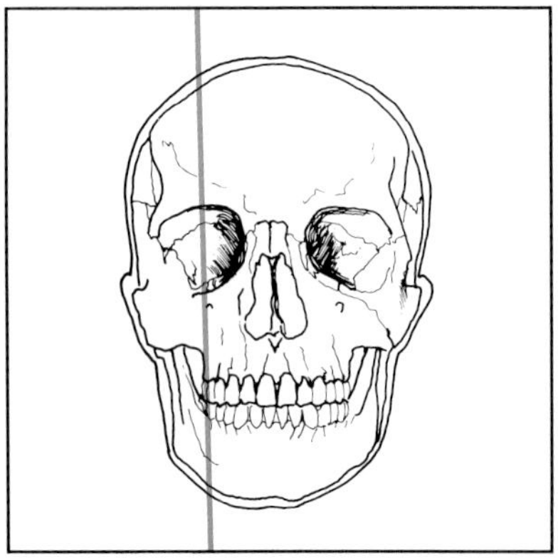

Figures 5–10 and 5–11 are identical in that both are MRI sagittal sections from the same area of the head. Figure 5–10 emphasizes by labels the musculoskeletal and neurovascular unit structures, whereas Figure 5–11 highlights by labels the visceral and enclosing unit structures.

A. Superior rectus muscle of eye
B. Optic nerve
C. Inferior rectus muscle of eye
D. Lateral pterygoid muscle
E. Maxillary sinus
F. Area of medial pterygoid muscle
G. Area of choroid plexus of third ventricle
H. Area of internal carotid artery (petrous segment)
I. Area of transverse sinus
J. Area of cervical portion of internal carotid artery

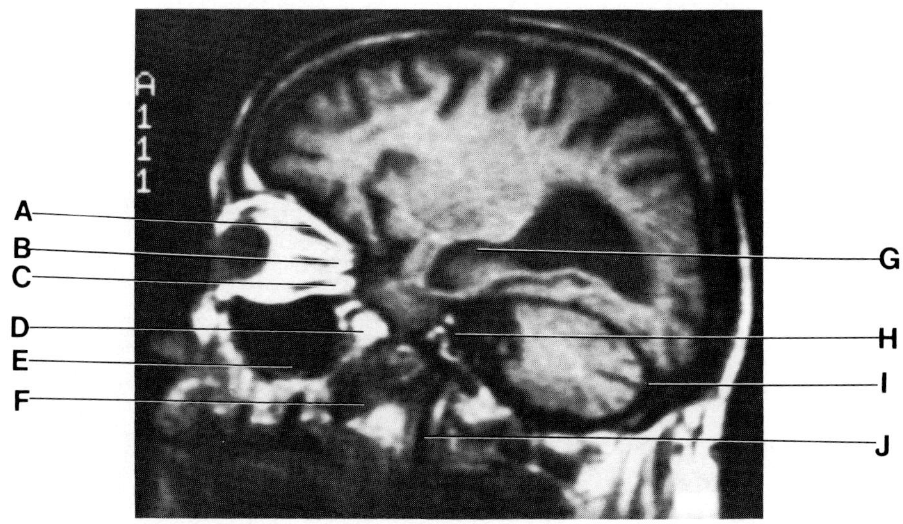

Figure 5–10
MRI sagittal section of the head, highlighting both musculoskeletal and neurovascular unit structures.

A. Frontal lobe
B. Amygdala
C. Orbital globe
D. Orbital fat
E. Temporal horn of lateral ventricle
F. Area of scalp
G. Parietal lobe
H. Pulvinar of thalamus
I. Occipital lobe
J. Occipital horn of lateral ventricle
K. Parahippocampal gyrus
L. Area of tentorium cerebelli
M. Cerebellum
N. Hippocampus

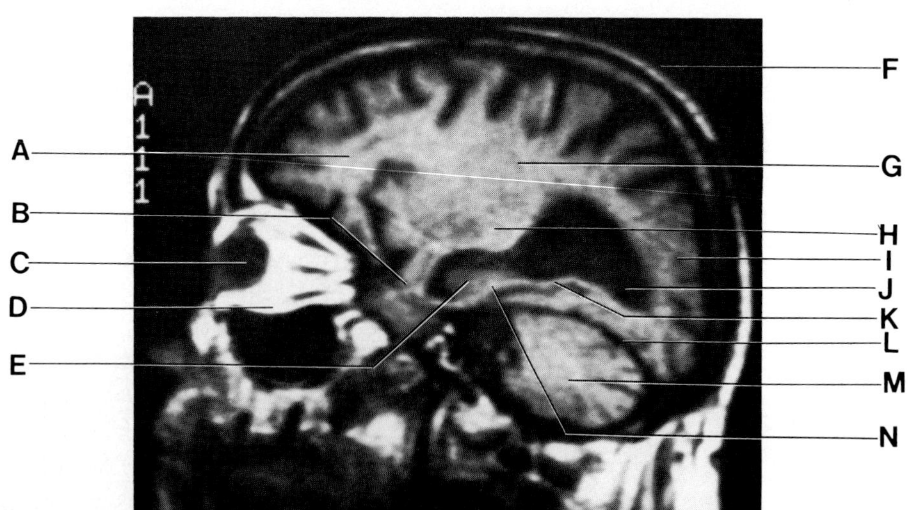

Figure 5–11
MRI sagittal section of the head, highlighting both visceral and enclosing unit structures.

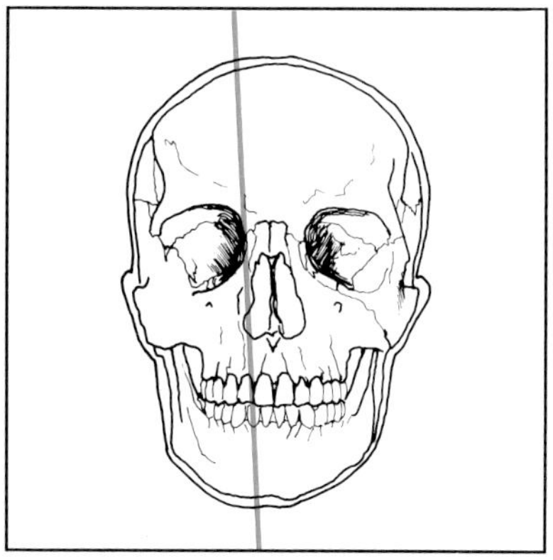

Compare Figures 5–12 and 5–13. Both are identical pictures of sagittal sections from the same area of the head. Figure 5–12 emphasizes musculoskeletal and neurovascular unit structures, and Figure 5–13 highlights visceral and enclosing unit structures.

A. Area of middle cerebral artery within sylvian fissure
B. Superior rectus muscle
C. Optic nerve
D. Maxilla
E. Choroid plexus
F. Area of transverse sinus

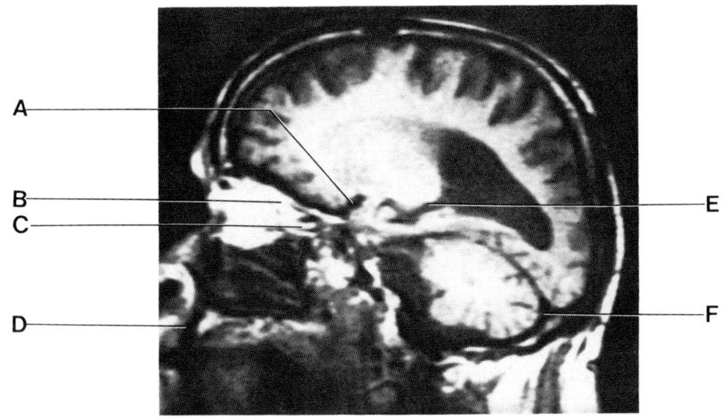

Figure 5–12
MRI sagittal section of the head, highlighting both musculoskeletal and neurovascular unit structures.

A. Frontal lobe
B. Pulvinar of thalamus
C. Hippocampus
D. Orbital fat
E. Parietal lobe
F. Lateral ventricle
G. Choroidal fissure
H. Hippocampal gyrus
I. Occipital horn (posterior horn) of lateral ventricle
J. Area of tentorium cerebelli
K. Occipital lobe
L. Cerebellum

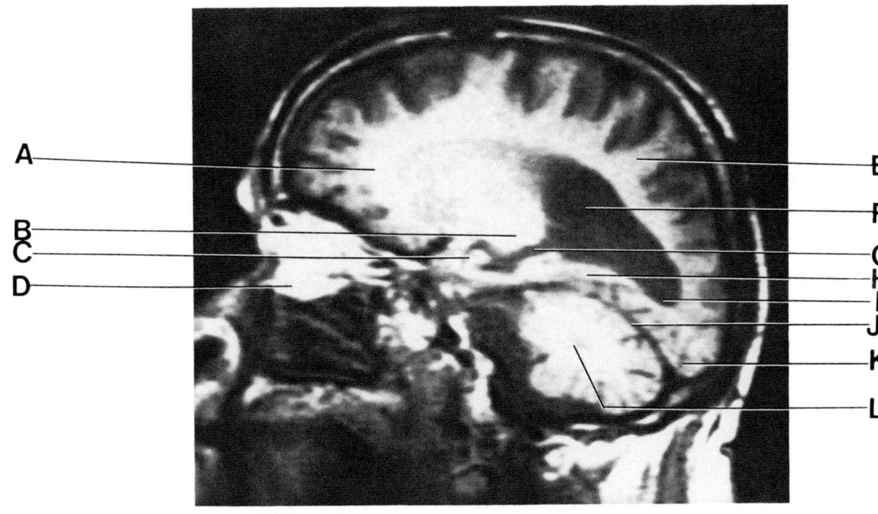

Figure 5–13
MRI sagittal section of the head, highlighting both visceral and enclosing unit structures.

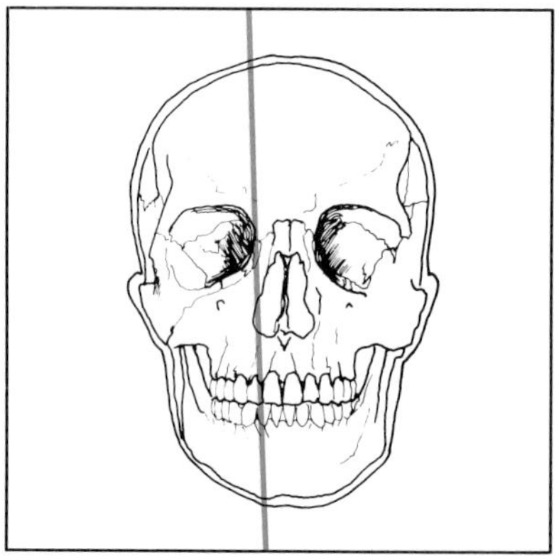

Figure 5–14 is an ultrasound sagittal view of fetal brain, which shows parts of the lateral ventricle. The lateral and third ventricles, as well as other brain structures, are depicted in Figure 5–15. This ultrasound picture of the fetal brain is in the sagittal plane. Figure 5–16 is another ultrasound picture that shows a sagittal view of the fetal brain. Normal brain structures are seen, as well as a cerebrovascular hemorrhage inferior to the central part (body) of the lateral ventricle.

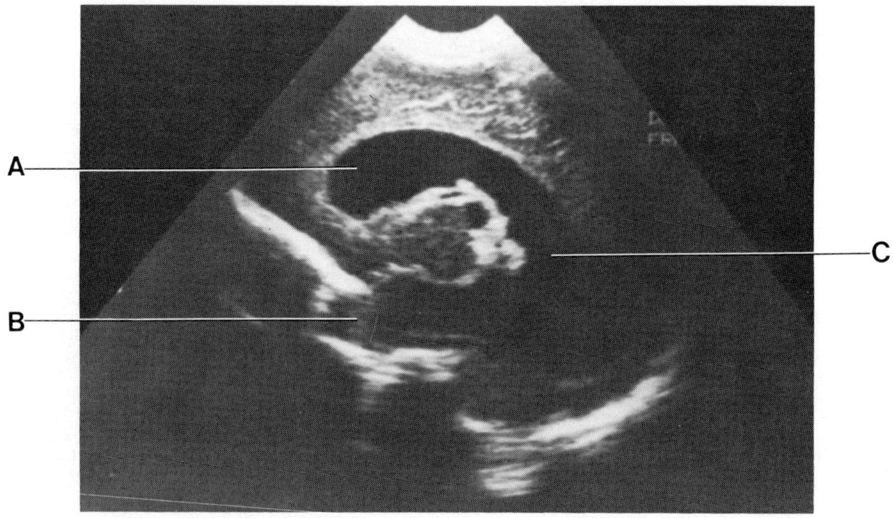

A. Anterior horn of lateral ventricle
B. Posterior horn of lateral ventricle
C. Temporal (inferior) horn of lateral ventricle

Figure 5–14
Ultrasound sagittal view of fetal brain, highlighting parts of the lateral ventricle.

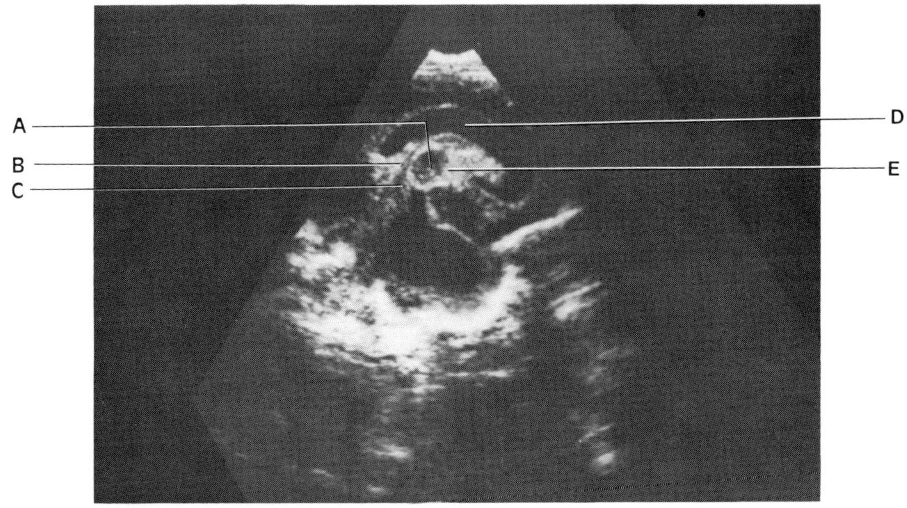

A. Third ventricle
B. Caudate nucleus
C. Blood
D. Lateral ventricle
E. Choroid plexus of lateral ventricle

Figure 5-15
Ultrasound sagittal view of fetal brain, highlighting lateral and third ventricles.

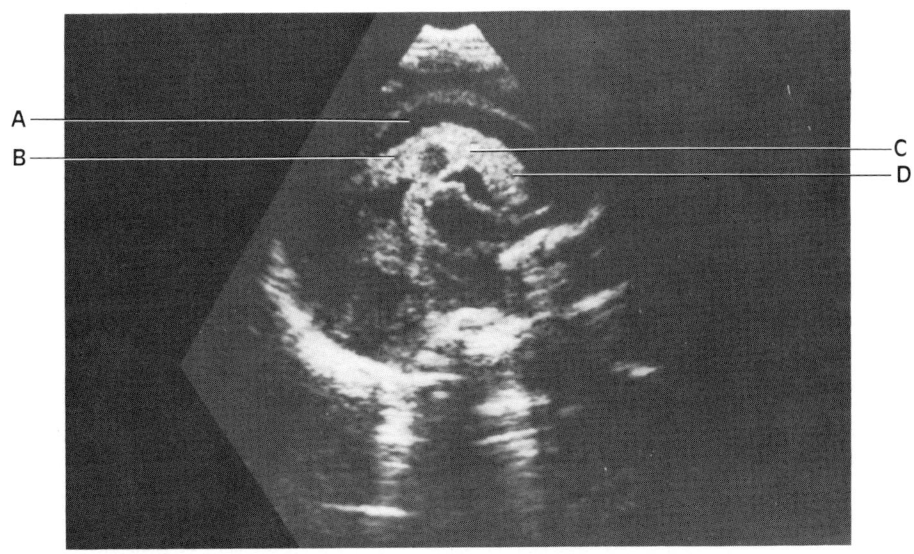

A. Anterior horn of lateral ventricle
B. Head of caudate nucleus
C. Cerebrovascular hemorrhage seen inferior to central part (body) of lateral ventricle
D. Choroid plexus

Figure 5-16
Ultrasound sagittal view of fetal brain, highlighting the lateral ventricle and the head of the caudate nucleus.

Appendix to Part II: Clinical Correlations for the Head Region

CASE ONE (Figs. AII–1 and AII–2)

History. The patient was examined in follow-up for a right frontal lobe lesion.

Findings. Compared with the previous computed tomography (CT) survey, the magnetic resonance image (MRI)* reveals a very small, 3 to 4 mm lesion in the cortex of the right frontal lobe anteriorly, with surrounding edema. However, the degree of edema has significantly decreased since the previous survey, and the mass effect on the right frontal horn is no longer identified. There is no evidence of a midline shift or extra-axial fluid collections. Mild asymmetry of the size of the lateral ventricles is considered a normal variation.

The middle cerebral artery shows somewhat ectatic vessels, and a small aneurysm in this area cannot be excluded by this survey. However, a previously suspected vascular malformation in the left basal ganglia region could not be identified by MRI.

Impression. Slight decrease in the size of the cortical lesion in the right frontal lobe anteriorly, with central low-signal intensity consistent with necrosis. Decrease in surrounding edema since the previous CT survey. No evidence of vascular malformations identified in the region of the basal ganglia, as described. No new lesions are noted.

*Magnetic resonance images in this Appendix are only samples of the hard copy.

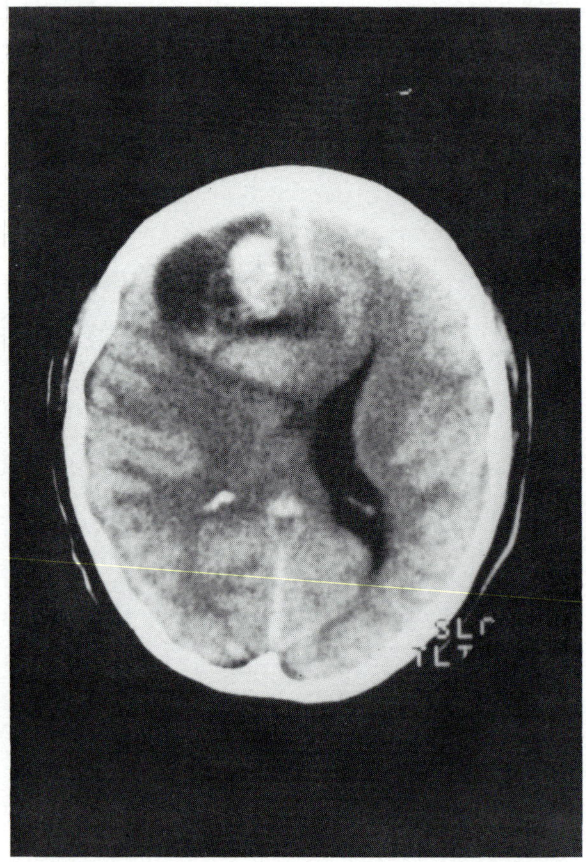

Figure AII–1.
A CT transverse image of the head, highlighting the primary tumor

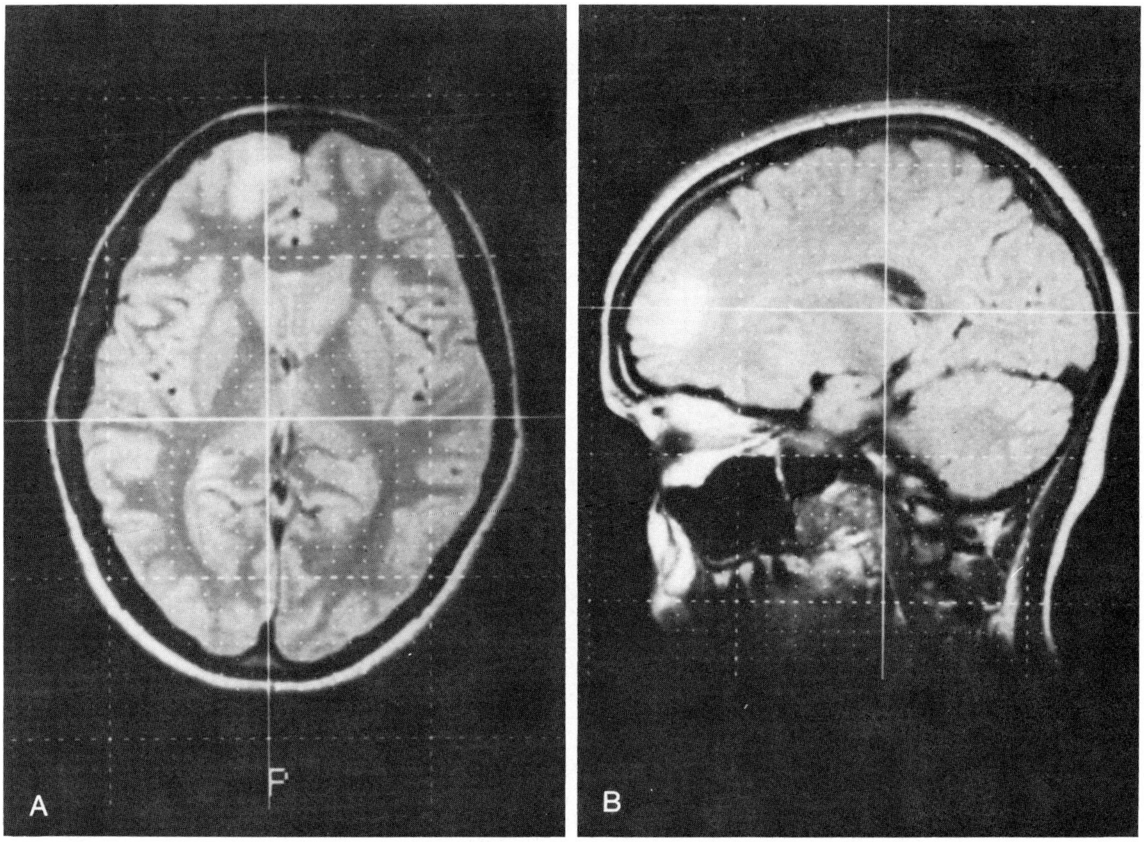

Figure AII–2.
Magnetic resonance images, highlighting a lesion in the right frontal lobe. *A*, Transverse image; *B*, sagittal image.

CASE TWO (Fig. AII–3)

History. The patient was seen in follow-up.

Findings. An extra-axial lesion, measuring 3 × 4 × 4 cm, was identified in the anterior left temporal fossa area. This has signal intensity in all T1, protein density, and T2-weighted images similar to that of cerebrospinal fluid (CSF). For this reason, it is felt to represent an arachnoid cyst. Comparison with previous MRI study is strongly recommended to see whether there is any change in the size of the lesion.

A minimal, bilateral maxillary and ethmoid sinusitis was noted incidentally.

The ventricular size is normal, with no midline shift. No large aneurysms or vascular malformation is noted. The pituitary gland is within normal limits.

There is mild thickening of the eighth cranial nerve bilaterally. This has a slightly higher signal intensity compared with CSF in the proton density images. For this reason, this area needs further evaluation with the previous study or with gadolinium-enhanced images.

Impression. A 3 × 4 × 4 cm extra-axial lesion anterior to the left temporal lobe is most likely an arachnoid cyst, as described. Previous MRI imaging study is not available for comparison. There is mild, nonspecific thickening of the eighth cranial nerve bilaterally. Comparison with prior MRI study or gadolinium-enhanced imaging of the eighth cranial nerve is suggested for further evaluation. Mild bilateral maxillary sinusitis.

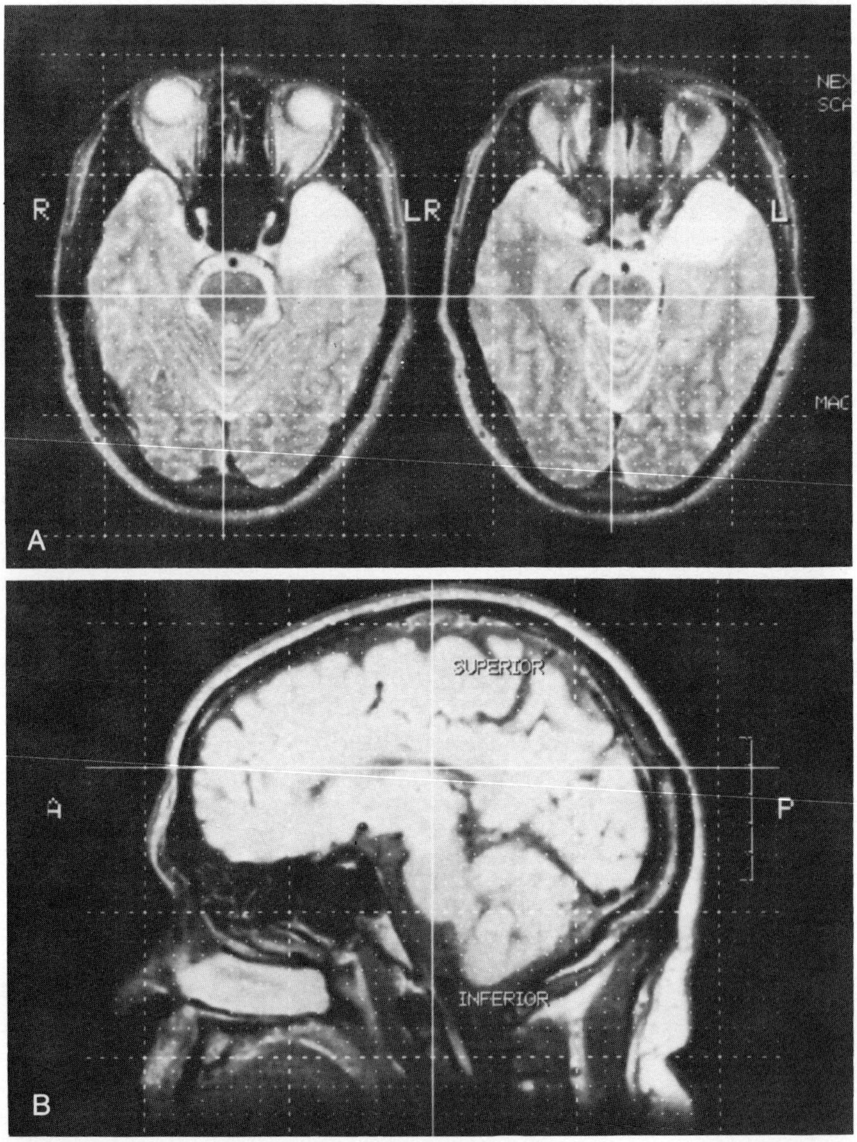

Figure AII–3.
Magnetic resonance images, highlighting an arachnoid cyst and mild maxillary sinusitis. *A,* Transverse image; *B,* sagittal image.

CASE THREE (Fig. AII–4)

History. The patient reported sinusitis and facial abscess.

Findings. A large mass was identified in the right maxillary sinus, extending laterally into the right infratemporal fossa. Soft tissue thickening was also identified anterior to the right maxillary sinus, with partial destruction of the anterior wall of the right maxillary sinus. The mass extends into the inferior aspect of the right orbit and into the sphenoid sinus posteriorly. There is definite extension into bilateral ethmoid sinuses and also into the superior aspect of the left maxillary sinus. No definite intracranial extension is identified. This mass involves the retropharyngeal space. A tiny pocket of air is identified within the right maxillary sinus. However, foci of high signal intensity are identified within this mass and could represent hemorrhage. The mass still could represent an abscessed neoplastic process, and this should be strongly considered. Further correlation with a CT scan is recommended.

Impression. Inflammation or neoplastic mass from the right maxillary sinus, extending across the nasopharynx to the left maxillary sinus. Involvement of the bilateral ethmoidal sinuses, sphenoid sinus, and right orbit. Tiny pocket of air in the right maxillary sinus. Edema or soft tissue mass extending into the soft tissues lateral to the right maxillary sinus.

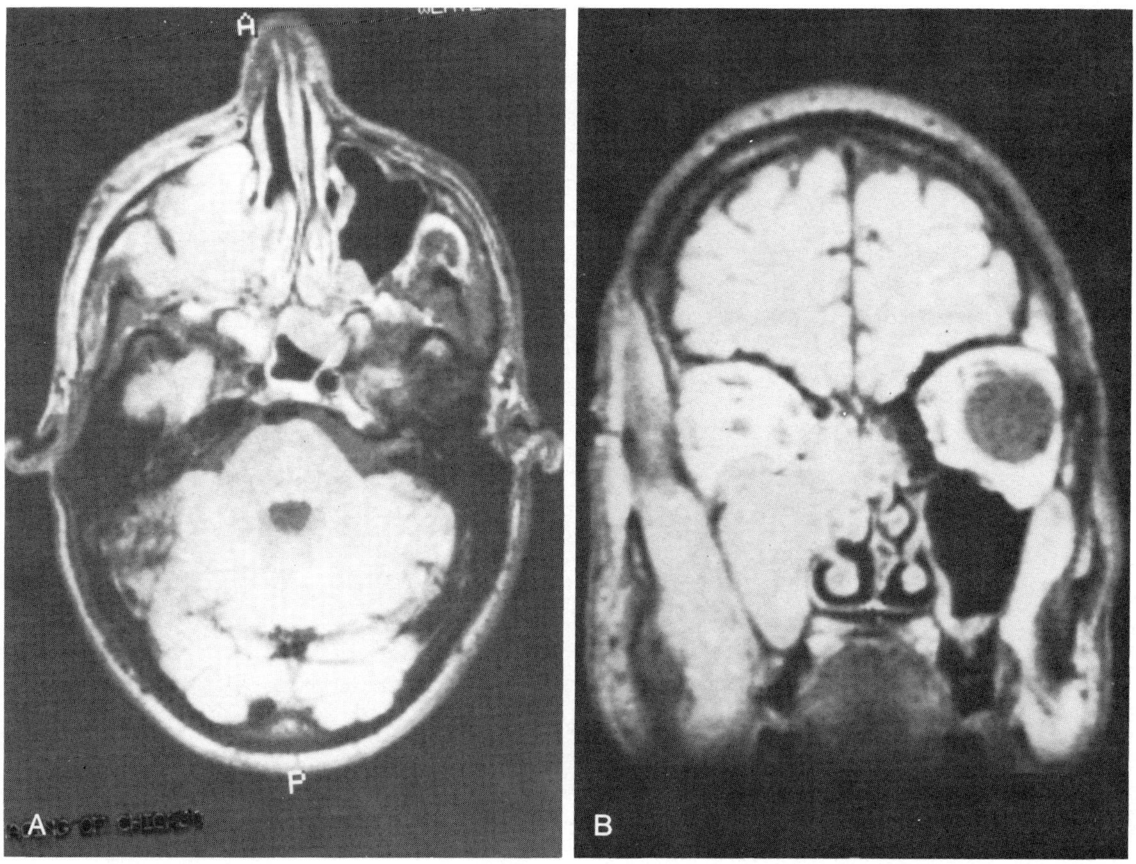

Figure AII–4.
Magnetic resonance images of the head, highlighting a pocket of air in the right maxillary sinus. *A,* Transverse (axial) image; *B,* coronal (frontal) image.

CASE FOUR (Fig. AII–5)

History. The patient has had left-sided headaches for the past 3 to 4 months. Acromegaly is noted.

Findings. A soft tissue mass is seen arising from the pituitary gland to the left of the midline, with extension into the sphenoid sinus and the left cavernous sinus. As this mass has a signal intensity very similar to that of the pituitary gland, the exact measurement of the mass is not possible. The approximate measurement is 2 cm in the vertical dimension. The cavernous carotid is displaced slightly laterally on the left side. However, the mass does not seem to encircle the carotid.

The pituitary stalk deviates slightly to the right of the midline. No definite evidence of hemorrhage is identified within this mass. No evidence of optic nerve compression is noted.

The axial images reveal mucosal thickening in the left maxillary sinus. The eighth cranial nerves are within normal limits. Minimal fluid is identified in the left mastoid air cells, consistent with a mild degree of mastoiditis, either acute or chronic. Mucosal thickening is also identified in the left ethmoidal air cells.

Impression. A 2-cm mass in the left side of the pituitary gland, with extension into the left cavernous sinus and left side of the sphenoid sinus and deviation to the right of the left internal carotid laterally and the pituitary stalk. No focal hemorrhages. The findings are consistent with left maxillary and ethmoid sinusitis. Questionable fluid was identified in the left mastoid air cells.

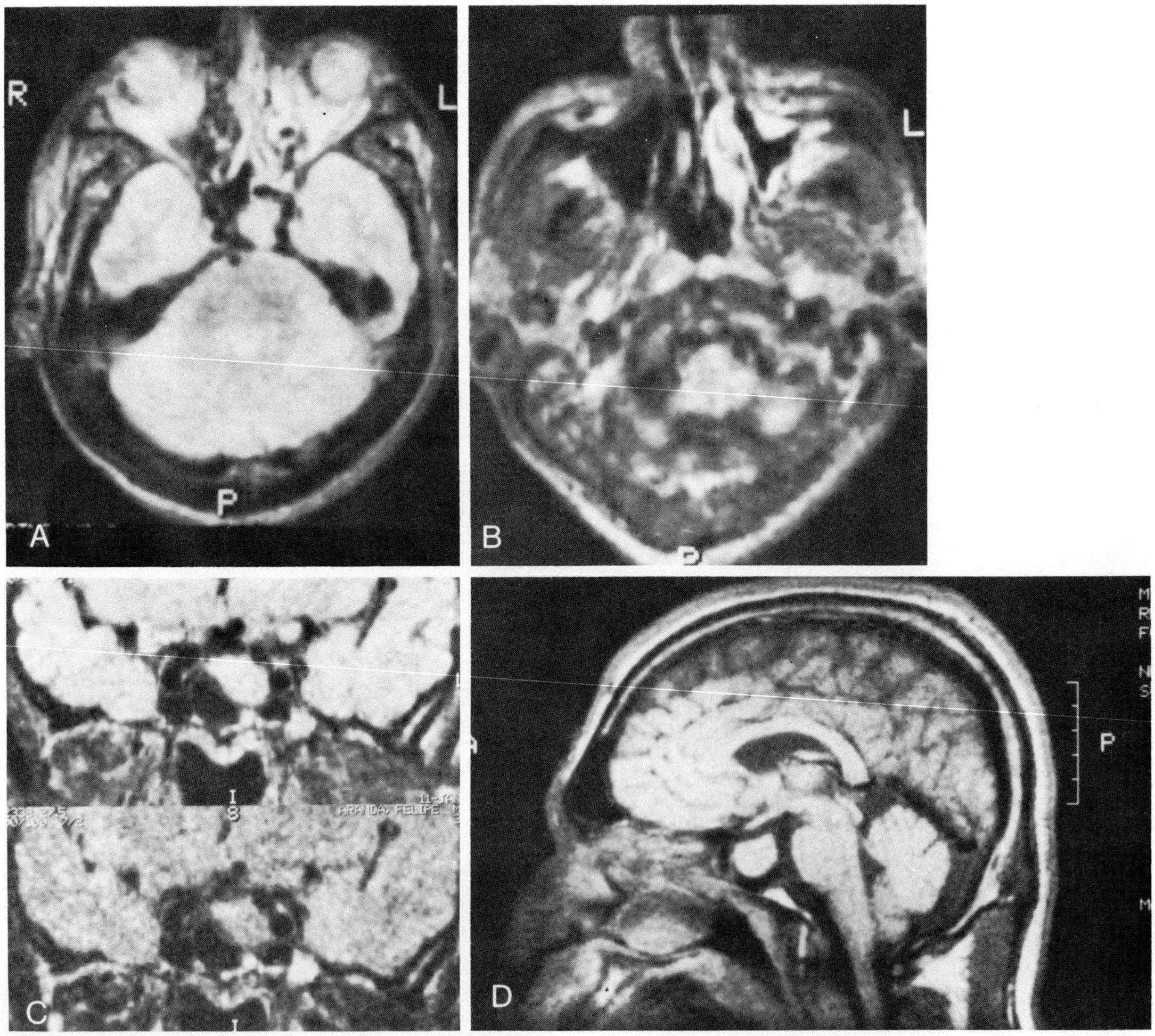

Figure AII–5.
Magnetic resonance images, highlighting left maxillary and ethmoid sinusitis and a mass in the left side of the pituitary gland. *A* and *B*, transverse images; *C*, coronal image; *D*, sagittal image.

PART III

THE NECK

INTRODUCTION

The seven cervical vertebrae of the neck represent the skeletal element. Typically, a cervical vertebra has a body, two pedicles, two transverse processes with a transverse foramen in each of the transverse processes, two laminae, one spinous process, two superior articular surfaces, and two inferior articular surfaces. Each transverse process has both anterior and posterior tubercles. As in other vertebrae, a spinal foramen transmits the cervical portion of the spinal cord. The three atypical cervical vertebrae include the atlas, the axis, and the seventh cervical vertebra.

The skeletal muscles of the neck serve in many capacities; some elevate the hyoid bone (suprahyoid muscles), some depress the hyoid bone and thyroid cartilage (infrahyoid muscles). Other muscles acting on the neck bring about abduction, rotation, flexion (sternocleidomastoid), and extension (trapezius, splenius capitis, semispinalis capitis, and a few other muscles). For further detail, consult the muscle tables in this part.

The vasculature of the neck consists of four main vessels for afferent cranial blood flow (the right and left carotid arteries, the right and left vertebral arteries). The efferent cranial blood flow that passes through the neck includes the right and left internal jugular veins and the right and left vertebral veins. More superficially, the external jugular veins are seen adjacent to the outer surface of the sternocleidomastoid muscles.

The nerves that traverse the neck are of two major types: the cervical plexus and the vagus nerves. The phrenic nerves originate from cervical nerves three, four, and five and are part of the cervical plexus.

The viscera of the neck include the trachea, esophagus, thyroid and parathyroid glands, laryngopharynx, and larynx.

The connective tissue sheaths are well defined into superficial and deep fasciae. Other than the superficial fascia, there are five layers of deep fascia within the neck. (1) Investing fascia is the most superficial of the deep fascia and splits to envelop the sternocleidomastoid and trapezius muscles. (2) The muscular layer encloses the infrahyoid muscles. (3) The carotid sheath surrounds the major neck neurovascular unit (internal jugular veins, common carotid arteries, and vagus nerves). (4) Prevertebral fascia surrounds the prevertebral muscles (longus colli and longus capitis). (5) Visceral (pretracheal) fascia surrounds the visceral unit.

Part III of *Modern Sectional Anatomy* includes Chapters 6, 7, and 8. Chapter 6 presents the transverse section of the neck in four anatomic categories. Chapter 7 discusses and presents pictures of neck coronal sections. Chapter 8 describes sagittal planes of the neck. Chapters 6 and 8 include some clinical images of magnetic resonance imaging (MRI).

Part III also presents two pathologic clinical images, which include radiologic reports. One of these images shows a mass in the right parotid gland and the other shows a partially extruded C3–C4 disk.

Table III-1 Suprahyoid and Extrinsic Muscles of Tongue

Muscle	Origin	Insertion	Innervation	Vasculature	Action
Digastric	Posterior belly: notch of mastoid process (temporal bone); anterior belly: mandible (digastric fossa)	Hyoid bone	Posterior belly: facial nerve branches: anterior belly: inferior alveoli (mylohyoid branch)	Posterior belly: posterior auricular (muscular branches), occipital artery (muscular branches)	Elevates hyoid bone, lowers jaw
Stylohyoid	Styloid process of temporal bone near its proximal end	Body of hyoid bone	Stylohyoid (branch of posterior trunk of facial)	Facial and muscular branches of occipital	Raises hyoid bone and base of tongue
Mylohyoid	Mylohyoid line of mandible	Median raphe (extends from chin to hyoid bone)	Inferior alveolar (mylohyoid branch) of trigeminal nerve 3	Mylohyoid branch; sublingual branch; submental branch of inferior alveolar, lingual, and facial, respectively	Raises hyoid bone and base of tongue, elevates floor of mouth
Geniohyoid	Mental spine of mandible	Front part of hyoid bone	Cranial nerve 1, a nerve incorporated in the hypoglossal nerve	Lingual (sublingual branch)	Raises hyoid bone and tongue
Hyoglossal	Hyoid bone	Back half of lateral parts of tongue	Hypoglossal nerve	Lingual (sublingual branch); facial (submental branch)	Depresses and retracts tongue

Table III-2 Infrahyoid Muscles

Muscle	Origin	Insertion	Innervation	Vasculature	Action
Omohyoid	Inferior belly: superior border of scapular notch and suprascapular ligament. The inferior belly passes posterior to sternocleido-mastoid and ends in central tendon. Superior belly: extends superiorly from central tendon	Inferior edge of body of hyoid bone	Ansa cervicalis, derived from cervical ventral (anterior) primary rami 1–3	Lingual (suprahyoid branch) and superior thyroid (sternocleido-mastoid branch)	Depresses hyoid bone
Sternohyoid	Posterior lateral edge of manubrium sterni, back of sternoclavicular joint, and clavicle (medial end)	Inferior edge of body of hyoid bone (medial section)	Ansa cervicalis, derived from cervical ventral (anterior) primary rami 1–3	Branches of sternocleido-mastoid and hyoid arteries of superior thyroid, also lingual	Depresses hyoid bone and larynx
Thyrohyoid	Anterior lamina of thyroid cartilage	Greater horn of hyoid bone and inferior border of hyoid body	Anterior branch of cervical nerve 1	Infrahyoid branch of superior thyroid artery	Raises and changes form of larynx
Sternothyroid	Back of manubrium sterni	Anterior thyroid cartilage	Ansa cervicalis, derived from cervical ventral (anterior) primary rami 1–3	Cricothyroid branch of superior artery	Depresses thyroid cartilage

Table III-3 Suboccipital Muscles

Muscle	Origin	Insertion	Innervation	Vasculature	Action
Straight (rectus) posterior major muscle of head (capitis)	Spinous process of axis	Occipital bone (side portion of inferior nuchal line)	First cervical nerve (posterior primary ramus)	Right and left vertebral arteries (muscular branches) and occipital artery (descending branch)	Head extension, abduction, and rotation
Straight (rectus) posterior minor muscle of head (capitis)	Posterior part of atlas (tubercle)	Back part of cranial bone (occipital bone) inferior to the inferior nuchal line	First cervical (posterior primary ramus)	Right and left vertebral arteries (muscular branches) and occipital artery (descending branch)	Head extension, abduction, and rotation
Oblique inferior muscle of head	Spinous process of axis	Atlas (transverse process)	Cervical nerves 1 and 2 (posterior primary rami)	Right and left vertebral arteries (muscular branches) and occipital artery (descending branch)	Rotation of head and atlas
Oblique superior muscle of head	Atlas (transverse process)	Superior to the inferior nuchal line and inferior to the superior nuchal line	First cervical (posterior primary ramus)	Right and left vertebral arteries (muscular branches) and occipital artery (descending branch)	Head rotation and extension

Table III-4 Prevertebral Muscles

Muscle	Origin	Insertion	Innervation	Vasculature	Action
Long muscle of neck (longus colli)	Anterior tubercles of transverse processes of cervical vertebrae 3–5, bodies of thoracic vertebrae 1–3, and bodies of the 3 upper thoracic and 3 lower cervical vertebrae	Anterior tubercles of transverse processes of cervical vertebrae 5 and 6, bodies of cervical vertebrae 2–4, and anterior tubercles of atlas	Cervical nerves 2–8 (anterior primary rami)	Ascending pharyngeal (prevertebral branches), ascending cervical (muscular branches), right and left vertebral arteries (muscular branches)	Head and neck flexion; synergist for neck rotation and abduction of neck (acting singly)
Long muscle of head (longus capitis)	Transverse processes (anterior tubercles) of cervical vertebrae 3–6	Lower part of basilar surface of occipital bone	Cervical nerves 1–4 (anterior primary rami)	Inferior thyroid (ascending cervical branch), pharyngeal (prevertebral branches), right and left arteries (muscular branches)	Head and neck flexion; synergist for neck rotation

Table III-5 Muscles Moving the Neck

Muscle	Origin	Insertion	Innervation	Vasculature	Action
Trapezius	Medial part of superior nuchal line and external occipital protuberance; spinous processes of 7th cervical vertebra and all thoracic vertebrae	Clavicle, acromion process, and base of spine of scapula	Spinal accessory and posterior primary rami of cervical nerves 3 and 4	Transverse cervical (descending scapula)	Pulls scapula toward the median, pulls acromion posteriorly, slants chin, and produces rotation of scapula
Sternocleidomastoid	Sternum and clavicle	Mastoid process and outer part of superior nuchal line of occipital bone	Spinal accessory nerve	A branch of superior thyroid (sternocleidooid), a branch of suprascapular artery (occipital muscular branch), and a branch of posterior auricular (occipital branch)	Flexes vertebral column, rotates head
Splenius	Spinous processes of thoracic vertebrae 3–6	Side portion of upper nuchal line and mastoid process of temporal bone	Posterior primary rami (lateral branches) of middle and inferior cervical nerves	Occipital artery (muscles and descending branches) and transverse cervical (superficial branch)	Neck extension, abduction of head and neck, slight rotation of head
Semispinal	Transverse processes of superior thoracic vertebrae	Surface between upper and lower nuchal lines near median of occipital bone	Cervical nerves 1–6 (posterior primary ramus)	Posterior intercostal (muscular branches), occipital artery (descending branch), and costocervical (deep cervical branch)	Head extension and some abduction
Longissimus muscle of neck (longissimus cervicis)	Thoracic vertebrae 4 and 5 (transverse processes)	Cervical vertebrae 2–6 (transverse processes)	Inferior cervical and upper thoracic nerves (posterior primary ramus)	Intercostal and lumbar arteries (posterior primary rami), occipital artery (muscular branches), and costocervical trunk (deep cervical branch)	Neck extension and abduction of head
Longissimus muscle of head (longissimus capitis)	Thoracic vertebrae 4 and 5 (transverse processes)	Temporal bone (mastoid process)	Cervical spinal nerves (posterior primary rami)	Occipital artery (muscular branches) and costocervical trunk (deep cervical branch)	Head extension and abduction
Diocostal	Angles of ribs 1–6	Transverse processes of cervical vertebrae 4–6	Posterior branches of cervical nerves	Intercostal (posterior primary rami)	Head extension, neck abduction, and some rotation of neck

Table III-6 Muscles of Respiration Primarily Located in Neck

Muscle	Origin	Insertion	Innervation	Vasculature	Action
Scalenus anterior	Cervical vertebrae 3–6 (anterior tubercles of transverse processes)	Upper part of first rib (scalene tubercle)	Cervical nerves 5–8 (anterior branches)	Inferior thyroid (ascending cervical branch)	Raises first rib
Scalenus medius	Cervical vertebrae 2–7 (posterior tubercles of transverse processes)	Superior surface of first rib lateral to tubercle of rib and medial to groove of subclavian artery	Cervical nerves 3 and 4 (anterior primary rami and lateral muscular branches)	Ascending cervical	Neck rotation, abduction, and flexion; rib elevation
Scalenus posterior	Cervical vertebrae 4–6 (posterior tubercles of transverse processes)	Lateral part of second rib posterior to attachment of serratus anterior	Cervical nerves 3 and 4 (both anterior and lateral muscular branches)	Ascending cervical and transverse cervical	Cervical vertebrae rotation, abduction of neck, second rib elevation, head and neck flexion

CHAPTER 6

Transverse (Axial) Cadaver Sections and Images of the Neck

There are five modules in this chapter. Each module represents a different level of the neck. The first, second, third, fourth, and fifth modules present studies of sections at the level of cervical vertebrae 2, 3, 4, 5, and 6, respectively. Module 2 includes sheet plastinated human sections. These plastinated specimens are 2.5 to 4 mm thick and can be projected on a screen by an overhead projector. Module 5 has well-illustrated transverse sections of the neck at the level of cervical vertebra 6. The illustrations at this level show the typical structures of the neck and their relationship in a transverse plane.

Several MRI transverse clinical slices of the neck are presented that show many of the normal structures in exquisite detail.

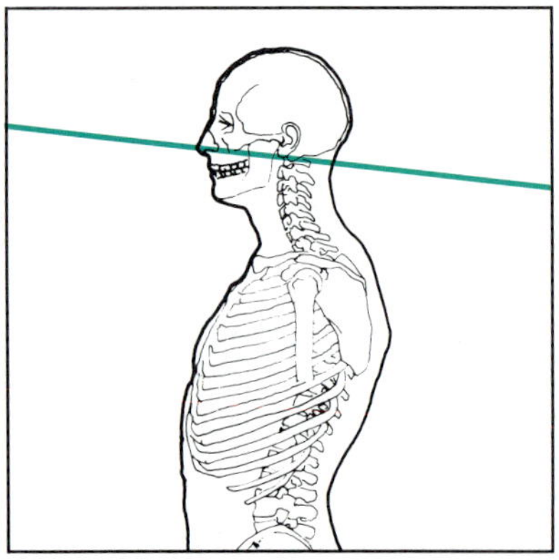

MODULE 1

This is a transverse (axial) section of the upper neck and the lower face at cervical vertebra 2, representing the images at all cervical vertebrae.

MUSCULOSKELETAL UNIT (Fig. 6–1)

I. Skeletal Portion

Ramus of Mandible (*M*). A bilateral, oval, posterior end of the mandible slice as seen in this neck section. This part of the mandible is anterior to the sternocleidomastoid muscle and the retromandibular vein.

Mandible (*L*). This bilateral section of the mandible appears oblong. The anterior part is thicker than the posterior part immediately medial to the masseter muscle and lateral to the transverse muscle of the tongue.

Transverse Foramen (*I*). This bilateral hole appears rounded in this neck section and is located within the transverse process of the cervical vertebra.

Nasal aperture (*J*). An opening seen in the median of this section.

Transverse muscle of tongue (*K*). Muscle fibers seen perpendicular to the median.

Oropharynx (*N*). Opening seen in the median and posterior to the transverse fibers on the tongue.

Body of the Cervical Vertebra (*P*). A thick aggregate of bone tissue anterior to the spinal foramen (vertebral cavity).

Transverse Process of the Cervical Vertebra (*O*). A lateral projection of the cervical vertebra. The most lateral projection on each bilateral prominence shows the anterior and posterior tubercles.

Spinal Foramen of the Cervical Vertebra (*Q*). A ring-shaped hole in the median of the cervical vertebra. Note the spinal cord within this foramen.

Lamina of the Cervical Vertebra. A diagonally placed, bilateral bone section that extends from the transverse process to the spinous process. The posterior end of this feature is more medial than the anterior end.

Spinous Process of the Cervical Vertebra. A posterior, bifid, median projection immediately medial to the multifidus muscle and posterior to the spinal cord.

Odontoid Process (Dens). A rather oval, oblong bone section located on the anterior part of the body of the axis.

Septal Nasal Cartilage. A plate of cartilaginous tissue between the right and left nasal apertures.

II. Muscular Portion

Masseter Muscle (*C*). A bulky section of bilateral muscle lateral to the mandible and anterior to the retromandibular vein.

Sternocleidomastoid Muscle (*H*). A bilateral, moderately large, oval muscle aggregate immediately lateral to the internal jugular vein and posterior to the retromandibular vein.

Levator Scapulae Muscle (*S*). A bilateral, elongated (anterior-to-posterior) muscle aggregate posterior and lateral to the cervical vertebra. The anterior end of this muscle section is more lateral than the posterior end and is medial to the sternocleidomastoid muscle.

Longissimus Capitis Muscle (*R*). A bilateral, elongated (anterior-to-posterior) muscle aggregate medial to the levator scapulae and lateral to the cervical vertebra. The anterior part of this muscle is in a more lateral position than its posterior part.

Longus Capitis Muscle (*G*). This is seen here bilaterally as a small slice of muscle anterior to the lateral part of the body of cervical vertebra 2. This muscle is immediately medial to the internal carotid artery.

Medial Pterygoid Muscle (*D*). A small, bilateral muscle slice seen medial to the mandible at this level. This muscle is located medial to the lateral pterygoid muscle and anterior to the levator muscle of the velum palatini.

Lateral Pterygoid Muscle (*E*). This bilateral muscle section is seen immediately medial to the mandible and lateral to the medial pterygoid muscle. This muscle is located anterior to the levator muscle of the velum veli palatini.

Alar Part of the Nasalis Muscle (*A*). This muscle mass is seen on the alar part of the nose.

Origin. Edge of the maxillary nasal notch and lesser alar cartilage.

Insertion. Skin near the lateral edge of the nose.

Levator Muscle of the Velum Palatini (Levator Palati Muscle) (*F*). A rounded, bilateral muscle section anterior and medial to the sternocleidomastoid muscle.

Facial Muscle (*B*). A superficial, sickle-shaped, embedded slice of bilateral muscle in the superficial fascia on the lateral anterior face.

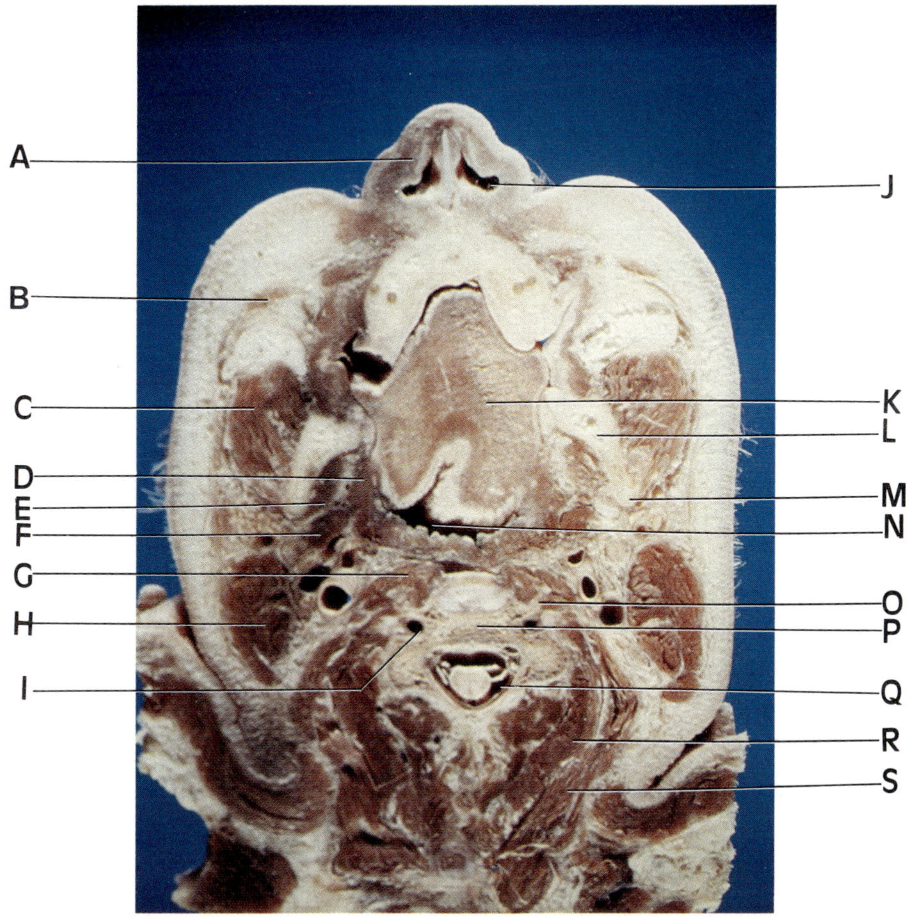

A. Alar part of nasalis muscle
B. Facial muscle
C. Masseter muscle
D. Medial pterygoid muscle
E. Lateral pterygoid muscle
F. Levator muscle of the velum palatini (levator palati muscle)
G. Longus capitis muscle
H. Sternocleidomastoid muscle
I. Transverse foramen
J. Nasal aperture
K. Transverse muscle of tongue
L. Mandible
M. Ramus of mandible
N. Oropharynx
O. Transverse process of cervical vertebra
P. Body of cervical vertebra
Q. Spinal foramen of cervical vertebra
R. Longissimus capitis muscle
S. Levator scapulae muscle

Figure 6-1
Transverse (axial) section of the upper neck and the lower face at cervical vertebra 2, highlighting musculoskeletal unit structures.

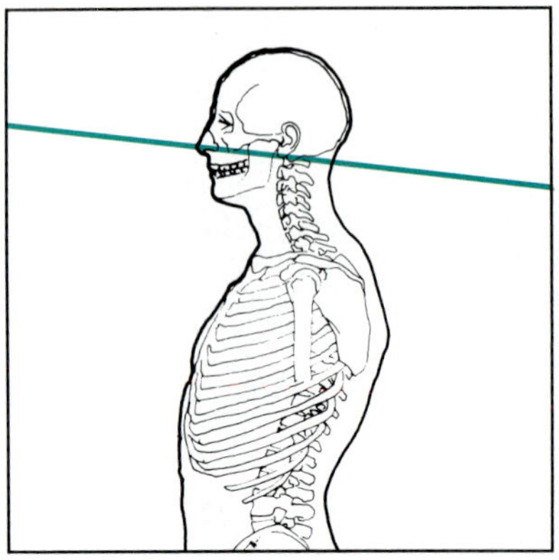

NEUROVASCULAR UNIT (Fig. 6–2)

I. Arteries

Internal Carotid Artery (*F*). A rounded, bilateral vascular structure anterior and medial to the internal jugular vein and posterior to the external carotid artery. It is formed from the bifurcation of the common carotid artery. The bilateral internal carotid artery terminates in the circle of Willis. Each artery is anterior and medial to its corresponding internal jugular vein.

Vertebral Artery (*B, H*). A bilateral, rounded structure within the transverse foramen of the cervical vertebra. This bilateral artery arises from its respective subclavian artery. Each ascends the neck through the transverse foramens of cervical vertebra 6 through cervical vertebra 1. Each vertebral artery enters the skull through the foramen magnum and terminates in a vessel they both form, the basilar artery.

External Carotid Artery (*E*). This is bilateral and is seen here as a rounded, hollow structure immediately anterior and slightly medial to the internal carotid artery.

Origin. This artery is formed by the bifurcation of its corresponding common carotid artery. Although each gives rise to eight branches, the vessel is the superficial temporal artery.

Distribution. This artery disperses blood to eight branches; from inferior to superior, they are the superior thyroid, ascending pharyngeal, lingual, facial, occipital, posterior auricular, maxillary, and superior temporal arteries.

Course. This artery starts adjacent to the superior edge of the thyroid cartilage, which corresponds to the disk between the third and fourth cervical vertebrae. It ascends superiorly, is slightly curved, and passes posterior to the neck of the mandible, one half the distance between the mastoid process and the angle of the jaw.

II. Veins

Internal Jugular Vein (*G*). A bilateral, oval, hollow structure posterior and lateral to its corresponding internal carotid artery. It is medial to the sternocleidomastoid muscle.

Retromandibular (Posterior Facial) Vein (*A*). A rounded, hollow, bilateral structure immediately anterior to the sternocleidomastoid muscle and posterior to the slice of the mandible seen in this section.

Origin. It arises from the union of the superficial temporal and maxillary veins. The bilateral retromandibular veins descend from their union through the substance of the parotid gland. At the inferior part of the gland, each vein joins the posterior auricular vein to form the external jugular vein.

Termination. External jugular vein.

Course. It runs downward within the parotid gland, superficial to the external carotid artery and medial to the facial nerve.

III. Nerves

Hypoglossal (12th Cranial) Nerve (*D*). A rounded, white section of bilateral nerve tissue anterior to the internal carotid artery. The hypoglossal nerve originates from the medulla oblongata of the brain and ends as it innervates all muscles of the tongue, including the hypoglossus and geniohyoid muscles.

Lingual Nerve (*C*). A rounded, white aggregate of bilateral nerve tissue anterior and medial to the hypoglossal nerve near the lateral surface of the medial (internal) pterygoid muscle.

Vagus Nerve. A rather small, bilateral, oval nerve section immediately medial to the internal jugular vein and posterior to the internal carotid artery.

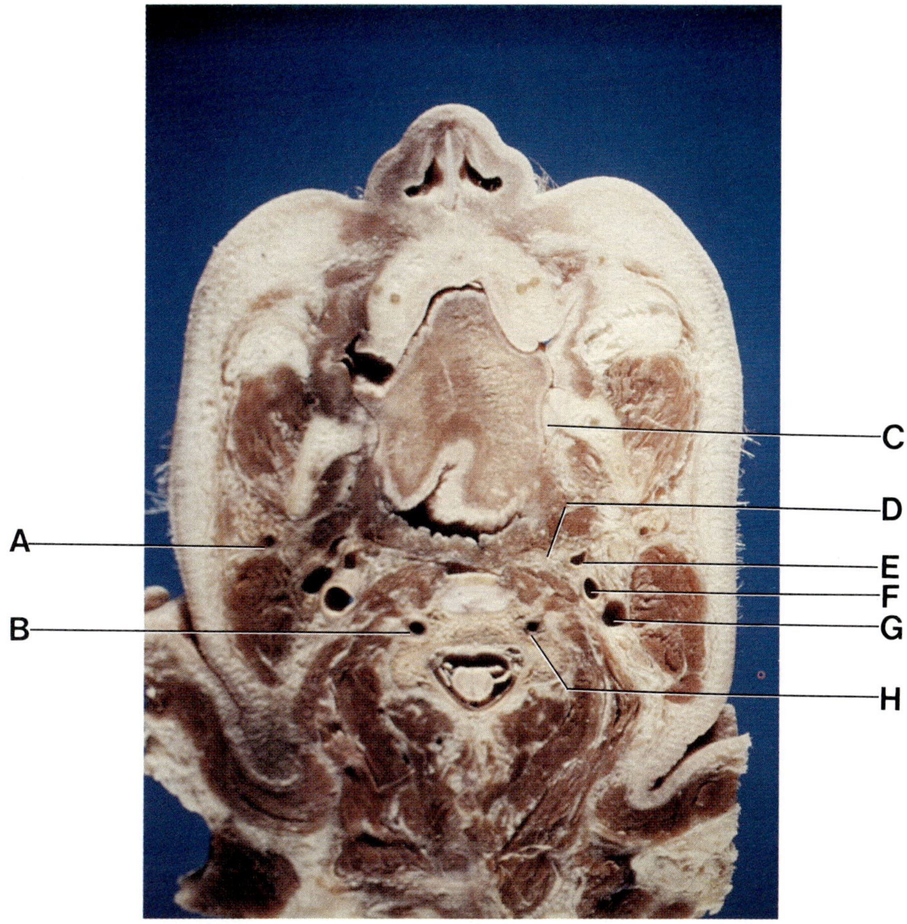

A. Retromandibular (posterior facial) vein
B. Right vertebral artery
C. Lingual nerve
D. Hypoglossal (12th cranial) nerve
E. External carotid artery
F. Internal carotid artery
G. Internal jugular vein
H. Left vertebral artery

Figure 6–2
Transverse section of the upper neck and the lower face at cervical vertebra 2, highlighting neurovascular unit structures.

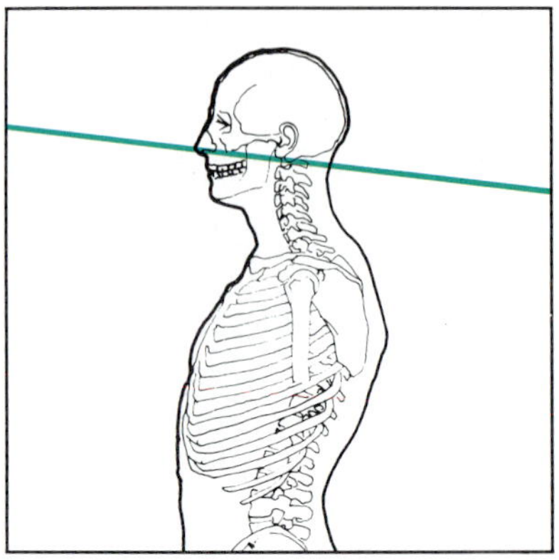

terminates at the level between lumbar vertebra 1 and lumbar vertebra 2 in the definitive condition.

VISCERAL UNIT (Fig. 6–3)

Septal Cartilage (*A*). A median strip of cartilaginous tissue. This structure runs anterior to posterior in this section.

Spinal Cord (*D*). A large cluster of nerve tissue within the spinal foramen of a cervical vertebra. This structure arises from the medulla oblongata and terminates at the level between lumbar vertebra 1 and lumbar vertebra 2 in the definitive condition.

Parotid Gland (*E*). A bilateral salivary gland, seen here as white glandular epithelial tissue within the subcutaneous layer (tissue), anterior and inferior to the ear.

Tongue (*B*). The tongue is seen here in the anterior median portion of this section. It lies within the mouth proper. The tongue is attached to the hyoid bone.

Oropharynx (*C*). A transversely placed slit posterior to the tongue and anterior to the retropharyngeal cleft.

Nasal Aperture. A bilateral opening within the nasal lumen. Septal cartilage is seen between the openings.

ENCLOSING UNIT (see Fig. 6–3)

Dura Mater (*F*). A thick sheet of tissue lining the inner wall of the spinal foramen.

Retropharyngeal Cleft. A rather small, transversely placed slit anterior to the body of the cervical vertebra and posterior to the oropharynx.

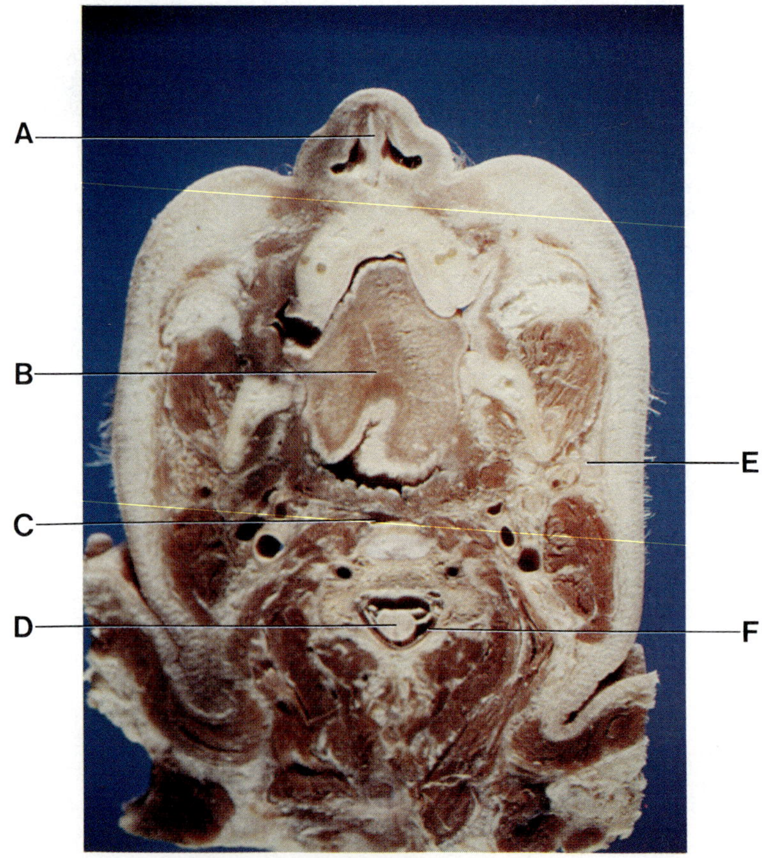

A. Septal cartilage
B. Tongue
C. Oropharynx
D. Spinal cord
E. Parotid gland
F. Dura mater

Figure 6–3
Transverse section of the upper neck and the lower face at cervical vertebra 2, highlighting both visceral and enclosing unit structures.

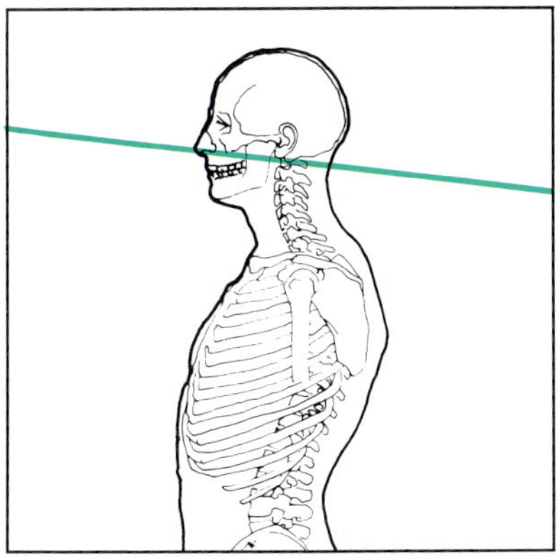

Figure 6–4 shows the transverse section of the upper neck (cervical vertebra 2). The enlargement of this transverse cervical slice provides a clear view of structures in musculoskeletal, neurovascular, visceral, and enclosing unit structures.

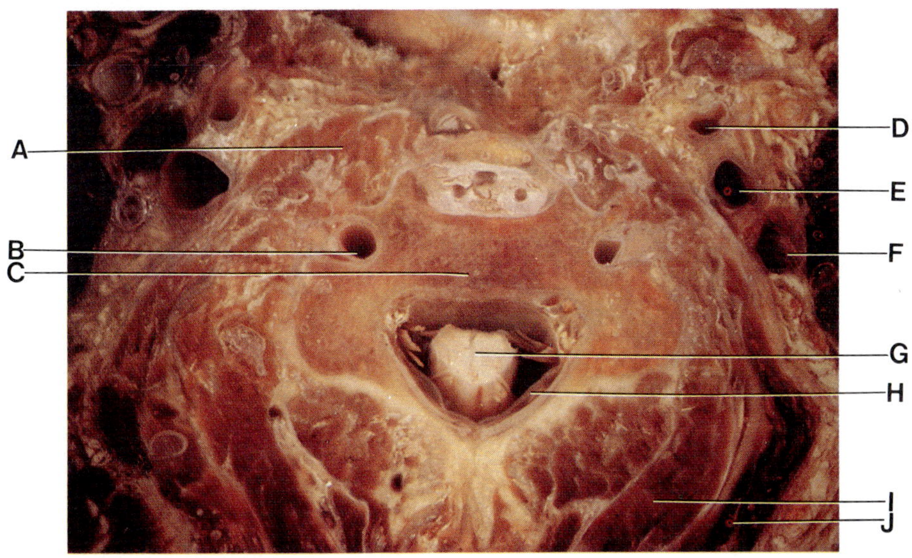

A. Longus capitis muscle
B. Vertebral artery
C. Body of cervical vertebra
D. External carotid artery
E. Internal carotid artery
F. Internal jugular vein
G. Spinal cord
H. Dura mater
I. Longissimus capitis muscle
J. Levator scapulae muscle

Figure 6–4
Enlargement of the transverse section of the upper neck at the level of cervical vertebra 2.

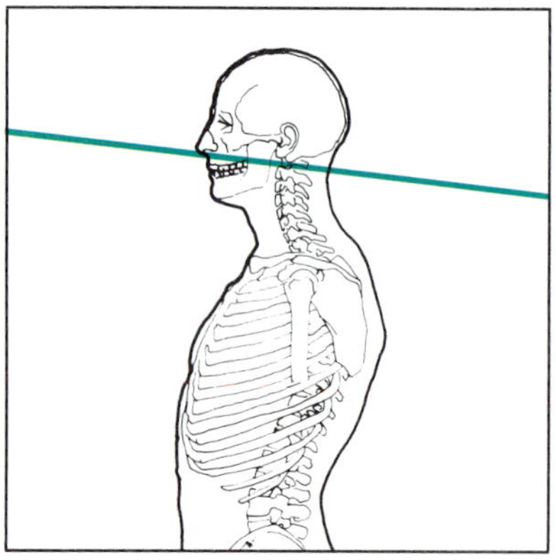

A transverse MRI image of the neck (Fig. 6–5) emphasizes structures of the musculoskeletal, neurovascular, visceral, and enclosing units. Twenty-one structures are clearly seen and labeled in this MRI picture. Another MRI image (Fig. 6–6) is a transverse slice of the neck. This section shows two structures (tongue and vagus nerve) that are not seen in Figure 6–5.

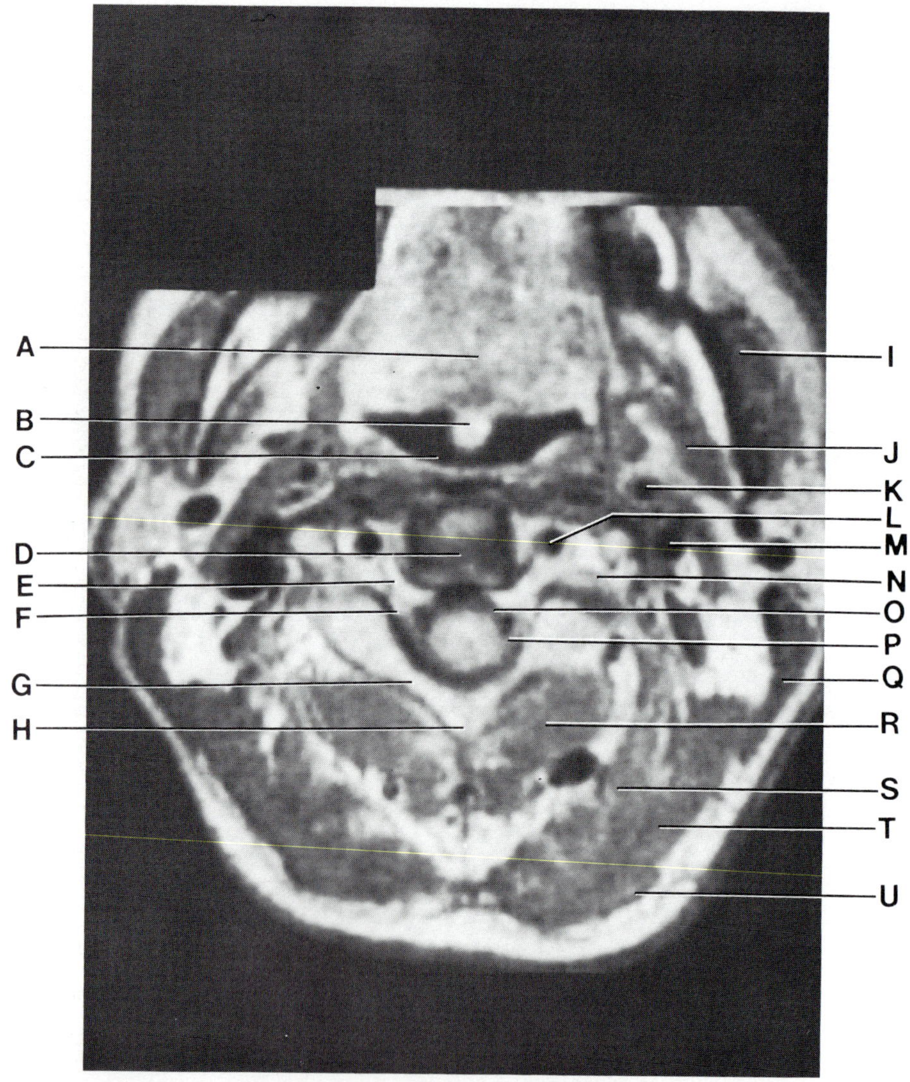

A. Soft palate
B. Uvula
C. Oropharynx
D. Body of cervical vertebra
E. Cervical spinal nerve
F. Spinal ganglion
G. Lamina of cervical vertebra
H. Spinous process of cervical vertebra
I. Masseter muscle
J. Medial pterygoid muscle
K. Carotid artery
L. Transverse foramen
M. Internal jugular vein
N. Transverse process of cervical vertebra
O. Anterior nerve root
P. Posterior nerve root
Q. Sternocleidomastoid muscle
R. Semispinalis cervicis muscle
S. Semispinalis capitis muscle
T. Splenius capitis muscle
U. Trapezius muscle

Figure 6–5
MRI transverse image of the neck, highlighting musculoskeletal, neurovascular, visceral, and enclosing unit structures.

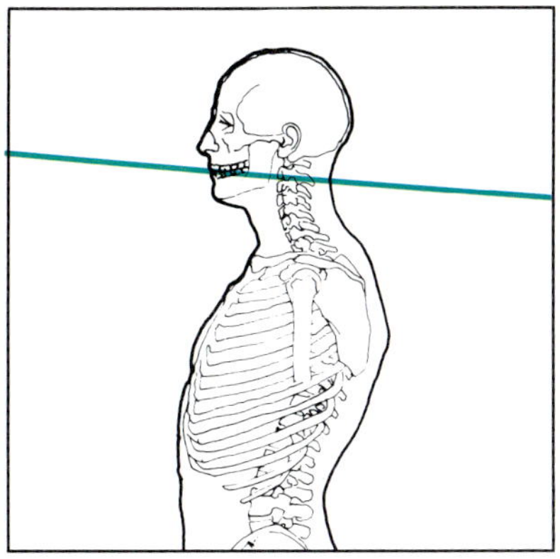

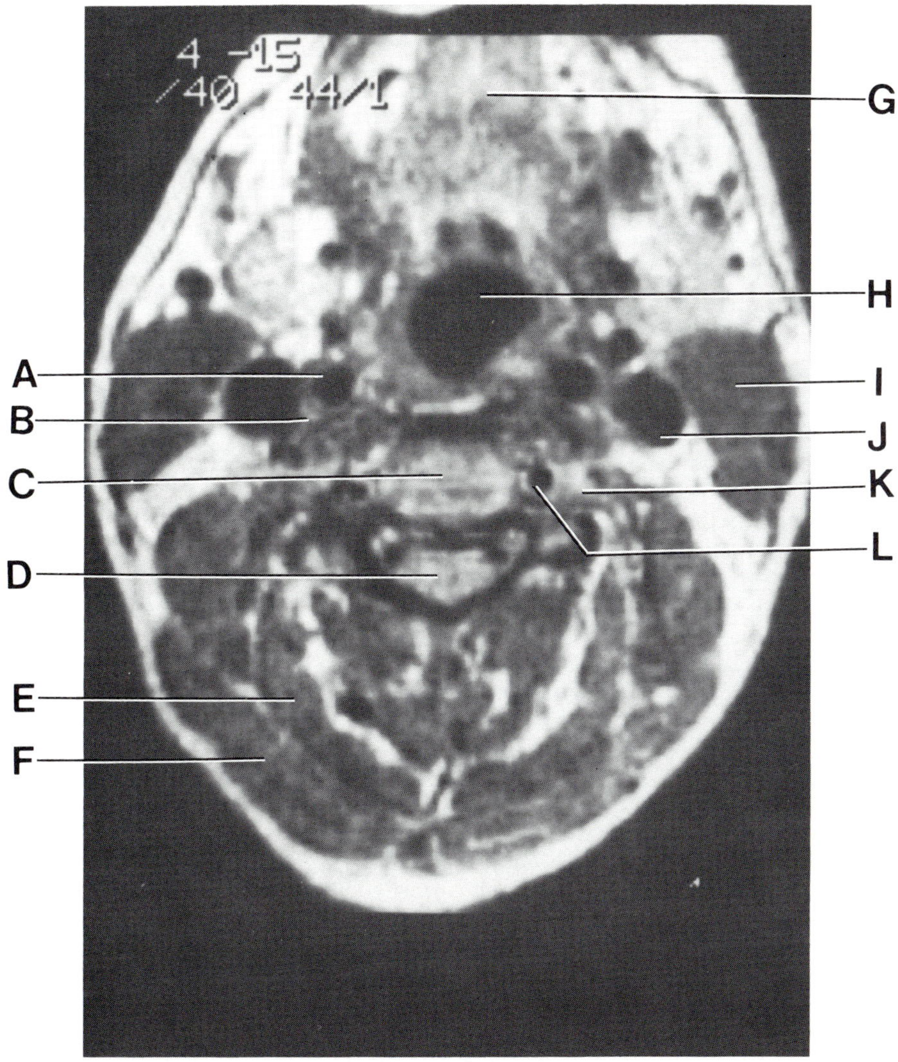

A. Common carotid artery
B. Vagus nerve
C. Body of cervical vertebra
D. Spinal cord
E. Semispinalis capitis muscle
F. Splenius capitis muscle
G. Tongue
H. Oropharynx
I. Sternocleidomastoid muscle
J. Internal jugular vein
K. Transverse process of cervical vertebra
L. Transverse foramen

Figure 6–6
MRI transverse image of the neck, highlighting musculoskeletal, neurovascular, visceral, and enclosing unit structures.

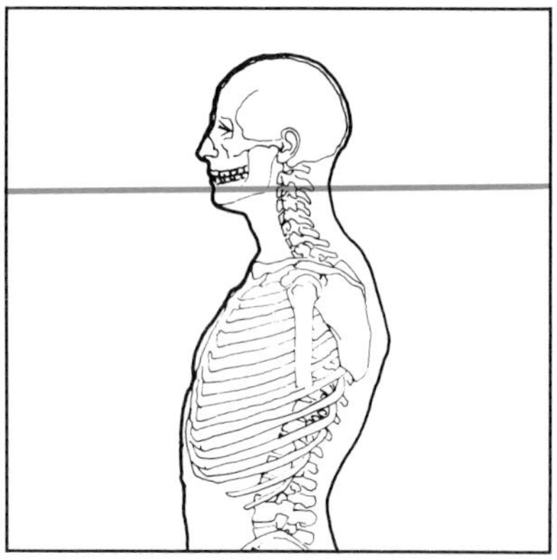

MODULE 2

This is a sheet plastinated transverse (axial) section of the neck and lower face at cervical vertebra 3, representing the images at all cervical vertebrae.

MUSCULOSKELETAL UNIT (Fig. 6–7)

I. Skeletal Portion

Mandible (*B*). A curved, rather thick bone section with tapering posterior ends. The concave surface is on its posterior surface. This feature is anterior to the longitudinal muscles of the tongue, as seen in this section, and posterior to the mentalis muscle.

Body of the Cervical Vertebra (*E*). A circular bone section anterior to the spinal foramen and posterior to the oropharynx. This bone feature is the most anterior part of this vertebra.

Transverse Process of the Cervical Vertebra (*I*). Bilateral lateral projections that extend from the sides of the vertebral body anterior to the lamina. The front lateral projection is rather sharp and immediately anterior to the vertebral artery. The posterior lateral projection is blunt and oval and posterior to the vertebral artery.

Lamina of the Cervical Vertebra (*J*). An oblique, bilateral, rather thick bone slice immediately posterior to the transverse process. The posterior end of each slice is more medial (joins at the median) than the anterior end.

Spinous Process of the Cervical Vertebra (*L*). A median bone slice immediately posterior to the spinal foramen and the laminae. This structure is immediately medial to the multifidus muscle.

Pedicle of the Cervical Vertebra (*H*). A bilateral, flat, rather thick bone feature immediately medial to the transverse foramen and posterior to the body of the vertebra.

Transverse Foramen (*G*). A bilateral hole that passes between the anterior and posterior parts of the transverse process.

II. Muscular Portion

Mentalis Muscle (*A*). A thin, curved strip of muscle tissue anterior to the mandible. The concave surface is adjacent to the mandible.

Platysma Muscle (*M*). A thin strip of curved, cutaneous muscle that extends from the sternocleidomastoid muscle on the right to the sternocleidomastoid muscle on the left. This muscle is in a semicircular arrangement and passes anterior to the mandible.

Anterior Belly of the Digastric Muscle (*C*). A rather thick elongated (anterior-to-posterior), bilateral muscle slice that extends from the medial pterygoid muscle to the mandible.

Mylohyoid Muscle (*D*). A rather small, bilateral, elongated muscle section immediately lateral to the intrinsic muscle of the tongue and immediately medial to the anterior belly of the digastric muscle.

Medial Pterygoid Muscle (*N*). A bilateral muscle section posterior to the anterior belly of the digastric muscle, oblong in shape and medial to the masseter muscle.

Masseter Muscle (*O*). An oval, bilateral muscle slice adjacent to the medial pterygoid muscle on its medial side. This muscle is anterior to the sternocleidomastoid muscle and medial to the posterior part of the platysma muscle.

Sternocleidomastoid Muscle (*Q*). An oblong, bilateral muscle section immediately lateral to the internal carotid artery and internal jugular vein. This muscle is posterior to the masseter, and its posterior end is more lateral than its anterior end.

Longus Colli Muscle (*P*). A small, bilateral strip of muscle tissue anterior to the lateral part of the body of the vertebra and medial to the longus capitus muscle.

Longus Capitis Muscle (*F*). A strip of bilateral, oblong (medial-to-lateral) muscle tissue, anterior to the transverse process of the cervical vertebra and lateral to the longus colli muscle.

Levator Scapulae Muscle (*R*). A bilateral muscle section, showing three bulges that make this muscle rather elongated in this section. The anterior bulge is adjacent to the transverse process in a more medial position than the posterior bulge. The posterior bulge is immediately lateral to the anterior part of the splenius capitis muscle.

Multifidus Muscle (*S*). A rather oblong, bilateral muscle slice adjacent to the lamina of the cervical vertebra and medial to the semispinalis cervicis muscle.

Semispinalis Cervicis Muscle (*T*). A rather elongated, oblique, bilateral muscle slice lateral to the multifidus muscle and posterior to the transverse process. The posterior end of this muscle is more medial than the anterior part.

Semispinalis Capitis Muscle (*U*). An oblong, curved, bilateral muscle section with tapering ends. The anterior end is adjacent to the transverse process,

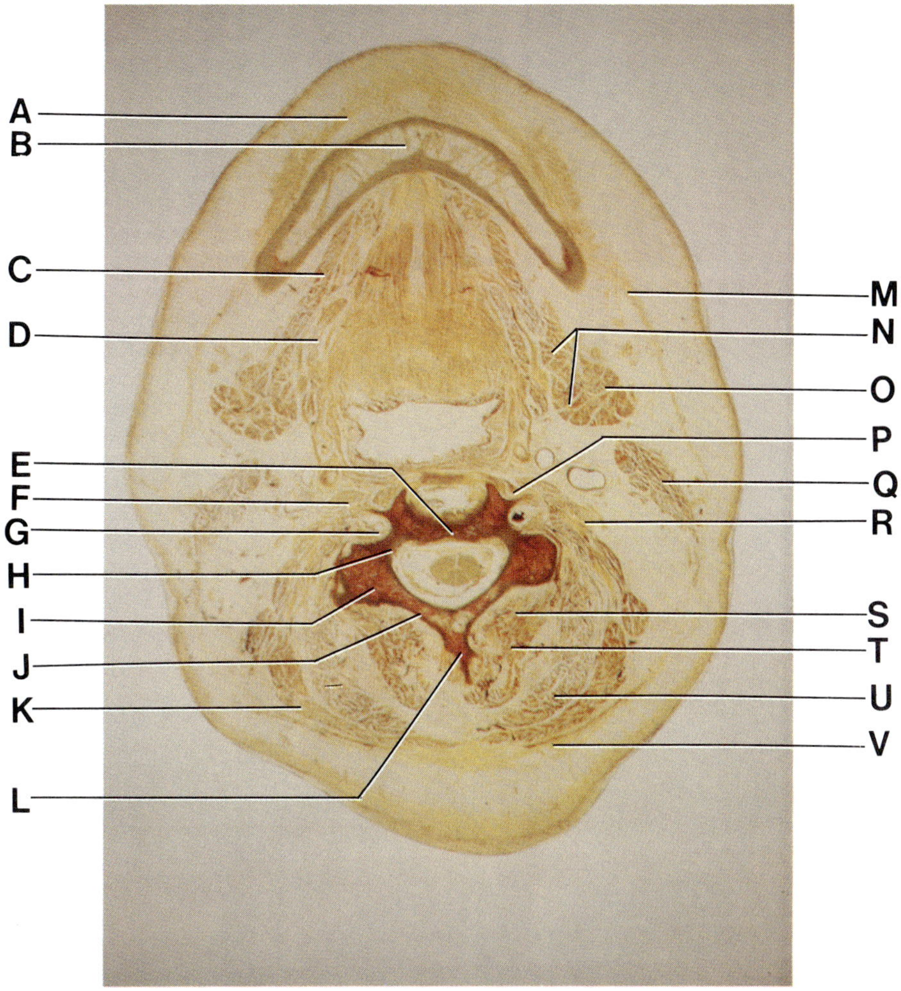

A. Mentalis muscle
B. Mandible
C. Anterior belly of digastric muscle
D. Mylohyoid muscle
E. Body of cervical vertebra
F. Longus capitis muscle
G. Transverse foramen
H. Pedicle of cervical vertebra
I. Transverse process of cervical vertebra
J. Lamina of cervical vertebra
K. Splenius capitis muscle
L. Spinous process of cervical vertebra
M. Platysma muscle
N. Medial pterygoid muscle
O. Masseter muscle
P. Longus colli muscle
Q. Sternocleidomastoid muscle
R. Levator scapulae muscle
S. Multifidus muscle
T. Semispinalis cervicis muscle
U. Semispinalis capitis muscle
V. Trapezius muscle

Figure 6-7
Transverse (axial) section of the neck and lower face at cervical vertebra 3, highlighting musculoskeletal unit structures.

and the posterior end is anterior to the medial part of the splenius capitis muscle. This muscle has a medial concave surface.

Splenius Capitis Muscle (*K*). An elongated, curved, bilateral muscle slice immediately deep to the trapezius muscle. The posterior end of this muscle section is in the median, and the anterior end is medial to the levator scapulae muscle. The medial part of this feature shows a concave arrangement.

Trapezius Muscle (*V*). A thin strip of bilateral muscle tissue immediately posterior and lateral to the splenius capitis muscle. It shows a fascial connection with the sternocleidomastoid muscle, which is anterior to it.

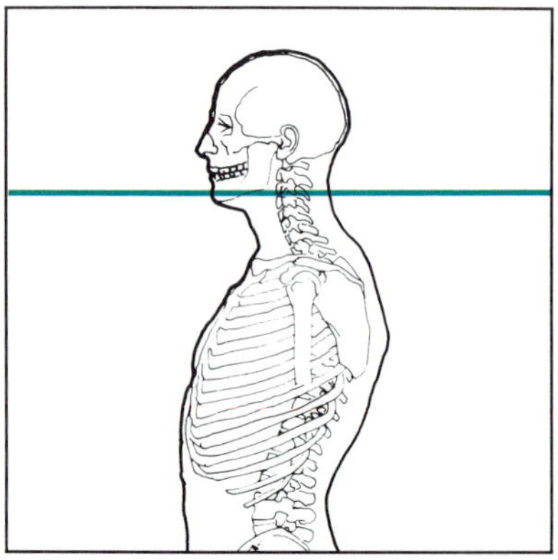

NEUROVASCULAR UNIT (Fig. 6–8)

I. Arteries

Vertebral Artery (*B*). A rounded, bilateral vascular structure within the transverse foramen of the transverse process.

External Carotid Artery (*A*). A rounded, bilateral structure anterior to the internal carotid artery and medial to the posterior part of the medial pterygoid muscle.

Internal Carotid Artery (*C*). A rather large, rounded, bilateral vascular structure anterior and medial to the internal jugular vein. It is posterior to the external carotid artery.

II. Veins

Internal Jugular Vein (*D*). A rather large, oblong, hollow, bilateral vascular structure posterior and lateral to the internal carotid artery. This vessel is medial to the sternocleidomastoid muscle.

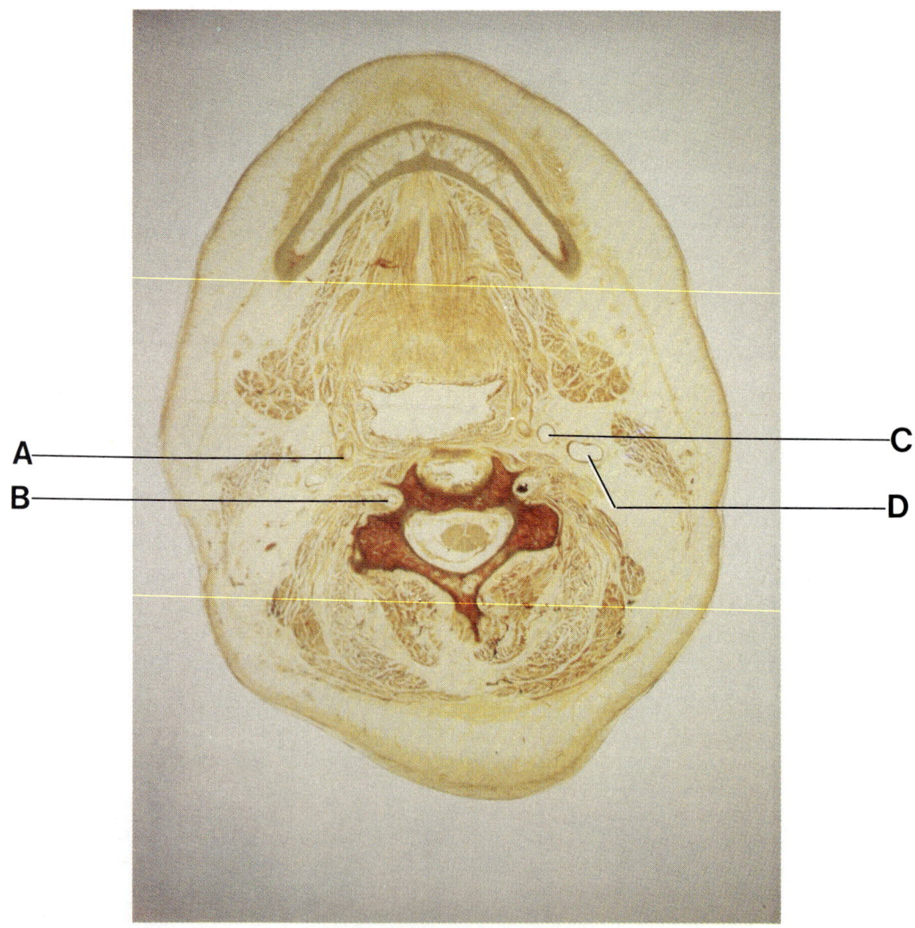

A. External carotid artery
B. Vertebral artery
C. Internal carotid artery
D. Internal jugular vein

Figure 6–8
Transverse section of the neck and lower face at cervical vertebra 3, highlighting neurovascular unit structures.

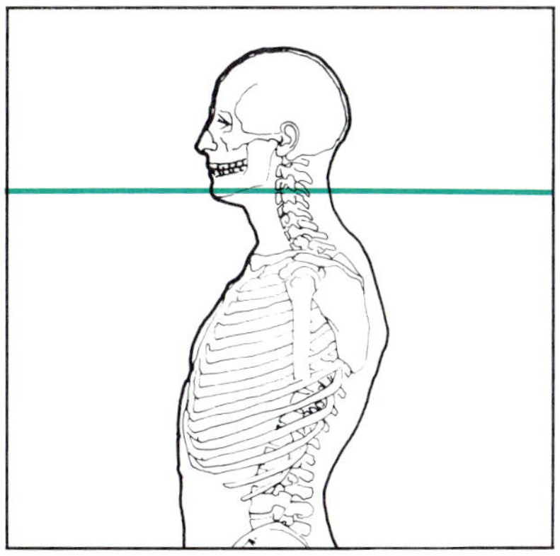

VISCERAL UNIT (Fig. 6–9)

Inferior Longitudinal Muscles of the Tongue (*A*). Two parallel, rather thick, elongated (anterior-to-posterior) muscle strips. These muscle sections extend from the oropharynx to the mandible.

Transverse Muscles of the Tongue (*B*). Curved muscle strips that extend across the tongue. These muscle strips have a posterior concave surface and appear to connect with muscle strips that surround the oropharynx.

Oropharynx (*C*). A rather rectangular space posterior to the extrinsic muscles of the tongue and anterior to the body of the vertebra.

Spinal Cord (*D*). An oval section of nerve tissue with an anterior median fissure located within the spinal foramen.

Lingual Tonsils (*E*). Aggregates of lymph tissue seen on posterior tongue.

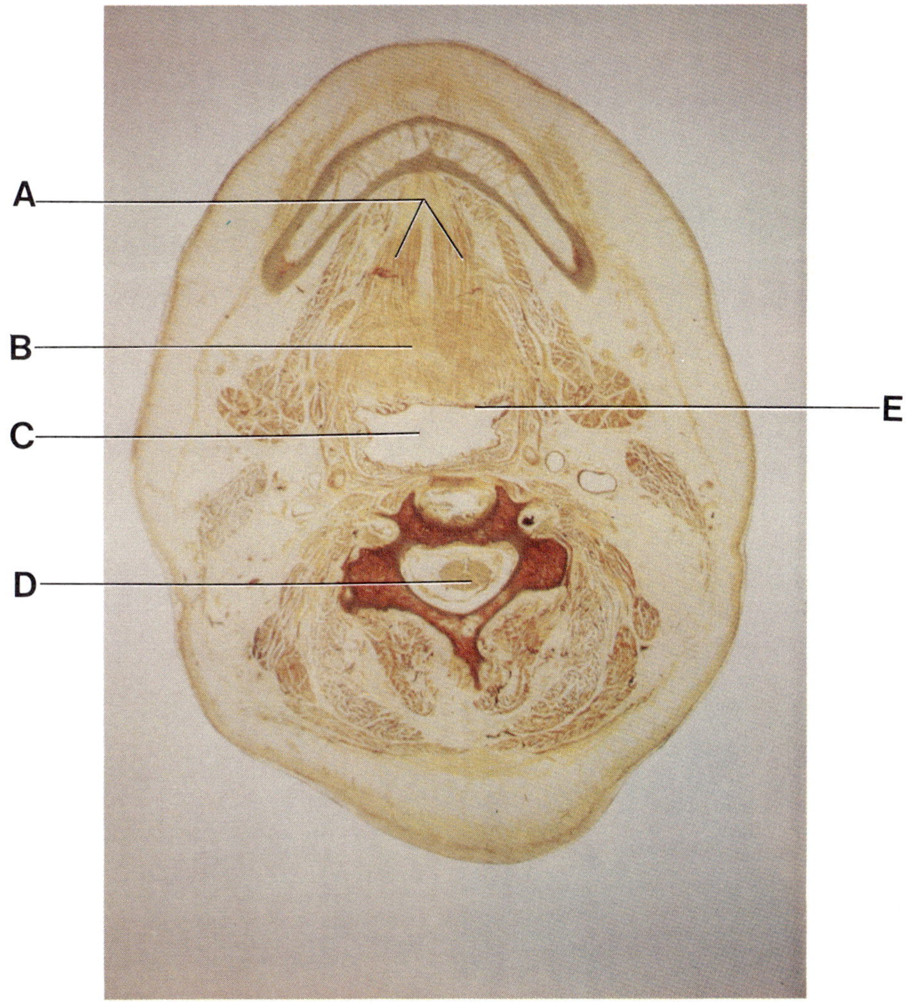

A. Inferior longitudinal muscles of tongue
B. Transverse muscles of tongue
C. Oropharynx
D. Spinal cord
E. Lingual tonsil

Figure 6–9
Transverse section of the neck and lower face at cervical vertebra 3, highlighting visceral unit structures.

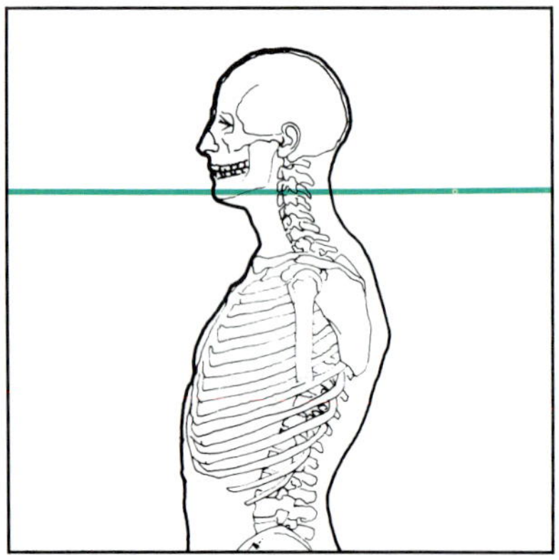

ENCLOSING UNIT (Fig. 6–10)

Mucosa (Mucous Membrane) of Oropharynx (*D*). A curved inner layer of the oropharynx. The anterior ends of this membrane are adjacent to the lateral parts of the tongue's intrinsic muscles.

Dura Mater (*F*). A rather thick membrane adjacent to the inner wall of the spinal foramen.

Subdural Space (*E*). An empty area immediately deep to the dura mater, which is superficial to the arachnoid meninx.

Arachnoid (*B*). A rather irregularly shaped membrane immediately deep to the subdural space and superficial to the subarachnoid space.

Subarachnoid Space (*A*). An empty area immediately deep to the arachnoid, which extends to the outer surface of the pia mater.

Pia Mater (*C*). A very thin sheet of tissue that encases the spinal cord.

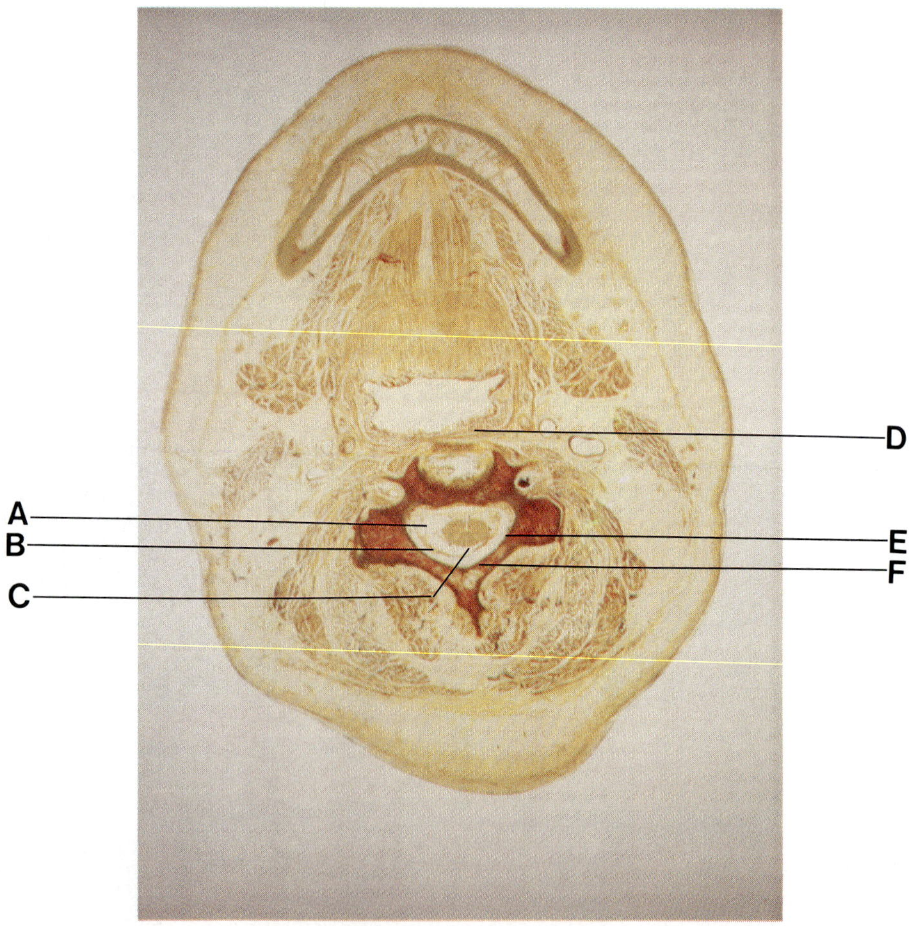

A. Subarachnoid space
B. Arachnoid
C. Pia mater
D. Mucosa (mucous membrane) of oropharynx
E. Subdural space
F. Dura mater

Figure 6–10
Transverse section of the neck and lower face at cervical vertebra 3, highlighting enclosing unit structures.

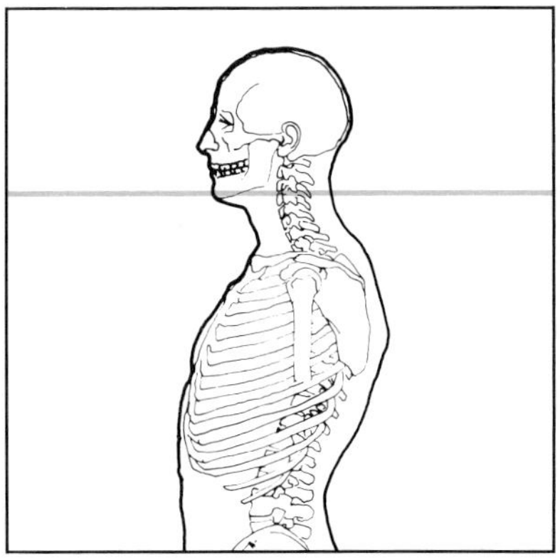

Figure 6–11 shows an MRI transverse image that highlights musculoskeletal, neurovascular, visceral, and enclosing unit structures.

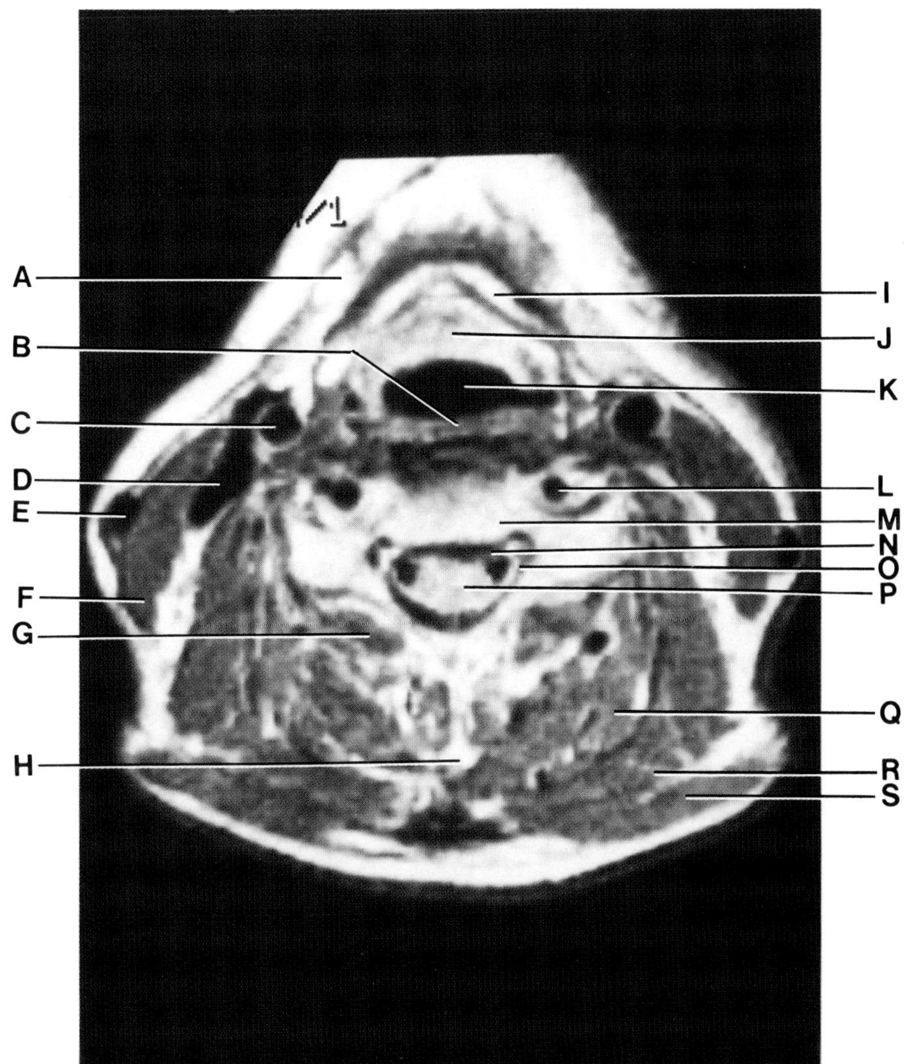

A. Mandible
B. Pharyngeal constrictor muscle
C. Carotid artery
D. Internal jugular vein
E. External jugular vein
F. Sternocleidomastoid muscle
G. Deep cervical muscle
H. Spinous process of cervical vertebra
I. Hyoid bone
J. Tongue
K. Laryngopharynx vestibule
L. Transverse foramen
M. Body of cervical vertebra
N. Anterior nerve root
O. Posterior nerve root
P. Spinal cord
Q. Semispinalis muscle
R. Splenius muscle
S. Trapezius muscle

Figure 6–11
MRI transverse image of neck, highlighting musculoskeletal, neurovascular, visceral, and enclosing unit structures.

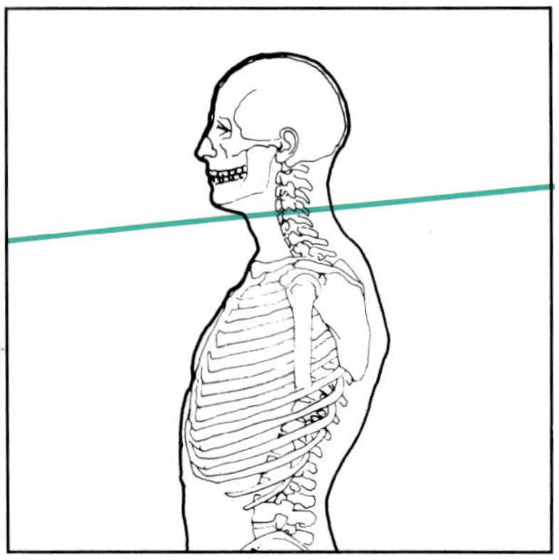

MODULE 3

This is a transverse (axial) section of the neck at the level of cervical vertebra 4.

MUSCULOSKELETAL UNIT (Fig. 6–12)

I. Skeletal Portion

Thyroid Cartilage (*B*). A curved strip of cartilaginous tissue. The posterior surface shows a concave arrangement, and the posterior end of this cartilage is lateral to the nasal pharynx. It is the largest cartilage of the larynx.

Body of the Cervical Vertebra (*C*). An oval mass of bone tissue immediately anterior to the spinal cord and posterior to the retropharyngeal cleft.

Laminae of the Cervical Vertebra (*D*). Two oblique sections of bone tissue. The posterior ends are more medial (join) than the anterior ends. The anterior ends of these features are posterior to the body of the vertebra.

Spinous Process of the Cervical Vertebra (*E*). A posterior median projection at the distal ends of the laminae.

II. Muscular Portion

Platysma Muscle (*A*). A thin strip of cutaneous muscle extending across the anterior part of this neck section.

Sternohyoid Muscle (*G*). A small, transversely curved, bilateral muscle section anterior to the thyroid cartilage. This muscle is medial to the sternothyroid muscle.

Sternothyroid Muscle (*H*). A thin strip of bilateral muscle immediately lateral to the sternohyoid muscle and anterior to the omohyoid muscles.

Omohyoid Muscle. A rather large, bilateral muscle strip adjacent to the lateral part of the thyroid cartilage and posterior to the sternothyroid muscle.

Sternocleidomastoid Muscle (*I*). An oblong, bilateral muscle slice immediately lateral to the neurovascular unit of the neck (internal jugular vein, internal carotid artery, and vagus nerve). This muscle is anterior to the trapezius muscle.

Splenius Capitis Muscle (*J*). An oblong, curved, rather large, bilateral muscle strip with a posterior tapered and pointed end. This feature is deep to the sternocleidomastoid and trapezius muscles. The concave surface is toward the middle of the section.

Trapezius Muscle (*K*). A narrow strip of bilateral, superficial muscle tissue posterior to the sternocleidomastoid and superficial to the splenius capitis muscles.

Semispinalis Capitis Muscle (*L*). An oblong bilateral muscle section deep to the posterior part of the splenius capitis muscle. The posterior part of this feature is thicker than the anterior part.

Multifidus Muscle (*F*). An oval, bilateral muscle section immediately posterior to the lamina of the cervical vertebra and lateral to the spinous process. This muscle is medial to the anterior part of the semispinalis capitis muscle.

Semispinalis Cervicis Muscle. A rather small, oval, bilateral muscle section posterior to the medial part of the multifidus muscle and medial to the semispinalis capitis muscle.

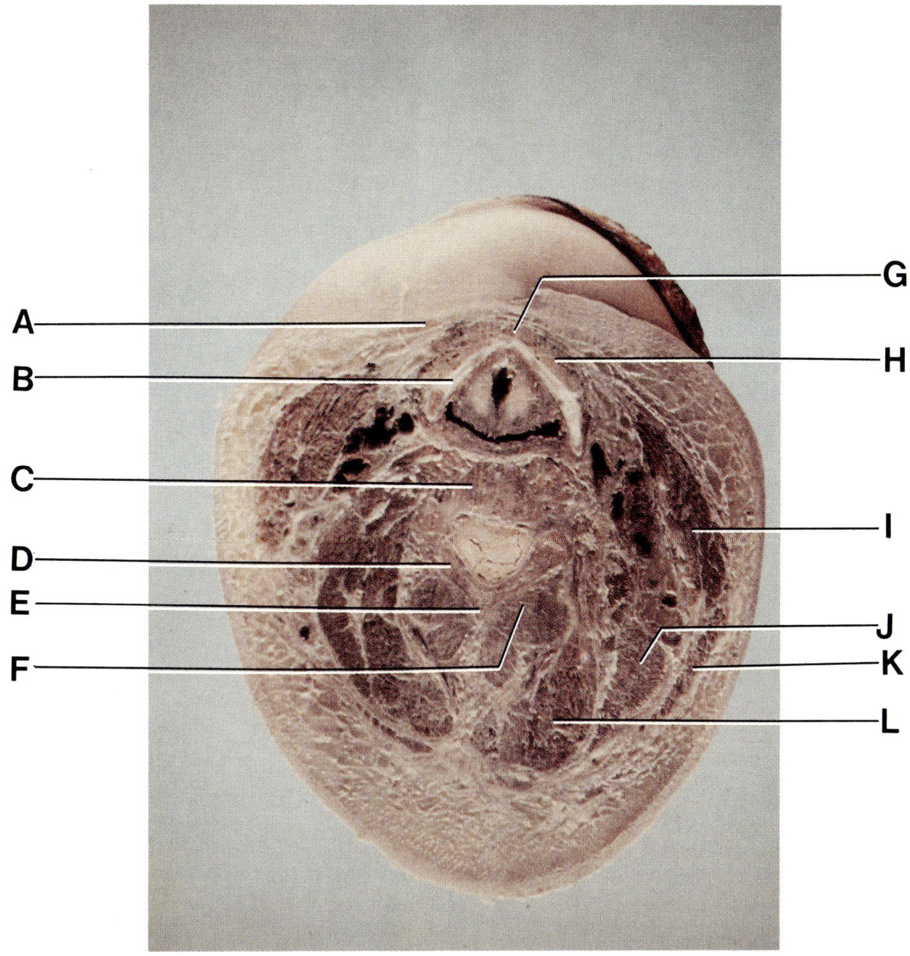

A. Platysma muscle
B. Thyroid cartilage
C. Body of cervical vertebra
D. Laminae of cervical vertebra
E. Spinous process of cervical vertebra
F. Multifidus muscle
G. Sternohyoid muscle
H. Sternothyroid muscle
I. Sternocleidomastoid muscle
J. Splenius capitis muscle
K. Trapezius muscle
L. Semispinalis capitis muscle

Figure 6–12
Transverse section of the neck at the level of cervical vertebra 4, highlighting musculoskeletal unit structures.

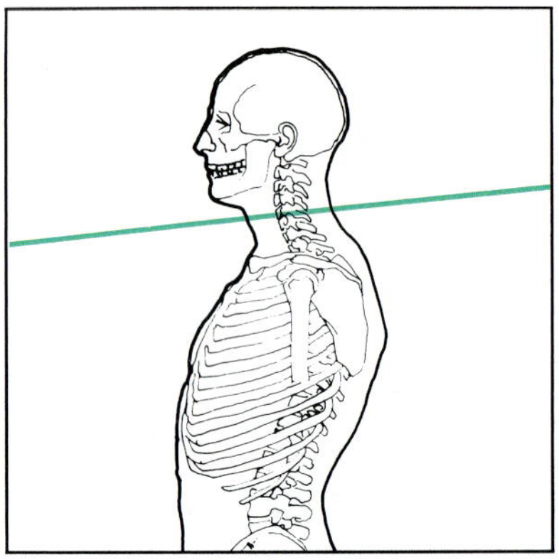

NEUROVASCULAR UNIT (Fig. 6–13)

I. Arteries

Common Carotid Artery (*A*). A rounded, hollow, bilateral vascular structure medial to the internal jugular vein and lateral to the posterior ends of the thyroid cartilage.

Origin. The left common carotid artery arises from the arch of the aorta, and the right common carotid artery originates from the brachiocephalic artery.

Distribution. Both arteries yield blood to their internal and external carotid arteries.

Course. The left common carotid artery leaves the arch of the aorta and ascends to the left of the trachea within the carotid sheath to reach the carotid triangle of the neck. This artery bifurcates in this triangle to form the left internal and external carotid arteries. The right common carotid artery is formed from the brachiocephalic artery posterior to the sternoclavicular joint and ascends the neck to the right of the trachea within its carotid sheath to reach the triangle of the neck. The artery bifurcates in this triangle to form the right internal and external carotid arteries.

II. Veins

Internal Jugular Vein (*B*). An irregular, hollow, bilateral structure immediately lateral to the common carotid artery and medial to the sternocleidomastoid muscle.

III. Nerves

Vagus (10th Cranial) Nerve (*C*). A small, oval, bilateral nerve structure posterior to and between the common carotid artery and the internal jugular vein.

Origin. Medulla oblongata.

Distribution. The vagus nerve is parasympathetic to the following viscera: neck viscera, thoracic viscera, and abdominal viscera (except the descending colon and sigmoid colon). The pelvic viscera receive their parasympathetic innervation from sacral nerves 2, 3, and 4.

Course. The vagus nerve exits from the skull through the jugular foramen and descends the neck posterior to the common carotid artery/internal jugular vein within the carotid sheath. This nerve enters the thorax lateral to the trachea through the superior thoracic aperture and passes posterior to the root of the lungs. The right and left vagus nerves form a vagal trunk on the anterior surface of the esophagus and then proceed through the diaphragm to reach the abdomen.

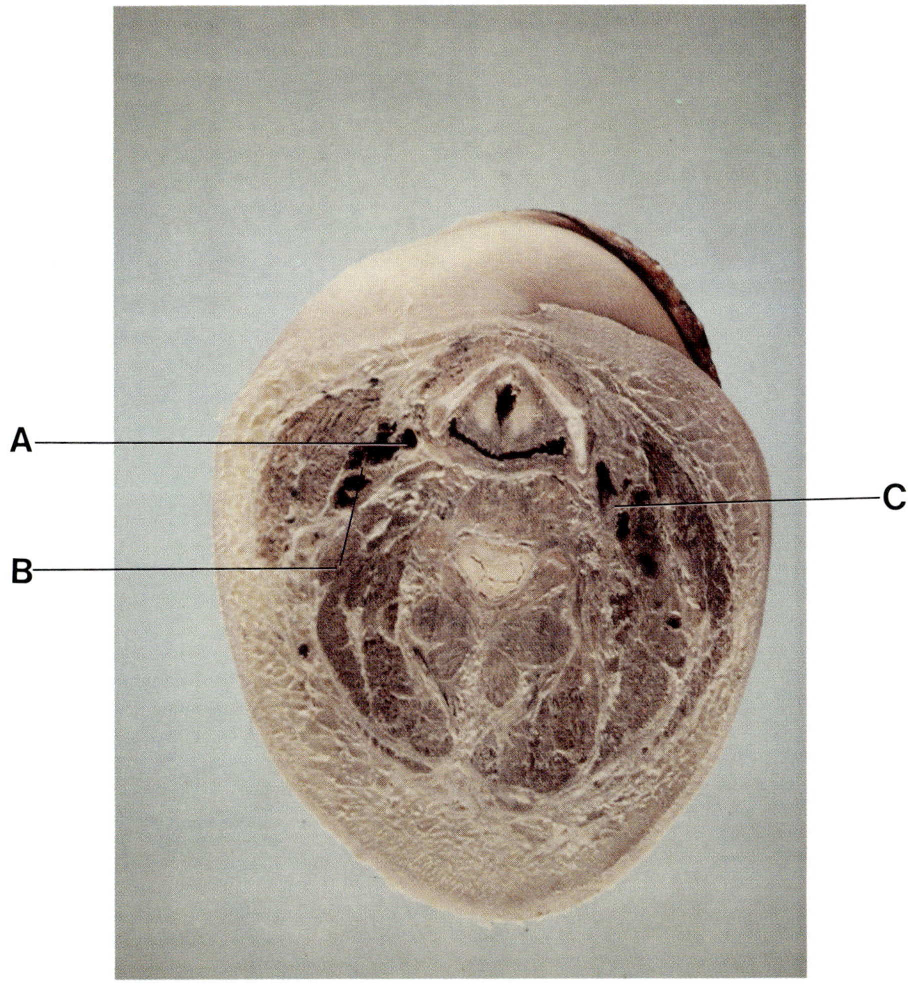

A. Common carotid artery
B. Internal jugular vein
C. Vagus nerve

Figure 6–13
Transverse section of the neck at the level of cervical vertebra 4, highlighting neurovascular unit structures.

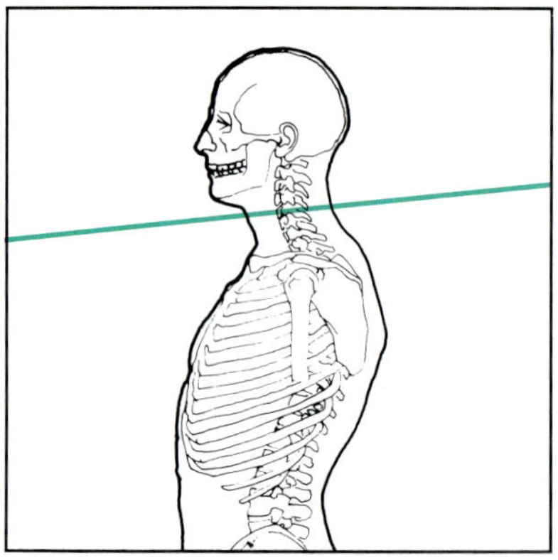

VISCERAL UNIT (Fig. 6–14)

Larynx (*A*). An oval mass of tissue with a rather wide slit in the median, covered on both the anterior and lateral parts by the thyroid cartilage, as seen in this section. This organ is anterior to the laryngopharynx and posterior to the infrahyoid muscles.

Course. The larynx starts at the level of cervical vertebra 3 (*note:* the larynx starts at a higher location in children and women, at approximately cervical vertebra 2) and extends as far as the superior part of cervical vertebra 6. In its course it passes anterior to the laryngopharynx and posterior to the infrahyoid muscles superiorly. This structure is attached to the oropharynx, and inferiorly this organ is attached to the trachea.

Laryngopharynx (*C*). A transversely placed cleft immediately posterior to the larynx.

Course. This hollow structure starts at about the level of cervical vertebra 2 in women and young children. In men, this structure starts at about the level of cervical vertebra 3. The Laryngopharynx extends to the upper level of cervical vertebra 6 and connects with the esophagus in the human organism. In its course it passes posterior to the larynx and anterior to the retropharyngeal cleft.

Spinal Cord (*B*). An oval nerve slice within the spinal foramen, posterior to the body of the vertebra and anterior to the laminae.

Origin. Medulla oblongata.

Termination. Lumbar vertebral disk between lumbar vertebrae 1 and 2.

Course. The spinal cord extends from the medulla oblongata of the brain to the level of the lumbar vertebral disk between lumbar vertebrae 1 and 2. The spinal cord exits from the skull through the foramen magnum and continues inferiorly within the spinal foramen to reach its inferior limit.

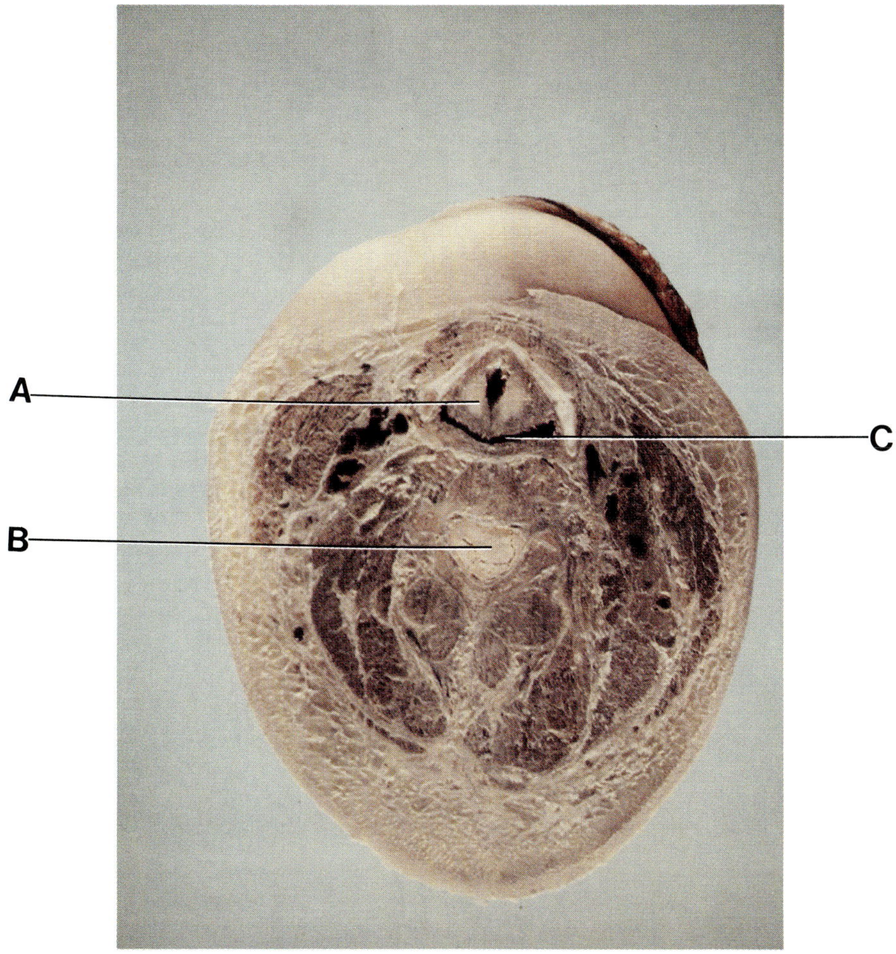

A. Larynx
B. Spinal cord
C. Laryngopharynx

Figure 6–14
Transverse section of the neck at the level of cervical vertebra 4, highlighting visceral unit structures.

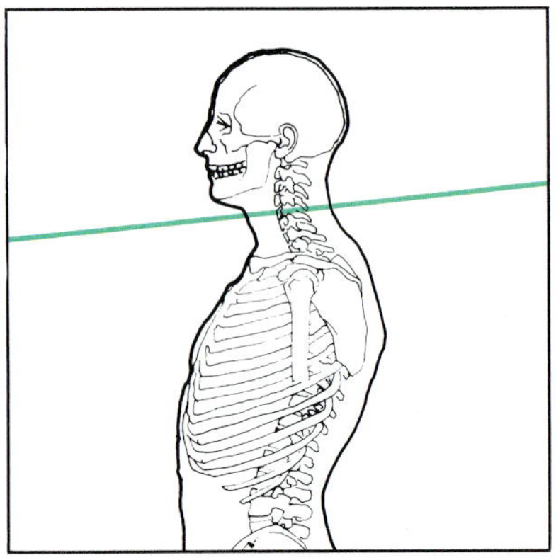

ENCLOSING UNIT (Fig. 6–15)

Investing Fascia (*A*). A whitish layer of connective tissue, seen here enclosing the sternocleidomastoid muscle and extending as a fascial sleeve to the trapezius muscle, which it also surrounds. This connective tissue sheath also continues anteriorly in front of the infrahyoid muscles.

Retropharyngeal Cleft (*C*). A tiny, transversely placed slit anterior to the prevertebral fascia and posterior to the mucosa of the laryngopharynx.

Course. This space starts at the base of the skull posterior to the pharynx and continues downward between the posterior walls of the pharynx/esophagus to reach the mediastinum below. In its course, the lateral structures adjacent to its lateral parts are the right and left carotid sheaths. Infections can spread from the neck region through this space to the upper part of the thorax.

Prevertebral Fascia (*B*). A sheath of connective tissue seen surrounding the anterior aspect of the prevertebral muscles (longus colli and longus capitis).

Dura Mater (*D*). A rather thick membrane that lines the inner wall of the spinal foramen.

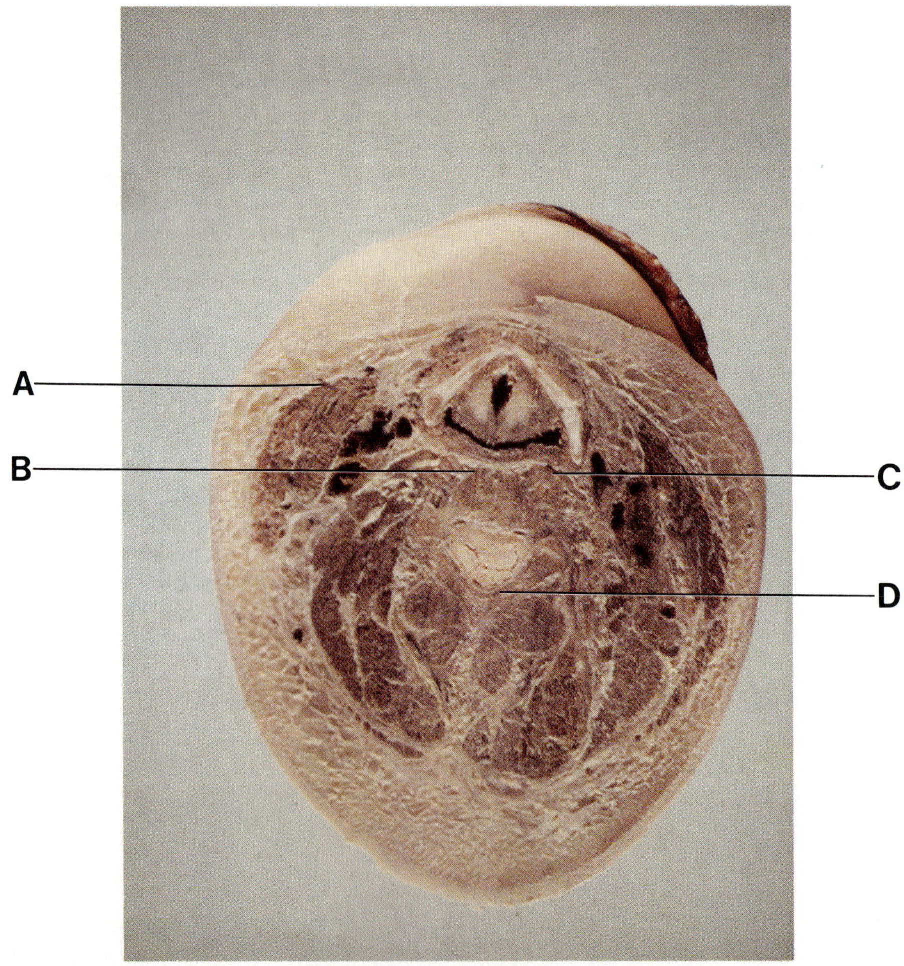

A. Investing fascia
B. Prevertebral fascia
C. Retropharyngeal cleft
D. Dura mater

Figure 6–15
Transverse section of the neck at the level of cervical vertebra 4, highlighting enclosing unit structures.

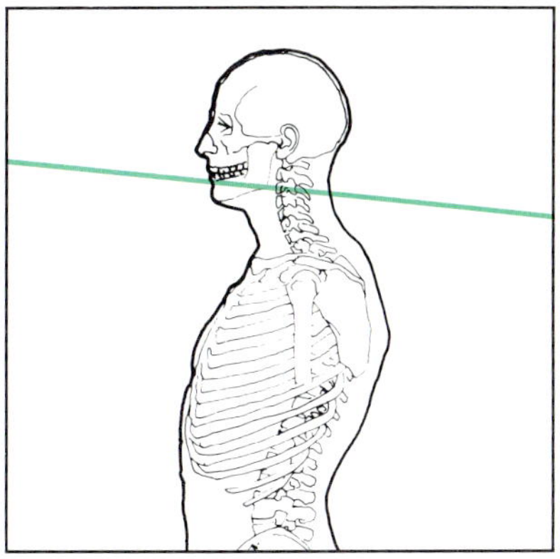

The structures in musculoskeletal, neurovascular, visceral, and enclosing unit are shown in this MRI slice (Fig. 6–16).

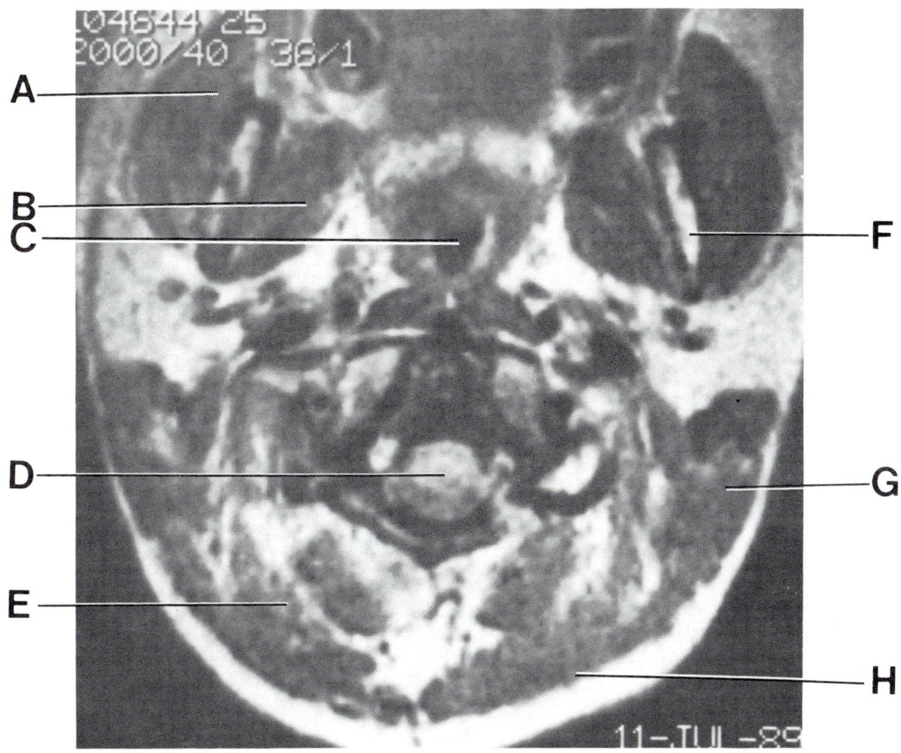

A. Masseter muscle
B. Medial pterygoid muscle
C. Oropharynx
D. Spinal cord
E. Splenius capitis muscle
F. Ramus of mandible
G. Sternocleidomastoid muscle
H. Trapezius muscle

Figure 6–16
MRI transverse image of the neck, highlighting musculoskeletal, visceral, and enclosing unit structures.

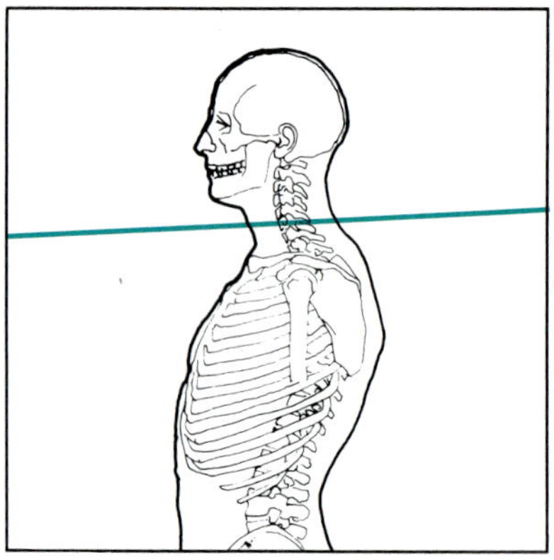

MODULE 4

This is a transverse (axial) section of the neck at the level of cervical vertebra 5.

MUSCULOSKELETAL UNIT (Fig. 6–17)

I. Skeletal Portion

Cricoid Cartilage (*A*). A ring-shaped, cartilaginous structure that surrounds the lumen of the larynx. The posterior part of this ring-shaped feature is thicker than the anterior portion. This is the only cartilage that surrounds the larynx.

Body of Cervical Vertebra 5 (*B*). An oval mass of bone tissue immediately posterior to the nasopharynx and anterior to the spinal cord.

Transverse Process (*F*). A lateral projection that extends from the part of the vertebra immediately posterior to the pedicle and anterior to the lamina. This bifid bone slice is bilateral.

Anterior Tubercle of the Transverse Process (*D*). A rather small, bilateral projection that extends from the anterior part of the transverse process.

Posterior Tubercle of the Transverse Process (*E*). A rather small, bilateral projection that extends from the posterior part of the transverse process.

Pedicle of Cervical Vertebra (*C*). A short and rather narrow bilateral strip of bone tissue between the lateral part of the body of the vertebra and the transverse process.

Lamina of the Cervical Vertebra (*G*). An oblique, narrow, bilateral strip of bone tissue immediately posterior to the transverse process and anterior to the spinous process. The posterior ends are more medial than the anterior ends. The posterior ends join in the median anterior to the spinous process.

Spinous Process of the Cervical Vertebra (*H*). A median, rather sharp, posterior projection between the posterior ends of the laminae.

II. Muscular Portion

Sternohyoid Muscle (*I*). Oblong, bilateral muscle section anterior to the sternothyroid muscle and cricoid cartilage.

Sternothyroid Muscle (*J*). An oblique, bilateral muscle strip posterior to the sternohyoid muscle and anterior to the cricothyroid muscle.

Cricothyroid Muscle (*K*). A rather large, bilateral, oblong (anterior-to-posterior) muscle section lateral to the cricoid cartilage. The anterior and posterior ends taper. The posterior end is immediately lateral to the nasopharynx.

Origin. The side part of the cricoid cartilage. This muscle splits into two groups of fibers (upper and lower).

Insertion. The upper group inserts into the inferior edge of the thyroid cartilage near its posterior surface. The lower group of fibers takes an oblique course to insert on the anterior border of the inferior horn of the thyroid cartilage.

Innervation. Superior laryngeal nerve (external laryngeal branch).

Vasculature. Superior thyroid (cricothyroid branch) and inferior thyroid (inferior laryngeal branch).

Action. Cricoid cartilage elevation and backward tilting of its border. This action causes tensing and elongation of the vocal cords.

Sternocleidomastoid Muscle (*L*). An oblong, bilateral muscle slice seen near the anterior and lateral parts of this section. The posterior part of this muscle is lateral to the neurovascular unit of the neck. The anterior part is lateral to the sternothyroid muscle.

Longus Colli Muscle (*M*). A rather small, oblique, bilateral muscle tissue anterior to the lateral part of the cervical vertebra (body). This muscle is medial to the longus capitis muscle, as seen in this section.

Longus Capitis Muscle (*N*). A rather small strip of bilateral muscle tissue immediately lateral to the longus colli muscle and lateral to the cervical vertebra (body).

Levator Scapulae Muscle (*O*). Two rather small, oblong strips of bilateral muscles. The lateral strip is longer than the medial strip, as seen in this section. These muscle strips are anterior to the splenius capitis muscle.

Semispinalis Capitis Muscle (*Q*). A bilateral, curved, rather thick, oblong muscle medial to the splenius capitis muscle. The concave part is medial to the convex part. This muscle extends from the area posterior to the spinous process to an area lateral to the transverse process.

Splenius Capitis Muscle (*R*). A curved, oblong, bilateral muscle strip with tapering anterior and posterior ends. The anterior end is medial to a part of the levator scapulae muscle. The posterior part of this muscle is deep to the trapezius muscle. The concave surface of this muscle is adjacent to the semispinalis capitis muscle.

Trapezius Muscle (*S*). A rather small, bilateral strip of muscle tissue lateral to the posterior part of the splenius capitis muscle.

Multifidus Muscle (*P*). A rather large, bilateral muscle mass immediately lateral to the spinous process and posterior to the lamina.

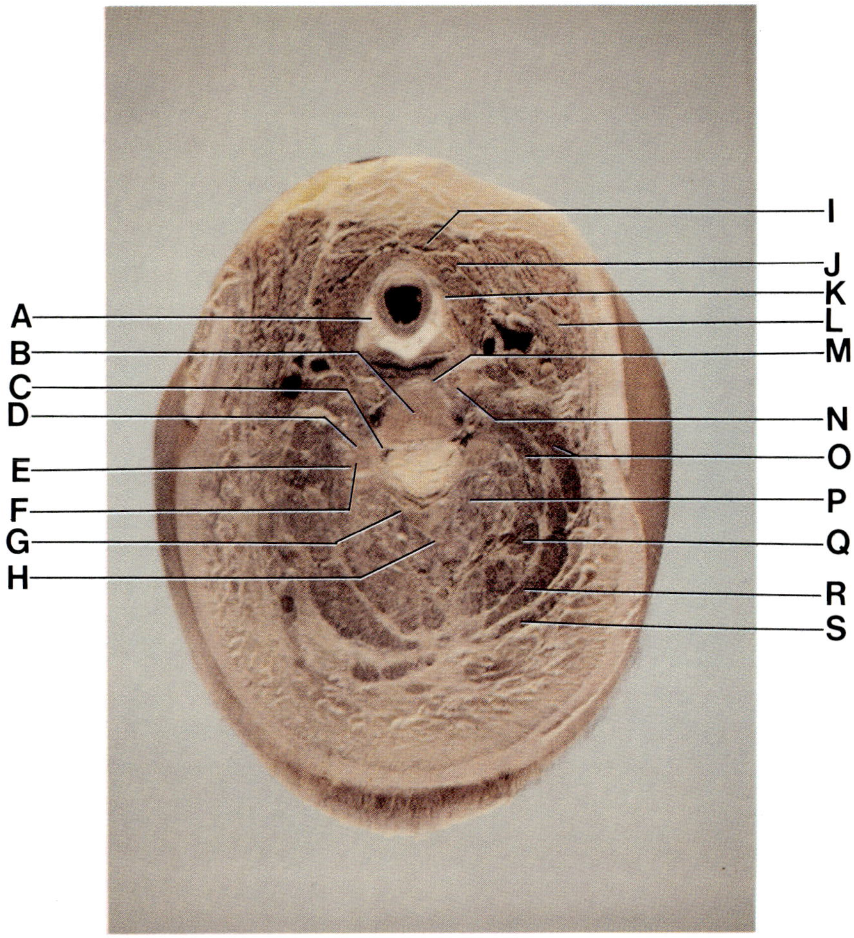

A. Cricoid cartilage
B. Body of cervical vertebra
C. Pedicle of cervical vertebra
D. Anterior tubercle of transverse process
E. Posterior tubercle of transverse process
F. Transverse process
G. Lamina of cervical vertebra
H. Spinous process of cervical vertebra
I. Sternohyoid muscle
J. Sternothyroid muscle
K. Cricothyroid muscle
L. Sternocleidomastoid muscle
M. Longus colli muscle
N. Longus capitis muscle
O. Levator scapulae muscle
P. Multifidus muscle
Q. Semispinalis capitis muscle
R. Splenius capitis muscle
S. Trapezius muscle

Figure 6-17
Transverse (axial) section of the neck at the level of cervical vertebra 5, highlighting musculoskeletal unit structures.

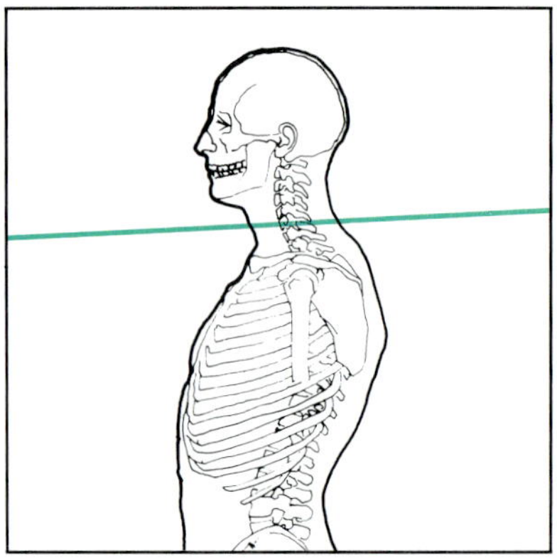

NEUROVASCULAR UNIT (Fig. 6–18)

I. Arteries

Common Carotid Artery (*A*). An oval, hollow, bilateral vascular structure immediately lateral to the posterior part of the cricothyroideus muscle and medial to the internal jugular vein.

II. Veins

Internal Jugular Vein (*B*). A rounded, somewhat irregular bilateral vascular structure immediately lateral to the common carotid artery and medial to the sternocleidomastoid muscle.

III. Nerves

Vagus Nerve (*C*). A small, rounded, bilateral nerve structure located in the posterior area of the carotid sheath. This nerve is posterior and lateral to the common carotid artery, as well as posterior and medial to the internal jugular vein.

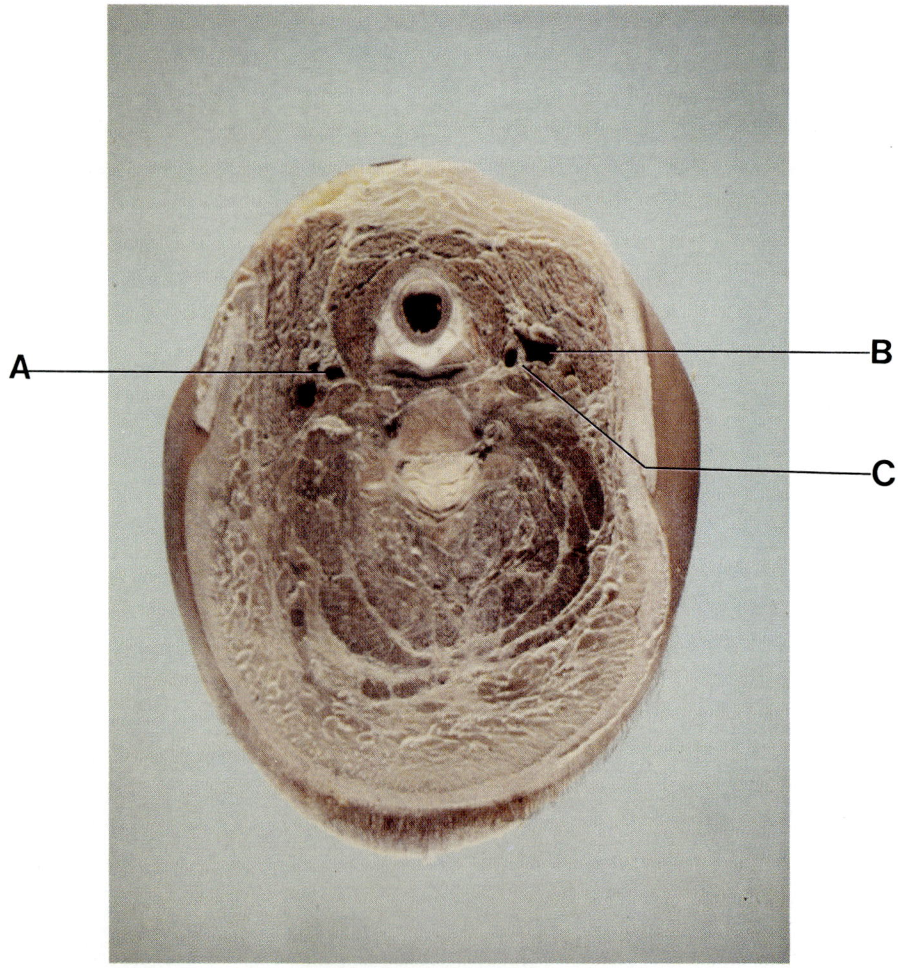

A. Common carotid artery
B. Internal jugular vein
C. Vagus nerve

Figure 6–18
Transverse section of the neck at the level of cervical vertebra 5, highlighting neurovascular unit structures.

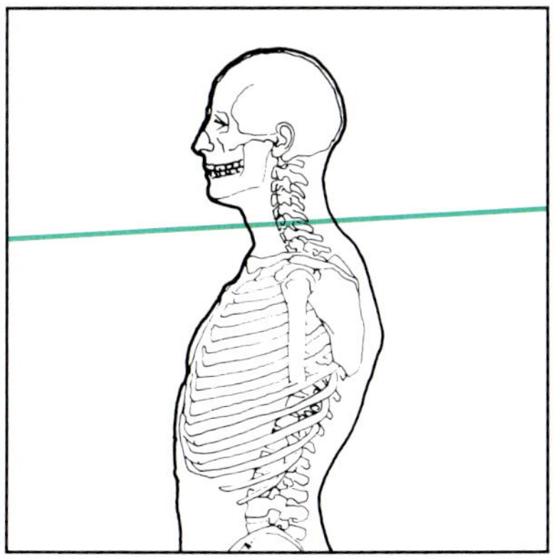

VISCERAL UNIT (Fig. 6–19)

Inferior Part of the Larynx (*A*). A rather large, rounded, hollow structure encased by its most inferior cartilage (cricoid cartilage). The cricothyroid muscles are found immediately lateral to this part of the larynx.

Laryngopharynx (*B*). A transversely placed slit immediately posterior to the cricoid cartilage of the larynx and anterior to the retropharyngeal cleft.

Spinal Cord (*C*). An oval assemblage of nerve tissue within the spinal foramen.

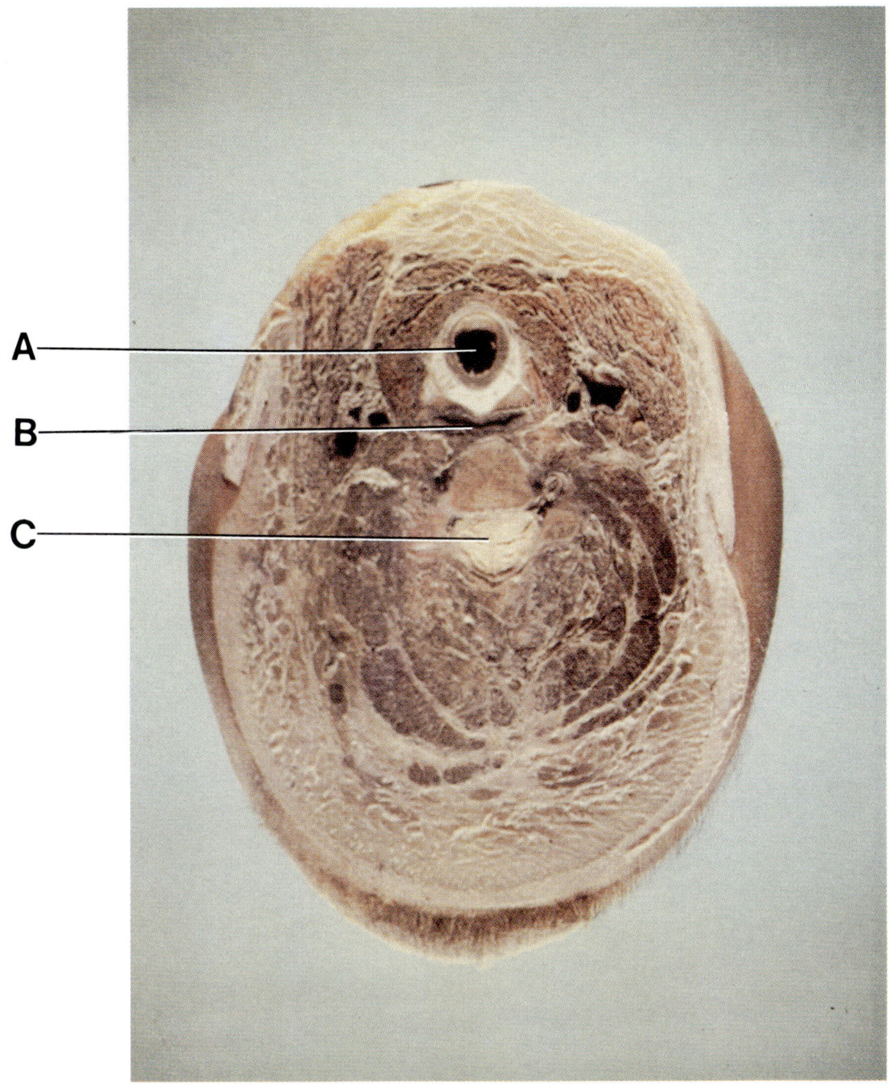

A. Inferior part of larynx
B. Laryngopharynx
C. Spinal cord

Figure 6–19
Transverse section of the neck at the level of cervical vertebra 5, highlighting visceral unit structures.

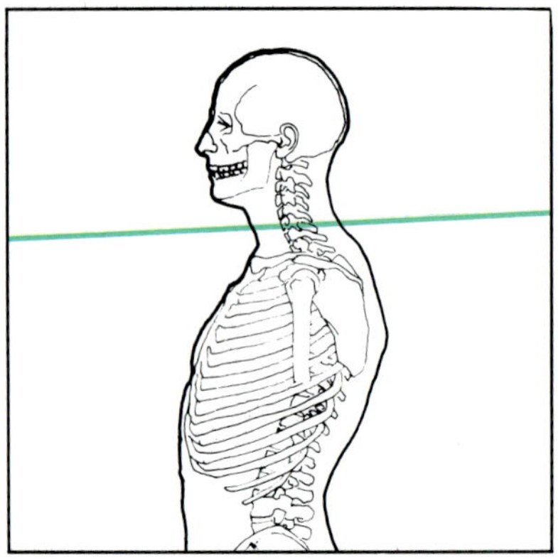

ENCLOSING UNIT (Fig. 6–20)

Carotid Sheath (*D*). A thin layer of connective tissue that surrounds the internal jugular vein, common carotid artery, and vagus nerve. This structure is bilateral.

Retropharyngeal Cleft (*A*). A very narrow, transversely placed space immediately posterior to the nasopharynx and immediately anterior to the body of the cervical vertebra.

Dura Mater (*B*). A moderately thick connective tissue membrane that lines the inner wall of the spinal foramen.

Investing Fascia (*C*). A fairly prominent sheet of connective tissue that surrounds the sternocleidomastoid muscle on both sides. This connective tissue structure also extends posteriorly and envelops the trapezius muscles on both sides of the section.

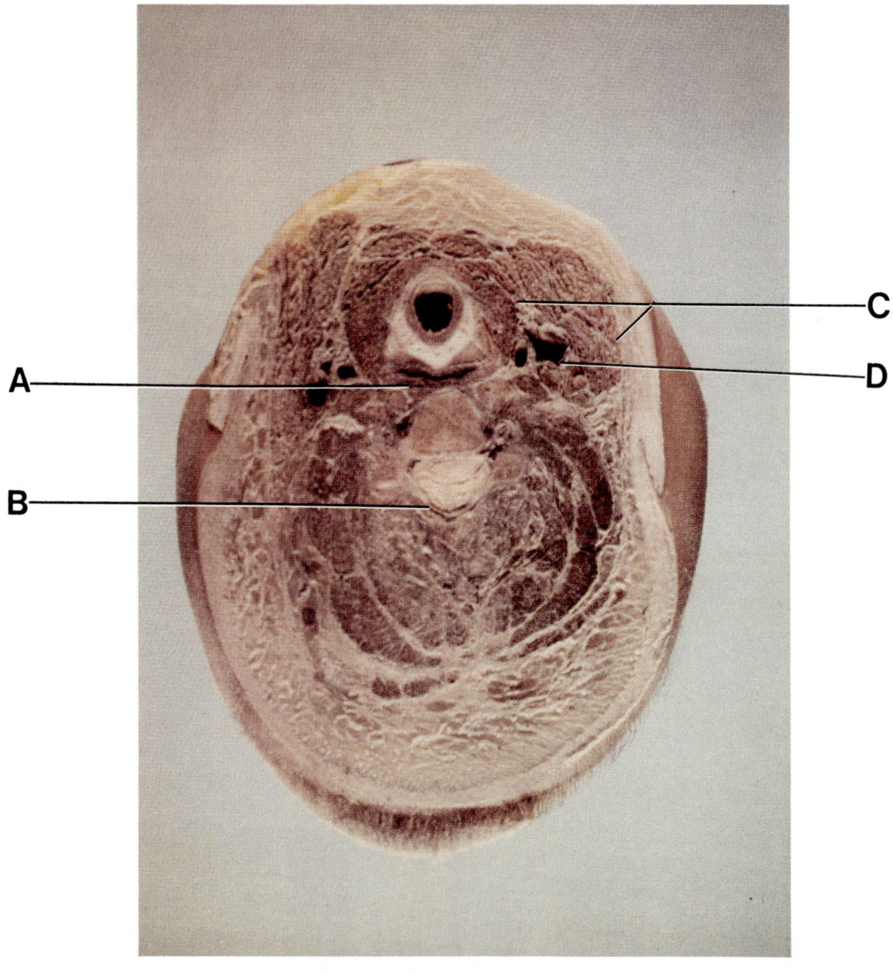

A. Retropharyngeal cleft
B. Dura mater
C. Investing fascia
D. Carotid sheath

Figure 6–20
Transverse section of the neck at the level of cervical vertebra 5, highlighting enclosing unit structures.

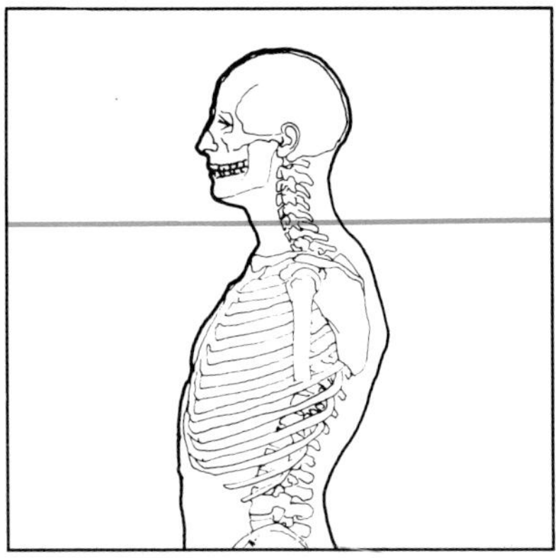

Figure 6–21 is a picture of an MRI scan that emphasizes musculoskeletal, visceral, and enclosing unit structures.

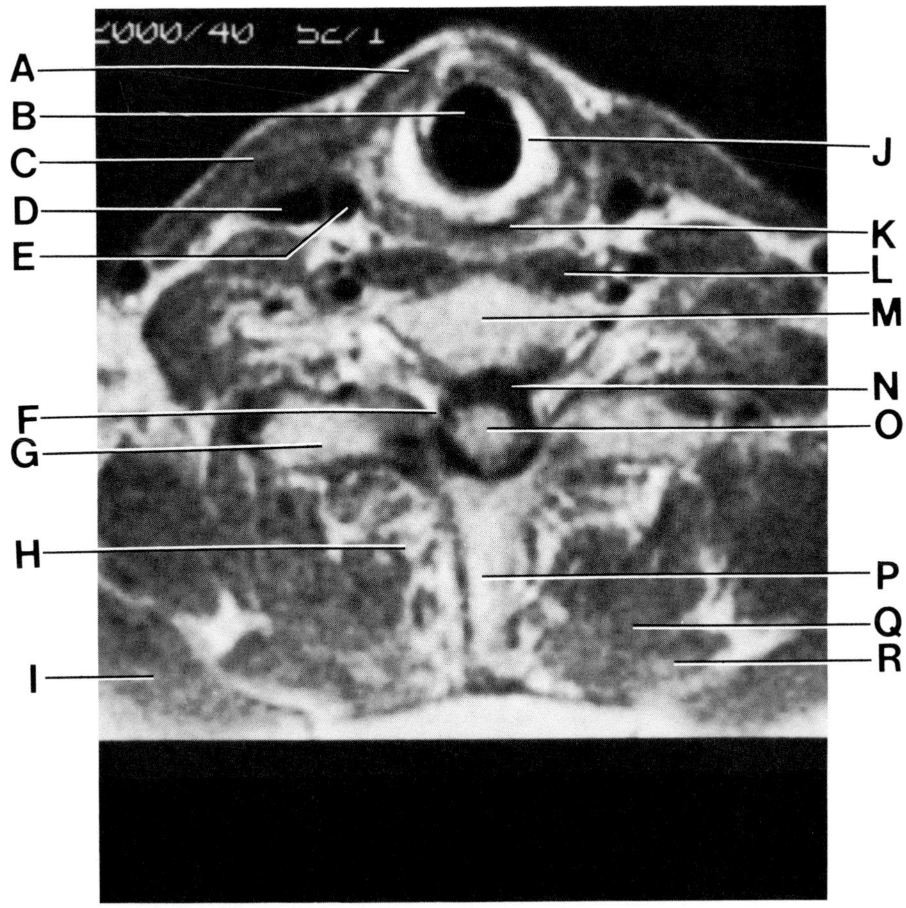

A. Infrahyoid muscle
B. Inferior part of larynx
C. Sternocleidomastoid muscle
D. Internal jugular vein
E. Right common carotid artery
F. Posterior nerve root
G. Transverse process of cervical vertebra
H. Deep cervical muscle
I. Trapezius muscle
J. Cricoid cartilage
K. Laryngopharynx
L. Longus colli and longus capitis muscles
M. Body of cervical vertebra
N. Spinal canal
O. Spinal cord
P. Spinous process of cervical vertebra
Q. Semispinalis capitis muscle
R. Splenius capitis muscle

Figure 6–21
MRI transverse image of the neck, highlighting musculoskeletal, neurovascular, and visceral unit structures.

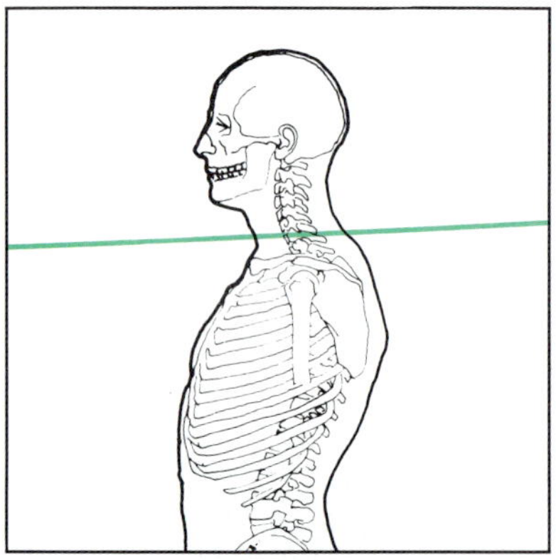

MODULE 5

This is a transverse (axial) section of the neck at cervical vertebral level 6.

MUSCULOSKELETAL UNIT (Fig. 6–22)

I. Skeletal Portion

Body of Sixth Cervical Vertebra (*D*). This is seen in cross-section. As in other typical cervical vertebrae, its features are:

1. **Body** (*D*). The anterior thickened portion of the vertebra.
2. **Pedicles.** Bilateral notched features that extend from the body of the vertebra to the transverse process.
3. **Transverse processes.** Bilateral projections that project from the sides of the vertebrae are seen, each with an anterior and posterior tubercle. Each transverse process contains a hole known as the transverse foramen.
4. **Laminae.** Bilateral flattened features that extend from the transverse process to the spinous process.
5. **Spinous process** (*K*). A posterior median projection of the vertebra is noted.

II. Muscular Portion

Sternohyoid Muscle (*A*). A transversely bilateral, oblong muscle immediately anterior to the sternothyroid muscle and posterior to the investing fascia.

Seventh Cervical Spinous Process (*E*). A posterior median projection of the seventh cervical vertebra.

Part of the Posterior Arch of Sixth Cervical Vertebra (*F*). A curved-back part of the sixth cervical vertebra. The concave surface is anterior to the convex surface.

Sternothyroid Muscle (*B*). A transversely oblong, bilateral muscle with tapering ends, immediately posterior to the sternohyoid muscle.

Sternocleidomastoid Muscle (*M*). A rather thick, bilateral muscle deep to the platysma muscle and superficial to the internal jugular vein.

Platysma Muscle (*N*). A thin strip of superficial muscle that extends from approximately the right midaxillary line to the left midaxillary line.

Scalenus Anterior Muscle (*O*). An oval, bilateral muscle slice anterior to the cervical plexus and medial to the posterior part of the sternocleidomastoid muscle. This muscle can be observed lateral to the vertebral vessels.

Scalenus Medius and Scalenus Posterior Muscles (*P*). These two bilateral muscles appear in the neck fused as one muscle section. They are located posterior to the cervical plexus and medial to the anterior part of the trapezius muscle. They are anterior to the longissimus cervicis and capitis muscles.

Longissimus Capitis Muscle (*Q*). A rounded, bilateral muscle section posterior to the scalenus medius and scalenus posterior muscles. This muscle is anterior to the levator scapulae and lateral to the longissimus cervicis muscle.

Longissimus Cervicis Muscle (*R*). A small, rounded, bilateral muscle feature medial to the longissimus capitis and anterior to the levator scapulae muscles.

Levator Scapulae Muscle (*S*). A bilateral, oblong, fairly small muscle feature immediately posterior to the longissimus capitis muscle and lateral to the anterior part of the semispinalis capitis muscle.

Splenius Capitis Muscle (*J*). A moderately thick, bilateral muscle anterior to the posterior part of the trapezius muscle and superficial to the medial portion of the semispinalis capitis muscle. The anterior end is more tapered than the medial end.

Semispinalis Capitis Muscle (*I*). A moderately thick muscle that extends from the spinous process to the transverse process. This curved muscle shows a concave surface adjacent to the semispinalis cervicis muscle.

Semispinalis Cervicis Muscle (*H*). A rather long, rounded, curved, bilateral muscle section. The concave surface is on its deep part. This muscle is anterior to the semispinalis capitis and posterior to the multifidus muscles.

Multifidus Muscle (*G*). A moderately oblong, bilateral muscle slice adjacent to the lamina and the spinous process. It is anterior to the semispinalis cervicis muscle.

Trapezius Muscle (*T*). A thick, bilateral, curved muscle section with tapering ends, located posterior to the sternocleidomastoid muscle. Its posterior part is posterior to the splenius capitis muscle. Its concave surface is on its deep part.

Longus Colli Muscle (*C*). A sickle-shaped, bilateral muscle slice anterior to the lateral part of the cervical vertebral body.

Omohyoid Muscle (*L*). A small, oblong, bilateral muscle section immediately posterior to the sternothyroid muscle and deep to the sternocleidomastoid muscle.

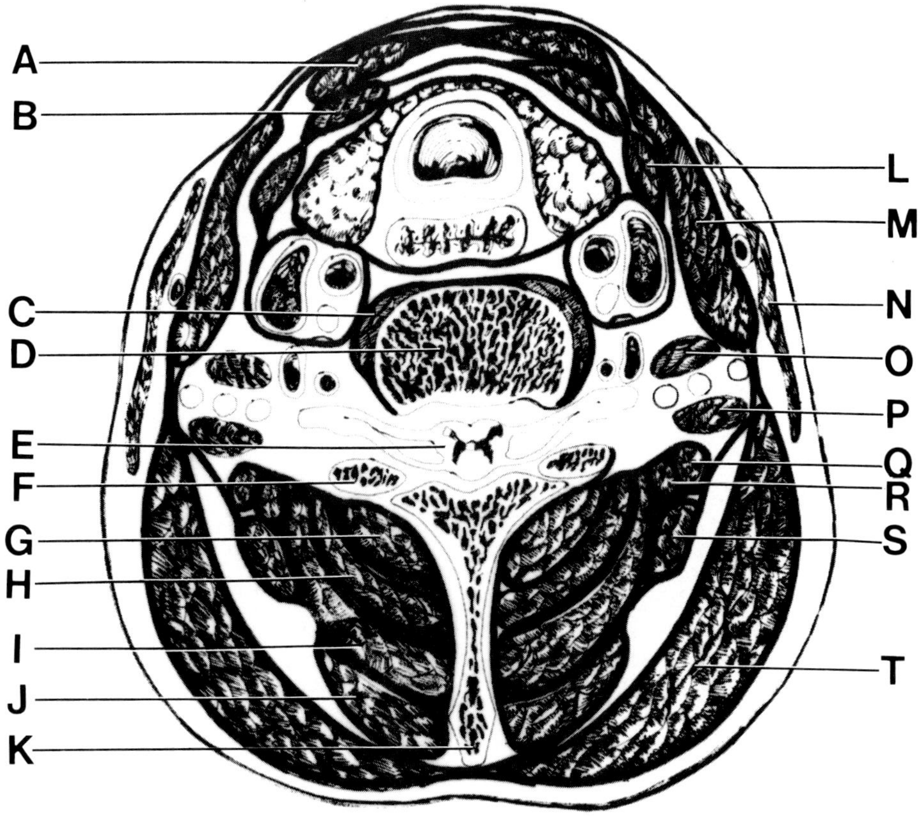

A. Sternohyoid muscle
B. Sternothyroid muscle
C. Longus colli muscle
D. Body of sixth cervical vertebra
E. Seventh cervical spinous process
F. Part of posterior arch of sixth cervical vertebra
G. Multifidus muscle
H. Semispinalis cervicis muscle
I. Semispinalis capitis muscle
J. Splenius capitis muscle
K. Spinous process of cervical vertebra
L. Omohyoid muscle
M. Sternocleidomastoid muscle
N. Platysma muscle
O. Scalenus anterior muscle
P. Scalenus medius and scalenus posterior muscles
Q. Longissimus capitis muscle
R. Longissimus cervicis muscle
S. Levator scapulae muscle
T. Trapezius muscle

Figure 6–22
Transverse section (illustration) of the neck at the level of cervical vertebra 6, highlighting musculoskeletal unit structures.

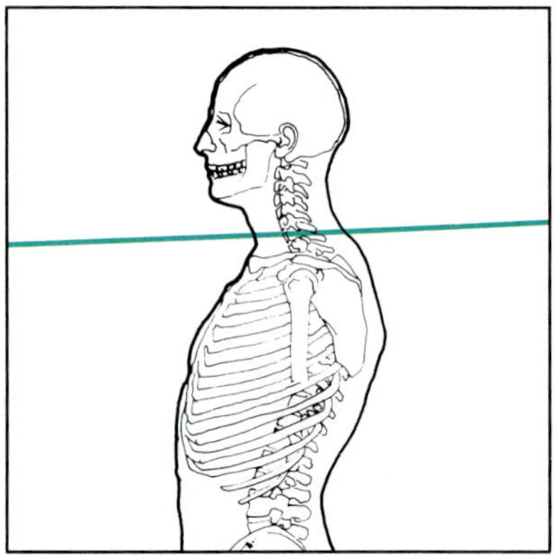

NEUROVASCULAR UNIT (Fig. 6–23)

I. Arteries

Common Carotid Artery (Left and Right) (*D*). There are two arteries, each one being situated on the side of the visceral unit, conducting blood (afferent) to the brain. This rounded vascular structure is seen within the carotid sheath, medial to the internal jugular vein and anterior to the vagus nerve.

II. Veins

Internal Jugular Veins (Left and Right) (*A*). There are two veins, each one being situated on the side of the visceral unit conducting blood (efferent) from the brain. This oblong, hollow vascular structure is seen within the carotid sheath lateral to both the common carotid artery and the vagus nerve.

External Jugular Vein (Left and Right) (*E*). A bilateral, hollow, rounded vascular structure seen immediately deep to the platysma muscle and superficial to the sternocleidomastoid muscle.

Origin. This vessel is formed by retromandibular and posterior auricular veins below the ear at the level of the mandibular angle.

Termination. Subclavian vein.

Course. This vessel descends the neck from the mandibular angle to reach the middle posterior part of the clavicle. In its course it crosses superficially and obliquely on the sternocleidomastoid muscle, penetrates the deep fascia below the clavicle, and terminates in the subclavian vein.

Vertebral Vein (Left and Right) (*G*). A rather oblong, vascular, hollow organ seen within the transverse foramen of the cervical vertebra.

Origin. Internal vertebral venous plexuses within the suboccipital triangle.

Termination. Posterior part of the brachiocephalic veins.

Course. This vessel descends the neck within the transverse foramens of the cervical vertebrae and exits before it enters the transverse process of the seventh cervical vertebra. This vessel courses downward after it exits from the transverse foramen of the sixth cervical vertebra on the front and side of the vertebral artery before it terminates in the superior and posterior portions of the brachiocephalic vein.

III. Nerves

Vagus Nerve (Left and Right) (*B*). A pair of nerves, each one located on the side of the visceral unit, containing the parasympathetic outflow from the medulla oblongata to the neck region and the thoracic and abdominal viscera. The vagus nerves are cranial nerves (cranial nerve X). This rounded nerve organ section is seen posterior to the common carotid artery and medial to the posterior part of the internal jugular vein.

Sympathetic Trunk (Chain) (Left and Right) (*F*). A bilateral, ganglionated nerve column that extends from the base of the skull to the coccyx.

Origin. It is formed by the sympathetic outflow of all the thoracic and the upper two or three lumbar nerves.

Course. This chain descends the neck on the posterior surface of the carotid sheath in front of the transverse processes of the cervical vertebrae; in the thorax it is in front of the rib heads. In the abdomen this structure is located adjacent to the anterolateral parts of the lumbar vertebral bodies. This structure passes in the pelvis in front of the sacrum medial to the sacral foramina. Near its termination in the front part of the coccyx, the two chains fuse with each other to form the ganglion impar (no pair).

Cervical Plexus (Left and Right) (*C*). Formed by the anterior primary rami of cervical nerves 1 through 4. It is situated at the level of cervical vertebrae 1 through 4, deep to the internal jugular vein, sternocleidomastoid muscle, anterior of the scalene medius muscle, and levator scapula muscle. This plexus is posterior to the anterior scalene muscle.

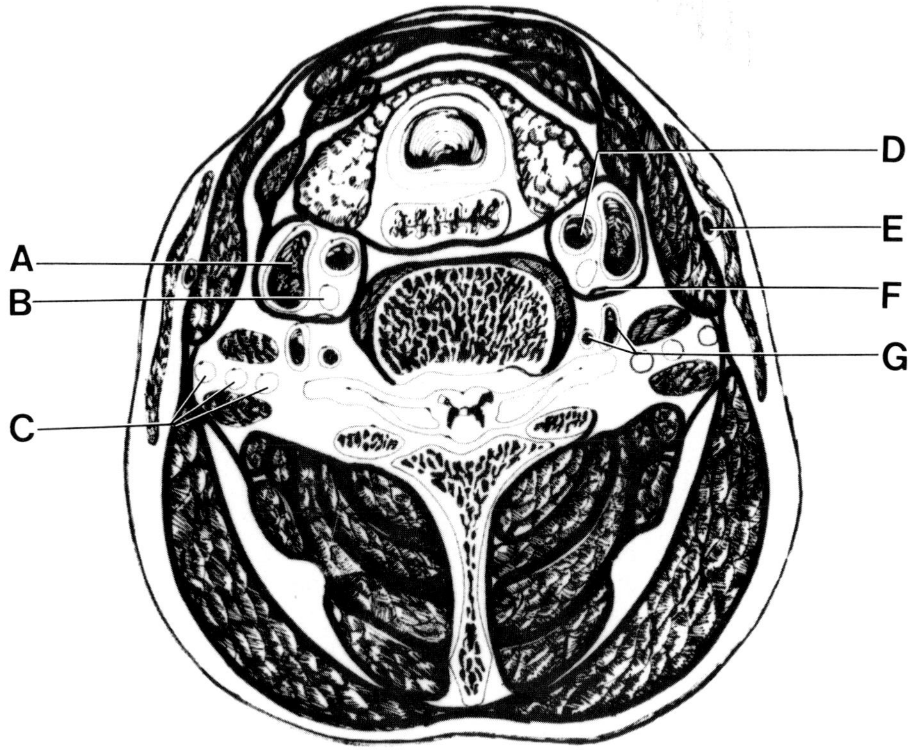

A. Right internal jugular vein
B. Right vagus nerve
C. Right cervical plexus
D. Left common carotid artery
E. Left external jugular vein
F. Left sympathetic trunk (chain)
G. Left vertebral artery and vein

Figure 6–23
Transverse section of the neck at the level of cervical vertebra 6, highlighting neurovascular unit structures.

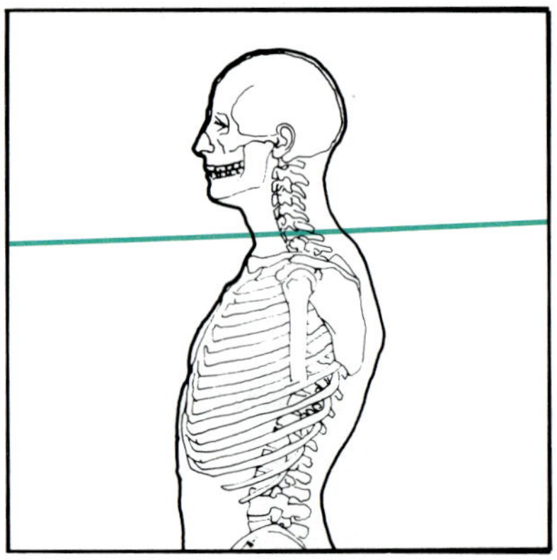

VISCERAL UNIT (Fig. 6–24)

Trachea (*B*). A tube that is anterior to the esophagus, is about 11 cm in length, and conducts air to the lungs.

Origin. Lower border of the cricoid cartilage opposite the sixth cervical vertebra.

Course. Descends from its origin inferiorly anterior to the esophagus and enters the superior thoracic aperture to reach the level between thoracic vertebrae 4 and 5 before it yields to primary bronchi. The bifurcation of the trachea at this level corresponds to an anterior sternal feature known as the sternal angle. At its bifurcation the trachea is slightly right of median, but during the rest of its course it is located primarily in the median plane.

Esophagus (*D*). A tube, posterior to the trachea, for the passage of food to the stomach. This hollow organ is approximately 25 cm (10 inches) long and connects with the nasopharynx at the level of cervical vertebra 6. It terminates at the cardiac opening of the stomach at the level of the 11th thoracic vertebra.

Course. It starts in the neck at the inferior border of the cricoid cartilage. The origin of this organ is at the level of the sixth cervical vertebra, where it is attached to the inferior part of the nasopharynx. From its origin this structure descends in front of the vertebral column through the superior thoracic aperture. As it descends through the thorax, it is found immediately anterior to the bodies of the vertebrae in approximately the median but lies slightly to the left side at some point of its course. The esophagus penetrates the diaphragm through the esophageal hiatus at the level of the 10th thoracic vertebra and terminates in the stomach at the level of the 11th thoracic vertebra.

Thyroid Gland (*C, E*). A two-lobed gland for the secretion of thyroid hormones. This gland is located in front of cervical vertebrae 5, 6, and 7 as well as the first thoracic vertebra.

Parathyroid Glands. The four parathyroid glands are located on the posterior surface of the thyroid gland, between the posterior edges of the thyroid lobes and the capsule enclosing the thyroid gland. There are two on each lobe of the thyroid. The upper parathyroid gland on each lobe is known as the superior parathyroid gland and is located at the level of the middle part of the posterior edge of the thyroid lobe. The inferior parathyroid gland has a variable position: it may lie near the bottom pole of the thyroid gland and also it may lie outside the capsule immediately superior to the inferior thyroid artery. Sometimes the inferior parathyroid gland is embedded within the posterior inferior part of the lobe of the thyroid gland.

Isthmus of Thyroid Gland (*A*). A transverse structure that connects the inferior parts of the right and left thyroid lobes. Usually this structure crosses the front of the trachea at the level of the second and third rings of the trachea.

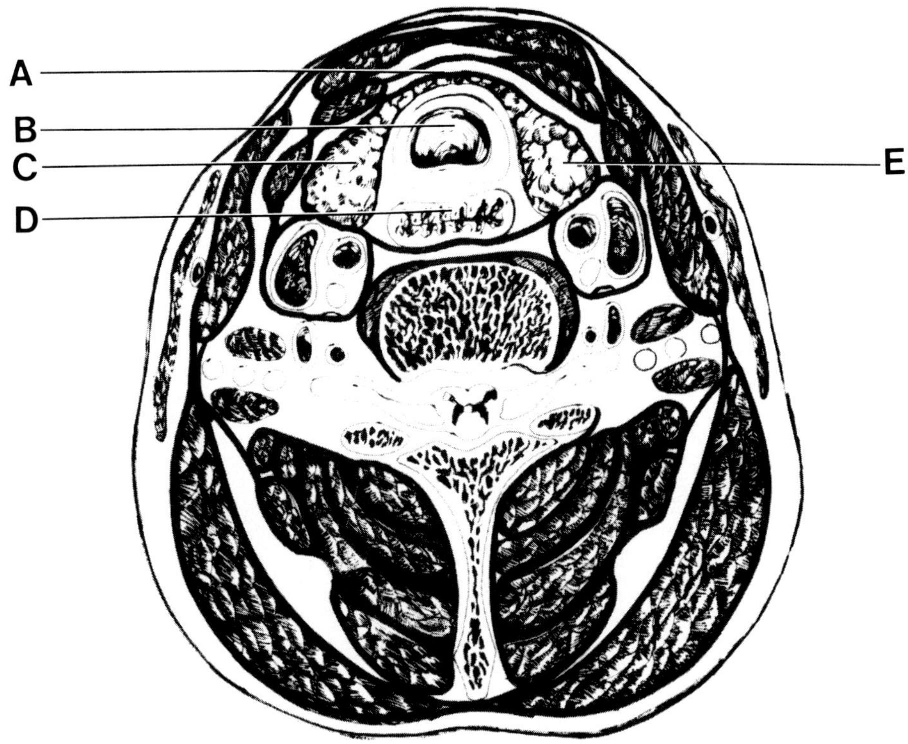

A. Isthmus of thyroid gland
B. Trachea
C. Right lobe of thyroid gland
D. Esophagus
E. Left lobe of thyroid gland

Figure 6–24
Transverse section of the neck at the level of cervical vertebra 6, highlighting visceral unit structures.

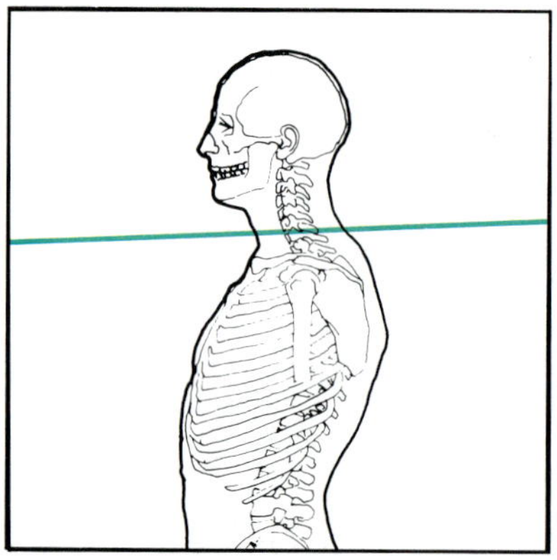

ENCLOSING UNIT (Fig. 6–25)

Skin. The most peripheral structure of the body.

Superficial Fascia. Also known as the hypodermis or subcutaneous tissue. This connective tissue layer attaches the skin to the superficial muscles.

Deep Fascia. Connective tissue with no fat cells. There are four layers of deep fascia.

Investing Fascia (*G*). This layer is the most superficial of the deep fascia and forms a complete collar around the neck. It splits to envelop the sternocleidomastoid and trapezius muscles and covers the posterior triangles of the neck.

Muscular Fascia (*B*). This layer encloses the infrahyoid muscles.

Carotid Sheath (*E*). Fascia formed as a tube, which surrounds the neurovascular unit of the neck. The bilateral neurovascular structures within this sheath are the internal jugular vein, common carotid artery, and vagus nerve.

Prevertebral Fascia (*I*). Surrounds the prevertebral muscles.

Visceral Fascia (*H*). Also known as the pretracheal fascia, this is a connective tissue sheath that encloses the visceral unit.

Premuscular Cleft (*A*). Seen as a transversely positioned anterior median slit posterior to investing fascia.

Previsceral Cleft (*C*). Seen in this section as a transversely curved slit that extends across median anterior to the thyroid gland. The concave surface is posterior to the convex surface.

Retropharyngeal Cleft (*D*). Seen as a transversely positioned space that crosses the median posterior to the esophagus.

Nuchal Fascia (*F*). Connective tissue between the right and left trapezius muscles.

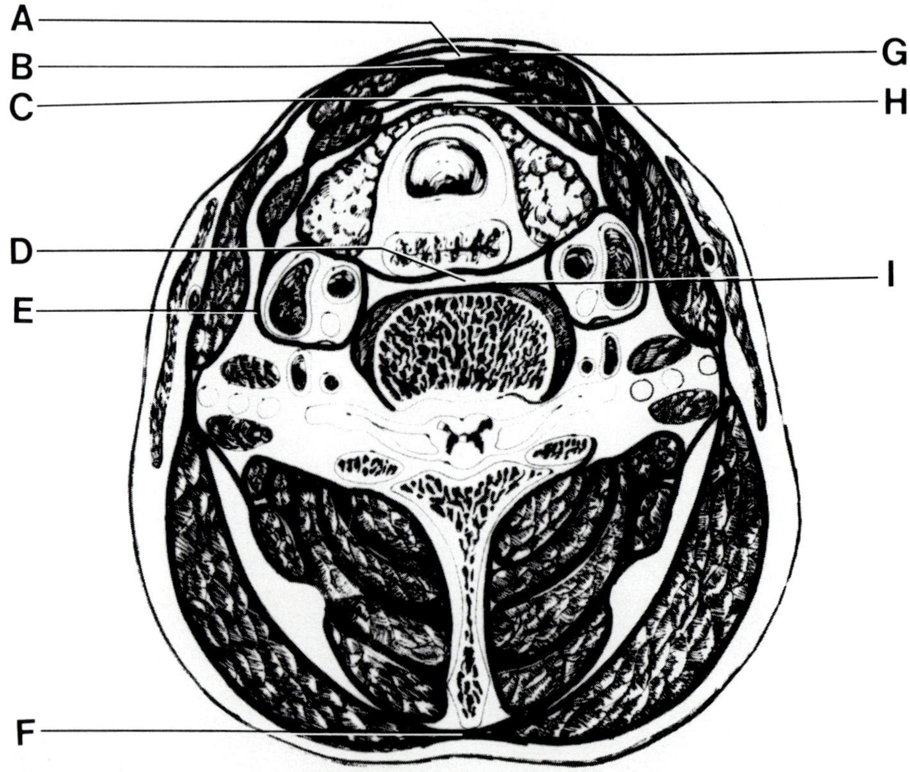

A. Premuscular cleft
B. Muscular fascia
C. Previsceral cleft
D. Retropharyngeal cleft
E. Carotid sheath
F. Nuchal fascia
G. Investing fascia
H. Visceral fascia
I. Area of prevertebral fascia

Figure 6–25
Transverse section of the neck at the level of cervical vertebra 6, highlighting enclosing unit structures.

CHAPTER 7

Coronal (Frontal) Cadaver Neck Sections

The two coronal illustrations of the neck sections combine the four anatomic categories. One illustration includes labels of the musculoskeletal unit, and the other has labels for neurovascular, visceral, and enclosing units. Structures are clearly seen in a longitudinal plane, and the width of the structures is clearly visible.

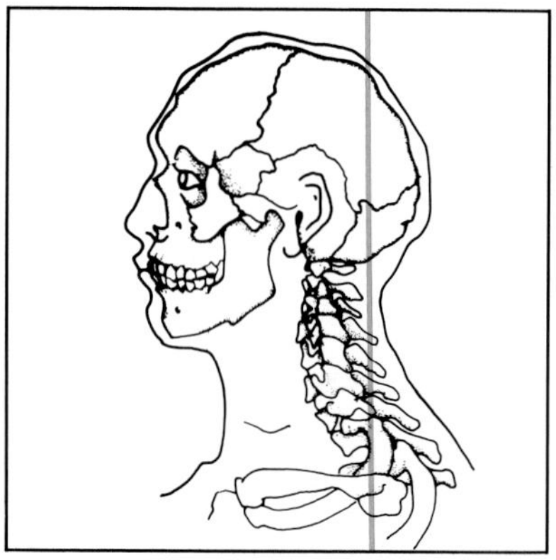

MODULE 1

The coronal (frontal) section of the neck is illustrated here. It passes through the posterior atlanto-occipital membrane, the spines of the cervical vertebrae, and the lower part of the cervical spinal cord.

MUSCULOSKELETAL UNIT (Fig. 7–1)

I. Skeletal Portion

Posterior Tubercle of the Atlas (*H*). An egg-shaped bone section seen in the median superior neck between the posterior venous vertebral plexuses. It is immediately inferior to the posterior atlanto-occipital membrane.

Spinous Process of the Axis (*J*). Three peripherally projected processes that represent the bifid spinous process. This bone section is seen immediately medial to the inferior oblique muscle of the head.

II. Muscular Portion

Sternocleidomastoid Muscle (*A*). A small, bilateral portion of muscle slice immediately lateral to the splenius capitis muscle and superior to the lesser occipital nerve.

Splenius Capitis Muscle (*B*). An elongated, vertical, sausage-shaped, bilateral muscle section that extends from the splenius cervicis muscle to the lateral part of the occipital bone. This muscle courses laterally to the longissimus capitis muscle. Its superior end is lateral to the occipital artery as seen in this section.

Longissimus Capitis Muscle (*C*). An elongated, ribbon-shaped, bilateral muscle slice immediately medial to the splenius capitis muscle and lateral to the semispinalis capitis muscle.

Inferior Oblique Muscle of the Head (Obliquus Capitis Inferior Muscle) (*I*). A triangular, rather large, bilateral muscle section obliquely lateral to the spinous process of the axis and medial to the superior part of the semispinalis capitis muscle as seen in this section. The apex is located at a higher level than the base of this triangular muscle section.

Semispinalis Capitis Muscle (*D*). A bulky, elongated, thick, bilateral muscle slice immediately medial to the longissimus capitis muscle. The superior part is immediately lateral to the inferior oblique muscle of the head.

Rectus Capitis Posterior Major Muscle (*G*). An oblong, bilateral, obliquely placed muscle section. The inferior end is more medial than the superior end. The superior end is seen attached to the occipital bone and medial to the superior oblique muscle of the head.

Superior Oblique Muscle of the Head (Obliquus Capitis Superior Muscle) (*F*). An obliquely placed, oblong, bilateral muscle section immediately inferior to the occipital bone and lateral to the rectus capitis posterior major muscle. Its inferior end is placed more medial than its superior end.

Semispinalis Cervicis Muscle (*K*). A rather elongated, bilateral muscle strip medial to the semispinalis capitis muscle and lateral to the multifidus muscle. Its inferior end is more tapered than its superior end.

Multifidus Muscle (*E*). A rather oblong, bilateral muscle slice immediately medial to the profundus cervicalis vein. Its superior end is placed more medially than its inferior end.

Splenius Cervicis Muscle (*M*). An elongated, bilateral muscle slice inferior to the longissimus capitis muscle and lateral to the inferior part of the semispinalis capitis muscle.

Levator Scapulae Muscle (*L*). A thick, elongated, bulky, bilateral muscle slice immediately lateral to the splenius cervicis muscle. The upper part of this muscle section is lateral to the inferior part of the longissimus capitis muscle.

Scalenus Posterior Muscle (*N*). A rather pyramid-shaped, bilateral muscle slice inferior to the levator scapulae muscle as seen in this neck section. This description applies better to the left splenius posterior muscle than to the right splenius posterior muscle. Both muscles are immediately lateral to their respective splenius cervicis muscles.

A. Sternocleidomastoid muscle
B. Splenius capitis muscle
C. Longissimus capitis muscle
D. Semispinalis capitis muscle
E. Multifidus muscle
F. Superior oblique muscle of the head (obliquus capitis superior muscle)
G. Rectus capitis posterior major muscle
H. Posterior tubercle of atlas
I. Inferior oblique muscle of the head (obliquus capitis inferior muscle)
J. Spinous process of axis
K. Semispinalis cervicis muscle
L. Levator scapulae muscle
M. Splenius cervicalis muscle
N. Scalenus posterior muscle

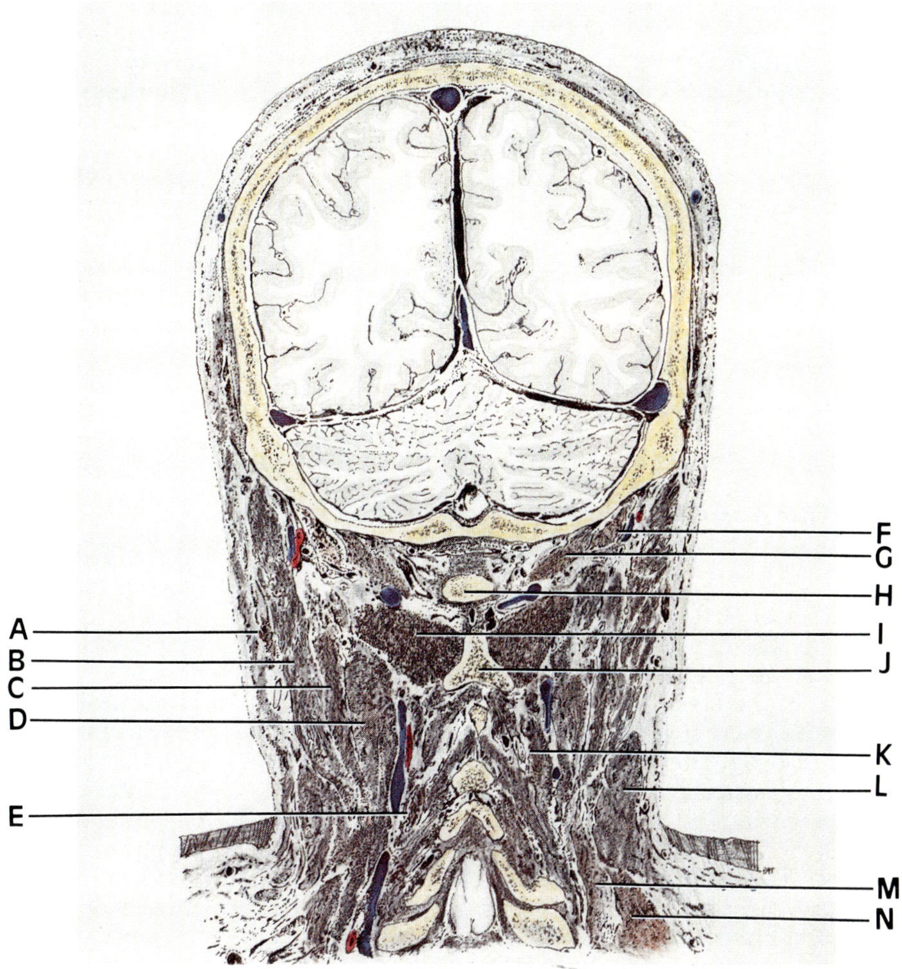

Figure 7–1
Coronal (frontal) section, which passes through the atlanto-occipital membrane, the spines of the cervical vertebrae, and the lower part of the cervical spinal cord, highlighting musculoskeletal unit structures.

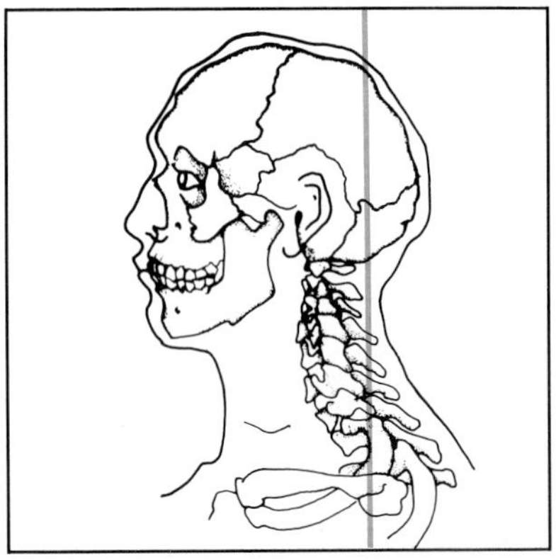

NEUROVASCULAR UNIT (Fig. 7–2)

I. Arteries

Occipital artery (*A*). An elongated, cylindrical, bilateral vascular structure immediately medial to the occipital vein and lateral to the superior oblique muscle of the head.

Origin. External carotid artery.

Distribution. Sternocleidomastoid, occipital, descending, auricular, mastoid, meningeal, and muscular branches.

Course. This artery courses superficially to the hypoglossal nerve, transverse process of the atlas, mastoid process, rectus capitis, squamous portion of the temporal bone, and rectus capitis lateralis muscle. It ascends and penetrates the fascia on the skull between the trapezius and sternocleidomastoid muscles. From this point the artery is found in a curving course in the superficial fascia of the posterior scalp.

Deep Cervical (Profunda Cervicalis) Artery (*D*). A vertical, hollow, bilateral vascular structure immediately lateral to the semispinal muscle of the neck and medial to the deep cervical vein in the superior part. The inferior part is lateral to the deep cervical vein.

Origin. Costocervical trunk muscles.

Distribution. Postvertebral muscles (erector spinae muscle of the neck).

Course. This artery passes posteriorly between the first rib and the transverse process of the seventh cervical vertebra following its place of origin (costocervical trunk) and ascends the neck medial to the semispinal muscle of the head and lateral to the semispinal muscle of the neck. It connects with inferior branches of the occipital artery.

II. Veins

Occipital Vein (*B*). A hollow, vertical, bilateral vascular structure immediately medial to the upper part of the splenius capitis muscle, as seen in this neck section, and lateral to the occipital artery.

Origin. Venous network in the posterior division of the scalp.

Termination. Deep cervical (profunda cervicalis) and vertebral vein (on rare occasions the occipital vein drains into the internal jugular vein, sometimes in the posterior auricular vein and sometimes in the external jugular vein).

Course. The occipital vein penetrates the superior part of the trapezius muscle and into the suboccipital triangle before it terminates in the vertebral and deep cervical veins.

Deep Cervical (Profunda Cervicalis) Vein (*E*). An elongated, hollow, bilateral vascular structure medial to the semispinal muscle of the head and lateral to the deep cervical artery in the case of its superior part. The inferior part seen in this coronal (frontal) body section shows that the vein is medial to the deep cervical artery.

Origin. It starts from a plexus in the suboccipital triangle.

Termination. The vertebral vein and sometimes the brachiocephalic vein.

Course. It descends the deep neck between the semispinal muscle of the head and the multifidus muscle.

Internal Dorsal Venous Vertebral Plexus (*J*). A hollow, bilateral vascular structure seen immediately superior to the inferior oblique muscle of the head and inferior to the medial part of the rectus capitis posterior major muscle. This vessel is lateral to the posterior tubercle of the atlas.

Origin. The basilar and occipital sinuses join on either side of the foramen magnum and empty into the vertebral venous system (internal dorsal venous vertebral plexus).

Termination. Vertebral vein.

III. Lymphatic Structures

Cervical Lymph Nodes (*F*). Oval, superficial, nodular, bilateral structures seen immediately lateral to the inferior part of the levator scapulae muscle.

IV. Nerves

Lesser Occipital Nerve (*C*). Elongated, vertical, bilateral sensory nerve tissue seen lateral to the splenius capitis muscle and inferior to the sternocleidomastoid muscle.

Origin. Posterior primary rami.

Distribution. Skin (scalp) posterior to the ears.

Course. This nerve courses upward on the lateral surface of the sternocleidomastoid muscle, following its penetration of deep fascia, to reach the scalp of the occipital area of the head.

VISCERAL UNIT (see Fig. 7–2)

Spinal Cord (*L*). An aggregate vertically placed column-like nerve tissue in the median of the section located within the spinal foramen.

A. Occipital artery
B. Occipital vein
C. Lesser occipital nerve
D. Deep cervical artery (profunda cervicalis artery)
E. Deep cervical (profunda cervicalis) vein
F. Cervical lymph nodes
G. Deep cervical (profunda cervicalis) vein
H. Deep cervical (profunda cervicalis) artery
I. Posterior atlanto-occipital membrane
J. Internal dorsal venous vertebral plexus
K. Ligamentum flavum
L. Spinal cord

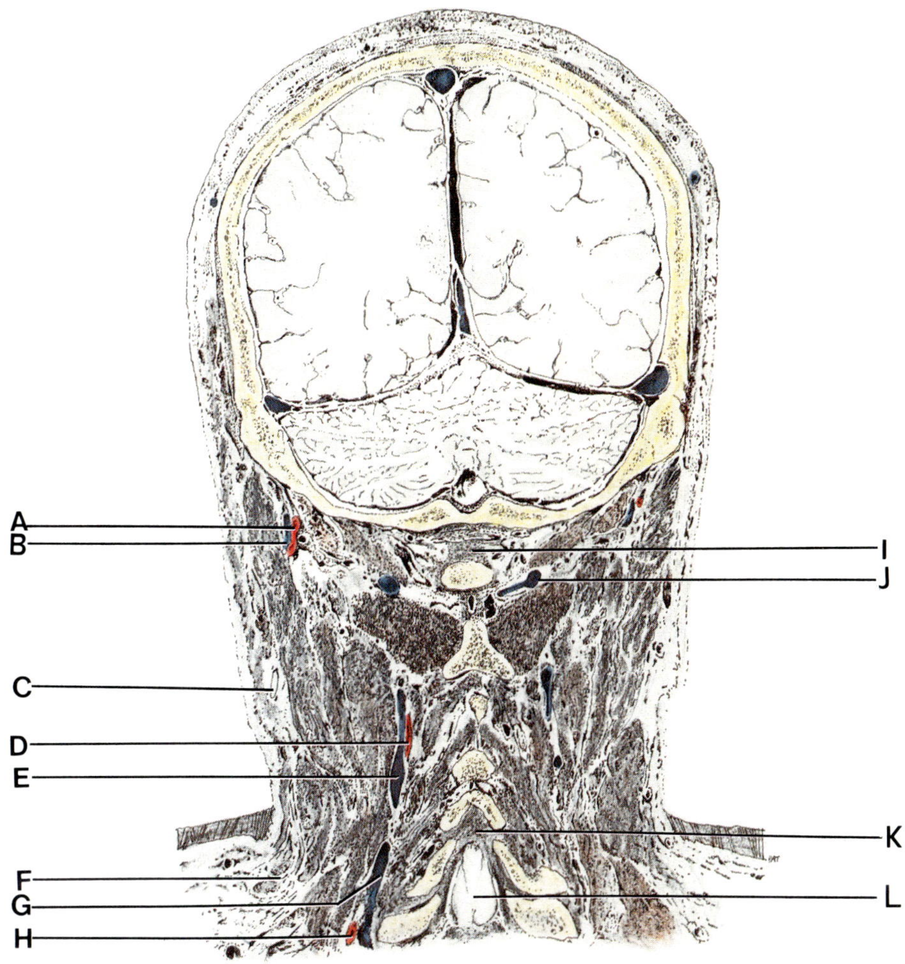

Figure 7-2
Coronal section, which passes through the posterior atlanto-occipital membrane, the spines of the cervical vertebrae, and the lower part of the cervical spinal cord, highlighting neurovascular, visceral, and enclosing unit structures.

ENCLOSING UNIT (see Fig. 7-2)

Posterior Atlanto-Occipital Membrane (*I*). A sheet of connective tissue seen adjacent to the inferior surface of the occipital bone in the median and superior surface of the posterior tubercle of the atlas. This membrane is wider on its superior end than on its inferior end.

Attachments. This is a ligament that attaches the vertebral arch (lamina) of the atlas to the occipital bone at the posterior margin of the foramen magnum.

Ligamentum Flavum (*K*). A rather wide sheet of ligament attached to the laminae of the vertebrae. This ligament is seen from the inferior surface of the posterior part of the vertebral canal (spinal foramen).

Attachments. Each ligament is attached to the lamina of a vertebra above and the lamina of a vertebra below. There are right and left ligamenta flava between each right and left lamina. There are 23 pairs of ligamenta flava.

CHAPTER 8

Sagittal Cadaver Sections and Clinical Images of the Neck

These midsagittal sections (Figs. 8–1 to 8–5), of the neck show labels on one for the musculoskeletal unit and the other combines both the visceral and enclosing units. The sagittal sections of the neck show the longitudinal view of the structures as well as the anterior-to-posterior extent of the structures. Note the orientation of each structure in a longitudinal plane. For example, the spinal cord is seen as a vertical, slightly curved slice of nerve tissue within the spinal foramen of the neck vertebrae.

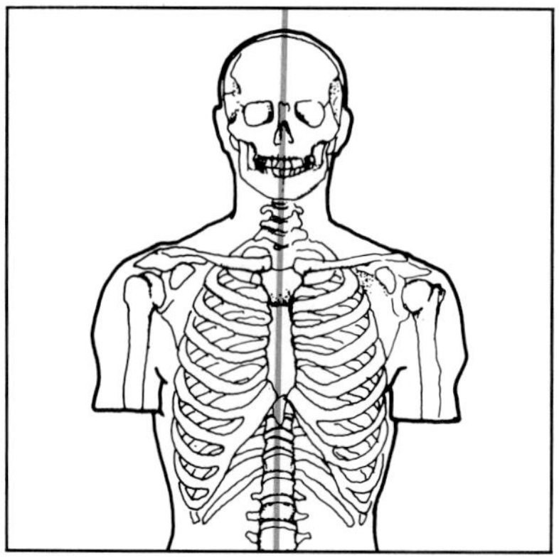

MODULE 1

This is a midsagittal section of the head, neck, and upper thorax, showing the epiglottis, cricoid cartilage, trachea, vocal cords, and thyroid cartilage. The neck structures are discussed and described in this module.

MUSCULOSKELETAL UNIT (Fig. 8–1)

I. Skeletal Portion

Body of the cervical vertebra (*H*). A rectangular bone slice anterior to the spinal cord and posterior to the anterior longitudinal ligament. The intervertebral disk is between the bodies of the vertebrae.

Odontoid Process (Dens) (*F*). A vertical slice of bone tissue that represents the superior projection of the anterior part of the axis. Its superior part is posterior to the anterior inner surface of the atlas. This entire structure is seen anterior to the superior part of the spinal cord. The inferior end of this bone slice is slightly more anterior than the superior part.

Median Atlantoaxial Joint (*E*). A pivot joint, seen in this sagittal section between the superior anterior part of the odontoid process and the posterior part of the front portion of the atlas.

Fovea Dentis (*A*). A shallow depression on the posterior part of the anterior portion of the atlas.

Spinous Process of the Cervical Vertebra (*I*). An anterior-to-posterior oblique bone slice posterior to the spinal cord. The posterior end is more inferior than the anterior end.

Intervertebral Disk (*C*). A plate of cartilaginous tissue seen between the bodies of the vertebrae.

Anterior Longitudinal Ligament (*D*). An anterior, vertical strip of regular connective tissue immediately anterior to the bodies and intervertebral disks of the vertebral column. This structure extends from the basilar part of the occipital bone to the anterior superior part of the sacrum. This ligament is attached to the basilar part of the occipital bone and along its entire course to the anterior part of the vertebral column.

Posterior Longitudinal Ligament (*B*). A posterior, vertical strip of regular connective tissue within the spinal canal, immediately posterior to the bodies and intervertebral disks of the vertebral column. This structure extends from the body of the axis to the superior part of the sacrum. This ligament is attached to the body of the axis and along its entire course to the posterior parts of the vertebral bodies and intervertebral disks.

Ligamenta Flava (*G*). Strips of connective tissue within the spinal canal, posterior to the spinal cord and anterior to the spinous process. Each ligamentum flavum is located between adjacent laminae. Each ligamentum flavum is attached to the articular capsule and the inner surface of the lamina.

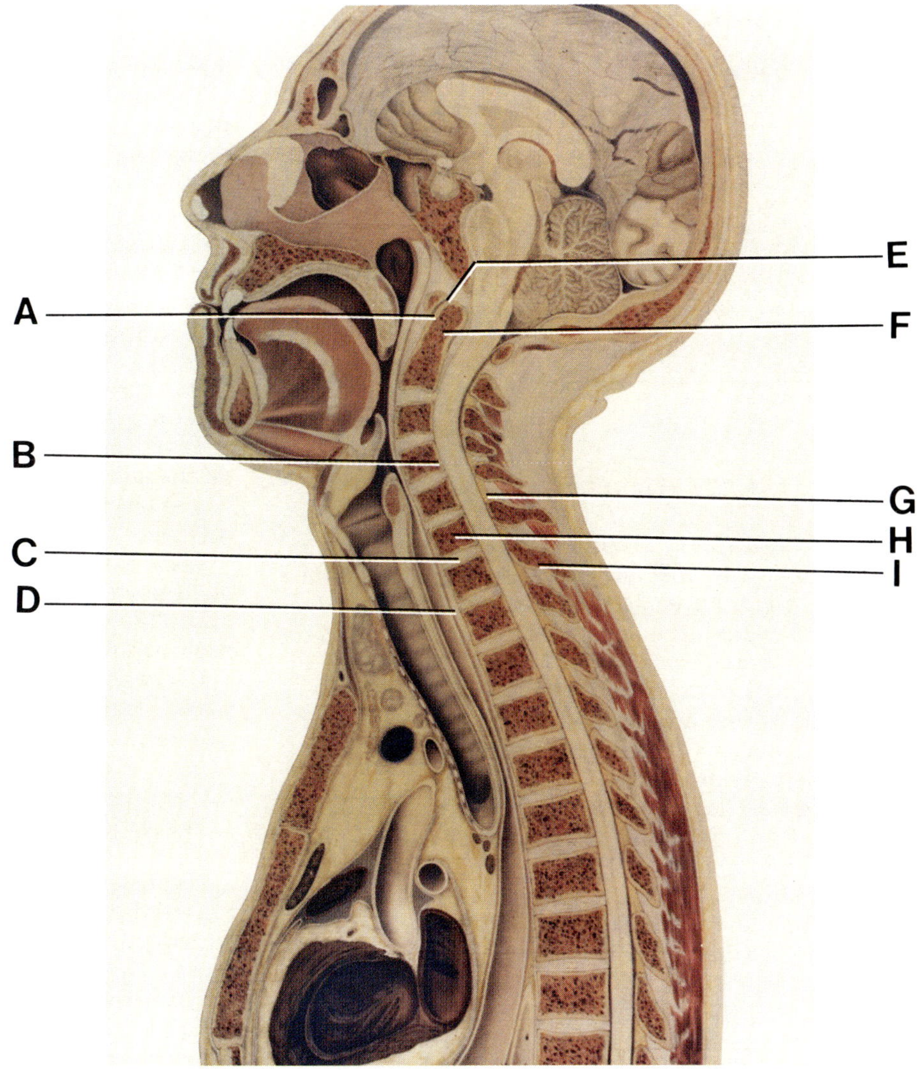

A. Fovea dentis
B. Posterior longitudinal ligament
C. Intervertebral disk
D. Anterior longitudinal ligament
E. Median atlantoaxial joint
F. Odontoid process (dens)
G. Ligamenta flava
H. Body of cervical vertebra
I. Spinous process of cervical vertebra

Figure 8–1
Midsagittal section of the neck highlighting musculoskeletal unit structures.

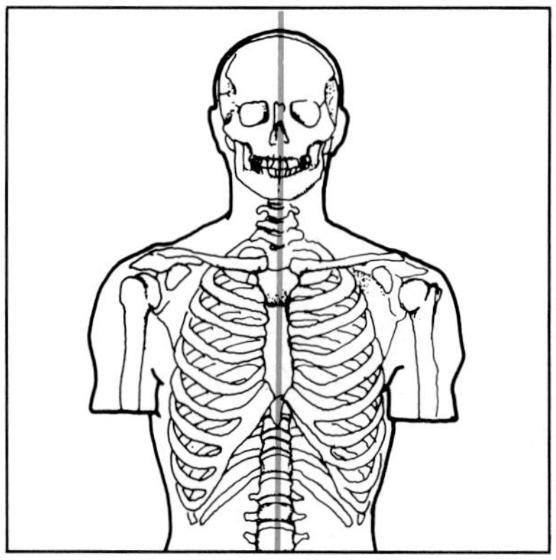

VISCERAL UNIT (Fig. 8–2)

Spinal Cord (*I*). A vertical slice of curved nerve tissue attached to the inferior part of the medulla oblongata that follows a course within a spinal canal. In the neck region, the spinal cord shows a concave surface on the posterior part.

Epiglottis (*H*). A vertical, oblique structure inferior to the oropharynx and anterior to the upper part of the laryngopharynx. The superior part is more posterior than its inferior part.

Thyroid Cartilage (*D*). An oblique strip of cartilaginous tissue anterior to the vocal cords and inferior to the hyothyroid membrane. It is superior to the cricothyroid ligament. This structure is known as the Adam's apple in men and Eve's apple in women.

Cricoid Cartilage (*K*). This is seen in this sagittal section as two portions of cartilaginous tissues. The anterior part of this structure is immediately between the cricothyroid ligament and the anterior part of the trachea. The posterior part is rather elongated (superior-to-inferior) and immediately superior to the posterior part of the trachea.

Trachea (*M*). A vertical, elongated, hollow, cartilaginous ring structure anterior to the esophagus and inferior to the cricoid cartilage of the larynx. The inferior end of this structure is slightly more posterior than the superior end.

Laryngopharynx (*J*). A vertical, hollow, canal-like structure posterior to the larynx and superior to the esophagus to which it attaches. The laryngopharynx is between the oropharynx and the esophagus.

Esophagus (*L*). A vertical, hollow tube posterior to the trachea, which continues into the thorax to reach the abdomen after passing through the diaphragm. The cervical portion of the esophagus lies posteriorly as it descends.

Thyroid Gland (*G*). An aggregate of glandular tissue anterior to the trachea and inferior to the anterior portion of the cricoid cartilage.

True Vocal Cord (*E*). An anterior-to-posterior fold on the lateral wall of the larynx, immediately inferior to the laryngeal ventricle and superior to the cricoid cartilage.

False Vocal Cord (Vestibular Fold) (*A*). An anterior-to-posterior, rather small fold immediately superior to the laryngeal ventricle and posterior to the protuberance of the thyroid cartilage.

Laryngeal Ventricle (*C*). An anterior-to-posterior slitlike feature, as seen in this sagittal section, between the true and false vocal cords.

ENCLOSING UNIT (see Fig. 8–2)

Hyothyroid Membrane (*B*). A vertical, oblique, rather thin strip of connective tissue between the superior part of the thyroid cartilage and the hyoid bone. The inferior end of this structure is more anterior than the superior end.

Cricothyroid Ligament (*F*). A vertical, oblique, rather thin connective tissue strip between the inferior part of the thyroid cartilage and the anterior part of the cricoid cartilage. The inferior end of this structure is more posterior than its superior end.

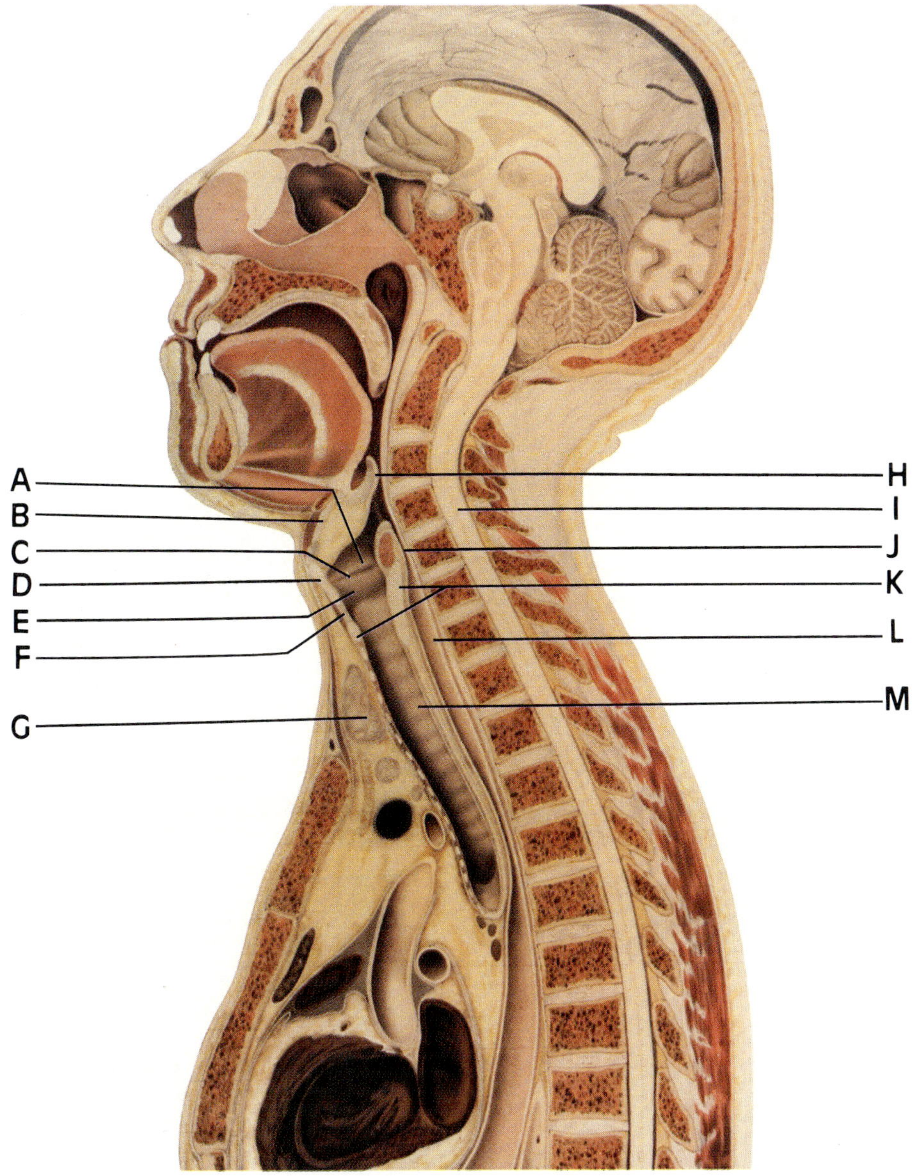

A. False vocal cord (vestibular fold)
B. Hyothyroid membrane
C. Laryngeal ventricle
D. Thyroid cartilage
E. True vocal cord
F. Cricothyroid ligament
G. Thyroid gland
H. Epiglottis
I. Spinal cord
J. Laryngopharynx
K. Cricoid cartilage
L. Esophagus
M. Trachea

Figure 8–2
Midsagittal section of the neck highlighting visceral and enclosing unit structures.

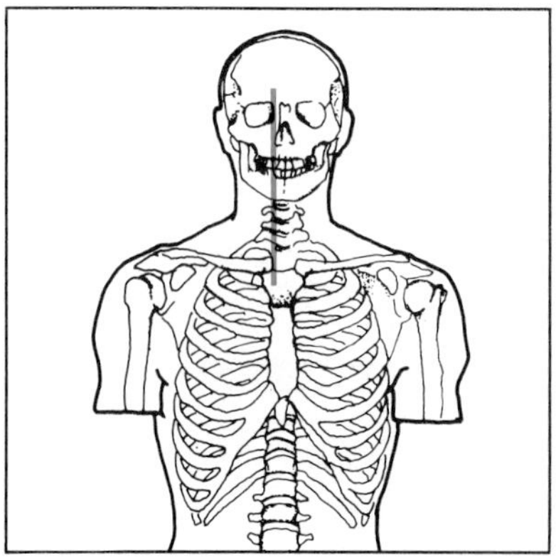

This sagittal section is an MRI picture (Fig. 8–3) that shows musculoskeletal, neurovascular, visceral, and enclosing unit structures.

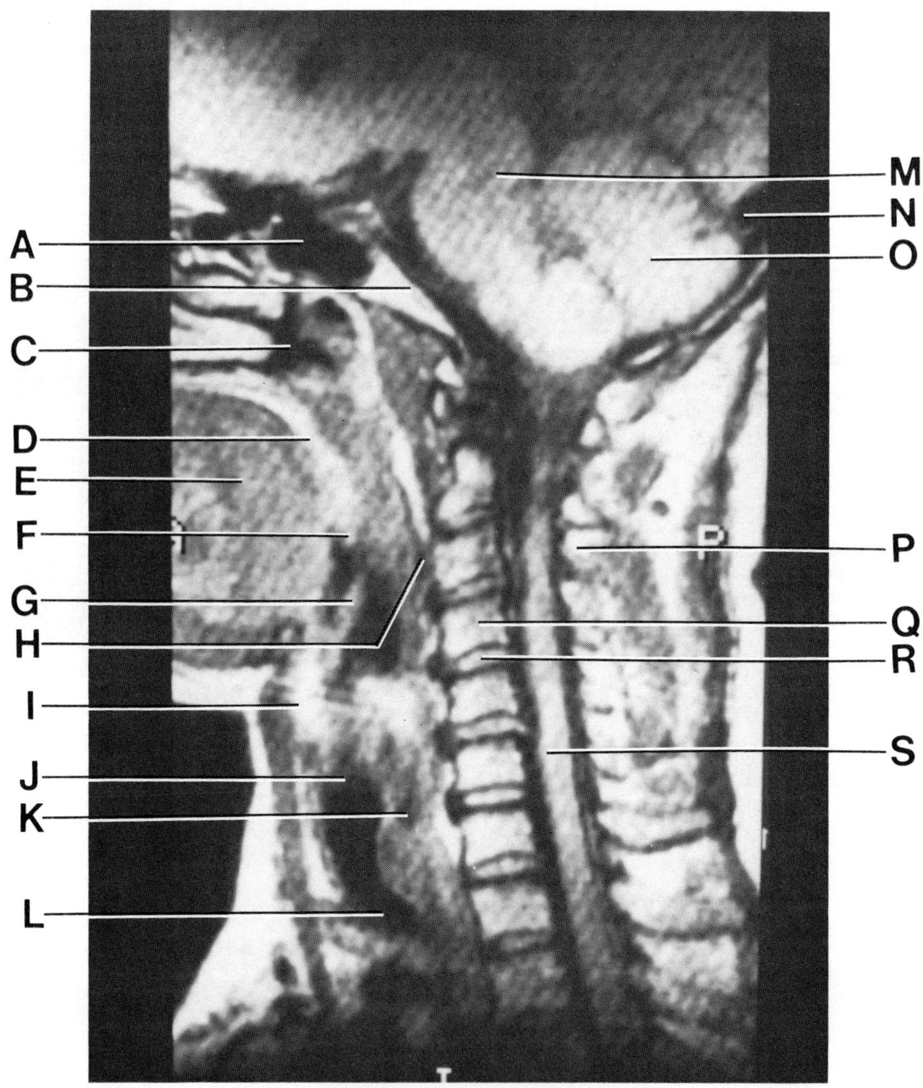

A. Sphenoid sinus
B. Clivus
C. Nasopharynx
D. Soft palate
E. Tongue
F. Uvula
G. Epiglottis
H. Oropharynx
I. Thyroid cartilage
J. Larynx
K. Esophagus
L. Trachea
M. Pons
N. Transverse sinus
O. Cerebellum
P. Spinous process of cervical vertebra
Q. Body of cervical vertebra
R. Intervertebral disk
S. Spinal cord

Figure 8–3
MRI sagittal image of neck, highlighting musculoskeletal, neurovascular, visceral, and enclosing unit structures.

Figure 8–4 is an MRI scan. This sagittal slice highlights musculoskeletal, neurovascular, visceral, and enclosing unit structures.

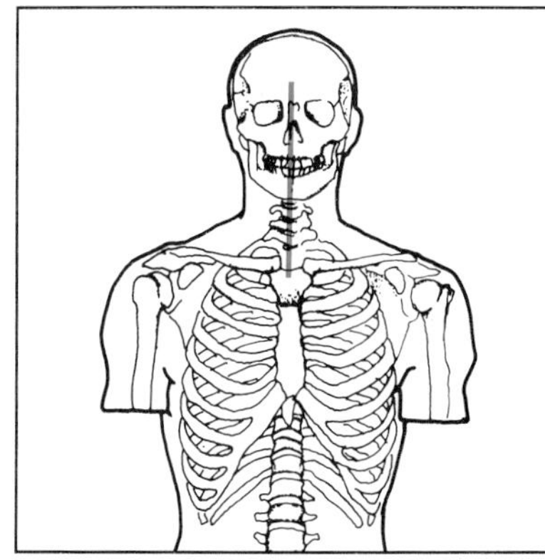

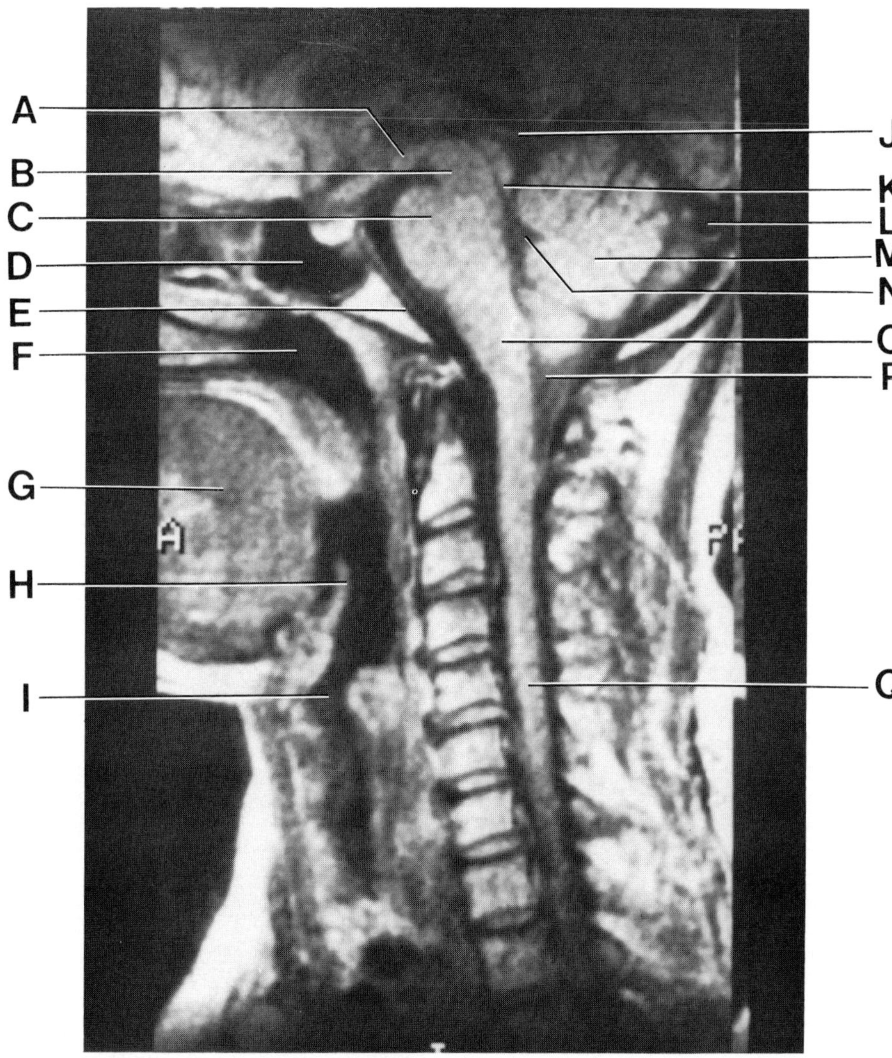

A. Mamillary body
B. Midbrain
C. Pons
D. Sphenoid sinus
E. Clivus
F. Nasopharynx
G. Tongue
H. Epiglottis
I. Larynx
J. Pineal gland
K. Aqueduct of Sylvius
L. Transverse sinus
M. Cerebellum
N. Fourth ventricle
O. Medulla oblongata
P. Cisterna magna
Q. Spinal cord

Figure 8–4
MRI sagittal image of neck, highlighting musculoskeletal, neurovascular, visceral, and enclosing unit structures.

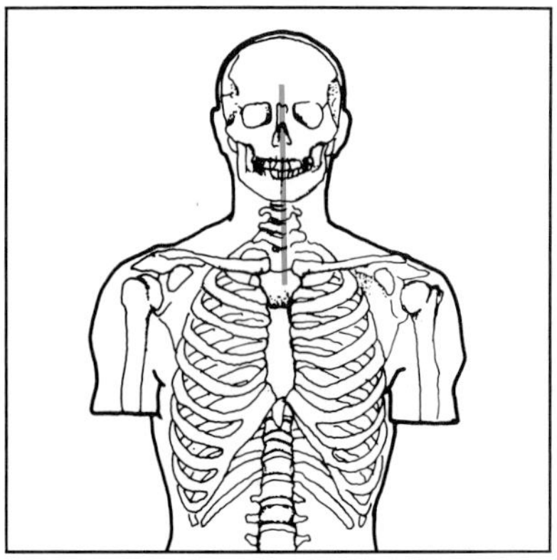

Figure 8–5 is another MRI slice showing structures in the sagittal plane. Twenty-one structures can be seen clearly, including head and neck structures.

A. Nasopharynx
B. Hard palate
C. Soft palate
D. Uvula
E. Oropharynx
F. Tongue
G. Epiglottis
H. Larynx
I. Thyroid cartilage
J. Cricoid cartilage
K. Trachea
L. Pons
M. Fourth ventricle
N. Clivus
O. Cerebellum
P. Cisterna magna
Q. Medulla oblongata
R. Body of cervical vertebra
S. Spinal cord
T. Intervertebral disk
U. Esophagus

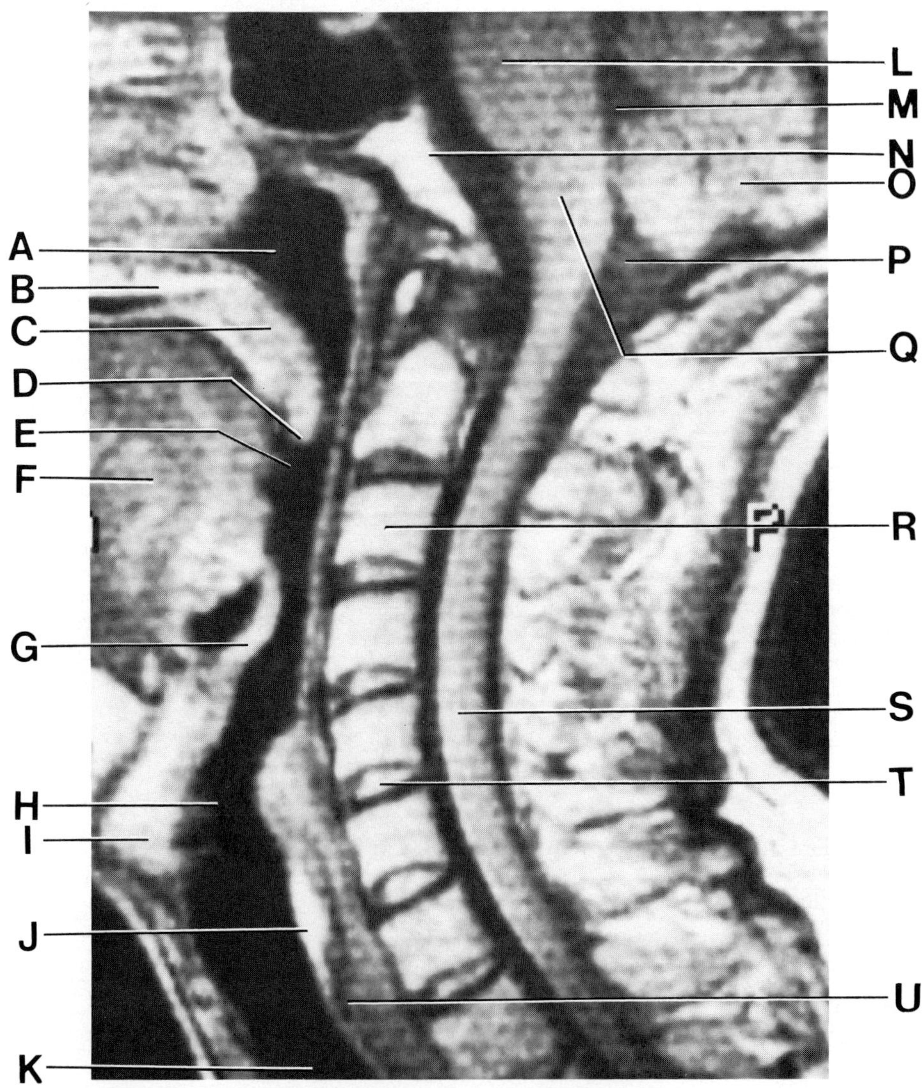

Figure 8–5
MRI sagittal image of neck, highlighting musculoskeletal, visceral, and enclosing unit structures.

Appendix to Part III: Clinical Correlations for the Neck Region

CASE ONE (Fig. AIII–1)

History. The patient has had a palpable mass in the region of the right parotid gland for the past 2 months.

Findings. In the T1-weighted images, a medium signal intensity lesion approximately 1.5 cm in diameter, with a somewhat ill-defined posterior border, is seen approximately 4 cm below the level of the right external auditory canal. Its signal intensity is different from the normal signal intensity of the adjacent right parotid gland. T2-weighted images show increased signal intensity within this mass. Inferior to the level of this mass is a suggestion of slightly enlarged lymph nodes in the anterior portion of the neck.

No other definite mass or lymphadenopathy is visualized.

The aforementioned signal characteristic of the right parotid mass suggests a solid rather than cystic structure, and therefore malignancy definitely should be considered.

Impression. A 1.5 cm mass in the inferior aspect of the right parotid gland. Its margins are ill defined posteriorly. Its signal characteristic suggests a solid mass. There is a suggestion that adjacent lymph nodes are slightly enlarged.

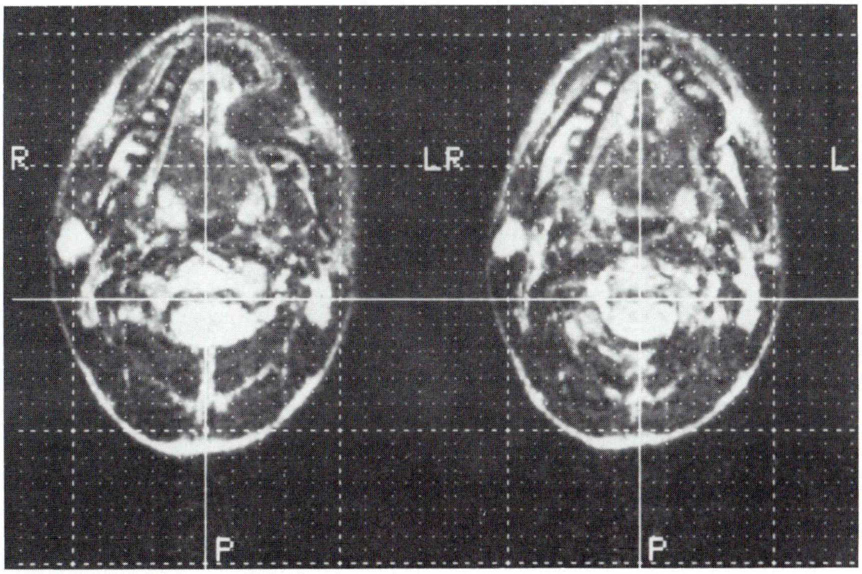

Figure AIII–1
MRI transverse image, highlighting a mass in the inferior aspect of the right parotid gland.

CASE TWO (Fig. AIII-2)

History. The patient reported numbness of the third and fourth fingers of the right hand and pain in the right forearm for nearly 1 year.

Findings. A partially extruded C3–C4 disk is seen that results in an indentation on the central portion of the cervical spinal cord.

Mild bulging of the C5–C6 disk is also seen centrally encroaching on the anterior aspect of the cervical spinal cord. In the axial images only, there is also a suggestion of a slight right lateral bulge of the C4–C5 disk. This last finding is equivocal.

No discrete evidence of cervical spinal stenosis or abnormalities involving the cerebellar tonsil or cervical spinal cord itself are seen in this study.

Impression. Partially extruded C3–C4 disk. Moderately centrally bulging C5–C6 disk. Possible right lateral bulge of the C4–C5 disk as described.

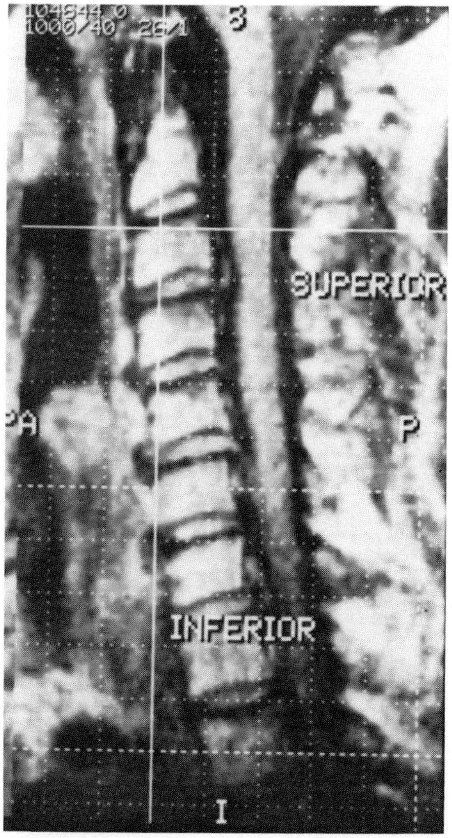

Figure AIII–2
Magnetic resonance sagittal image, highlighting a partially extruded C3–C4 disk.

PART

IV

THE THORAX

INTRODUCTION

There are 12 thoracic vertebrae and 12 pairs of ribs. Five per cent of the population has cervical ribs, which may appear as a bilateral and symmetric pair, as a bilateral and asymmetric pair, or as a unilateral cervical rib. There is one sternum divided into three segments. From superior to inferior, these are the manubrium, the body, and the xiphoid process.

The thorax has six bilateral intrinsic muscles: (1) external intercostal (11 pairs), (2) internal intercostal (11 pairs), (3) innermost intercostal (11 pairs or may be variable), (4) subcostal (4 pairs), (5) transversus thoracis (5 pairs), and (6) levator (levatores costarum). For more information, see the muscle tables of the thorax at the end of this section.

A typical pattern of the arterial blood supply to the thorax is shown in Figures IV–1 and IV–2. The first intercostal artery on both sides arises from the costocervical trunk. The costal cervical trunk on both sides divides at the neck of the first rib to form two posterior intercostal arteries to supply blood to the first two intercostal spaces. All other posterior intercostal segments arise from the descending thoracic

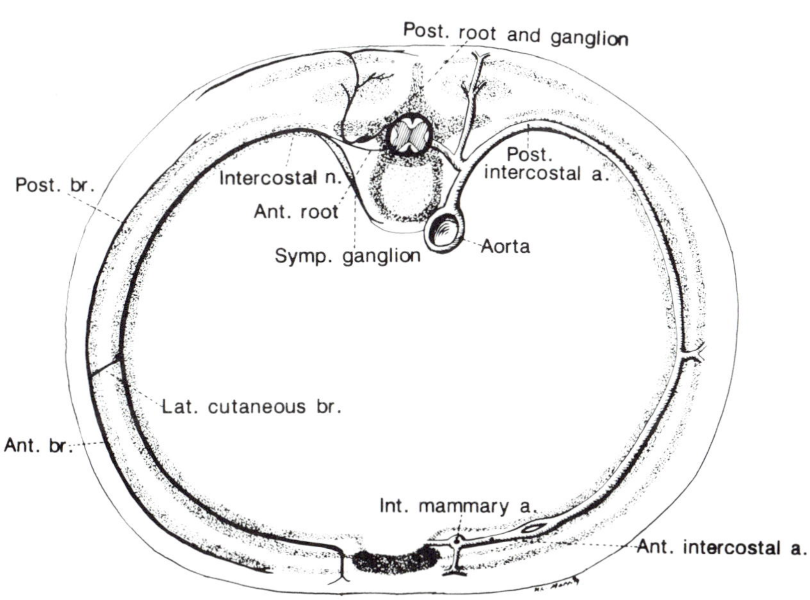

Figure IV–1
Arterial supply of the thorax seen in transverse section.

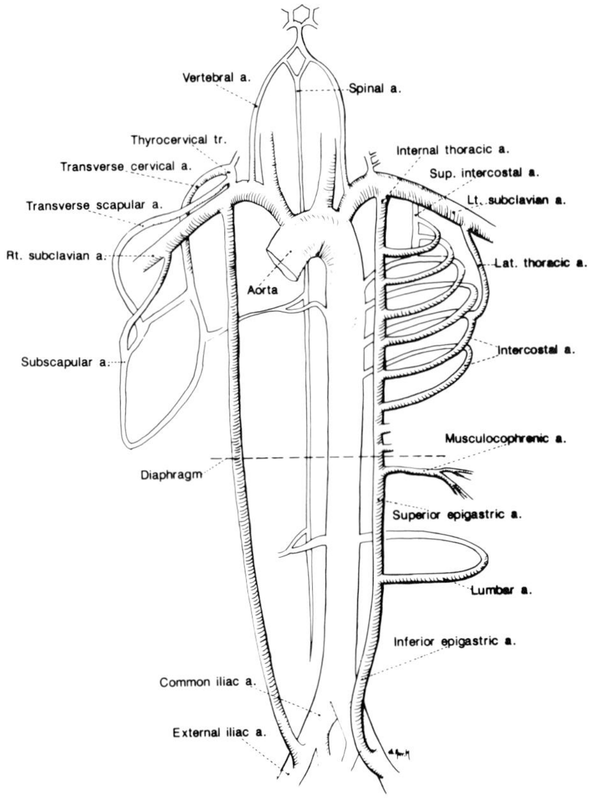

Figure IV-2
Arterial supply of the thorax.

aorta, as is clearly shown in Figure IV-2. Each anterior segment of the intercostal arteries arises from the internal thoracic artery and anastomoses with a corresponding posterior segment to complete one intercostal artery. There are 11 pairs of intercostal arteries, each formed by anterior and posterior segments.

The venous drainage (Fig. IV-3) is accomplished by the azygos system. The typical pattern of this system on the left side is the intercostal vein connected to both the azygos vein posteriorly and the internal thoracic vein anteriorly. The back part of the intercostal vein is referred to as the posterior intercostal segment, and the front half is known as the anterior intercostal part. Each vein anastomoses to form one intercostal vein. These typical intercostal veins drain intercostal spaces 5 through 11.

The first intercostal spaces on both the right and left sides drain by their respective supreme intercostal vein. Both these veins terminate in the respective brachiocephalic veins. Intercostal spaces 2, 3, and 4 on both sides are drained by the tributaries of the superior intercostal veins. Blood from the superior intercostal vein on the left side flows into the brachiocephalic vein. On the left side intercostal veins 5, 6, 7, and 8 are drained into the accessory hemiazygos vein. The remainder of the intercostal veins on the left side drain into the hemiazygos vein. The intercostal veins on the right side of intercostal spaces 5 and below drain into the azygos vein. The blood from both the accessory hemiazygos and the hemiazygos drains into the azygos by an arching venous structure. The blood in the azygos vein then drains into the superior vena cava.

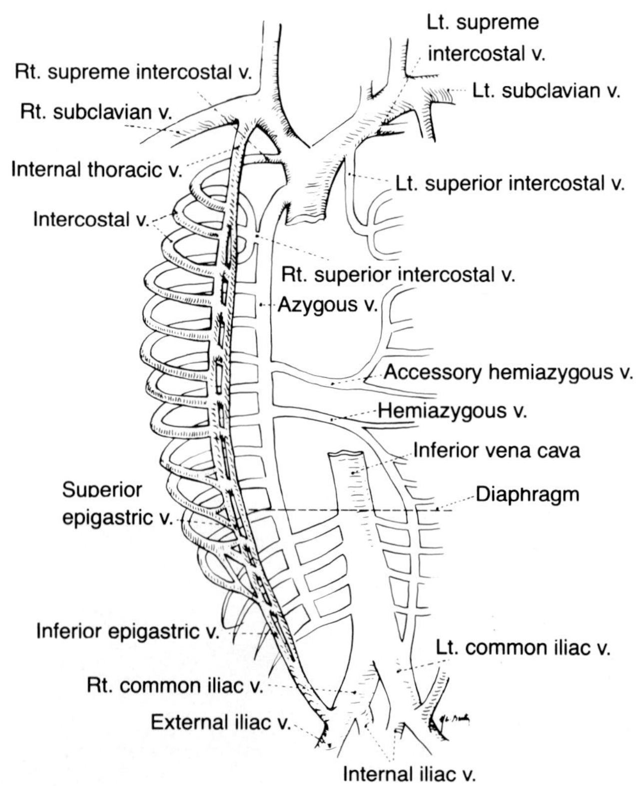

Figure IV-3
Venous drainage of the thorax.
Rt. supreme intercostal v.
Rt. subclavian v.
Internal thoracic v.
Intercostal v.
Superior epigastric v.
Inferior epigastric v.
Rt. common iliac v.
External iliac v.
Lt. supreme intercostal v.
Lt. subclavian v.
Lt. superior intercostal v.
Rt. superior intercostal v.
Azygos v.
Accessory hemiazygos v.
Hemiazygos v.
Inferior vena cava
Diaphragm
Lt. common iliac v.
Internal iliac v.

There are 12 pairs of thoracic nerves; the first eleven are also known as the intercostal nerves. Typically, each intercostal nerve supplies the muscles and other structures between the ribs, as well as some cutaneous branches of the thorax. Each nerve consists of two main branches: the posterior primary ramus and the anterior primary ramus. The 12th thoracic nerve is known as the subcostal nerve and frequently takes part in formation of the lumbar plexus.

This Part of *Modern Sectional Anatomy* includes Chapters 9, 10, and 11. Chapter 9 presents transverse (axial) sections of the thorax in four anatomic categories. Chapter 10 includes coronal (frontal) sections, and Chapter 11 discusses the sagittal sections. Two pathologic clinical images showing lung cancer, one from magnetic resonance imaging (MRI) and the other from computed tomography (CT) are presented in the Appendix to this Part.

Table IV-1 Anterior Muscles Associated with Thorax That Function Elsewhere

Muscle	Origin	Insertion	Innervation	Vasculature	Action
Subclavius	Costochondral joint area of first rib	Groove on inferior surface of clavicle	Nerve to subclavius muscle	Clavicular branch of thoracoacromial artery	Scapular depression and protraction
Greater pectoral (pectoralis major)	Medial third of clavicle (clavicular origin), sternum, and first 7 costal cartilages (sternal origin)	Side lip of intertubercular sulcus of humerus	Lateral and medial pectoral nerves	Pectoral branch of thoracoacromial artery	Clavicular origin: Flexion, adduction, and medial rotation of arm. Sternal origin: Adduction, extension, and medial rotation of arm
Smaller pectoral (pectoralis minor)	Superficial surface of superior margin of ribs 3–5	Coracoid process of scapula	Medial pectoral nerve	Intercostal branches of internal thoracic, lateral thoracic, and thoracoacromial arteries	Depresses and protracts scapula; rib elevation if shoulder is fixed
Serratus anterior	Superficial surface of ribs 1–9	Vertebral border of scapula on deep surface	Long thoracic nerve	Lateral thoracic artery	Scapular elevation when shoulder is fixed and scapular abduction

Table IV-2 Posterior Muscles Associated with Thorax That Function Elsewhere

Muscle	Origin	Insertion	Innervation	Vasculature	Action
Levator m. of scapula (levator scapulae)	Transverse processes, cervical vertebrae 1–4	Superior angle of scapula	Dorsal scapular and posterior primary rami of cervical nerves 3–4	Transverse cervical artery	Scapular elevation; bends neck to sides if scapula is fixed
Rhomboid major	Spinous processes of thoracic vertebrae 2–5	Inferior part (below spine) of vertebral border of scapula	Dorsal scapular nerves	Transverse cervical artery	Adduction and rotation of scapula
Rhomboid minor	Inferior part of ligamentum nuchae and spinous processes of 7th cervical to 1st thoracic vertebrae	Medial portion of spine of scapula	Dorsal scapular nerves	Transverse cervical artery	Adduction and rotation of scapula

Table IV-3 Intrinsic Muscles of Thorax

Muscle	Origin	Insertion	Innervation	Vasculature	Action
External intercostal	Inferior edge of rib except 12th rib	Superior edge of inferior rib	Intercostal (thoracic) nerves (anterior primary branch)	Intercostal arteries and their branches. Costocervical trunk and internal thoracic artery's intercostal branches	Elevation of rib cage
Internal intercostal	Inferior edge of costal cartilage and body of rib	Superior edge of adjacent inferior rib	Intercostal (thoracic) nerves (anterior primary branch)	Intercostal arteries and their branches. Internal thoracic artery's intercostal branches. Musculophrenic and costocervical trunk	Elevation of rib cage. Prevents evagination and invagination of intercostal space during inspiration and expiration
Transverse m. of thorax (transversus thoracis)	Posterior surface of xiphoid process and inferior half of body of sternum	Anterior medial costal cartilages 2–6	Intercostal (thoracic) nerves T1–T6 (anterior primary branch)	Intercostal arteries and sternal branches of internal thoracic arteries	Muscle of expiration and depresses costal cartilage
Levator m. of ribs (levatores costarum)	Transverse processes of T1–T12	Between tubercle and angle of adjacent inferior rib	Intercostal (thoracic) nerves (anterior primary branch)	Intercostal arteries—main back branches	Aids in elevation of ribs in respiration

Table IV-3 Intrinsic Muscles of Thorax (*Continued*)

Muscle	Origin	Insertion	Innervation	Vasculature	Action
Serratus posterior superior	Spinous processes of inferior 3 cervical vertebrae and spinous processes of superior 4 thoracic vertebrae	Superior edge of ribs 2–5	Anterior primary rami of intercostal nerves 1–3	Supreme intercostal and posterior intercostal arteries	Ribs elevation, to assist in inspiration
Serratus posterior inferior	Spinous processes of inferior 2 thoracic vertebrae and spinous processes of superior 2 lumbar vertebrae	Inferior edge of ribs 9–12	Anterior primary rami of intercostal nerves 10–12	Posterior intercostal arteries	Ribs depression, to assist in expiration
Subcostal	These 4 pairs of muscles arise separately from deep surfaces of ribs near their angles	Superior edge of rib inferior to rib of origin	Anterior primary rami of intercostal nerves 9–12	Posterior intercostal arteries and musculophrenic artery	Ribs depression
Diaphragm	Anterior: posterior surface of xiphoid process. Anterolateral: 7–10 costal cartilages and 11–12 ribs. Lateral: 12th rib. Posterior: transverse processes of 12th thoracic vertebra, 1st lumbar vertebra, and crura of diaphragm attach to bodies of lumbar vertebrae 1–2	Central tendon	Phrenic	Branches of musculophrenic, superior and inferior phrenic, and pericardiacophrenic arteries	Respiration; contractions increase vertical area of thorax

CHAPTER 9

Transverse (Axial) Cadaver Sections and Images of the Thorax

This chapter contains seven modules. Each module represents a different level of the thorax. The first, second, third, fourth, fifth, sixth, and seventh modules present studies of sections at the levels of thoracic vertebrae 1, 2, 4, 5, 6, 7, and 8, respectively. The first and fourth modules include sheet plastinated human sections. These plastinated specimens are 2.5 to 4 mm thick. All other sections are commonly known as thick sections (approximately 20 mm). Each transverse section is shown from its inferior surface, regardless of its thickness.

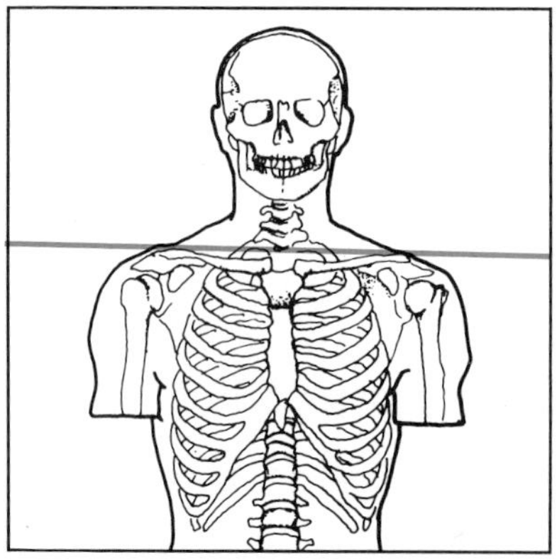

MODULE 1

This is a sheet plastinated (2.5 mm thick), transverse (axial) section of the thorax at the level of thoracic vertebra 1.

MUSCULOSKELETAL UNIT (Fig. 9–1)

I. Skeletal Portion

Body of the Thoracic Vertebra (*K*). A rectangular bone tissue with irregular lateral borders immediately medial to the head of the left and right first ribs.

Lamina of the Thoracic Vertebra (*G*). An oblique, bilateral bone section immediately posterior and lateral to the spinal cord and anterior to the multifidus muscle. The posterior ends are placed more medial than the anterior ends and join at the spinous process.

Spinous Process of the Thoracic Vertebra (*I*). A median posterior, rather long bone slice immediately posterior to the spinal cord and medial to the erector spinae muscles.

Costovertebral Capitular Joint (*B*). A bilateral gap between the head of the rib and the body of the vertebra.

Head of the Rib. Medial, rather oval, bilateral rib end lateral to the body of the vertebra.

Neck of the Rib (*E*). A constricted, bilateral portion of the rib immediately lateral to the rib head.

Body (Shaft) of the Rib (*F*). A bilateral bone slice immediately lateral to the neck of the rib.

Clavicle (*C*). A small, bilateral slice of the clavicle immediately medial to the anterior part of the deltoid muscle and anterior to the subscapularis muscles. The left slice is a little longer than the right slice in this transverse section (thoracic vertebra 1).

Vertebral Border of the Scapula. A sickle-shaped, bilateral bone section lateral to the rhomboid minor muscle and medial to the supraspinatus muscle. This bone slice is immediately posterior to the subscapularis muscle.

Spine of the Scapula (*T*). A transversely elongated, bilateral bone slice immediately posterior to the supraspinatus muscle and lateral to the trapezius muscle.

Head of the Humerus (*Q*). A rounded, bilateral bone section seen on the left side of the picture. This feature is immediately lateral to the supraspinatus muscle and medial to the deltoid muscle.

Glenohumeral Joint (*D*). An oval, bilateral space seen on the right side of this picture. The lateral wall of this joint is an articular capsule. The inner lining of this wall (adjacent to the space) shows a synovial membrane.

II. Muscular Portion

Deltoid Muscle (*R*). A curved, bilateral muscle section lateral, anterior, and posterior to the features of the humerus seen in this section. The medial portion shows the concave surface.

Trapezius Muscle (*V*). A transversely elongated, sickle-shaped, bilateral muscle slice immediately medial to the spine of the scapula and lateral to the spinous process of the thoracic vertebra. The medial end is immediately posterior to the rhomboid muscle.

Supraspinatus Muscle (*S*). A rather wide, bilateral muscle slice immediately lateral to the features of the humerus seen in this section. This muscle is anterior to the spine of the scapula and medial to the vertebral border. The medial end is wider and more irregular than the lateral end.

Subscapularis Muscle (*P*). In this transverse section, the subscapularis muscle is seen as a small, rather oblong, bilateral muscle slice anterior to the vertebral border of the scapula.

Rhomboid Minor Muscle (*U*). A bilateral muscle slice immediately anterior to the medial part of the trapezius and posterior to the erector spinae muscles. The lateral end is larger than the medial end, making this muscle slice appear cone-shaped.

Transversospinalis Muscle (*H, J*). An oval, rather large, bilateral aggregate muscle slice lateral to the lamina and spinous process of the thoracic vertebra in this section. This muscle system includes the semispinalis thoracis, semispinalis cervicis, multifidus, and rotator muscles. The semispinalis muscle of the thorax originates from the transverse processes of thoracic vertebrae 7, 8, 9, 10, 11, and 12 and terminates into the spinous processes of thoracic vertebrae 1, 2, 3, and 4, as well as cervical vertebrae 6 and 7. The semispinal muscle of the neck originates from the transverse process of thoracic vertebrae 1, 2, 3, 4, 5, and 6 and the articulate processes of cervical vertebrae 4, 5, 6, and 7. This muscle inserts into the spinous processes of cervical vertebrae 2, 3, 4, and 5. The multifidus muscle originates continuously from (1) the sacrum deep to the tendon of the sacrospinalis muscle; (2) the posterior sacroiliac ligaments; (3) the mamillary processes of the lumbar vertebrae; (4) the transverse processes of the thoracic vertebrae; and (5) the articular processes of cervical vertebrae 4, 5, 6, and 7. This muscle terminates successively into the spinous processes as far up as the axis (second cervical vertebra).

There are 11 pairs of rotator muscles. These are small muscles located in the vertebral groove of the

A. Sternohyoid muscle
B. Costovertebral capitular joint
C. Clavicle
D. Glenohumeral joint
E. Neck of rib
F. Body (shaft) of rib
G. Lamina of thoracic vertebra
H. Right transversospinalis muscle
I. Spinous process of thoracic vertebra
J. Left transversospinalis muscle
K. Body of thoracic vertebra
L. Sternothyroid muscle
M. Longus colli muscle
N. Scalenus anterior muscle
O. Serratus anterior muscle
P. Subscapularis muscle
Q. Head of humerus
R. Deltoid muscle
S. Supraspinatus muscle
T. Spine of scapula
U. Rhomboid minor muscle
V. Trapezius muscle
W. Serratus posterior superior muscle

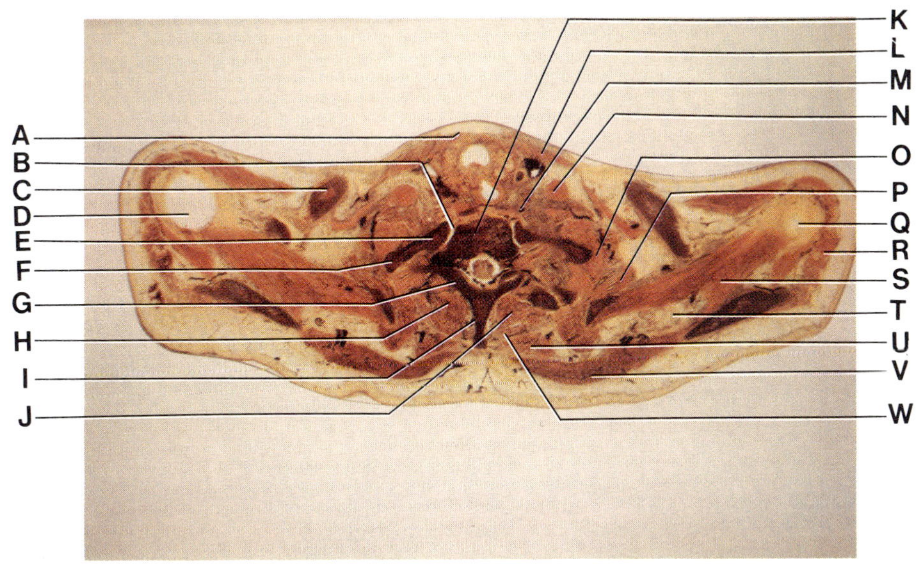

Figure 9–1
Transverse (axial) section of the thorax at the level of thoracic vertebra 1, highlighting musculoskeletal unit structures.

thoracic region under the multifidus muscle. Rotator muscles are divided into a longus and a brevis. Each rotator longus originates successively from the root of the transverse process to the junction of the transverse process and terminates into the lamina of the second vertebra above. For example, the one that arises from the transverse process of thoracic vertebra 12 terminates into the lamina of thoracic vertebra 10. Each rotator brevis muscle arises successively from the root of the transverse process to the junction of the transverse process and terminates into the lamina of the vertebra immediately above. For example, the one that arises from the transverse process of thoracic vertebra 12 terminates into the lamina of thoracic vertebra 11.

Serratus Posterior Superior Muscle (*W*). A rather thin, transversely placed, bilateral muscle deep to the rhomboid minor muscle and superficial to the transversospinalis muscle system. This thin, membranous muscle originates continuously from the nuchal ligament/spines of cervical vertebra 7 and thoracic vertebrae 1, 2, and sometimes 3. This muscle inserts into the superior edge of ribs 2, 3, 4, and 5 (immediately lateral to the angles). The sources of innervation are the anterior primary rami of thoracic nerves 1, 2, and 3. The highest intercostal muscle and the posterior intercostal muscles are the sources for the blood supply. The movement capability is to raise the ribs during inspiration.

Scalenus Anterior Muscle (*N*). A rounded, bilateral muscle slice posterior and lateral to the internal jugular vein. It is posterior and medial to the lateral part of the sternothyroid muscle.

Sternothyroid Muscle (*L*). A transverse, oblique, bilateral muscle slice anterior to the omohyoid muscle and lateral to the trachea. The lateral end of this muscle is slightly posterior to its medial end.

Sternohyoid Muscle (*A*). A rather small, transverse, bilateral muscle anterior to the thyroid gland and medial to the sternothyroid muscle.

Serratus Anterior Muscle (*O*). A bilateral muscle aggregate adjacent to the lateral part of the rib, as seen in this section, posterior to the anterior scalene and anterior to the subscapularis muscle.

Longus Colli Muscle (*M*). An irregularly shaped, bilateral muscle slice anterior to the lateral part of the thoracic vertebra and head of rib in this section.

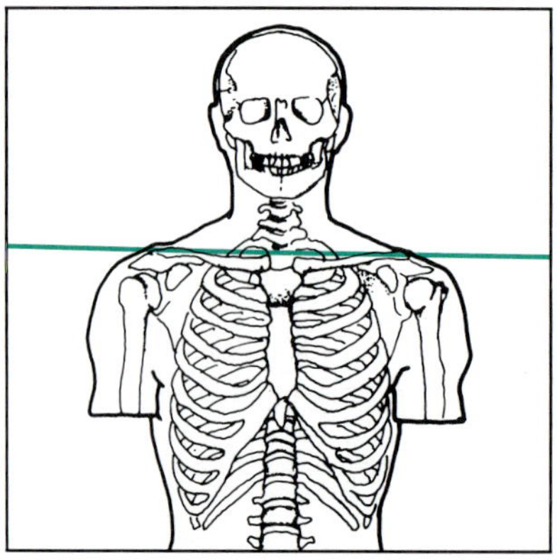

NEUROVASCULAR UNIT (Fig. 9–2)

I. Arteries

Common Carotid Artery (*A*). An oval, bilateral, hollow vascular structure lateral to the thyroid gland and medial to the internal jugular vein. The right common carotid artery arises from the brachiocephalic artery, and the left common carotid artery originates from the arch of the aorta. Both end in the carotid triangle of the neck by forming internal and external carotid arteries.

Vertebral Artery (*C*). A rather small, rounded, hollow, bilateral vascular structure anterior and lateral to the head of the rib in this section. It is lateral to the longus colli muscle and posterior to the inferior thyroid artery.

Inferior Thyroid Artery (*B*). A small, rounded, bilateral, hollow vascular structure anterior and lateral to the vertebral artery. It is located lateral to the esophagus and medial to the anterior scalene muscle.

Origin. Thyrocervical trunk.

Distribution. Infrahyoid, longus colli, and scalenus anterior muscles, inferior constrictor of the throat, thyroid gland, and esophagus.

Course. Ascends superiorly anterior to the medial border of the anterior scalene muscle. It courses medial and anterior to the vertebral arteries and posterior to the carotid sheath to reach its distribution.

II. Veins

Right Internal Jugular Vein (*D*). A rounded, hollow, bilateral vessel lateral to the common carotid artery, posterior to the sternothyroid muscle, and lateral to the thyroid gland.

III. Nerves

Dorsal Root of a Spinal Nerve (*G*). A thin strip of nerve tissue that extends from the posterior part of the spinal cord to join a common nerve structure between the body and transverse process of the vertebra.

Ventral Root of a Spinal Nerve (*F*). A thin, bilateral nerve strip that extends from the anterior part of the spinal cord to join a common nerve structure between the body and transverse process of the vertebra.

Spinal Nerve (*E*). A transversely elongated, curved, bilateral nerve structure lateral to the vertebra and posterior to the rib. The concave surface of this nerve is anterior.

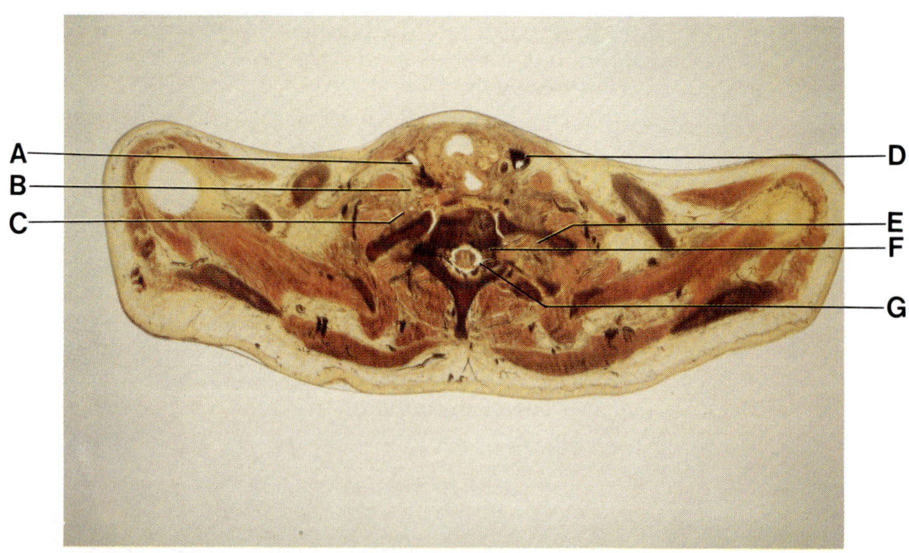

A. Right common carotid artery
B. Inferior thyroid artery
C. Vertebral artery
D. Left internal jugular vein
E. Spinal nerve
F. Ventral root of spinal nerve
G. Dorsal root of spinal nerve

Figure 9–2
Transverse (axial) section of the thorax at the level of thoracic vertebra 1, highlighting neurovascular unit structures.

VISCERAL UNIT (Fig. 9–3)

Right Lobe of the Thyroid Gland (*B*). An oval, rather large, curved aggregate of glandular tissue immediately lateral to the trachea and anterior part of the esophagus. The concave part of this lobe is on the medial surface.

Left Lobe of the Thyroid Gland (*E*). This is similar to the right lobe but smaller in this section. It is located on the side of the trachea and the anterior part of the left side of the esophagus.

Isthmus of the Thyroid Gland (*D*). A rather thin strip of glandular tissue anterior to the trachea, extending from the right lobe to the left lobe.

Trachea (*A*). A C-shaped ring of cartilaginous tissue that forms the anterior and lateral walls of this hollow structure. On the posterior part is soft tissue, which includes some skeletal muscles and connective tissue as well as epithelial tissue.

Esophagus (*F*). A rather oval, thick-walled hollow structure posterior and slightly to the left of the trachea. This organ is immediately anterior to the retropharyngeal cleft and medial to the common carotid arteries.

Spinal Cord (*C*). An oval aggregate of nerve tissue located within the spinal foramen. This structure extends from the medulla oblongata within the inferior part of the skull to the level between lumbar vertebra 1 and lumbar vertebra 2. This organ is a place for attachment of 31 pairs of spinal nerves.

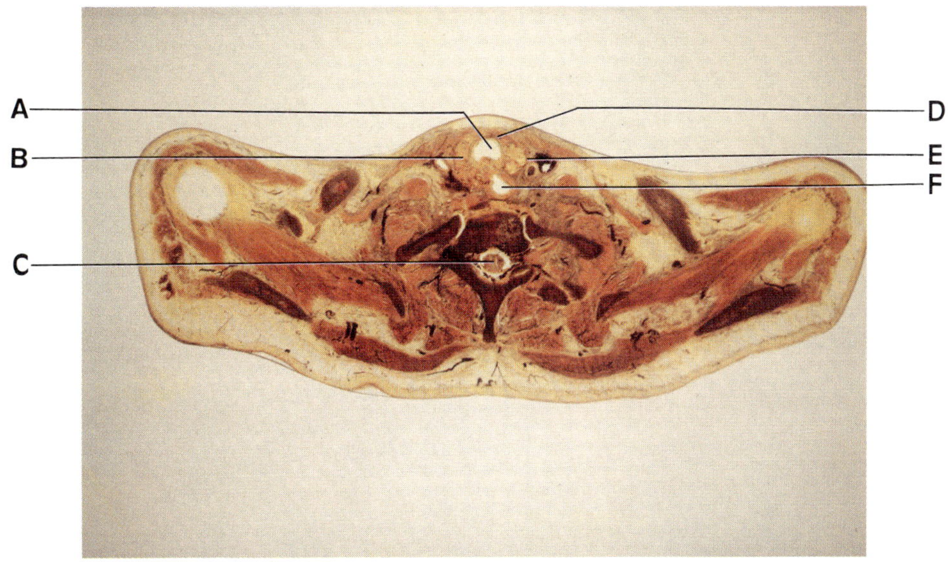

A. Trachea
B. Right lobe of thyroid gland
C. Spinal cord
D. Isthmus of thyroid gland
E. Left lobe of thyroid gland
F. Esophagus

Figure 9–3
Transverse (axial) section of the thorax at the level of thoracic vertebra 1, highlighting visceral unit structures.

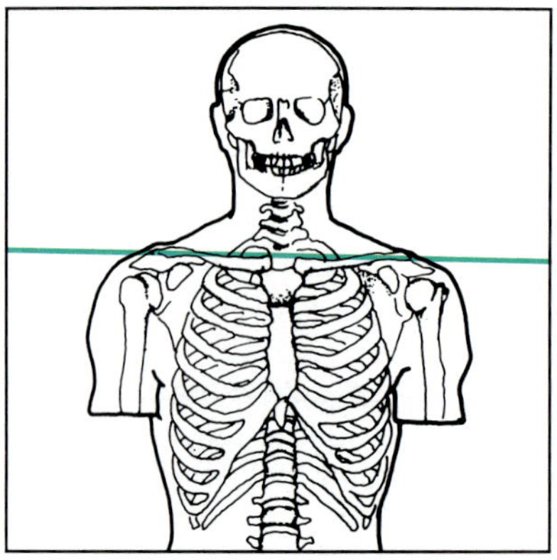

ENCLOSING UNIT (Fig. 9–4)

Carotid Sheath (*C*). An encapsulated enclosure that surrounds the internal jugular vein, common carotid artery, and vagus nerve.

Retropharyngeal Cleft (*D*). A transversely placed, fissure-like feature immediately posterior to the esophagus and anterior to the body of the vertebra.

Synovial Membrane (*A*). A bilateral connective tissue membrane seen lining the articular capsule of the glenohumeral joint (shoulder joint).

Articular Capsule (*B*). A rather thick connective tissue appears in this circular, bilateral structure that surrounds the shoulder joint cavity.

Shoulder Joint Cavity. A bilateral circular space surrounded immediately by synovial membrane and more peripherally by the articular capsule.

Dura Mater (*E*). A sheet of connective tissue seen in this section lining the internal wall of the spinal foramen.

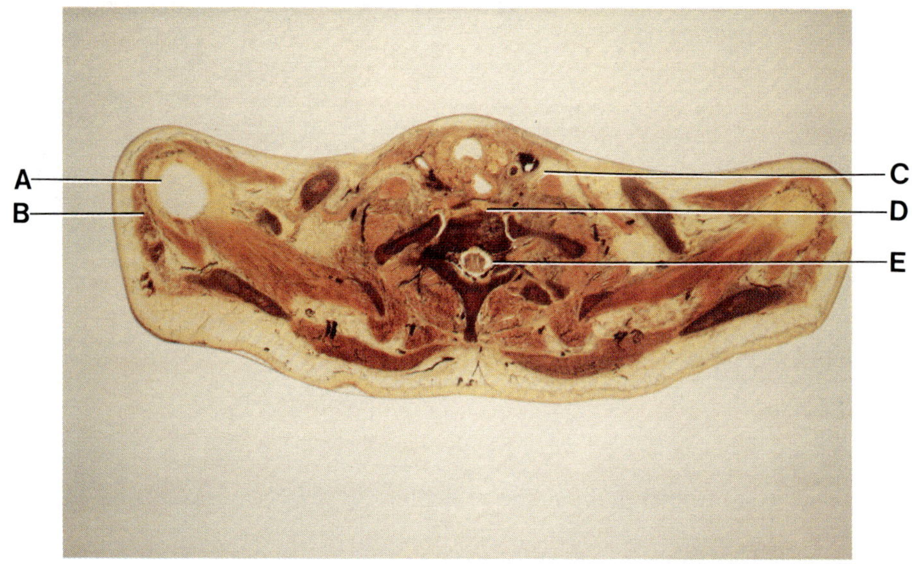

A. Synovial membrane
B. Articular capsule
C. Carotid sheath
D. Retropharyngeal cleft
E. Dura mater

Figure 9–4
Transverse section of the thorax at the level of thoracic vertebra 1, highlighting enclosing unit structures.

Figure 9–5 is the picture of an MRI transverse section of the thorax, which highlights the musculoskeletal, neurovascular, and visceral unit structures. This MRI slice is across the upper part of the superior mediastinum.

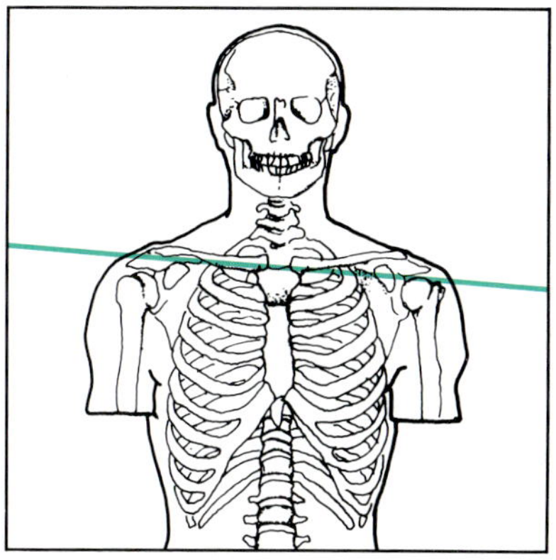

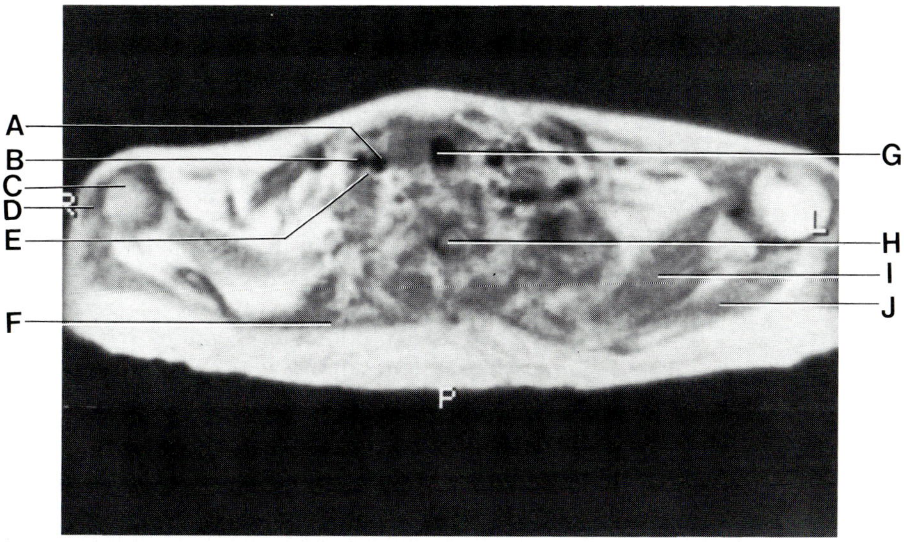

A. Right common carotid artery
B. Right internal jugular vein
C. Head of humerus
D. Deltoid muscle
E. Right vagus nerve
F. Trapezius muscle
G. Trachea
H. Spinal cord
I. Subscapularis muscle
J. Infraspinatus muscle

Figure 9–5
MRI transverse section of the thorax, highlighting musculoskeletal, neurovascular, and visceral unit structures.

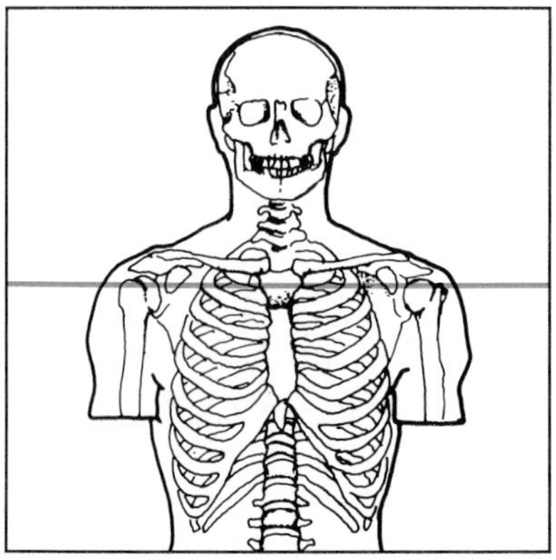

MODULE 2

This is a transverse (axial) section at the level of thoracic vertebra 2.

MUSCULOSKELETAL UNIT (Fig. 9–6)

I. Skeletal Portion

Body of a Thoracic Vertebra (*L*). This is seen here as a half moon–shaped feature posterior to the esophagus and anterior to the spinal cord. The body is the largest and most anterior feature of a vertebra. The anterior part of this feature is covered with a portion of an intervertebral disc.

Spinous Process of a Thoracic Vertebra (*W*). A triangular feature posterior to the spinal cord. The base of the triangle is adjacent to the posterior part of the spinal cord.

Lamina of a Thoracic Vertebra (*V*). A flattened, layered feature between the transverse process and the spinous process. There are two laminae in this section, one each on the right and left sides of the spinous process.

Transverse Process of a Thoracic Vertebra (*M*). An oblong feature between the pedicle and the lamina. There are right and left transverse processes of the vertebra in this section.

Clavicle (*A*). A bilateral oval slice posterior to the medial part of the pectoralis major muscle.

Rib (*N*). An elongated, oblique bone slice lateral to both right and left lungs.

Vertebral Border of the Scapula (*T*). A thin edge on the medial side of the scapula near the vertebral column and lateral to the rhomboid major muscle.

Spine of the Scapula (*F*). A posterior, irregularly shaped bone feature immediately lateral to the medial end of the teres minor muscle. It is anterior and lateral to the trapezius muscle.

Scapula (*E*). A transversely placed, elongated, thin bone section seen between the infraspinatus and subscapularis muscles.

Jugular Notch of the Sternum (*K*). A depression of the sternum seen between the medial ends of the clavicles. This feature is posterior and medial to the medial part of the pectoralis major muscle.

II. Muscular Portion

Infraspinatus Muscle (*S*). This is seen on both right and left sides of the section posterior to the scapula as a triangular feature.

Subscapularis Muscle (*P*). This is seen bilaterally as a bulky muscle slice with narrow medial ends anterior to the scapula and medial to the long head of the triceps brachii muscle.

Teres Major Muscle (*R*). An oblong, bilateral muscle slice posterior to the long head of the triceps brachii muscle and lateral to the infraspinatus muscle.

Triceps Brachii Muscle (Long Head) (*Q*). This appears bilaterally as a narrow, oblong muscle anterior to the teres major muscle.

Trapezius Muscle (*H*). This is seen bilaterally as an elongated, sickle-shaped feature posterior to the rhomboid major muscle and lateral to the spinous process of the thoracic vertebra.

Pectoralis Major Muscle (*B*). A transversely placed, elongated, bilateral muscle slice with each end tapered. The medial part is anterior to the lateral part of the sternum and the sternal end of the clavicle. The lateral part is anterior to the pectoralis minor muscle.

Pectoralis Minor Muscle (*C*). A transversely positioned, bilateral muscle section with medial and lateral tapered ends posterior to the lateral part of the pectoralis major muscle.

Rhomboid Major Muscle (*U*). A transversely placed, bilateral muscle section anterior to the trapezius muscle and lateral to the spinous process. The medial part is posterior to the multifidus muscle.

Multifidus Muscle (*J*). An oval, bilateral muscle aggregate immediately lateral to the lamina and spinous process of a thoracic vertebra. This muscle is anterior to the medial part of the rhomboid major muscle.

Erector Spinae (Sacrospinalis) Muscle Group (*I*). A thin, bilateral muscle section immediately lateral to the multifidus muscle and anterior to the rhomboid muscle.

Serratus Anterior Muscle (*D, O*). A rather thick, bilateral muscle slice anterior and medial to the subscapularis muscle. It is lateral to the rib section.

Teres Minor Muscle (*G*). A transversely placed, elongated, bilateral muscle strip posterior to the infraspinatus muscle. The medial end is narrower than the lateral end.

A. Clavicle
B. Pectoralis major muscle
C. Pectoralis minor muscle
D. Serratus anterior muscle (right)
E. Scapula
F. Spine of scapula
G. Teres minor muscle
H. Trapezius muscle
I. Erector spinae (sacrospinalis) muscle group
J. Multifidus muscle
K. Jugular notch of sternum
L. Body of thoracic vertebra
M. Transverse process of thoracic vertebra
N. Rib
O. Serratus anterior muscle (left)
P. Subscapularis muscle
Q. Triceps brachii muscle (long head)
R. Teres major muscle
S. Infraspinatus muscle
T. Vertebral border of scapula
U. Rhomboid major muscle
V. Lamina of thoracic vertebra
W. Spinous process of thoracic vertebra

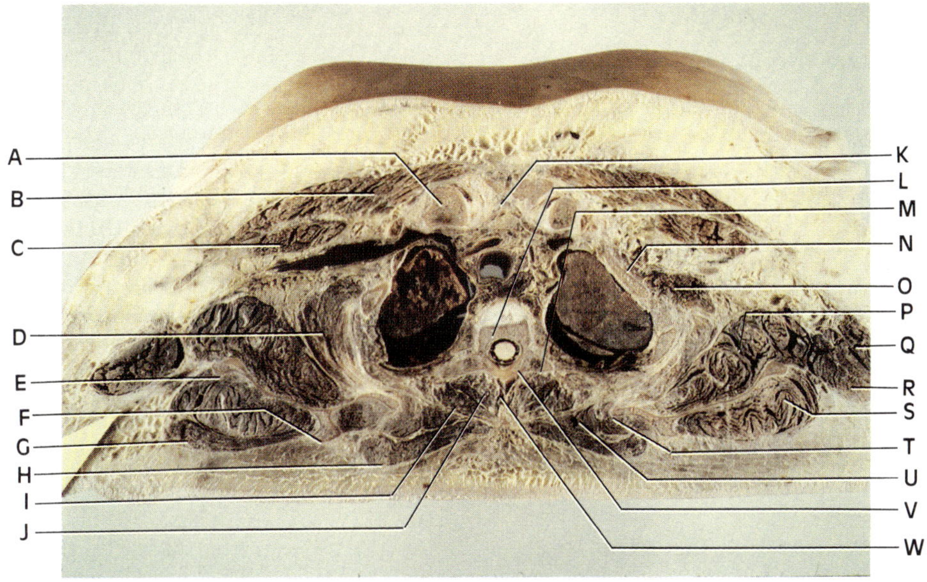

Figure 9–6
Transverse (axial) section of the thorax at the level of thoracic vertebra 2, highlighting musculoskeletal unit structures.

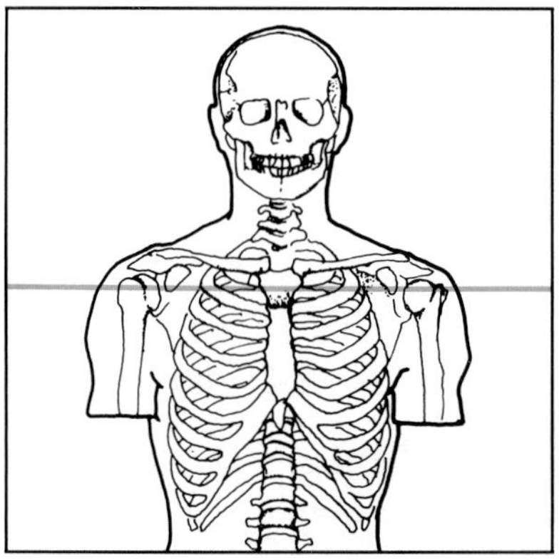

NEUROVASCULAR UNIT (Fig. 9–7)

I. Arteries

Brachiocephalic Artery (*A*). A slit-shaped, hollow vascular organ anterior to the right lateral part of the trachea.

Origin. Arch of the aorta.

Distribution. Right common carotid and right subclavian arteries.

Course. This vessel arises from the right portion of the arch of the aorta and ascends posterior to the inferior part of the manubrium to reach the level of the right sternoclavicular joint, where it splits into the right common carotid and right subclavian arteries.

Right Axillary Artery (*C*). A transversely placed, hollow vascular structure seen posterior to the pectoralis minor and anterior to the serratus anterior muscles. This vessel is seen as a continuation of the right subclavian artery.

Origin. Right subclavian artery.

Distribution. Brachial and six named branches but sometimes seven branches.

Course. This artery arises from the right subclavian artery as the right subclavian artery crosses the superior surface of the first rib. This artery courses lateral to the axillary vein and extends from the superior surface of the first rib to the level in the arm for the attachment of the teres major muscle.

Right Subclavian Artery (*B*). A transversely placed, hollow vascular structure immediately anterior to the pleural cavity and lateral to the brachiocephalic artery. It is seen as a continuous structure with the right axillary artery.

Origin. Brachiocephalic artery.

Distribution. Right axillary artery and branches of the right subclavian artery.

Course. This artery arises from the brachiocephalic artery posterior to the right sternoclavicular joint, courses laterally to reach the superior surface of the first rib, and connects with the right axillary artery.

Left Subclavian Artery (*F*). A partially collapsed, hollow vascular structure medial to the anterior part of the pleural cavity and posterior as well as lateral to the left common carotid artery.

Origin. Left portion of the aortic arch.

Distribution. Left axillary artery and branches of the left subclavian artery.

Course. This artery arises from the left portion of the aortic arch and proceeds deep to the scapula toward the left superior rib. The origin of this vessel is at the approximate level of the intervertebral disk between the third and fourth thoracic vertebrae.

Left Common Carotid Artery (*D*). A nearly collapsed vascular structure anterior and medial to the left subclavian artery. It is posterior and medial to the left brachiocephalic vein.

II. Veins

Left Brachiocephalic Vein (*E*). A triangular, partially collapsed vascular structure anterior to the left pleural cavity and anterior/lateral to the left common carotid artery.

Origin. Arises by the union of the left internal jugular and left subclavian veins.

Termination. Superior vena cava.

Course. This vein is formed posteriorly to the left sternoclavicular joint and courses vertically across the superior mediastinum in an oblique plane. The inferior end, which drains into the superior vena cava, is more toward the right than its superior end.

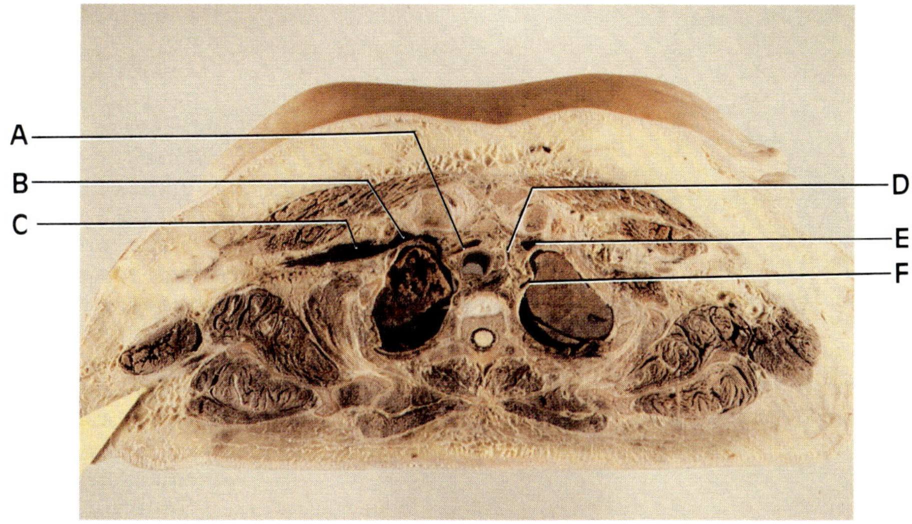

A. Brachiocephalic artery
B. Right subclavian artery
C. Right axillary artery
D. Left common carotid artery
E. Left brachiocephalic vein
F. Left subclavian artery

Figure 9-7
Transverse section of the thorax at the level of thoracic vertebra 2, highlighting neurovascular unit structures.

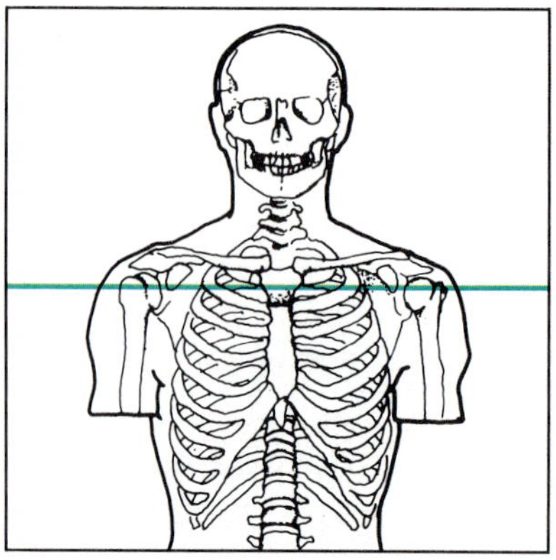

VISCERAL UNIT (Fig. 9–8)

Trachea (*A*). A hollow, C-shaped, cartilaginous structure. The anterior and lateral parts of this hollow organ are formed by cartilage, whereas the posterior part is formed by muscle and other soft tissues. The trachea is known as the windpipe and is about 11 cm long.

Esophagus (*D*). This structure has a partially collapsed, hollow, slitlike appearance posterior to the left part of trachea and anterior to the left part of a thoracic vertebra. The esophagus transports food from the throat to the stomach by a process known as peristalsis.

Superior Lobe of the Left Lung (*E*). A rather large, irregularly circular feature within the pleural cavity. The posterior part is lateral to both the body of a thoracic vertebra and the left inferior lobe of the left lung. The anterior part of this lobe is lateral to the left subclavian and left common carotid arteries.

Inferior Lobe of the Left Lung (*F*). An oblique, narrow strip of lung tissue adjacent to the posterior medial surface of the superior lobe of the left lung.

Superior Lobe of the Right Lung (*B*). An irregularly circular slice of lung tissue within the right pleural cavity, posterior to the right subclavian artery and medial to the serratus anterior muscle.

Spinal Cord (*C*). A white, spherical nerve feature within the spinal foramen, surrounded by dura mater.

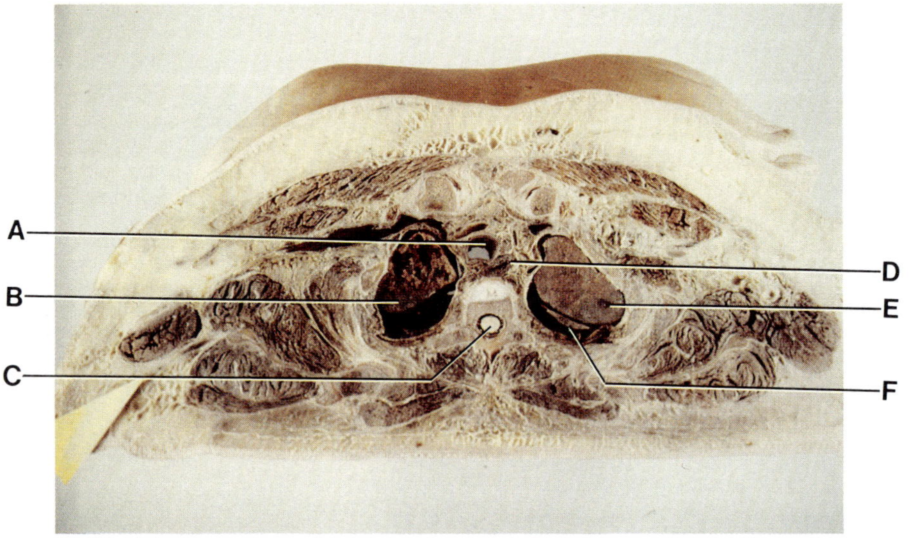

A. Trachea
B. Superior lobe of right lung
C. Spinal cord
D. Esophagus
E. Superior lobe of left lung
F. Inferior lobe of left lung

Figure 9–8
Transverse section of the thorax at the level of thoracic vertebra 2, highlighting visceral unit structures.

ENCLOSING UNIT (Fig. 9–9)

Interpleural Space (Pleural Cavity) (*C, D*). A narrow, circular, empty area around each lung part.

Visceral Pleura (*A*). A thin sheet of tissue around each lung section.

Parietal Pleura (*B*). The portion of the pleura that extends from the roots of the lungs and lines the walls of the thoracic cavity.

Oblique Fissure of the Left Lung (*E*). A slit separating the superior and inferior lobes of the left lung.

The musculoskeletal, neurovascular, and visceral unit structures are emphasized in Figure 9–10. This is an MRI transverse section of the upper thorax.

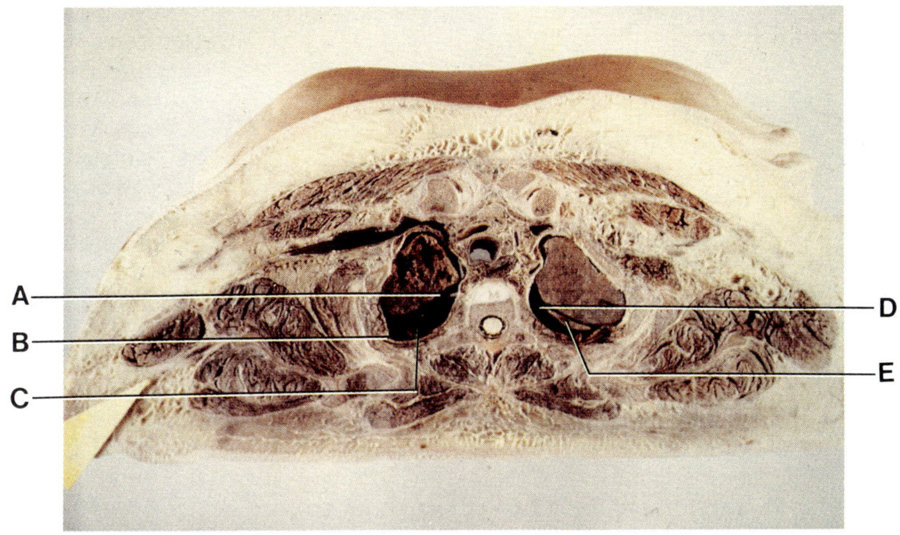

A. Visceral pleura
B. Parietal pleura
C. Right interpleural space (pleural cavity)
D. Left interpleural space (pleural cavity)
E. Oblique fissure of left lung

Figure 9–9
Transverse section of the thorax at the level of thoracic vertebra 2, highlighting enclosing unit structures.

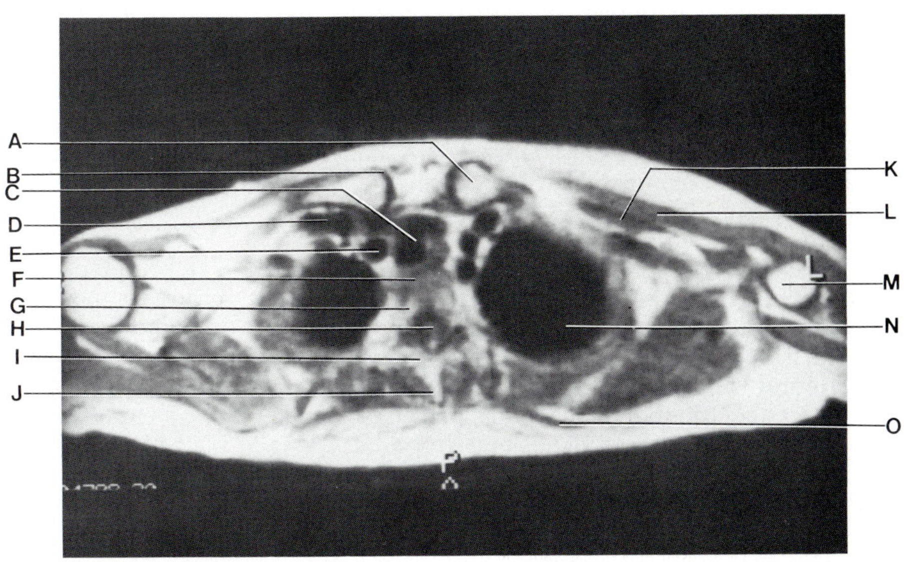

A. Sternal end of clavicle
B. Sternoclavicular joint
C. Trachea
D. Right internal jugular vein
E. Right common carotid artery
F. Esophagus
G. Body of thoracic vertebra
H. Spinal cord
I. Lamina of thoracic vertebra
J. Spinous process of thoracic vertebra
K. Pectoralis minor muscle
L. Pectoralis major muscle
M. Head of humerus
N. Superior lobe of left lung
O. Trapezius muscle

Figure 9–10
MRI transverse section of the thorax highlighting musculoskeletal, neurovascular, and visceral unit structures.

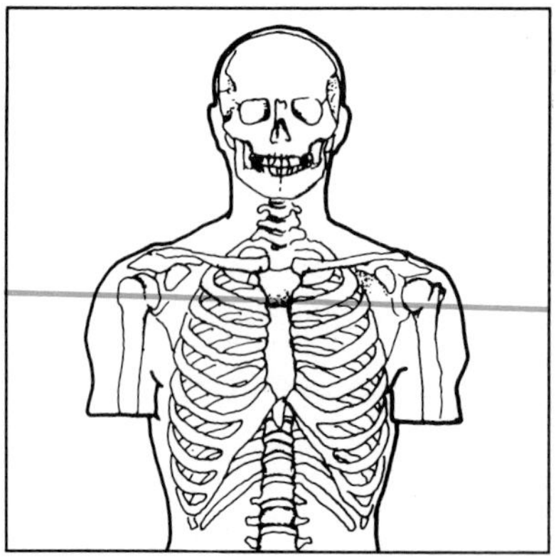

MODULE 3

This is a transverse (axial) section at the vertebral level of thoracic vertebra 4.

MUSCULOSKELETAL UNIT (Fig. 9–11)

I. Skeletal Portion

Sternum (*A*). A flat, semicircular bone anterior to the mediastinum and posterior to the medial ends of the pectoralis major muscles.

Body of a Thoracic Vertebra (*D*). A bony cup-shaped feature located in the posterior median of this thoracic section.

Rib (*Q*). Curved rib sections lateral to the lungs. The concave surfaces are adjacent to the lungs. Between each rib section, muscles are seen.

Body (Shaft) of a Rib (*W*). There are three rib slices on each side of this thoracic section. The anterior one joins with cartilage to form a joint.

Spinous Process of a Thoracic Vertebra (*L*). The posterior median projection of a thoracic vertebra, seen between deep back muscles.

Lamina of a Thoracic Vertebra (*E*). A flattened, layered, oblique feature located between the transverse process and the spinous process. Right and left laminae are seen in this section.

Scapula (*T*). A narrow, bilateral strip of bone tissue seen between the infraspinatus and subscapularis muscles. The subscapularis muscle is anterior to it, and the infraspinatus muscle is posterior to it.

Vertebral Border of the Scapula (*V*). A thin, bilateral edge on the medial part of the scapula, adjacent to the rhomboid major muscle.

Sternocostal Joint (*M*). This is found between the sternum and the rib cartilage. It is posterior to the medial part of the pectoralis major muscle.

Costal Cartilage (*N*). Hyaline cartilage that extends from a body (shaft) of a rib to the sternum.

Costochondral Joint (*O*). A joint between costal cartilage and its corresponding body of a rib.

II. Muscular Portion

Infraspinatus Muscle (*U*). A transversely placed, oblong, bilateral muscle section posterior to the scapula, as seen in this thoracic section. The medial end is narrower than the lateral end. The lateral end is adjacent to the teres minor muscle.

Subscapularis Muscle (*F*). A bulky, oblique, and transversely placed bilateral muscle anterior to the scapula. The medial end is narrower and more posterior than the lateral end.

Teres Minor Muscle (*G*). A small, oblong, bilateral muscle section seen between the infraspinatus and teres major muscles.

Teres Major Muscle (*S*). An oblong feature posterior and medial to the latissimus dorsi muscle. It is lateral to the infraspinatus muscle at this level.

Latissimus Dorsi Muscle (*R*). A bulky, short, anterior and posterior bilateral muscle lateral to the teres major muscle.

Trapezius Muscle (*J*). A transversely positioned, bilateral muscle strip posterior to the rhomboid major muscle. This muscle is the most posterior muscle in this thoracic axial section.

Rhomboid Major Muscle (*I*). This is seen bilaterally, anterior to the trapezius muscle, as a transversely positioned muscle slice. Its medial end is narrower than its lateral end and is posterior to the spinalis thoracis muscle.

Pectoralis Major Muscle (*P*). A narrow strip of curved, bilateral muscle tissue seen superficial to the pectoralis minor muscle and deep to the mammary glands.

Pectoralis Minor Muscle (*B*). Rather thick, bilateral muscle bundles seen deep to the posterior part of the pectoralis major muscle.

Serratus Anterior Muscle (*C*). An elongated, bilateral, curved feature lateral to the rib bodies. The medial surface shows a concave part.

Iliocostalis Thoracis Muscle (*H*). A small, bilateral, rounded muscle bundle lateral to the longissimus thoracis muscle and anterior to the medial part of the rhomboid major muscle.

Longissimus Thoracis Muscle. A bilateral aggregate of muscle bundles located between the iliocostalis thoracis and spinalis thoracis muscles.

Spinalis Thoracis Muscle (*K*). A bilateral muscle medial to the longissimus thoracis muscle and lateral to the posterior part of the spinous process of a vertebra.

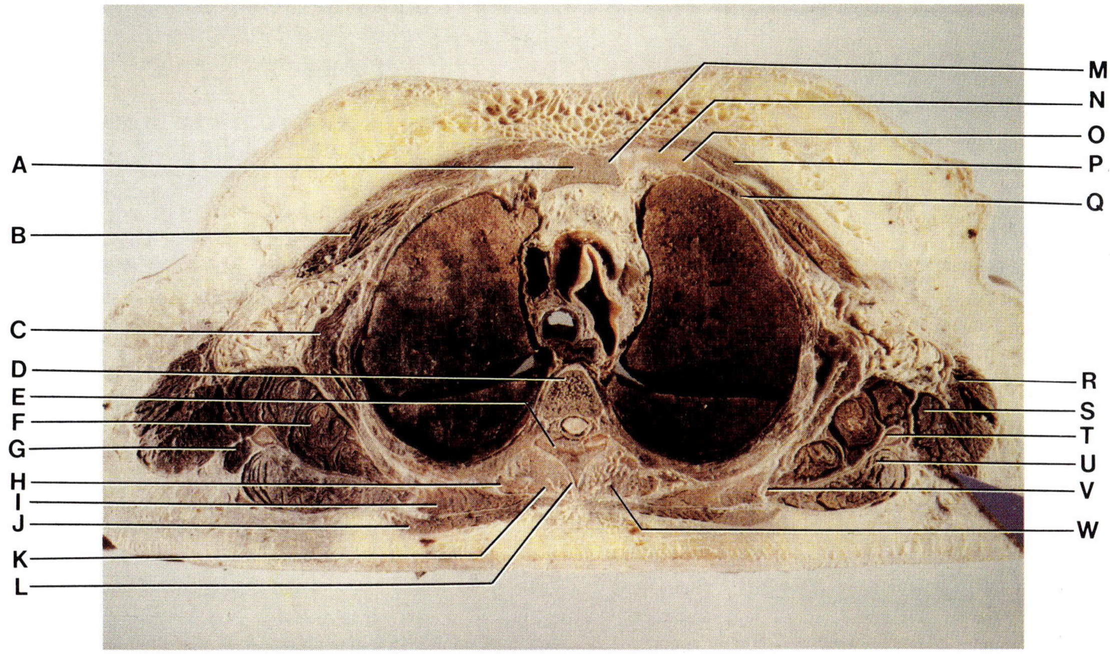

A. Sternum
B. Pectoralis minor muscle
C. Serratus anterior muscle
D. Body of thoracic vertebra
E. Lamina of thoracic vertebra
F. Subscapularis muscle
G. Teres minor muscle
H. Iliocostalis thoracis muscle
I. Rhomboid major muscle
J. Trapezius muscle
K. Spinalis thoracis muscle
L. Spinous process of thoracic vertebra
M. Sternocostal joint
N. Costal cartilage
O. Costochondral joint
P. Pectoralis major muscle
Q. Rib
R. Latissimus dorsi muscle
S. Teres major muscle
T. Scapula
U. Infraspinatus muscle
V. Vertebral border of scapula
W. Body (shaft) of rib

Figure 9-11
Transverse (axial) section of the thorax at the level of thoracic vertebra 4, highlighting the musculoskeletal unit structures.

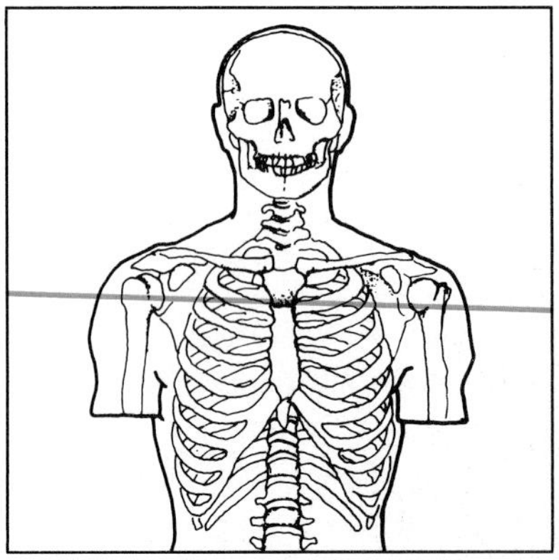

NEUROVASCULAR UNIT (Fig. 9–12)

I. Arteries

Internal Thoracic (Internal Mammary) Artery (*A*). A hollow, bilateral vascular structure lateral to the internal thoracic vein and sternum.

Origin. Subclavian artery.

Distribution. Musculophrenic and superior epigastric arteries, as well as perforating branches to the anterior thorax.

Course. This artery arises from the inferior aspect of the first part of the subclavian artery. It courses downward anteriorly and toward the median posterior to the sternocleidomastoid muscle and subclavian and internal jugular veins. This artery is adjacent to the parietal pleura and passes obliquely toward the median as it courses through the thorax. As this artery descends the thorax, it is found posterior to the upper six costal cartilages and the muscles between, immediately lateral to the sternum. This vessel terminates at the level of the sixth intercostal space by splitting into the superior epigastric and musculophrenic arteries.

Descending Thoracic Aorta (*D*). A portion of the aorta that extends from the arch of the aorta to the diaphragm. At this level the descending abdominal aorta begins.

Origin. Aortic arch.

Distribution. Its branches and the descending abdominal aorta.

Course. It courses to the left of the thoracic vertebrae within the thorax, anterior to the lateral left part of the bodies below its origin covered by parietal pleura. It enters the abdomen through the aortic hiatus of the diaphragm and becomes the descending abdominal aorta.

Ascending Aorta (*B*). This portion of the aorta terminates at this level and appears as a slitlike vascular structure.

Origin. Left ventricle.

Distribution. Arch of the aorta and its branches.

Course. Ascends within the anterior portion of the interatrial sulcus to reach the arch of the aorta above.

Arch of the Aorta (*C*). An elongated (anterior-to-posterior), hollow, vascular, slitlike structure between the ascending aorta and the descending thoracic aorta. This structure is located left of the superior vena cava.

Origin. Ascending aorta.

Distribution. Brachiocephalic artery, left common carotid artery, and left subclavian artery.

Course. The aortic arch is a sagittally placed portion of the aorta in the inferior part of the superior mediastinum, posterior to the inferior part of the manubrium.

II. Veins

Internal Thoracic (Internal Mammary) Vein (*E*). A small, oval, bilateral venous structure approximately 10 mm lateral to the sternocostal joint and 5 mm medial to the internal thoracic artery.

Origin. Union of the superior epigastric and musculophrenic veins.

Termination. Right and left brachiocephalic veins, depending upon its corresponding internal thoracic vein.

Course. This vein courses upward immediately medial to the internal thoracic artery and lateral to the sternum. It terminates into its corresponding brachiocephalic vein.

Superior Vena Cava (*F*). A large, hollow vascular vein right of the arch of the aorta and medial to the superior lobe of the right lung. It is located on the right of the median, anterior to the midaxillary line.

Origin. Right and left brachiocephalic veins.

Termination. Right atrium of the heart.

Course. It courses adjacent to the right side of the ascending aorta and receives the azygos vein before it terminates in the right atrium.

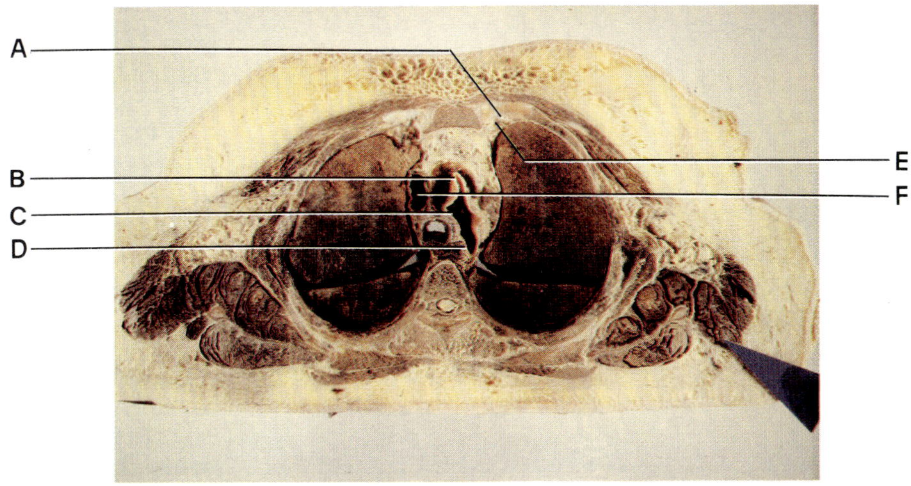

A. Left internal thoracic artery
B. Ascending aorta
C. Arch of aorta
D. Descending thoracic aorta
E. Left internal thoracic vein
F. Superior vena cava

Figure 9–12
Transverse section of the thorax at the level of thoracic vertebra 4, highlighting neurovascular unit structures.

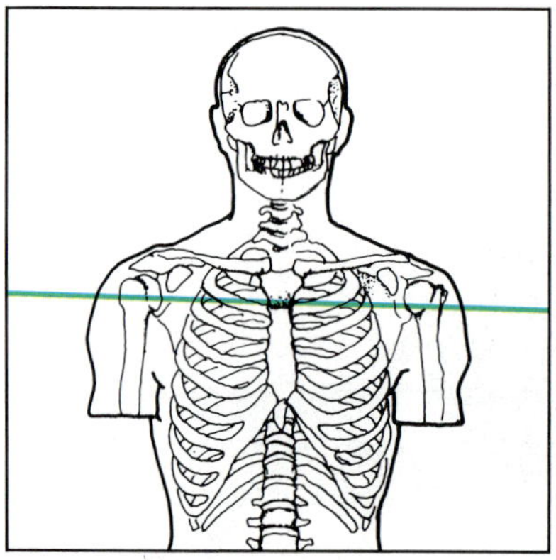

VISCERAL UNIT (Fig. 9–13)

Superior Lobe of the Right Lung (*A*). A large, brown, spongy, triangular lung tissue. The base of the triangular feature is posterior and anterior to the oblique fissure of the right lung.

Inferior Lobe of the Right Lung (*C*). A half moon–shaped lung tissue posterior to the oblique fissure.

Superior Lobe of the Left Lung (*D*). A large, triangular, spongy lung tissue anterior to the oblique fissure of the left lung. The base of the triangular feature is adjacent to the oblique fissure of the left lung.

Inferior Lobe of the Left Lung (*F*). A half moon–shaped lung slice posterior to the oblique fissure of the left lung. The convex surface of this part of the lung is on its posterior surface.

Trachea (*B*). A C-shaped, hollow, cartilaginous organ posterior and medial to the superior vena cava. It is to the right of the arch of the aorta and anterior to the right part of the esophagus.

Esophagus (*E*). A spherical feature with stellate projections in the lumen. It is located posterior to the right part of the trachea; on the right side, it is adjacent to the descending thoracic aorta.

Spinal Cord. This is seen here as a whitish, spherical nerve organ within the spinal foramen.

ENCLOSING UNIT (Fig. 9–14)

Oblique Fissure of the Right Lung (*C*). A slit that partially separates the superior lobe of the right lung from the inferior lobe. This oblique feature is positioned with its medial end slightly anterior to its lateral end.

Oblique Fissure of the Left Lung (*E*). A slit that separates the superior lobe of the left lung from the inferior lobe. This feature is positioned in such a way that the medial end is slightly anterior to its lateral end.

Interpleural Space (*B*). A narrow, bilateral, curved space between the visceral and parietal pleurae. The concave surface is deep to the convex surface. This space is seen surrounding the lungs.

Visceral Pleura. A thin, bilateral sheet of tissue adjacent to and covering each lung.

Parietal Pleura. A thin, bilateral, membranous lining for the inner wall of the thorax. This membrane is seen superficial to the interpleural space.

Mediastinal Pleura (*A*). A part of the parietal pleura adjacent to the mediastinum. This pleura is on the right and mediastinal pleura (*D*) is on the left of the section.

Mediastinal Pleura (*D*). A sheet of tissue adjacent to the mediastinal structures. This membrane is a part of the parietal pleura and medial to the lung.

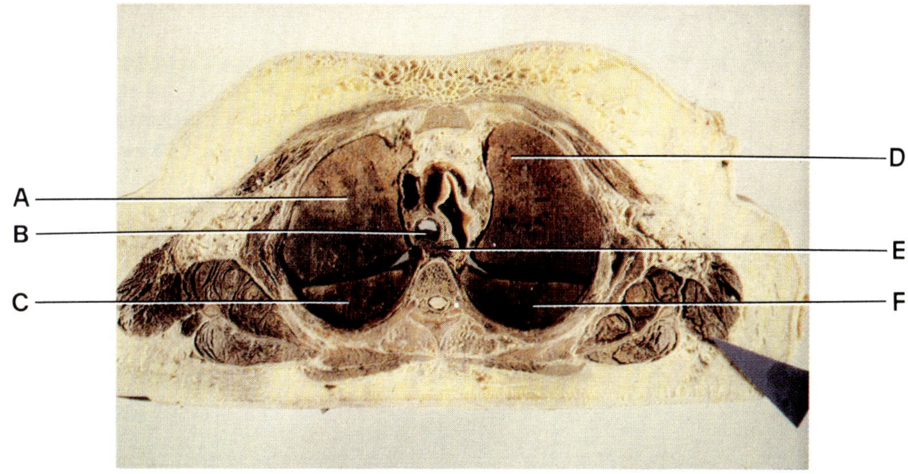

A. Superior lobe of right lung
B. Trachea
C. Inferior lobe of right lung
D. Superior lobe of left lung
E. Esophagus
F. Inferior lobe of left lung

Figure 9–13
Transverse section of the thorax at the level of thoracic vertebra 4, highlighting visceral unit structures.

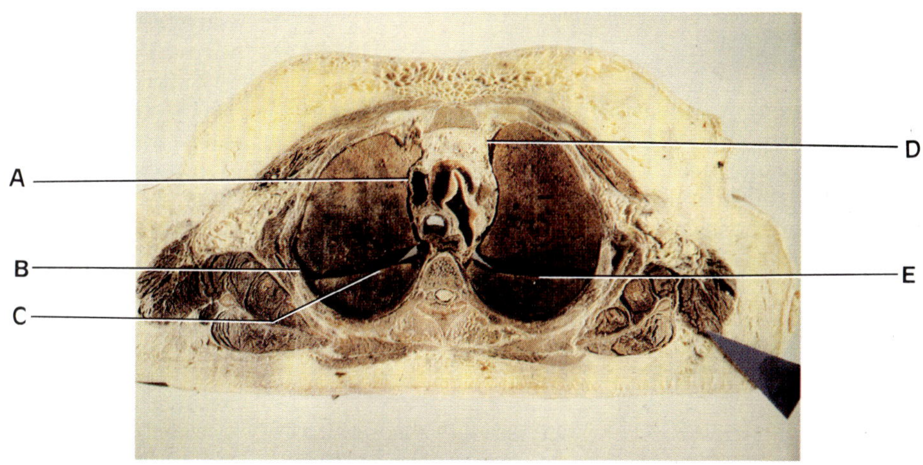

A. Mediastinal pleura (right)
B. Interpleural space
C. Oblique fissure of right lung
D. Mediastinal pleura (left)
E. Oblique fissure of left lung

Figure 9–14
Transverse section of the thorax at the level of thoracic vertebra 4, highlighting enclosing unit structures.

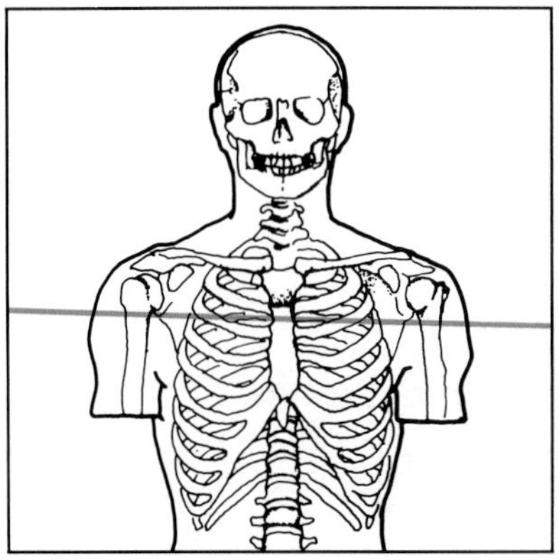

MODULE 4

This is a sheet plastinated transverse (axial) cadaver section at the level of thoracic vertebra 5.

MUSCULOSKELETAL UNIT (Fig. 9–15)

I. Skeletal Portion

Sternum (*A*). A transversely placed bone section in the anterior median of this section. This bone slice is rather thick and medial to costal cartilages.

Sternocostal Joint (*B*). A bilateral junction between the sternum and a costal cartilage.

Costal Cartilage (*C*). A transversely elongated slice of hyaline cartilage between the sternum and a rib body.

Costochondral Joint (*M*). A connection between the body of a rib and the lateral end of a costal cartilage.

Body (Shaft) of a Rib (*E*). There are five slices of rib bodies on each side of this section, most appear sickle-shaped—that is, each body of a rib section has two narrow ends.

Body of a Thoracic Vertebra (*F*). A large, rather triangular aggregate of bone tissue. It is located to the right of the descending thoracic aorta and posterior to the azygos vein.

Costovertebral Capitular Joint (*O*). A bilateral junction between the head of the rib and the lateral posterior part of the vertebral body.

Head of a Rib (*G*). A posterior, somewhat enlarged portion of a rib immediately lateral to the posterior lateral part of the vertebral body.

Neck of a Rib (*H*). A portion of a rib immediately lateral to the rib head, which shows some constriction.

Tubercle of a Rib (*J*). A small, bilateral, medial projection seen immediately posterior and lateral to the neck of the rib.

Lamina of a Thoracic Vertebra (*K*). A bilateral, oblique bone section seen connecting posteriorly at the spinous process of the vertebra. The posterior end is more medial than the anterior end.

Spinous Process of a Thoracic Vertebra. A median, rather thin bone section immediately posterior to the lamina and medial to the deep back muscles.

Scapula. A transversely placed, elongated, bilateral bone slice. The lateral end is more anterior and is thicker than the medial end.

II. Muscular Portion

Pectoralis Major Muscle (*N*). A transversely oblique, elongated, rather thin muscle section. The posterior end is thicker than the anterior and medial ends. This muscle is superficial to the pectoralis minor muscle.

Pectoralis Minor Muscle (*D*). A sickle-shaped, bilateral muscle slice immediately deep to the pectoralis major and superficial to rib body sections.

Serratus Anterior Muscle (*P*). A thin, curved strip of bilateral muscle tissue superficial to rib body slices. The medial part of this muscle strip is concave. This muscle section extends from the scapula to the posterior part of the pectoralis minor muscle.

External Intercostal Muscle (*R*). Bilateral muscle slices on the outer area between rib body sections, as seen at this level. The fibers of this muscle are obliquely placed, with their anterior ends more lateral than their posterior ends.

Internal Intercostal Muscle (*Q*). Bilateral muscle slices on the inner area between rib body sections, as seen at this level. The fibers of this muscle are obliquely placed, with their anterior ends more medial than their posterior ends.

Subscapularis Muscle (*I, S*). A bilateral, transversely oblique muscle slice seen immediately anterior to the scapula at this level. Its lateral end is thicker and more anterior than its medial end.

Teres Major Muscle (*T*). A rather thick, triangular, bilateral muscle section, seen lateral and posterior to the lateral part of the scapula. The base of this muscle is seen more medially than the apex.

Infraspinatus Muscle (*U*). A transversely submarine-shaped, oblique, bilateral muscle section posterior to the medial part of the scapula, as seen at this level. Each end is moderately thin.

Rhomboid Major Muscle (*W*). A transversely placed, bilateral muscle slice immediately deep to the trapezius and medial to the spinous process of the thoracic vertebra.

Trapezius Muscle (*V*). A transversely elongated, fairly thick, bilateral muscle slice posterior to the rhomboid major muscle.

Spinalis Thoracis Muscle (*L*). An irregularly shaped, bilateral muscle aggregate seen medial to the neck of the rib and lateral to the spinous process of the vertebra. This muscle slice is seen immediately anterior to the rhomboid major muscle.

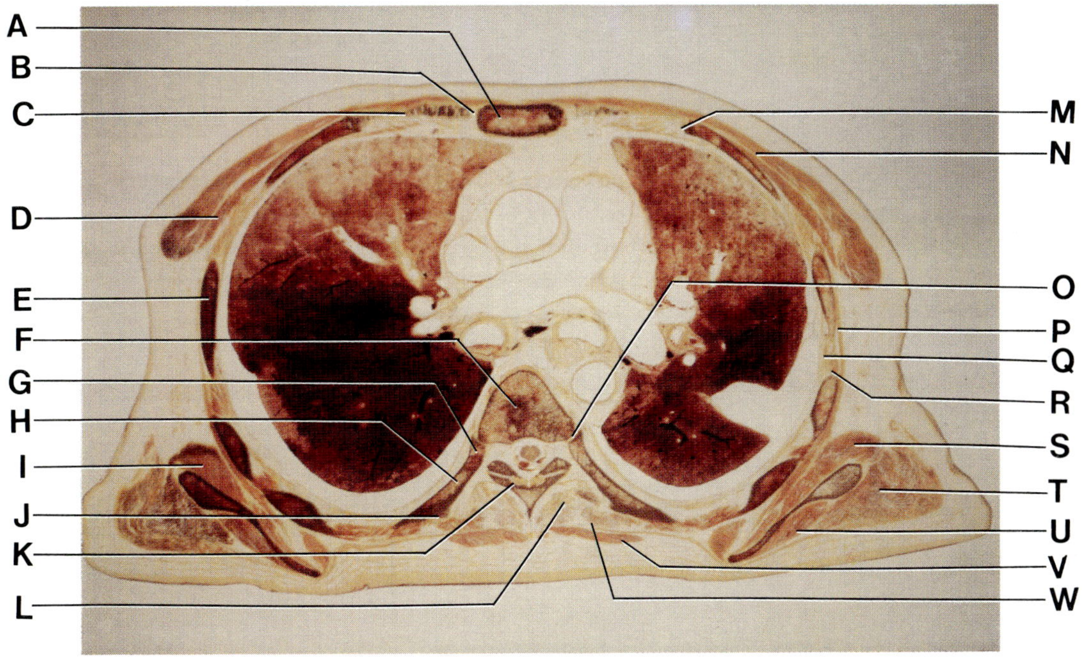

A. Sternum
B. Sternocostal joint
C. Costal cartilage
D. Pectoralis minor muscle
E. Body (shaft) of rib
F. Body of thoracic vertebra
G. Head of rib
H. Neck of rib
I. Subscapularis muscle (right)
J. Tubercle of rib
K. Lamina of thoracic vertebra
L. Spinalis thoracis muscle
M. Costochondral joint
N. Pectoralis major muscle
O. Costovertebral capitular joint
P. Serratus anterior muscle
Q. Internal intercostal muscle
R. External intercostal muscle
S. Subscapularis muscle (left)
T. Teres major muscle
U. Infraspinatus muscle
V. Trapezius muscle
W. Rhomboid major muscle

Figure 9–15
Transverse (axial) section of the thorax at the level of thoracic vertebra 5, highlighting musculoskeletal unit structures.

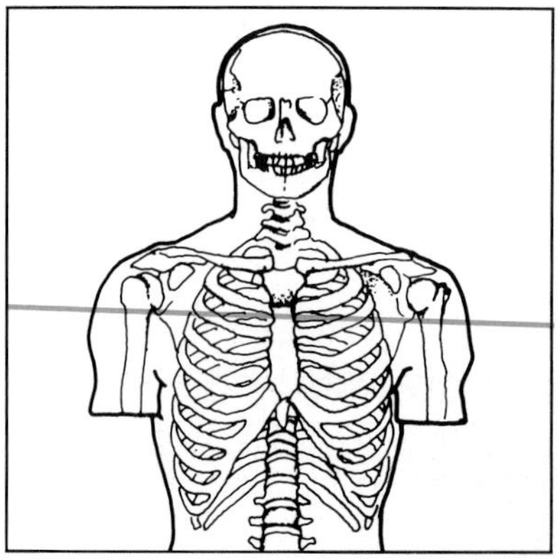

NEUROVASCULAR UNIT (Fig. 9–16)

I. Arteries

Ascending Aorta (*A*). A large, rounded vascular organ seen posterior to the sternum and anterior to the pulmonary trunk. It is to the left of the superior vena cava.

Pulmonary Trunk (*E*). A transversely oblong, curved, large, hollow vascular artery. The concave surface of this vessel is anterior to the convex surface. The anterior part shows a slight enlargement compared with the other part. The ascending aorta is anterior to this vessel, as seen in this section.

Origin. Right ventricle.

Distribution. Pulmonary circulation from the right ventricle to the right and left pulmonary arteries.

Course. It ascends from the right ventricle within the interatrial sulcus and courses deep to the arch of the aorta before dividing into the right and left pulmonary arteries.

Right Pulmonary Artery (*C*). A moderately narrow, transversely placed, hollow vascular structure that appears to be a continuation of the pulmonary trunk.

Origin. Pulmonary trunk.

Distribution. Blood flows to many vessels within the right lung to reach the pulmonary capillaries around the alveoli.

Course. It courses toward the right lung to reach its hilum before branching into several vessels.

Left Pulmonary Artery (*F*). An oblong, hollow vascular structure that is a branch of the pulmonary trunk. It is immediately medial to the left lung hilum and lateral to the pulmonary trunk. The left primary bronchus is seen medial to this vessel.

Origin. Pulmonary trunk.

Distribution. Blood flows to many vessels within the left lung to reach the pulmonary capillaries around the alveoli.

Course. It courses toward the left lung to reach its hilum before branching into several vessels.

Descending Thoracic Aorta (*G*). A large, rounded, thick-walled vascular structure seen to the left of the body of a thoracic vertebra, medial to the posterior part of the left lung. This structure is seen posterior to the esophagus and slightly to the left but directly posterior to the left primary bronchus.

II. Veins

Superior Vena Cava (*B*). A rounded, hollow vascular vein anterior to the right pulmonary artery, to the right of the ascending aorta and medial to the right lung.

Azygos Vein (*D*). A rounded, hollow vascular vein anterior to the body of a thoracic vertebra as seen at this level. It is seen adjacent to the right side of the esophagus.

Origin. The union of the right subcostal and ascending lumbar veins is the most common origin but sometimes by a branch from the right renal vein and sometimes from the inferior vena cava.

Termination. Superior vena cava.

Course. It reaches the thorax after passing through the aortic hiatus in the diaphragm. It courses adjacent to the right side of the vertebral column to reach the fourth thoracic vertebra. It curves anteriorly over the root of the right lung and terminates in the superior vena cava before the superior vena cava penetrates the pericardium.

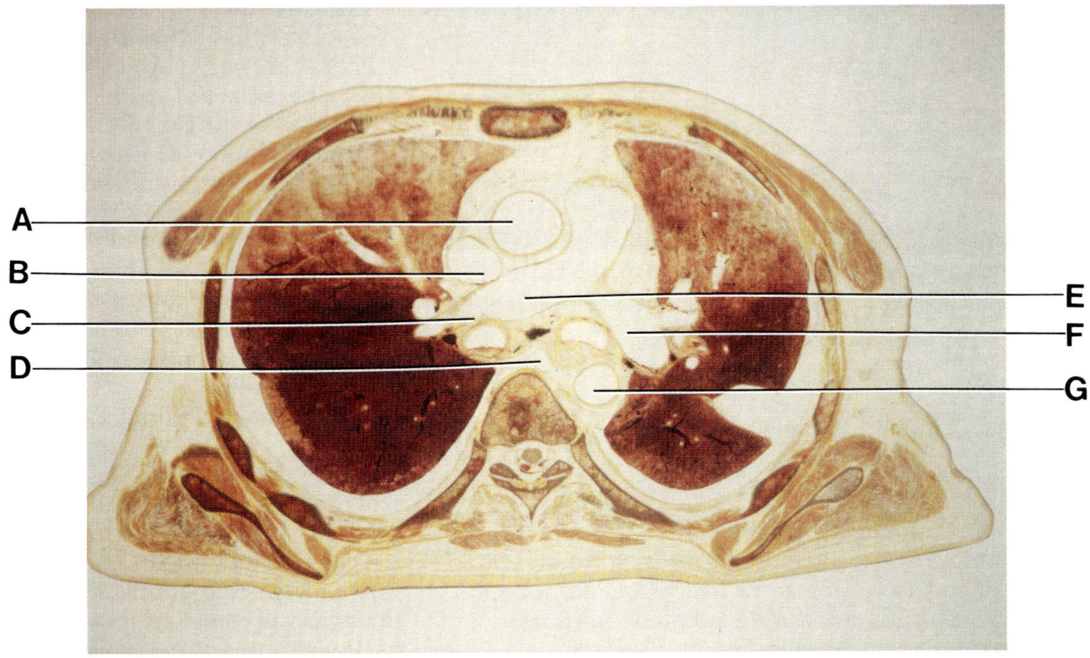

A. Ascending aorta
B. Superior vena cava
C. Right pulmonary artery
D. Azygos vein
E. Pulmonary trunk
F. Left pulmonary artery
G. Descending thoracic aorta

Figure 9–16
Transverse section of the thorax at the level of thoracic vertebra 5, highlighting neurovascular unit structures.

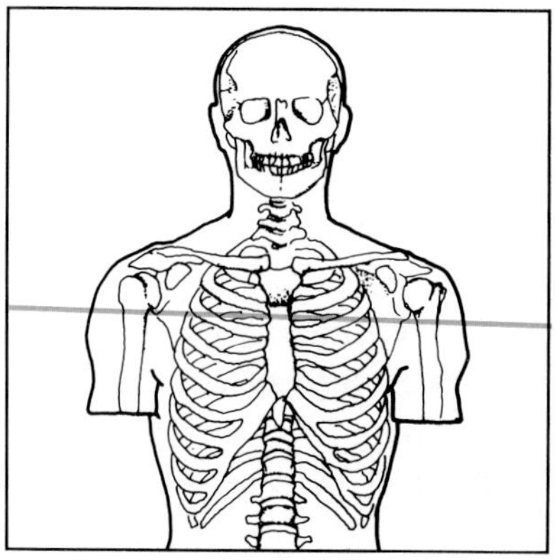

VISCERAL UNIT (Fig. 9–17)

Superior Lobe of the Right Lung (*A*). An anterior slice of lung tissue located on the right of the superior vena cava and ascending aorta. The most anterior and medial part of this structure is pointed and posterior to the right part of the sternum.

Inferior Lobe of the Right Lung (*C*). The posterior slice of lung tissue adjacent to the right side of the right primary bronchus.

Superior Lobe of the Left Lung (*E*). A triangular lung slice lateral to the pulmonary trunk and left pulmonary artery. The apex of the triangle is anterior and deep to a costal cartilage of a rib.

Inferior Lobe of the Left Lung (*H*). A posterior slice of lung tissue having a curved posterior part and a lateral anterior gap, as seen in this body section. This lobe is lateral to the descending thoracic aorta and left primary bronchus.

Right Primary Bronchus (*B*). A hollow, C-shaped ring of cartilaginous tissue and a fine mesh of tissue are seen projecting toward the lumen. This structure is medial to the inferior lobe of the right lung and lateral to the azygos vein.

Origin. Inferior part of the trachea.

Course. This 1-inch-long, hollow respiratory structure connects with the right lung adjacent to the fifth thoracic vertebra.

Left Primary Bronchus (*F*). A bean-shaped, hollow, thick-walled structure. The anterior and lateral parts of the wall consist of cartilaginous tissue, whereas the posterior part of the wall is soft tissue.

Origin. Inferior end of trachea.

Course. This 2-inch-long, hollow respiratory structure attaches to the lung at its hilus opposite the sixth thoracic vertebra. On its course it proceeds adjacent to the left inferior part of the aortic arch and anterior to the esophagus, thoracic duct, and descending thoracic aorta.

Esophagus (*G*). An oval, stellate organ posterior to the medial part of the left primary bronchus and adjacent to the right side of the azygos vein.

Spinal Cord (*D*). Oval nerve tissue within the spinal foramen.

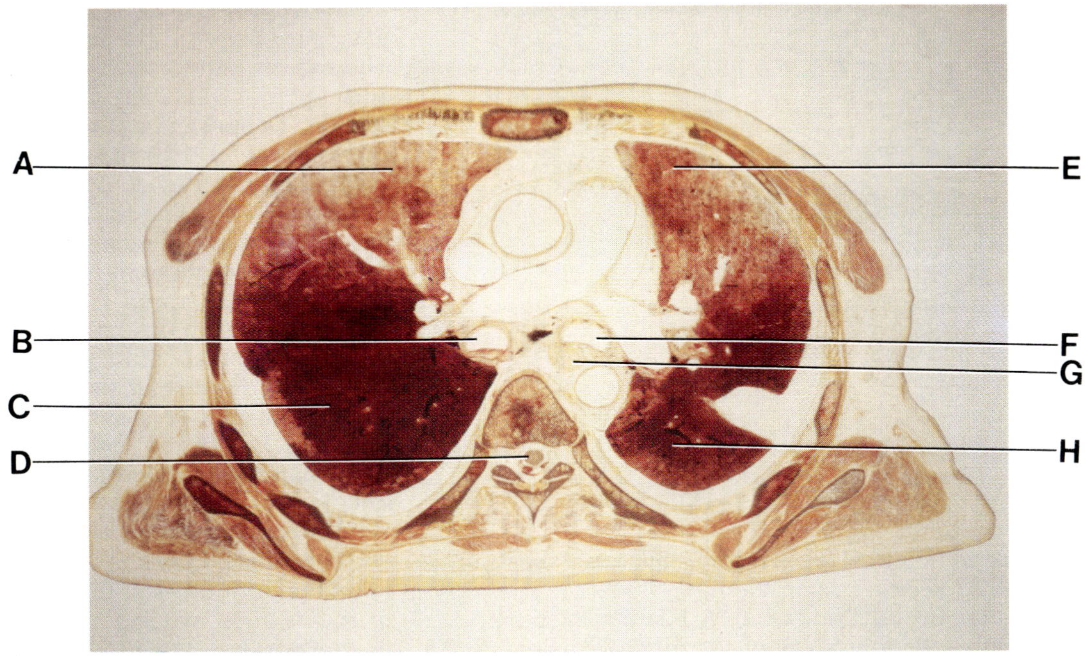

A. Superior lobe of right lung
B. Right primary bronchus
C. Inferior lobe of right lung
D. Spinal cord
E. Superior lobe of left lung
F. Left primary bronchus
G. Esophagus
H. Inferior lobe of left lung

Figure 9–17
Transverse section of the thorax at the level of thoracic vertebra 5, highlighting visceral unit structures.

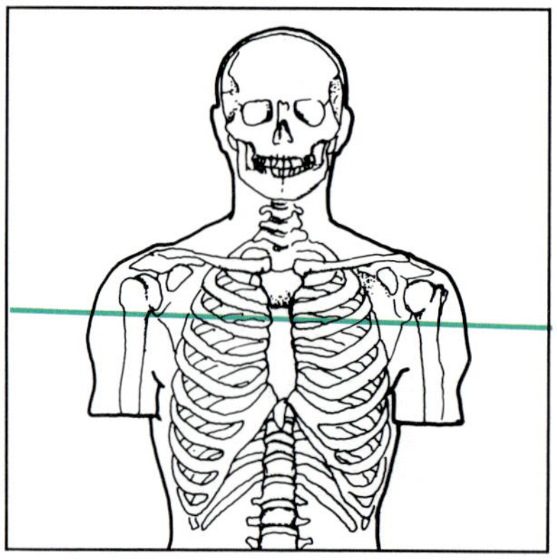

ENCLOSING UNIT (Fig. 9–18)

Visceral Pleura (*B*). A thin, bilateral sheet of membrane seen encasing the lung tissue.

Parietal Costal Pleura (*D*). A bilateral membrane adjacent to the inner wall of the thoracic cavity.

Interpleural Space (*E*). A curved, elongated, bilateral space between the visceral and parietal pleurae. The convex surface of the space is posterior, lateral, and anterior, as seen in this thoracic transverse section. The posterior part is wider in this plastinated cadaver section than the anterior part.

Mediastinal Pleura (*A*). A membranous sheet of tissue seen enclosing the superior vena cava, ascending aorta, and pulmonary trunk. This membrane crosses posteriorly to the right and left superior lobes as well as the sternum.

Dura Mater (*G*). A circular membrane seen adjacent to the spinal cord.

Oblique Fissure of the Right Lung (*C*). A transverse indication of a slit between the superior and inferior lobes of the right lung. This feature starts immediately lateral to the right pulmonary artery, as seen in this plastinated thoracic section.

Oblique Fissure of the Left Lung (*F*). A transversely oblique indication of a slit between the superior and inferior lobes of the left lungs. The lateral part of this feature is positioned more anterior than the medial part.

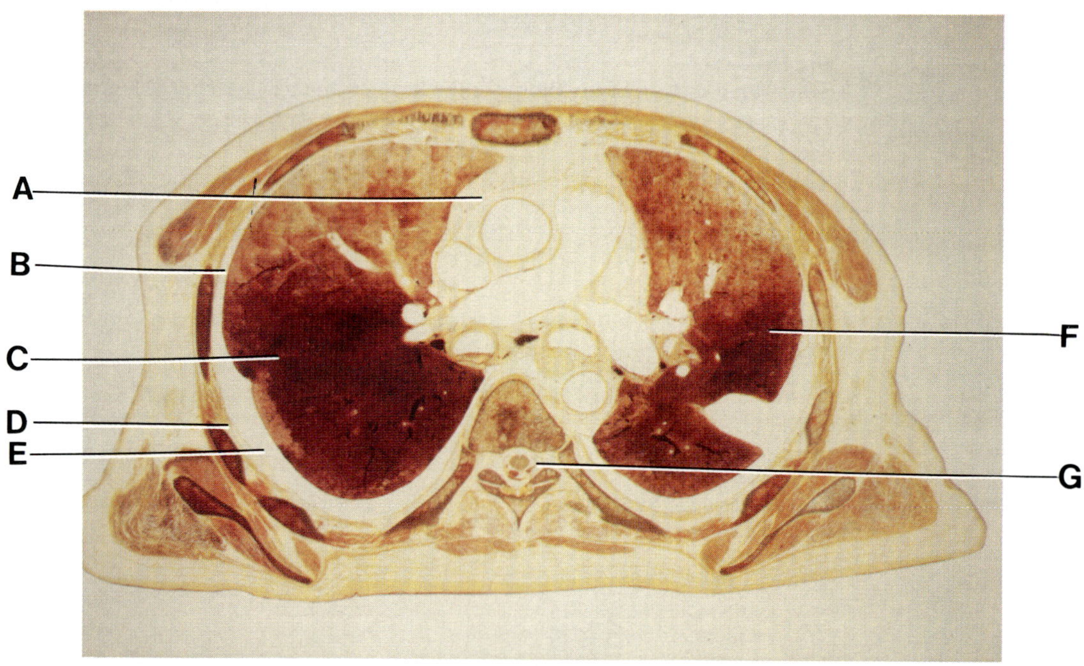

A. Mediastinal pleura
B. Visceral pleura
C. Oblique fissure of right lung
D. Parietal costal pleura
E. Interpleural space
F. Oblique fissure of left lung
G. Dura mater

Figure 9–18
Transverse section of the thorax at the level of thoracic vertebra 5, highlighting enclosing unit structures.

Figure 9–19 is an MRI transverse slice of the thorax. This section is across the upper part of the inferior mediastinum.

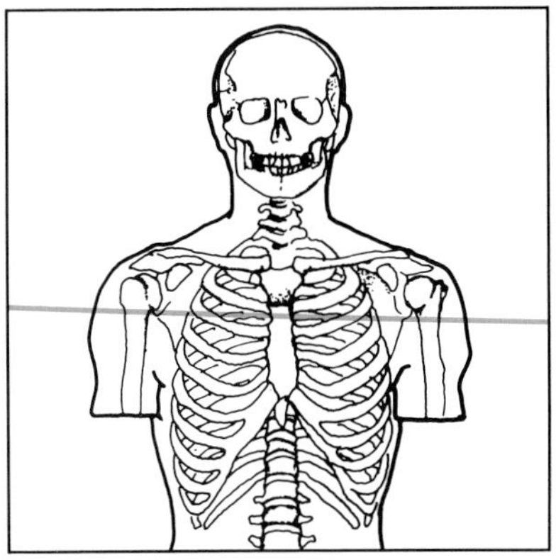

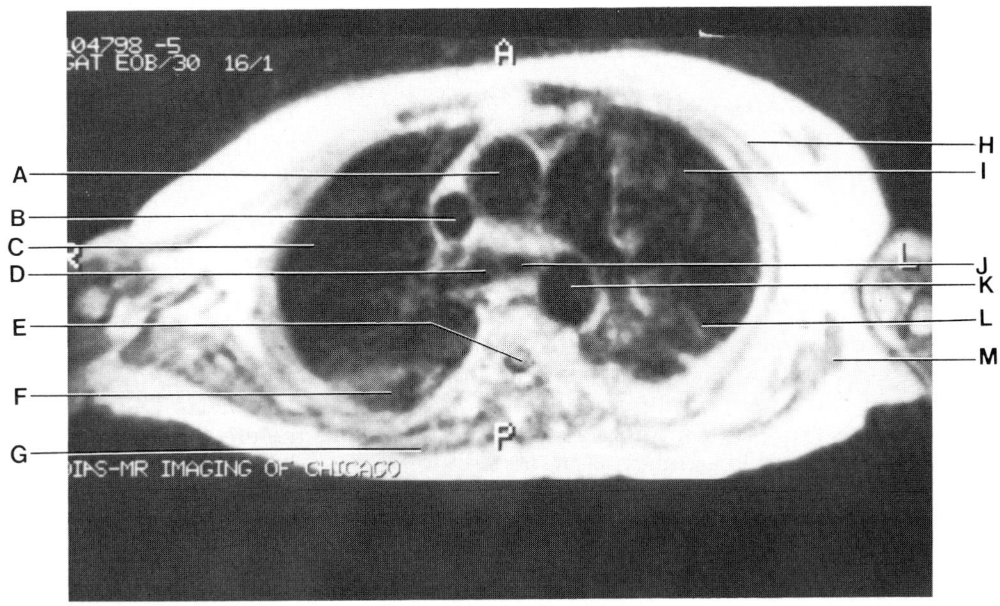

A. Ascending aorta
B. Superior vena cava
C. Superior lobe of right lung
D. Right primary bronchus
E. Spinal cord
F. Inferior lobe of right lung
G. Trapezius muscle
H. Pectoralis major muscle
I. Superior lobe of left lung
J. Left primary bronchus
K. Descending thoracic aorta
L. Inferior lobe of left lung
M. Latissimus dorsi muscle

Figure 9–19
MRI transverse section of the thorax, highlighting musculoskeletal, neurovascular, and visceral unit structures.

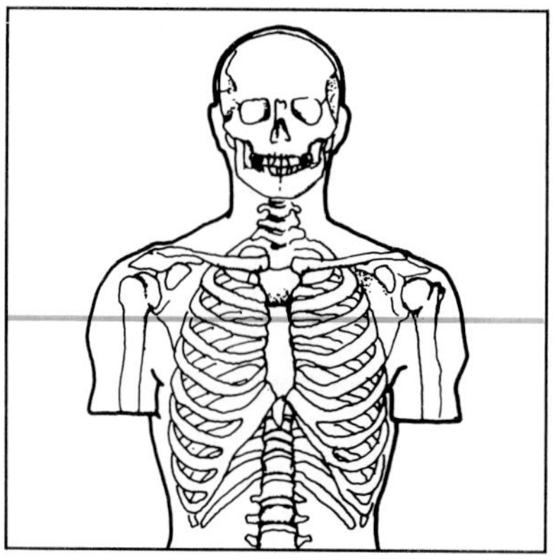

MODULE 5

This is a transverse (axial) section at the level of thoracic vertebra 6.

MUSCULOSKELETAL UNIT (Fig. 9–20)

I. Skeletal Portion

Body of a Thoracic Vertebra (*C*). An oval mass of bone tissue anterior to the spinal foramen. This is the largest part and anterior feature of a vertebra.

Transverse Process of a Thoracic Vertebra (*I*). A lateral projection of bilateral bone tissue anterior to the deep back muscles and posterior to the pedicle of a vertebra. This feature is between the pedicle and lamina of a vertebra.

Lamina of a Thoracic Vertebra (*J*). A flat, layered bilateral feature between the transverse process and the spinous process.

Spinous Process of a Thoracic Vertebra (*K*). A median posterior projection located between the deep back muscles. Its most posterior part is between the medial ends of the trapezius muscle.

Sternum (*B*). A rectangular bone structure in the anterior median part of this section.

Head of a Rib (*S*). A small, bilateral, rounded part of a rib, located adjacent to the posterior part of the body of a vertebra, forming the costovertebral capitular joints.

Tubercle of a Rib (*U*). A small, bilateral "swelling" on a rib, articulating with the transverse process of a thoracic vertebra to form the costotransverse joints.

Body (Shaft) of a Rib (*P*). A bony, elongated, bilateral slice of a rib part between the rib tubercle and the costochondral joint. These bony sections are seen on the peripheral part of the thorax.

Costal Cartilage (*N*). A bilateral slice of hyaline cartilage that connects the anterior part of the body of the rib to the sternum.

Sternocostal Joint (*A*). A simple, bilateral synovial joint formed by costal cartilage and the sternum.

Scapula (*T*). A thin, transversely oblique, bilateral bone slice between the infraspinatus and the subscapularis muscles. The medial end is slightly posterior to the lateral end.

II. Muscular Portion

Infraspinatus Muscle (*G*). A rounded, triangular, bilateral muscle section posterior to the scapula and medial to the teres major muscle. The medial end is the thinnest part of this slice.

Teres Major Muscle (*H*). A rounded, bilateral muscle tissue lateral to the infraspinatus and adjacent to the posterior medial part of the latissimus dorsi muscle in this thoracic section.

Latissimus Dorsi Muscle (*F*). An oblong, bilateral muscle structure. Its posterior part is immediately lateral to the teres major muscle, and its anterior part extends laterally to the serratus anterior muscle.

Trapezius Muscle (*W*). A transversely positioned, bilateral muscle strip posterior to the rhomboid major, longissimus thoracis iliocostalis, and spinalis thoracis muscles. Its medial end is thinner than its lateral end.

Rhomboid Major Muscle (*V*). A narrow, triangular, bilateral muscle tissue. The medial part is anterior to the trapezius muscle, and the lateral part is medial to the subscapularis muscle. The base of this triangular muscle slice is adjacent to the subscapularis muscle.

Spinalis Thoracis Muscle (*M*). The most medial of the erector spinae muscles; superficial to the multifidus muscle and lateral to the posterior part of the spinous process of a vertebra. This bilateral muscle is deep to the medial part of the trapezius muscle.

Longissimus Thoracis Muscle (*Y*). One of the erector spinae muscle group, located between the spinalis thoracis and the iliocostalis thoracis muscles. This muscle is bilateral.

Iliocostalis Thoracis Muscle (*X*). The most lateral member of the erector spinae muscle group, seen lateral to the longissimus thoracis muscle and deep to the trapezius muscle. This muscle is bilateral.

Multifidus Muscle (*L*). A bilateral muscle section of the second layer (transversospinalis) of deep back muscles, adjacent to the posterior surfaces, transverse process, and lamina of a vertebra.

Serratus Anterior Muscle (*R*). An elongated, irregularly shaped, bilateral muscle section lateral to the body of ribs anterior to the subscapularis muscle.

Pectoralis Major Muscle (*O*). A transversely placed, bilateral, thin muscle strip anterior to the costal cartilage and the anterior part of a rib body (shaft).

Pectoralis Minor Muscle (*Q*). A small, bilateral strip of muscle deep to the posterior part of the pectoralis major muscle.

Intercostal Muscles (*D*). Seen as a strip of muscle fibers closely associated with the ribs.

Subscapularis Muscle (*E*). A transversely oblique, bilateral muscle slice anterior to the scapula. The medial end is slightly posterior to the lateral end.

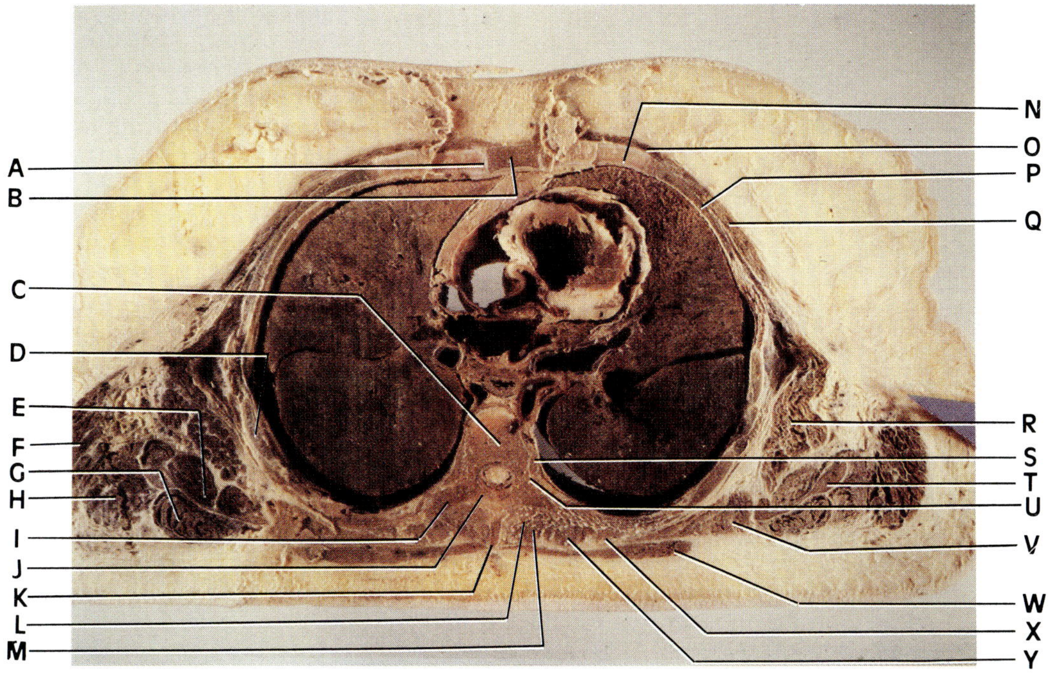

A. Sternocostal joint
B. Sternum
C. Body of thoracic vertebra
D. Intercostal muscles
E. Subscapularis muscle
F. Latissimus dorsi muscle
G. Infraspinatus muscle
H. Teres major muscle
I. Transverse process of thoracic vertebra
J. Lamina of thoracic vertebra
K. Spinous process of thoracic vertebra
L. Multifidus muscle
M. Spinalis thoracis muscle
N. Costal cartilage
O. Pectoralis major muscle
P. Body (shaft) of rib
Q. Pectoralis minor muscle
R. Serratus anterior muscle
S. Head of rib
T. Scapula
U. Tubercle of rib
V. Rhomboid major muscle
W. Trapezius muscle
X. Iliocostalis thoracis muscle
Y. Longissimus thoracis muscle

Figure 9–20
Transverse (axial) section of the thorax at the level of thoracic vertebra 6, highlighting musculoskeletal unit structures.

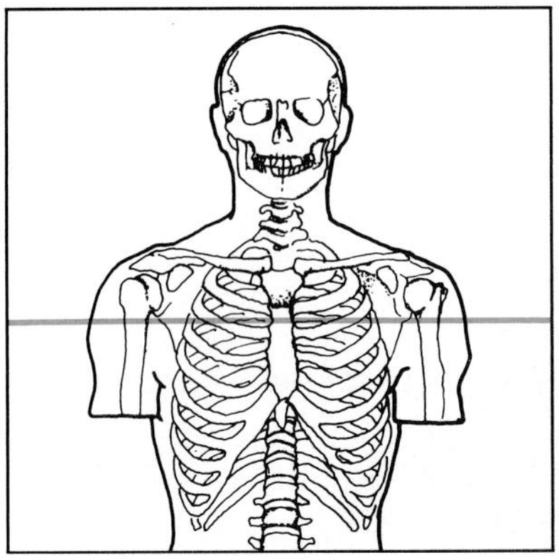

NEUROVASCULAR UNIT (Fig. 9–21)

I. Arteries

Descending Thoracic Aorta (*C*). A partially collapsed, hollow vascular organ anterior to the left part of the body of a thoracic vertebra. This vessel is to the left of the esophagus.

Intercostal Artery (*D*). A very small, circular feature posterior to the descending thoracic aorta and lateral to the body of a thoracic vertebra.

Origin. Descending thoracic aorta.

Distribution. Intercostal spaces, back muscles, vertebral column, and thoracic wall.

Course. This vessel courses within the intercostal space between the internal intercostal and innermost intercostal muscles. It lies between the intercostal vein above and the intercostal nerve (thoracic nerve) below.

Internal Thoracic Artery (*A*). A small, rounded, hollow, bilateral vascular structure immediately lateral to the internal thoracic vein.

Pulmonary Trunk (*B*). A transversely placed, hollow, wide vascular structure posterior to the pericardium and anterior to the esophagus, azygos vein, and descending thoracic aorta.

II. Veins

Accessory Hemiazygos Vein (*H*). A small, circular, venous structure posterior to the esophagus and the descending thoracic aorta.

Origin. A continuation of the fourth intercostal vein.

Termination. Azygos vein at the seventh cervical vertebra but sometimes in the hemiazygos vein to form a common trunk that empties into the azygos vein.

Course. It passes downward from its origin on the left side of the thoracic vertebrae to reach the level of approximately thoracic vertebra 7, where it tends to cross over and join the azygos vein.

Azygos Vein (*G*). A partially collapsed, rather small venous structure located on the anterior right part of a thoracic vertebra at this level.

Superior Vena Cava (*F*). A hollow, spherical venous structure medial to the superior lobe of the right lung and anterior to the right part of the pulmonary trunk.

Internal Thoracic Vein (*E*). A small, spherical, bilateral venous structure immediately medial to the internal thoracic artery and lateral to the sternocostal joint.

VISCERAL UNIT (Fig. 9–22)

Superior Lobe of the Right Lung (*A*). An irregular, triangular lung tissue slice. The base of this structure is adjacent to the oblique fissure of the right lung. The apex is anterior to the pericardium and posterior to the costal cartilage of a rib.

Inferior Lobe of the Right Lung (*D*). A spongy, large, half moon–shaped lobe posterior to the oblique fissure of the right lung.

Superior Lobe of the Left Lung (*E*). A triangular, curved lung tissue section. The base is adjacent to the oblique fissure of the left lung. The apex curves toward the median and is positioned posteriorly to the costal cartilage of a rib. Its most medial end is posterior to the lateral part of the sternum.

Inferior Lobe of the Left Lung (*H*). A semicircular lung tissue section posterior to the oblique fissure of the left lung.

Esophagus (*I*). A transversely short, thick-walled, hollow organ immediately posterior to the body of a vertebra. It is located between the azygos vein and the descending thoracic aorta.

Left Atrium of the Heart (*F*). A hollow, rounded chamber of the heart to the left of the right atrium. Its wall appears rather thick and is seen within the pericardial cavity. The left atrium receives oxygenated blood from the four pulmonary veins and pumps the blood into the left ventricle.

Right Atrium of the Heart (*B*). A transversely positioned, slitlike structure seen within the pericardial cavity posterior to the medial part of the left atrium. The right atrium receives blood from the vena cavae and coronary sinus and pumps blood into the right ventricle.

Left Primary Bronchus (*G*). A transversely oblique, hollow, elongated structure. The lateral end is slightly more anterior than the medial end. The primary bronchii (left and right) are the first generation of branchings from the trachea.

Right Primary Bronchus (*C*). A hollow, spherical feature anterior to the azygos vein and posterior to the medial part of the superior lobe of the right lung.

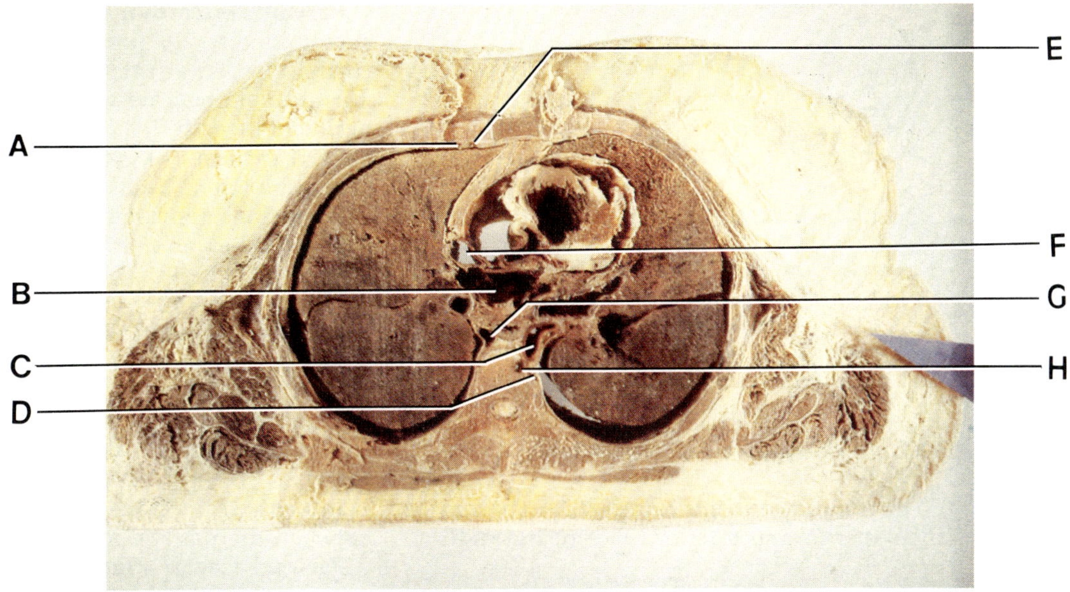

A. Right internal thoracic artery
B. Bifurcation of pulmonary trunk
C. Descending thoracic aorta
D. Intercostal artery
E. Right internal thoracic vein
F. Superior vena cava
G. Azygos vein
H. Accessory hemiazygos vein

Figure 9-21
Transverse section of the thorax at the level of thoracic vertebra 6, highlighting neurovascular unit structures.

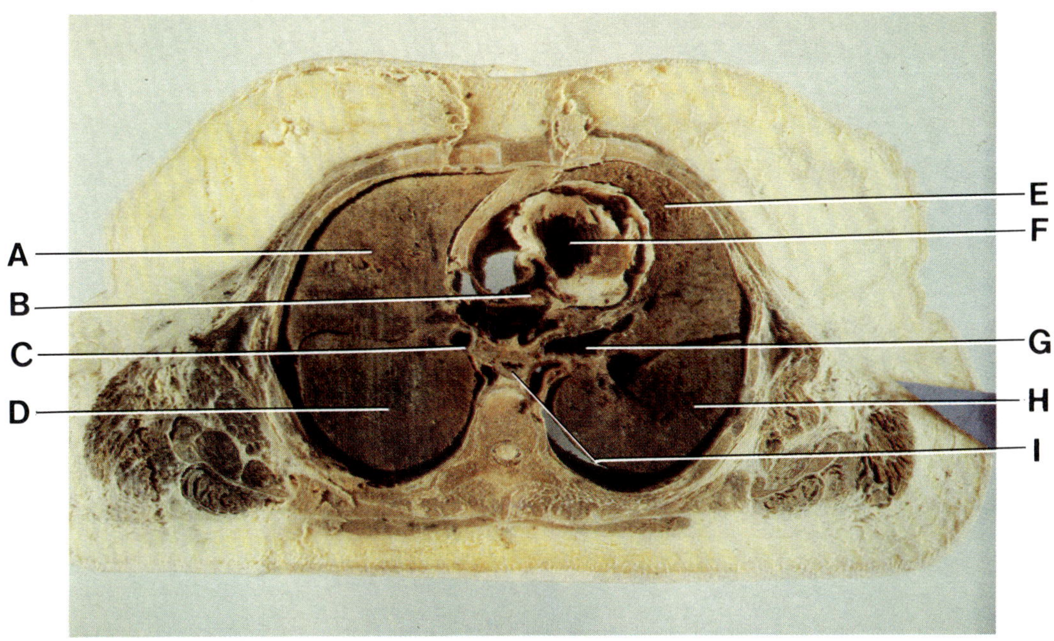

A. Superior lobe of right lung
B. Right atrium of heart
C. Right primary bronchus
D. Inferior lobe of right lung
E. Superior lobe of left lung
F. Left atrium of heart
G. Left primary bronchus
H. Inferior lobe of left lung
I. Esophagus

Figure 9-22
Transverse section of the thorax at the level of thoracic vertebra 6, highlighting visceral unit structures.

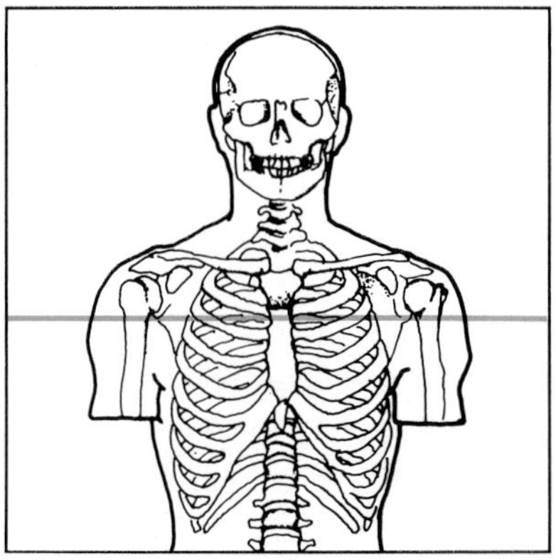

ENCLOSING UNIT (Fig. 9–23)

Oblique Fissure of the Right Lung (*D*). A slit that separates the superior and inferior lobes of the right lung.

Oblique Fissure of the Left Lung (*J*). A slit that separates the superior and inferior lobes of the left lung.

Part of Pericardial Cavity (*C, I*). A circular space surrounded by the pericardium. The right part of the cavity, as seen at this level, is wider than the other portions of the cavity. The heart is located in the pericardial cavity.

Pericardium (*B*). A rather thick membrane that forms the lateral walls of the pericardial cavity.

Posterior Mediastinal Area (*K*). An area that includes the descending thoracic aorta, the origin of the intercostal arteries, and the termination of the intercostal veins.

Middle Mediastinal Area (*H*). An area that includes the heart and the origin or termination of eight great vessels.

Anterior Mediastinal Pleura (*A*). An area that includes the lymph nodes and part of the thymus gland.

Interpleural Space (*E*). A circular, hollow area between the parietal and visceral pleurae.

Parietal Costal Pleura (*F*). A thin sheet of tissue lining the thoracic wall.

Visceral Pleura (*G*). A thin sheet of tissue surrounding both right and left lungs.

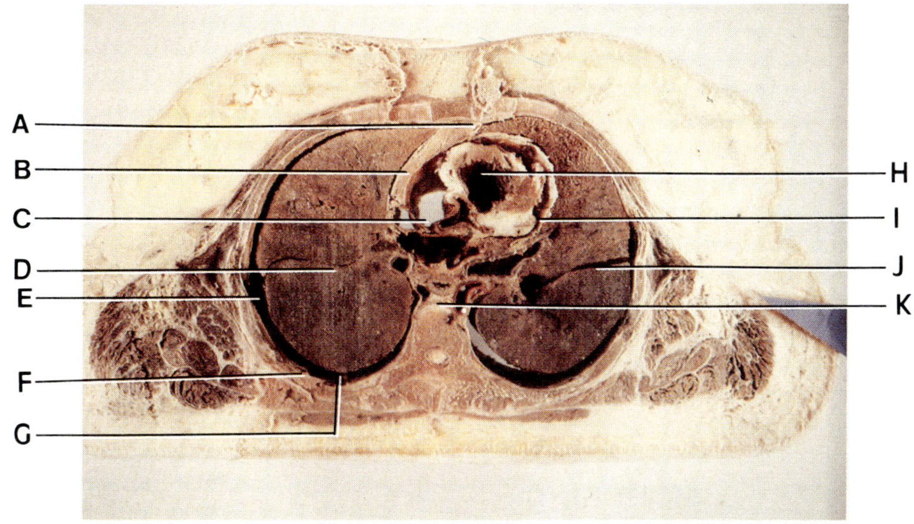

A. Anterior mediastinal pleura
B. Pericardium
C. Part of pericardial cavity
D. Oblique fissure of right lung
E. Interpleural space
F. Parietal costal pleura
G. Visceral pleura
H. Middle mediastinal area
I. Part of pericardial cavity
J. Oblique fissure of left lung
K. Posterior mediastinal area

Figure 9–23
Transverse section of the thorax at the level of thoracic vertebra 6, highlighting enclosing unit structures.

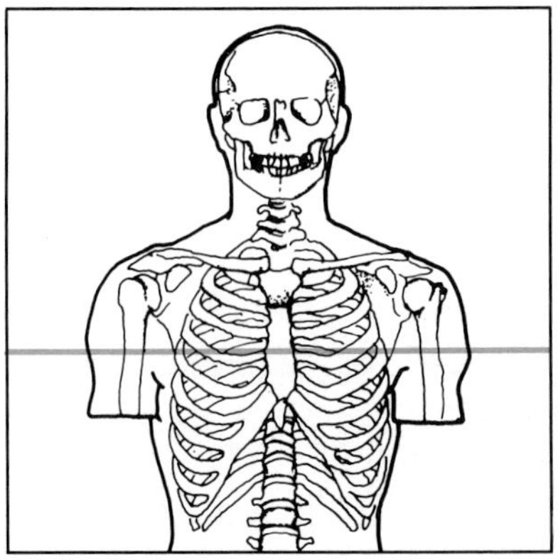

MODULE 6

This is a transverse (axial) section of the vertebral level of thoracic vertebra 7.

MUSCULOSKELETAL UNIT (Fig. 9–24)

I. Skeletal Portion

Body of a Thoracic Vertebra (*O*). An oval aggregate slice of bone tissue, which is the most anterior feature of a vertebra. This feature is also known as the centrum.

Pedicle of a Thoracic Vertebra (*B*). A bilateral strip of bone tissue posterior to the lateral part of the body (centrum) and anterior to the transverse process. Typically, the spinal nerve exits between two adjacent pedicles. For example, the pedicle of each vertebra is notched, and the superior and inferior adjacent vertebrae form a hole (intervertebral foramen) for the exits of the typical spinal nerve.

Lamina of a Thoracic Vertebra (*T*). A flattened, oblique, bilateral bone slice between the transverse process and the spinous process. The posterior end is more medial than the anterior end.

Body (Shaft) of a Rib (*M*). A bilateral slice of bone tissue located between the tubercle and the costochondral end.

Head of a Rib (*Q*). A bilateral, rounded, and most posterior feature of a rib, which forms the costovertebral capitular joint.

Neck of a Rib (*R*). A bilateral, constricted feature between the head and the tubercle of a rib.

Sternum (*A*). A transversely placed, rectangular bone section in the anterior median part of the section.

Costochondral Joint (*L*). A bilateral joint feature of a rib between the rib bone and the corresponding rib (costal) cartilage.

Sternocostal Joint (*J*). A bilateral, simple synovial joint formed by sternum and costal cartilage. There are seven pairs of these joints.

Costal Cartilage (*K*). A bilateral feature formed by hyaline cartilage, which connects the ribs to the sternum.

Costovertebral Capitular Joint (*P*). A bilateral joint between the head of a rib and the demifacet (costal facet) of a vertebra. There are 12 pairs of costovertebral capitular joints in a human body.

II. Muscular Portion

Trapezius Muscle (*S*). A transversely placed, thin, bilateral muscle strip immediately posterior to the rhomboid major muscle and lateral to the posterior part of the spinous process.

Latissimus Dorsi Muscle (*D*). A sickle-shaped, bilateral muscle slice lateral to the serratus anterior muscle. Both the anterior and posterior ends are thin compared with the rest of the muscle sections. The concave portion of this muscle is on its medial part.

Rhomboid Major Muscle (*F*). A triangular feature posterior and medial to the serratus anterior muscle. The base of the triangle is lateral, and the apex is toward the middle.

Spinalis Thoracis Muscle. A small, bilateral muscle section between the posterior part of the spinous process and the longissimus thoracis muscle. This muscle is superficial to the multifidus muscle.

Longissimus Thoracis Muscle (*H*). A small, bilateral muscle aggregate between the spinalis thoracis and iliocostalis thoracis muscles.

Iliocostalis Thoracis Muscle (*G*). A bilateral muscle slice and the most lateral member of the erector spinae group. This muscle is lateral to the longissimus thoracis muscle and anterior to the trapezius muscle.

Serratus Anterior Muscle (*C, N*). An irregularly shaped, bilateral muscle section lateral to the rib cage and anterior to the rhomboid major muscle.

Intercostal Muscle (*E*). There are three different intercostal muscles. The outermost intercostal muscle (external intercostal) shows a thin slice between rib bodies near the lateral edge of this feature. The adjacent medial slice is known as the internal intercostal and is seen between the external and innermost intercostal muscles. The third and deepest muscle between the ribs is sometimes variable and not clearly seen but when found is attached to the inferior medial surface of a rib and the superior medial edge of an adjacent rib. The neurovascular unit in the thoracic wall lies between the internal and innermost intercostal muscles.

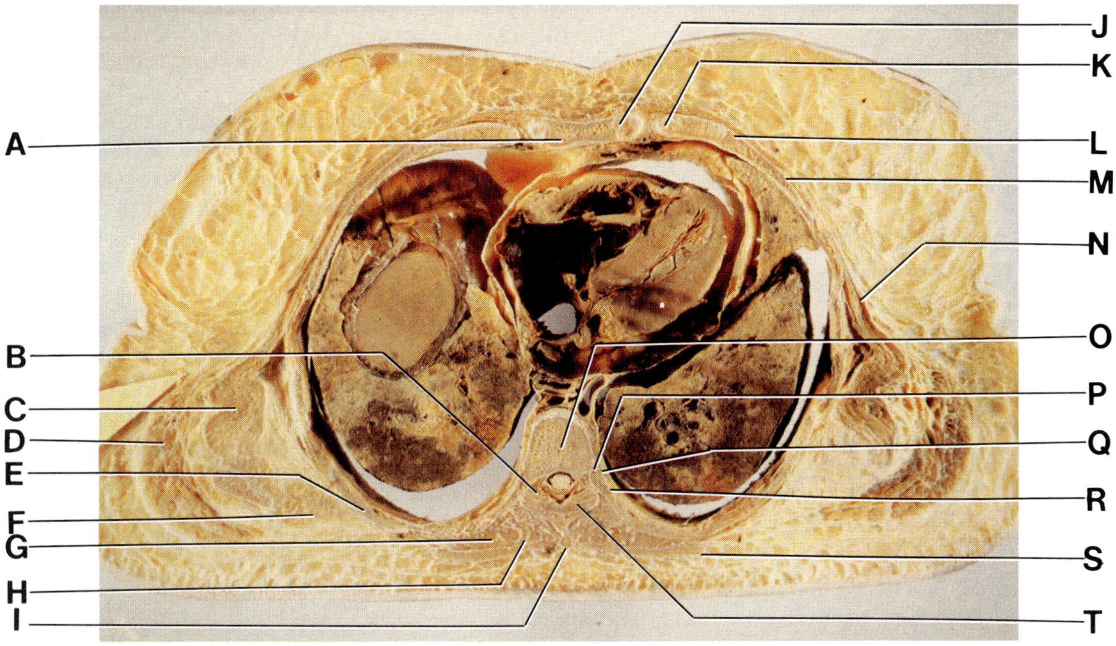

A. Sternum
B. Pedicle of thoracic vertebra
C. Serratus anterior muscle (right)
D. Latissimus dorsi muscle
E. Intercostal muscles
F. Rhomboid major muscle
G. Iliocostalis thoracis muscle
H. Longissimus thoracis muscle
I. Spinous process of thoracic vertebra
J. Sternocostal joint
K. Costal cartilage
L. Costochondral joint
M. Body (shaft) of rib
N. Serratus anterior muscle (left)
O. Body of thoracic vertebra
P. Costovertebral capitular joint
Q. Head of rib
R. Neck of rib
S. Trapezius muscle
T. Lamina of thoracic vertebra

Figure 9–24
Transverse (axial) section of the thorax at the level of thoracic vertebra 7, highlighting musculoskeletal unit structures.

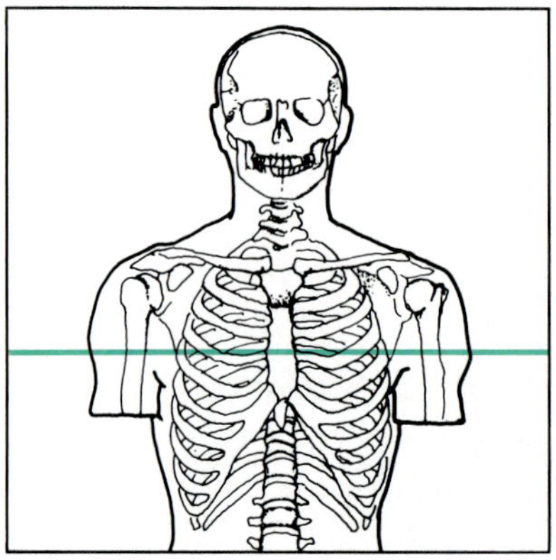

NEUROVASCULAR UNIT (Fig. 9–25)

I. Arteries

Ascending Aorta (*B*). A small, slitlike feature located at the superior part of the left ventricle. This vascular structure originates from the left ventricle.

Descending Thoracic Aorta (*C*). A partially collapsed, hollow vascular artery located to the left of the azygos vein and anterior to the left part of the vertebral body.

Pulmonary Trunk (*A*). A slitlike vascular structure at the superior part of the right ventricle near the interventricular septum. The pulmonary trunk arises from the right ventricle.

II. Veins

Azygos Vein (*E*). A partially collapsed venous structure anterior to the body of a thoracic vertebra and to the right of the descending thoracic aorta. It is posterior to the esophagus.

Inferior Vena Cava (*D*). A large, hollow, circular vascular structure within the right atrium, anterior to the esophagus and to the right of the ascending aorta.

A CT transverse image of the thorax in Figure 9–26 highlights musculoskeletal and neurovascular unit structures. This CT slice is across the inferior mediastinal area of the thorax.

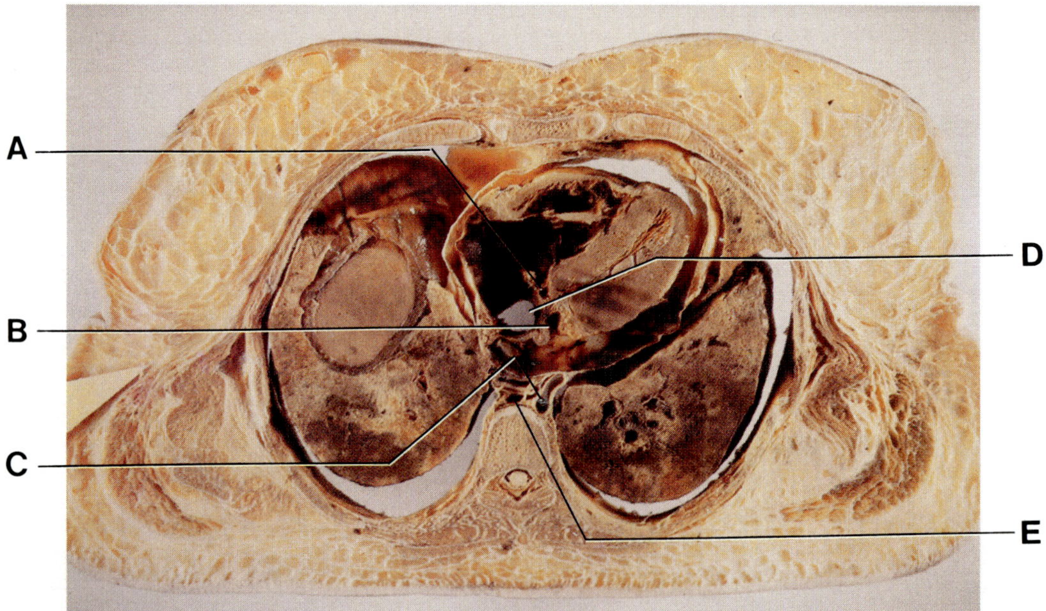

A. Pulmonary trunk
B. Ascending aorta
C. Descending thoracic aorta
D. Inferior vena cava
E. Azygos vein

Figure 9–25
Transverse section of the thorax at the level of thoracic vertebra 7, highlighting neurovascular unit structures.

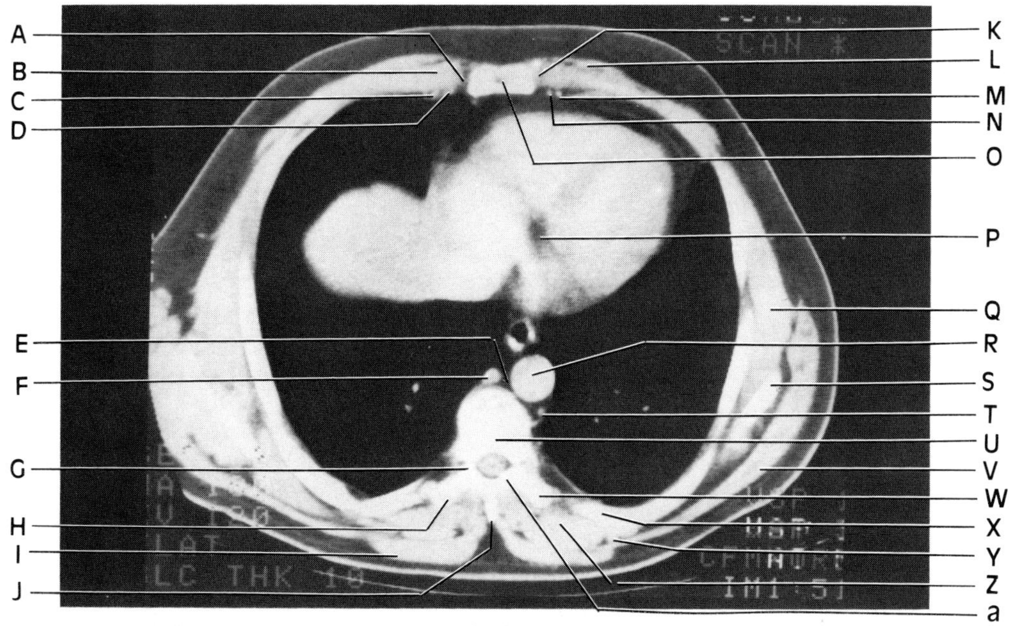

A. Sternocostal joint
B. Costal cartilage
C. Right internal thoracic artery
D. Right internal thoracic vein
E. Thoracic duct area
F. Azygos vein
G. Pedicle of thoracic vertebra
H. Transverse process of thoracic vertebra
I. Trapezius muscle
J. Spinous process of thoracic vertebra
K. Sternocostal joint
L. Pectoralis major muscle
M. Left internal thoracic artery
N. Left internal thoracic vein
O. Sternum
P. Inferior vena cava
Q. Serratus anterior muscle
R. Descending thoracic aorta
S. Scapula
T. Hemiazygos vein
U. Body of thoracic vertebra
V. Latissimus dorsi muscle
W. Transverse process of thoracic vertebra
X. Neck of rib
Y. Rib tubercle
Z. Erector spinae (sacrospinalis) muscle
a. Lamina of thoracic vertebra

Figure 9-26
CT (with contrast) transverse image of the thorax, highlighting both musculoskeletal and neurovascular unit structures.

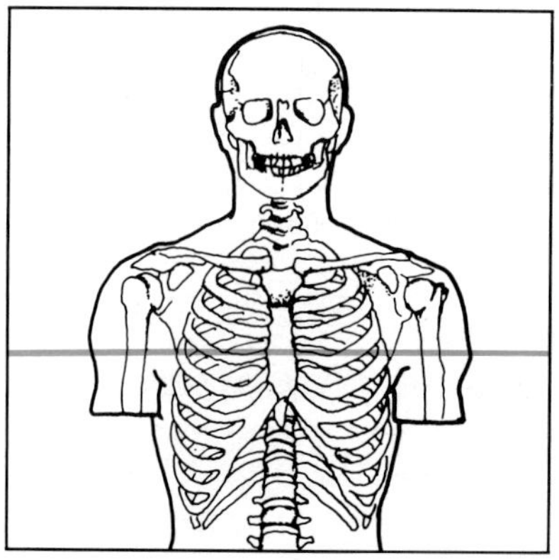

VISCERAL UNIT (Fig. 9–27)

Superior Lobe of the Right Lung (*B*). An anterior, oval lung slice seen lateral to the heart and adjacent to a portion of the diaphragm. A portion of the right liver is seen enclosed within a part of the diaphragm adjacent to the superior lobe of the right lung.

Inferior Lobe of the Right Lung (*F*). A rather large, oval part of the right lung, located in the posterior part of the thoracic cavity.

Superior Lobe of the Left Lung (*A*). A rather small, triangular lung slice. The base of this triangular feature is adjacent to the lateral part of the oblique fissure.

Inferior Lobe of the Left Lung (*K*). A large, sickle-shaped lung section immediately posterior to the oblique fissure of the left lung. The concave surface is adjacent to the pericardium.

Left Ventricle of the Heart (*I*). This is located in the pericardial cavity. From the left atrioventricular orifice, the left ventricle extends forward and to the left as far as the apex. At this level, the thickness of the wall of the chamber is seen to be three times that of the right ventricle. The left ventricle forms the apex of the heart and pumps blood into the ascending aorta, to be distributed to all organs in the body.

Apex of the Heart (*H*). A blunt, pointed feature of the left ventricle, seen within the pericardial cavity and the distal end of the left ventricle.

Right Ventricle of the Heart. A triangular chamber in the anterior part of the heart. The wall of this chamber is three times thinner than the wall of the left ventricle. This ventricle extends from the right atrium to the part of the heart immediately anterior to the apex of the heart.

Right Atrium of the Heart (*C*). A chamber seen at this level in the median and adjacent to the right ventricle. A rather large, oval vascular structure (inferior vena cava) is in the posterior part of this chamber.

Papillary Muscles of the Heart (*J*). These are large, fiber-like features in the left ventricle at this level. There are normally three papillary muscles in the right ventricle and two in the left ventricle. Papillary muscles shorten to sustain tension on the atrioventricular (A-V) valves, to inhibit these valves from protruding into the atrium during ventricular contraction.

Esophagus (*L*). A short, transversely placed, slitlike structure posterior to the right part of the heart, as seen at this level, and anterior to the azygos vein.

Mammary Gland (*G*). An anterior enlargement and thickening of the subcutaneous tissue anterior to the pectoralis major muscle. The lateral part of this organ extends as far back as the midaxillary line. This structure is bilateral.

Diaphragm (*D*). A circular slice of the diaphragm is seen at this level, which surrounds an oval slice of the right lobe of the liver. This slice of diaphragm is adjacent to the right lung.

Right Lobe of the Liver (*E*). An oval section of the right lobe of the liver is seen within a spherical sliced portion of the diaphragm.

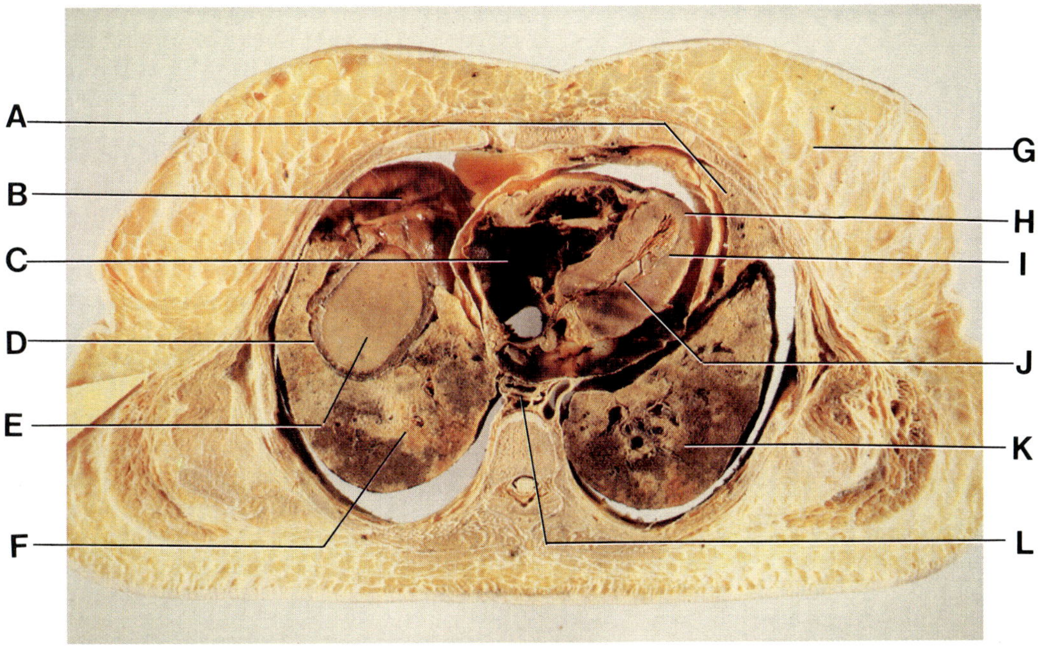

A. Superior lobe of left lung
B. Superior lobe of right lung
C. Right atrium of heart
D. Diaphragm
E. Right lobe of liver
F. Inferior lobe of right lung
G. Mammary gland
H. Apex of heart
I. Left ventricle of heart
J. Papillary muscles of heart
K. Inferior lobe of left lung
L. Esophagus

Figure 9–27
Transverse section of the thorax at the level of thoracic vertebra 7, highlighting visceral unit structures.

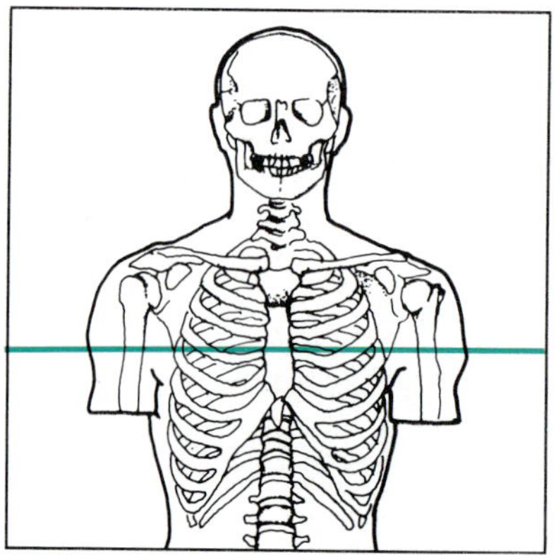

ENCLOSING UNIT (Fig. 9–28)

Interpleural Space (Pleural Cavity) (*F*). A bilateral, semicircular space between the visceral and parietal pleurae. The posterior part of this space on the right is much larger than the space on the left.

Visceral Pleura (*I*). A thin, bilateral membrane that encases the lungs.

Parietal Pleura (*E*). A thin, bilateral sheet of tissue that lines the inner wall of the thorax.

Pericardial Cavity (*B*). A semicircular space that surrounds the heart and is enclosed by the pericardium.

Pericardium (*G*). A rather thick membrane that surrounds the pericardial cavity. This is the parietal membrane of the pericardial cavity.

Anterior Mediastinal Area (*A*). The space in front of the heart that contains fat and lymph nodes.

Middle Mediastinal Area (*C*). The space surrounded by pericardium that contains the heart and is the origin or termination of eight great vessels.

Posterior Mediastinal Area (*D*). The space in back of the heart that contains the descending thoracic aorta, the origin of the intercostal arteries, and the termination of the intercostal veins.

Oblique Fissure of the Left Lung (*H*). A slitlike feature that separates the superior and inferior lobes of the left lung.

Figure 9–29 shows a CT (with contrast) image that highlights visceral and enclosing unit structures. This slice is across the mediastinal area of the thorax.

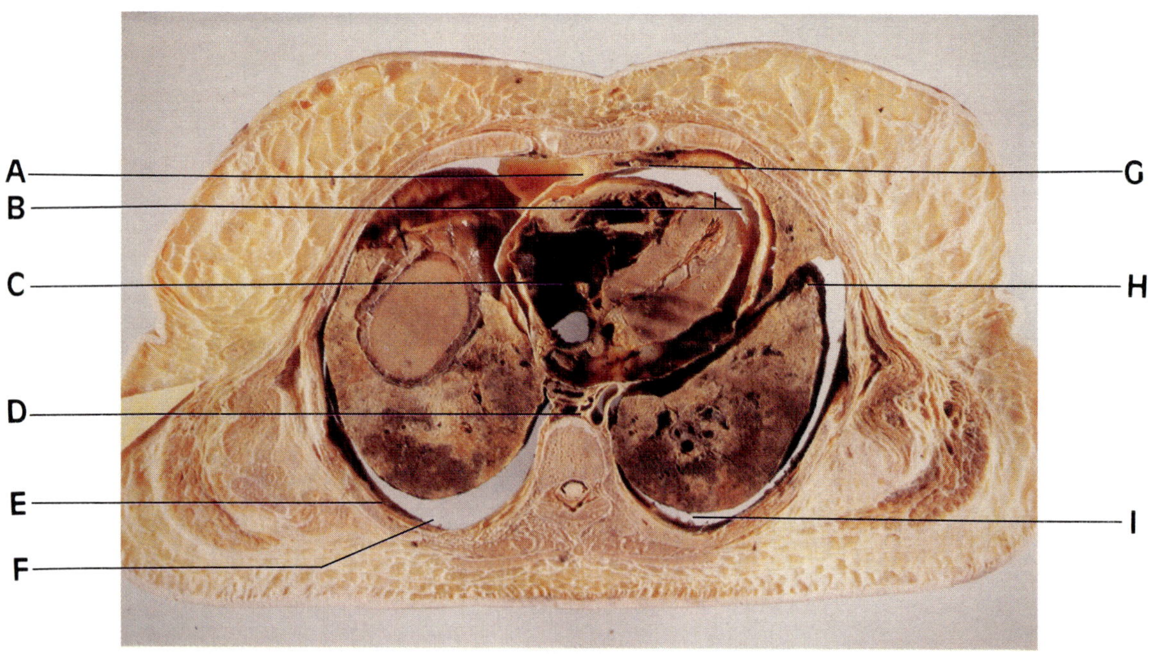

A. Anterior mediastinal area
B. Pericardial cavity
C. Middle mediastinal area
D. Posterior mediastinal area
E. Parietal pleura
F. Interpleural space (pleural cavity)
G. Pericardium
H. Oblique fissure of left lung
I. Visceral pleura

Figure 9–28
Transverse section of the thorax at the level of thoracic vertebra 7, highlighting enclosing unit structures.

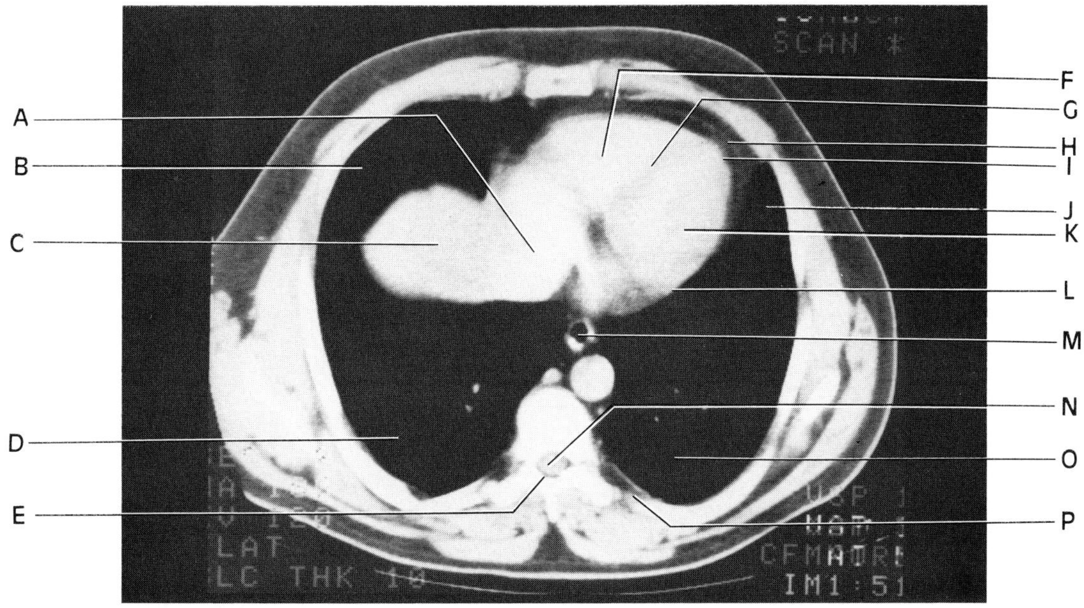

A. Right atrium of heart
B. Superior lobe of right lung
C. Hepatic dome of diaphragm (cupula of right diaphragm)
D. Inferior lobe of right lung
E. Dura mater
F. Right ventricle of heart
G. Interventricular septum
H. Pericardium
I. Apex of heart
J. Superior lobe of left lung
K. Left ventricle of heart
L. Coronary sulcus (atrioventricular sulcus)
M. Esophagus
N. Spinal cord
O. Inferior lobe of left lung
P. Parietal costal pleura

Figure 9–29
CT (with contrast) image of the thorax, highlighting both visceral and enclosing unit structures.

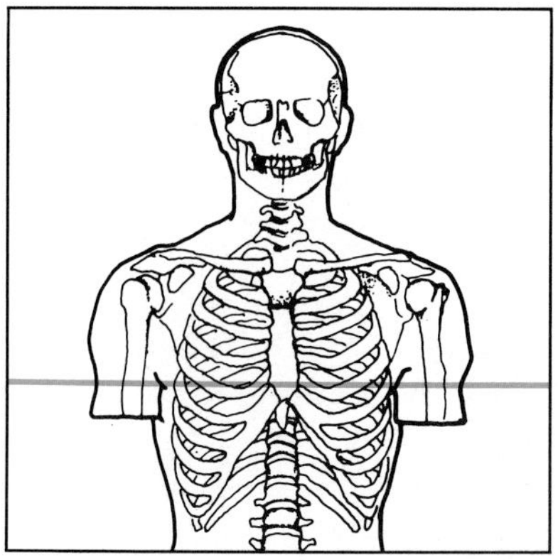

MODULE 7

This is a transverse (axial) section at the vertebral level of thoracic vertebra 8.

MUSCULOSKELETAL UNIT (Fig. 9–30)

I. Skeletal Portion

Body of a Thoracic Vertebra (*M*). An oval aggregate slice of bone tissue that is the most anterior feature of a vertebra. The lateral parts are directly anterior to the pedicles.

Pedicle of a Thoracic Vertebra (*E*). A rather thin, bilateral bone section immediately posterior to the lateral parts of the body and anterior to the transverse process.

Transverse Process of a Thoracic Vertebra (*P*). A lateral projection posterior to the pedicle. The lateral end of this bilateral feature is more posterior than its medial end.

Spinous Process of a Thoracic Vertebra (*Q*). An elongated (anterior-to-posterior) feature that projects posteriorly in the median line. This is the most posterior feature of a vertebra.

Head of a Rib (*G*). A rounded, bilateral bone section and the most posterior feature of a rib, which articulates with the demifacet (costofacet) on the body of a vertebra to form the costovertebral capitular joint.

Body (Shaft) of a Rib (*C*). A slightly elongated, bilateral bone slice located between the tubercle and the costochondral end.

Costal Cartilage (*A*). A strip of hyaline cartilage that connects the rib shaft (body) to the sternum.

Costotransverse Joint (*H*). A bilateral joint between the rib tubercle and the costotransverse facet. There are ten pairs of costotransverse joints in a human body.

Costovertebral Capitular Joint (*N*). A bilateral joint between the head of the rib and the demifacet (costofacet) of the vertebra. There are twelve pairs of costovertebral capitular joints in a human body.

Costochondral Joint (*B*). A bilateral joint between the anterior part of the rib shaft (body) and its corresponding costal cartilage. There are ten pairs of these joints in a human body.

Body of the Sternum (*K*). A rectangular bone slice in the anterior median part of this body section. The body of the sternum is located between the sternal angle and the xiphisternal joint.

Inferior Angle of the Scapula (*O*). A bilateral, thin, triangular bone section anterior to the medial part of the latissimus dorsi muscle.

II. Muscular Portion

Trapezius Muscle (*J*). The most posterior muscle at this level, seen as a thin, elongated (medial-to-lateral), bilateral muscle slice posterior to the rhomboid major muscle. The trapezius muscle is a superficial back muscle named on the basis of shape and is subdivided into superior and inferior portions.

Rhomboid Major Muscle (*I*). This is seen here bilaterally deep to the trapezius muscle as an elongated (medial-to-lateral), sickle-shaped structure. The medial end, as seen at this level, is thicker than the lateral end.

Latissimus Dorsi Muscle (*F*). This appears bilaterally as a curved, elongated, rather large muscle section posterior to the serratus anterior muscle, and the medial part is posterior to the inferior angle of the scapula. The concave surface is medial to the convex surface. It is a broad, flat, triangular, superficial back muscle with its terminal portion along the posterior axillary fold.

Serratus Anterior Muscle (*L*). A curved, bilateral muscle section lateral to the intercostal muscles and ribs. The concave surface of this muscle slice is medial to the convex surface.

Intercostal Muscles (*D*). These are seen between the ribs at this level. There are external and internal intercostal as well as innermost intercostal muscles.

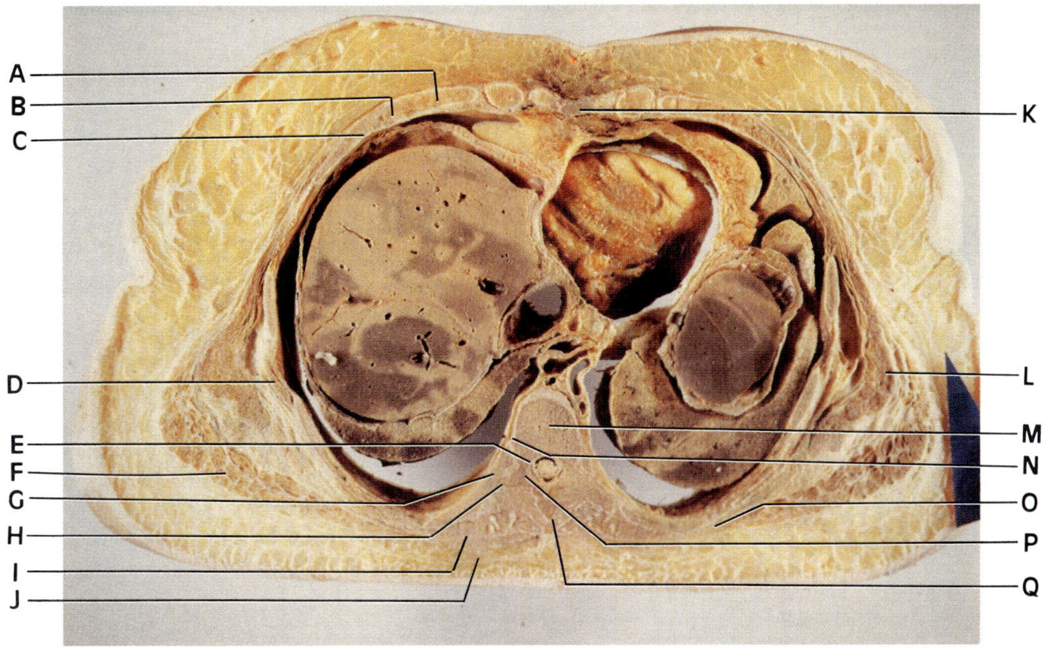

A. Costal cartilage
B. Costochondral joint
C. Body (shaft) of rib
D. Intercostal muscles
E. Pedicle of vertebra
F. Latissimus dorsi muscle
G. Head of rib
H. Costotransverse joint
I. Rhomboid major muscle
J. Trapezius muscle
K. Body of sternum
L. Serratus anterior muscle
M. Body of vertebra
N. Costovertebral capitular joint
O. Inferior angle of scapula
P. Transverse process of vertebra
Q. Spinous process of vertebra

Figure 9-30
Transverse (axial) section of the thorax at the level of thoracic vertebra 8, highlighting musculoskeletal unit structures.

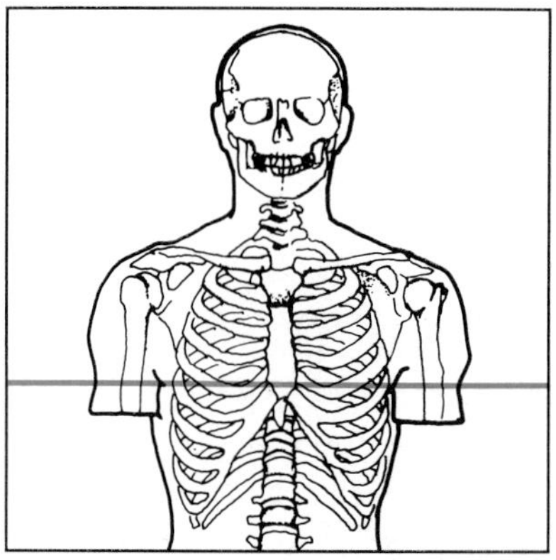

NEUROVASCULAR UNIT (Fig. 9–31)

I. Arteries

Descending Thoracic Aorta (*B*). A partially collapsed, hollow vascular structure seen anterior to the hemiazygos vein and left of the azygos vein as well as on the left side of the median.

Superior Epigastric Artery (*A*). A small, hollow, rounded, bilateral vascular feature lateral to the internal thoracic vein.

II. Veins

Superior Epigastric Vein (*C*). A small, rounded, bilateral, hollow venous structure medial to the internal thoracic vein and lateral to the edge of the sternum.

Azygos Vein (*E*). A small, hollow, spherical feature posterior to the esophagus and to the right of the descending thoracic aorta. It is anterior to the body of a vertebra.

Hemiazygos Vein (*F*). A small, hollow, spherical venous structure posterior to the descending thoracic aorta and to the left of the anterior part of the body of a thoracic vertebra.

Origin. The combination of the left ascending lumbar and left subcostal veins.

Termination. Azygos vein.

Course. It courses superiorly anterior to the vertebral column to reach the eighth thoracic vertebra, where it courses to arrive at the right side and terminates into the azygos vein.

Inferior Vena Cava (*D*). An oval, large, hollow vascular structure on the right side of the median, anterior to the esophagus and left of the right lobe of the liver.

Origin. Right and left iliac veins in front of the right part of lumbar vertebral body 5.

Termination. Right atrium of the heart.

Course. It ascends on the right anterior part of the vertebral column to reach a groove between the caudate lobe and the bare area on the deep surface of the right lobe of the liver. Then the inferior vena cava passes through the caval hiatus of the diaphragm to the right of the aorta, to pierce the fibrous pericardium, and passes posteriorly to enter the posterior inferior part of the right atrium.

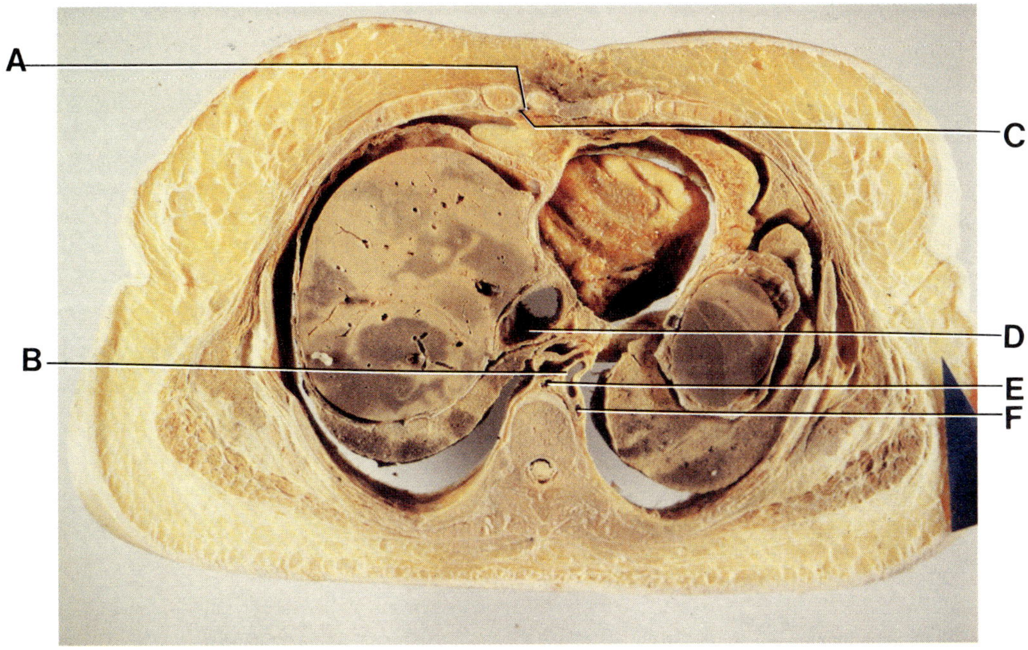

A. Right superior epigastric artery
B. Descending thoracic aorta
C. Right superior epigastric vein
D. Inferior vena cava
E. Azygos vein
F. Hemiazygos vein

Figure 9–31
Transverse section of the thorax at the level of thoracic vertebra 8, highlighting neurovascular unit structures.

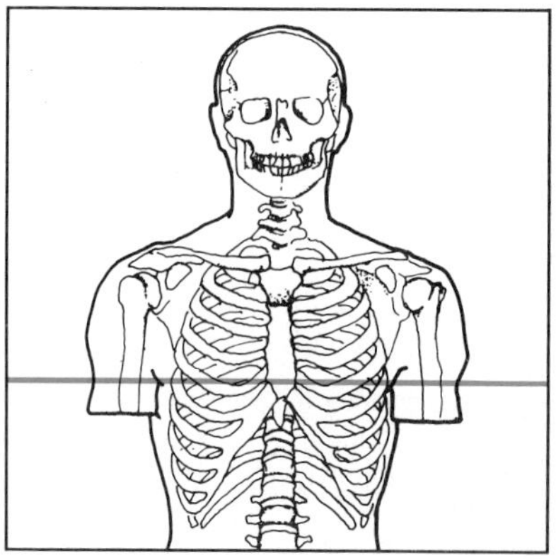

Inferior Lobe of the Right Lung (*C*). A curved feature, with its concave surface anterior and adjacent to a slice of liver. The medial end of this lung tissue is adjacent to the posterior part of the inferior vena cava.

Heart (*E*). This is seen here on the left side of the body section as a triangular structure surrounded by the pericardium located in the pericardial cavity. The pointed portion is more lateral than the other parts of the heart seen at this level.

Apex of the Heart (*F*). A narrow portion of the heart formed by the left ventricle and located toward the left anterior portion at this level.

Spinal Cord (*D*). A rounded, white nerve structure surrounded by meninges posterior to the body of a thoracic vertebra.

VISCERAL UNIT (Fig. 9–32)

Right Lobe of the Liver (*B*). A large, rounded liver section right of the inferior vena cava and anterior to the inferior lobe of the right lung. The anterior portion of this liver slice is adjacent to a part of the diaphragm. It is significant to note that the liver lies in a position corresponding approximately to a vertebral level between thoracic vertebra 8 and lumbar vertebra 2.

Spleen (*H*). A spherical structure surrounded by the diaphragm tissue, anterior to the inferior lobe of the left lung. The spleen is located primarily in the posterior part of the left hypochondriac region, which lies between the fundus of the stomach and the diaphragm.

Diaphragm (*G*). Two portions of the diaphragm are seen at this level. The right part is seen surrounding the right lobe of the liver, and the left part (*A*) is around the spleen slice. The diaphragm is located approximately at a vertebral level between thoracic vertebra 8 and lumbar vertebra 2.

Inferior Lobe of the Left Lung (*I*). A C-shaped lung section located in the posterior part of the pleural cavity. The concave surface is anterior and adjacent to a part of the diaphragm.

ENCLOSING UNIT (Fig. 9–33)

Interpleural Space (Pleural Cavity) (*B*). A bilateral, C-shaped, hollow area between the parietal and visceral pleurae. The concave portion is medial to the convex portion, and the medial ends are adjacent to the mediastinum.

Visceral Pleura (*F*). A thin, bilateral sheet of tissue (membrane) that encases the lungs.

Parietal Pleura (*G*). A bilateral, thin sheet of tissue (membrane) that lines the inner wall of the pleural cavity.

Pericardial Cavity (*E*). A circular space seen at this level to the left of the right lobe of the liver and anterior/medial to the spleen.

Pericardium (*D*). A rather thick sheet of tissue that encloses the pericardial cavity. The heart is located within this cavity.

Anterior Mediastinal Fat (*A*). An aggregate fatty slice in the anterior medial part of this body section. This fat is located anterior to the pericardium and part of the diaphragm, as seen at this level.

Dura Mater (*C*). A rather thick sheet of tissue lining the inner wall of the spinal foramen.

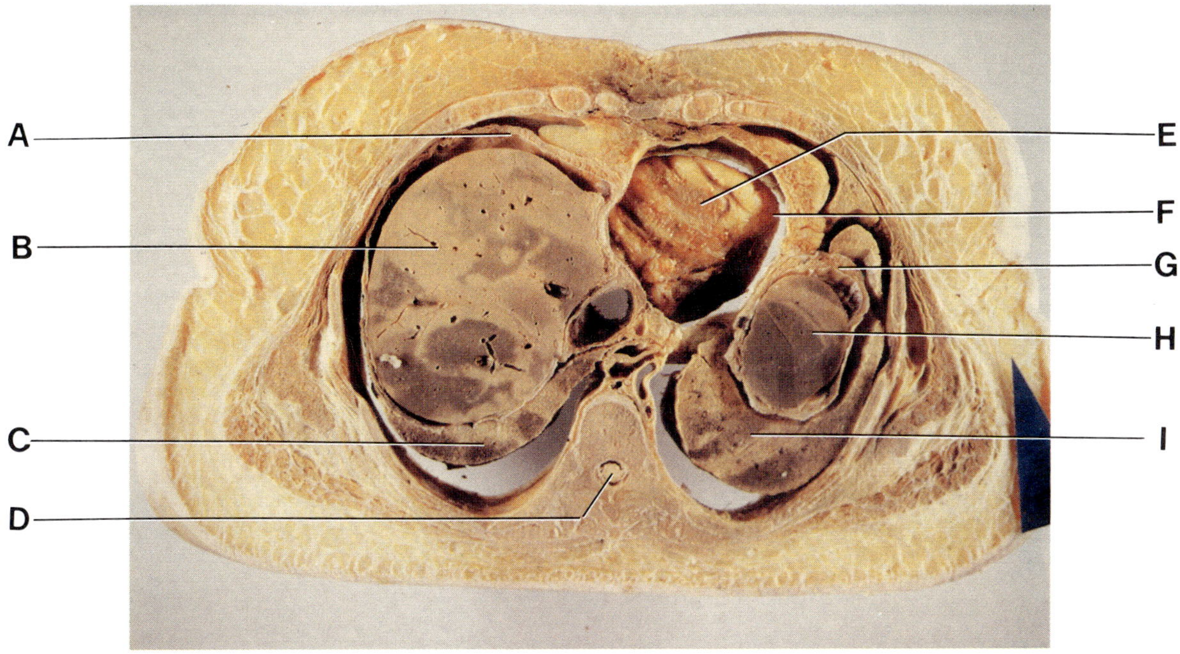

A. Diaphragm (right part)
B. Right lobe of liver
C. Inferior lobe of right lung
D. Spinal cord
E. Heart
F. Apex of heart
G. Diaphragm (left part)
H. Spleen
I. Inferior lobe of left lung

Figure 9–32
Transverse section of the thorax at the level of thoracic vertebra 8, highlighting visceral unit structures.

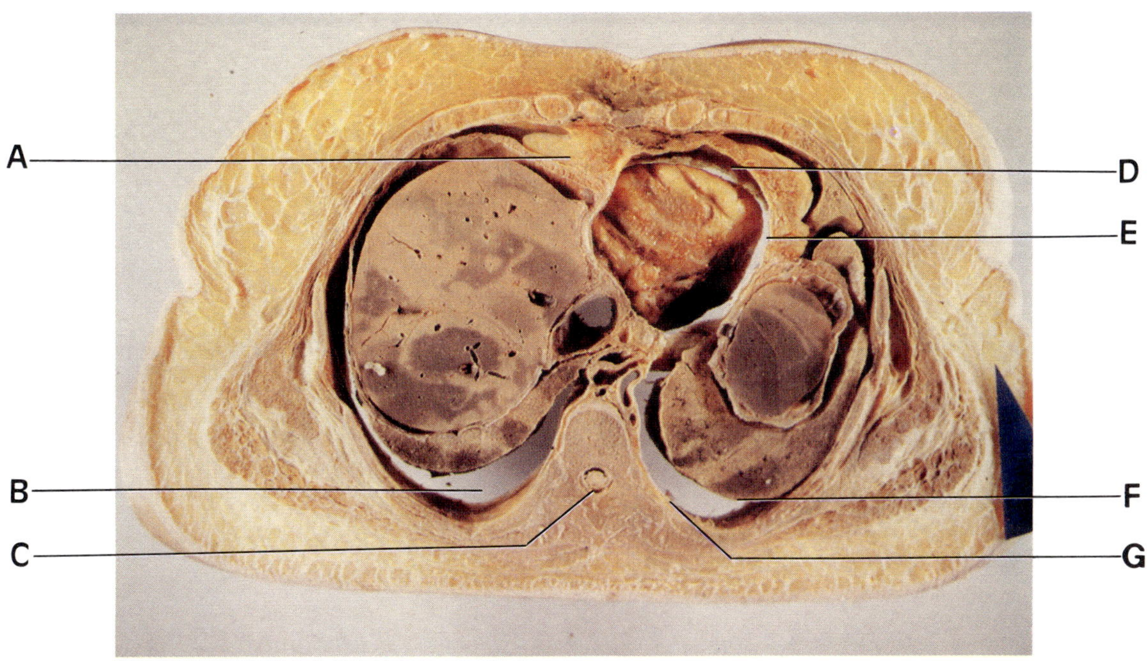

A. Anterior mediastinal fat
B. Interpleural space (pleural cavity)
C. Dura mater
D. Pericardium
E. Pericardial cavity
F. Visceral pleura
G. Parietal pleura

Figure 9–33
Transverse section of the thorax at the level of thoracic vertebra 8, highlighting enclosing unit structures.

CHAPTER 10

Coronal (Frontal) Cadaver Sections and Images of the Thorax

These cadaver coronal sections of the thorax combine the four anatomic categories on two identical photographs. One illustration includes the musculoskeletal and neurovascular structures, and the other, the visceral and enclosing units. Sections of most of the structures of the thorax are seen in these coronal views. Descriptions of these cadaver slices, seen approximately at the midaxillary line, relate to appearance, position, location, and relationship of adjacent structures. The origin, distribution, and termination of neurovascular structures are provided in many cases. A magnetic resonance image (MRI) of a coronal slice of the thorax completes the discussion.

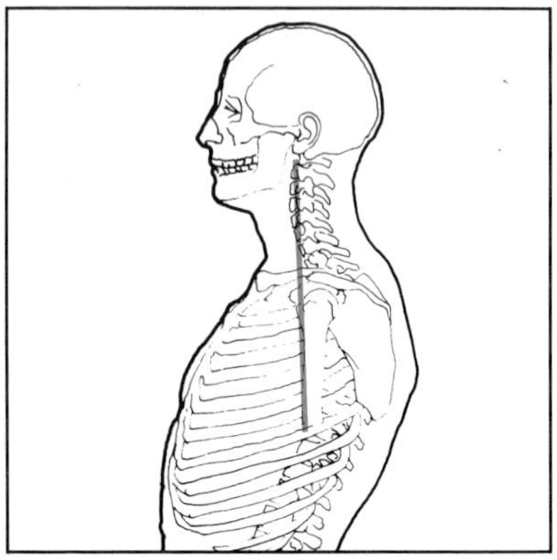

MODULE 1

This is a frontal (coronal) section of the thorax and upper parts of the arm (only the chest structures are labeled and described in this module), showing the arch of the aorta, the ascending aorta, the pulmonary trunk, the left ventricle, the superior vena cava, and the upper lobes of the lungs. *Note:* The upper limb structures are labeled and described in Chapter 19.

MUSCULOSKELETAL UNIT (Fig. 10–1)

I. Skeletal Portion

Rib (*Q*). An oblique, vertical bone slice adjacent to the intercostal muscles. The inferior ends of rib bones are more lateral than the superior ends.

Seventh Cervical Vertebra (*A*). A rectangular bone slice superior to the thyroid gland, as seen in this coronal section. This is the last cervical vertebra.

Clavicle (*D*). The right clavicle seen in this coronal plane is elongated (medial-to-lateral) superior to the coracoid process. The lateral end is narrower and slightly superior to the medial end. The left clavicle seen in this coronal view shows a circular slice of bone tissue medial to the deltoid muscle.

II. Muscular Portion

External Intercostal Muscle (*S*). Vertical, oblique strips of muscle tissue between the lateral parts of the ribs. The inferior end is more lateral than the superior end.

Internal Intercostal Muscle (*T*). Vertical, oblique strips of muscle tissue deep to the external intercostal muscles. The inferior end is more lateral than the superior end.

NEUROVASCULAR UNIT (see Fig. 10–1)

I. Arteries

Ascending Aorta (*L*). A large, hollow, rounded vascular structure seen at the superior part of the left ventricle and inferior to the arch of the aorta.

Arch of the Aorta (*J*). A curved, hollow, tubelike vascular structure extending from the ascending aorta to the descending thoracic aorta. The left surface of this vessel shows a concave part.

Descending Thoracic Aorta (*H*). A rounded, hollow vascular structure immediately inferior to the thyroid gland, as seen in this coronal view. This is the proximal portion of the descending thoracic aorta.

Pulmonary Trunk (*V*). A large artery adjacent to the concave surface of the arch of the aorta. This artery bifurcates to form right and left pulmonary arteries.

Left Pulmonary Artery (*U*). A rounded, hollow vascular structure posterior to the distal end of the pulmonary trunk and superior to the right pulmonary artery.

Right Pulmonary Artery (*W*). A rounded, hollow vascular structure posterior to the distal end of the pulmonary trunk and inferior to the left pulmonary artery.

Inferior Thyroid Artery (*B*). A moderately small artery, rounded in appearance, seen lateral to the thyroid glandular tissue in this coronal view.

Origin. Thyrocervical trunk of the subclavian artery.

Distribution. It disperses blood to six major branches: (1) esophageal, (2) thyroid terminal, (3) tracheal, (4) ascending cervical, (5) inferior laryngeal, and (6) muscular.

Course. This artery runs adjacent to the anterior surface of the scalenus anterior muscle. It takes a medial course between the carotid sheath and the vertebral artery to reach the thyroid gland. It ultimately anastomoses with the superior thyroid artery.

Right Subclavian Artery (*E*). An oval, hollow vascular structure seen medial to the first rib and the axillary artery. It is lateral to the proximal portion of the superior vena cava.

Origin. Brachiocephalic artery.

Distribution. It commonly disperses blood to four branches: (1) costocervical trunk, (2) thyrocervical trunk, (3) vertebral branch, and (4) internal thoracic (internal mammary) branch. The axillary artery is the continuation of this vessel.

Course. It arises from the brachiocephalic artery posterior to the sternoclavicular joint and curves laterally to reach the superior surface of the first rib, where its name changes to the axillary artery.

Brachiocephalic Artery (*F*). An oblong, hollow vascular structure adjacent to the left side of the superior vena cava and seen in this coronal view superior to the arch of the aorta. It is adjacent to the right part of the trachea.

Origin. Arch of the aorta.

Distribution. Right subclavian and right common carotid arteries.

Course. It starts from the arch of the aorta behind the center of the manubrium and ascends obliquely. As it climbs, it turns posteriorly to the right and becomes adjacent to the anterior part of the trachea, coursing to the right side of the trachea. Adjacent to the sternoclavicular joint, it divides into the right subclavian and right common carotid arteries.

Left Common Carotid Artery (*M*). A vertically elongated, curved vascular structure attached to the arch of the aorta. It courses to the left of the thyroid gland, as seen in this coronal section.

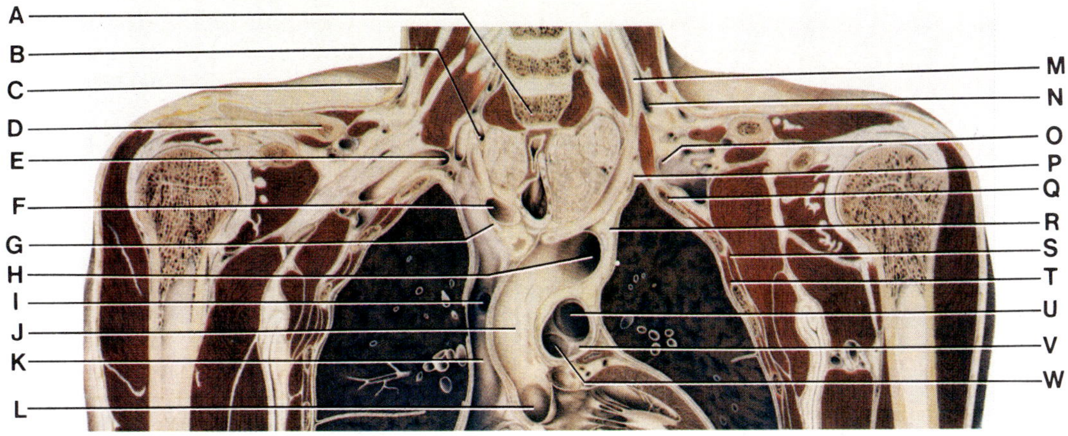

A. Seventh cervical vertebra
B. Inferior thyroid artery
C. Right external jugular vein
D. Clavicle
E. Right subclavian artery
F. Brachiocephalic artery
G. Right vagus nerve
H. Descending thoracic aorta
I. Azygos vein
J. Arch of aorta
K. Superior vena cava
L. Ascending aorta
M. Left common carotid artery
N. Left internal jugular vein
O. Left subclavian vein
P. Left phrenic nerve
Q. Rib
R. Left vagus nerve
S. External intercostal muscle
T. Internal intercostal muscle
U. Left pulmonary artery
V. Pulmonary trunk
W. Right pulmonary artery

Figure 10–1
Midcoronal (frontal) section of the thorax, highlighting both musculoskeletal and neurovascular unit structures.

Origin. Arch of the aorta.

Distribution. Internal and external carotid arteries.

Course. It starts on the left part of the aortic arch and climbs to the left sternoclavicular joint, where it becomes the cervical portion of the left common carotid artery. At the proximal part of the cervical portion, this artery is anterior to the trachea, but then the artery turns to the left and becomes adjacent to the left side of the trachea to reach the carotid triangle, where it divides into the external and internal carotid arteries.

II. Veins

Superior Vena Cava (*K*). A vertical, elongated, hollow venous structure adjacent to the right surface of the arch of aorta and the ascending aorta.

Azygos Vein (*I*). A rounded, hollow venous structure seen within the posterior wall of the superior vena cava.

Left Subclavian Vein (*O*). A tubelike vascular structure medial to the axillary vein and posterior to the phrenic nerve.

Origin. It is a continuation of the axillary vein.

Termination. Left brachiocephalic vein.

Course. This vein extends from the superior edge of the first rib to the medial margin of the scalenus anterior muscle. In its course the clavicle and the subclavius muscle are superior to this vessel.

Left Internal Jugular Vein (*N*). A vertical vascular slice, with the inferior part of this vessel as seen in this coronal view larger than the superior part. It is lateral to the left common carotid artery and medial to the sternocleidomastoid muscle.

Right External Jugular Vein (*C*). A moderately small, vertical venous structure lateral to the sternocleidomastoid muscle.

III. Nerves

Left Phrenic Nerve (*P*). A short, vertical mixed nerve tissue immediately to the left of the left common carotid artery and medial to the superior border of the first rib.

Origin. Cervical nerves 3, 4, and 5. Its chief origin is the fourth cervical nerve.

Distribution. Diaphragm, parietal and visceral pleurae, and abdominal serous membranes.

Course. It is formed adjacent to the lateral edge of the scalenus anterior muscle and takes a vertical course inferiorly. This nerve courses anterior to the scalenus anterior muscle and deep to the sternocleidomastoid muscle. It continues inferiorly posterior to the prevertebral fascia and accompanies the ascending cervical artery to reach a position between the left subclavian artery and vein, lateral to the left vagus nerve and the arch of the aorta. It runs anteriorly to the root of the left lung. As it crosses the front part of the root, this nerve is between the mediastinal pleura and the pericardium. From this point it runs toward the diaphragm, where it terminates.

Left Vagus Nerve (*R*). A vertically curved, mixed nerve tissue adjacent to the left part of the aortic arch and the descending thoracic aorta. The medial part shows a concave surface in this coronal plane.

Right Vagus Nerve (*G*). A vertically oblique, mixed nerve tissue adjacent to the lateral part of the brachiocephalic artery and medial to the superior vena cava.

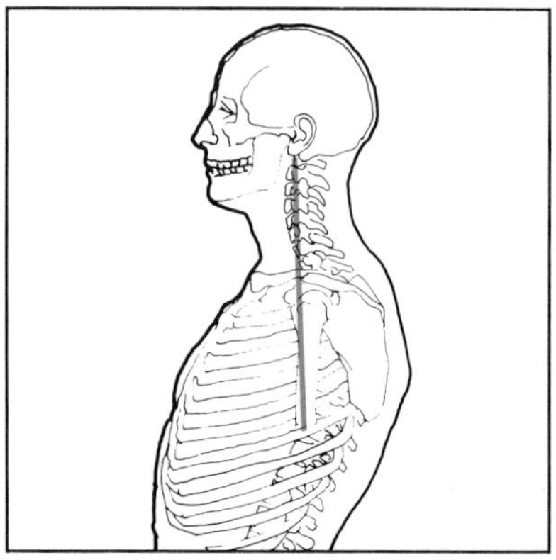

VISCERAL UNIT (Fig. 10–2)

Thyroid Gland (*K*). An aggregate glandular tissue on both sides of the trachea that represents a section of the left and right lobes. On this body section, the left lobe is larger than the right lobe. Each lobe is 5 cm in length, with the apex on its superior part.

Trachea (*J*). An oval, hollow respiratory structure, seen in this coronal plane between the inferior parts of the thyroid lobes and anterior to the esophagus.

Esophagus (*I*). A hollow, irregular digestive organ seen superior and posterior to the trachea. In this plane the esophagus is seen between the upper parts of the thyroid lobes.

Superior Lobe of the Right Lung (*B*). A pyramidal lung tissue superior to the transverse fissure and located in the upper part of the right pleural cavity. The medial surface of this lobe is adjacent to the superior vena cava.

Apex of the Right Lung (*A*). A vertical, rather sharp, superior part of the superior lobe of the right lung.

Middle Lobe of the Right Lung (*G*). A moderately wide respiratory structure immediately inferior to the transverse fissure. The superior part of this lobe is seen in this coronal view.

Superior Lobe of the Left Lung (*M*). An irregularly shaped upper part of the left lung with a bulge on its medial surface. This structure is above the oblique fissure and adjacent to the left side of the vessels in the superior mediastinum as well as the upper part of the heart.

Apex of the Left Lung (*L*). A sharp, superior feature on the superior lobe of the left lung.

Inferior Lobe of the Left Lung (*R*). A rather narrow portion of the left lung adjacent to the left lateral surface of the heart and below the oblique fissure.

Left Ventricle (*Q*). A cavity of the heart, with a thick wall of cardiac tissue adjacent to the medial surface of the left lung and inferior to the ascending aorta, as well as the arch of the aorta.

Papillary Muscles of the Left Ventricle (*O*). A moderately short, transversely placed structure within the left ventricle. There are two papillary muscles of the left ventricle and three such structures of the right ventricle.

Chordae Tendineae of the Left Ventricle (*S*). Tendinous cords are attached to the papillary muscles and bicuspid (mitral) valve.

Bicuspid (Mitral) Valve (*N*). A sheet of cusps with chordae tendineae attached at the inferior border of the superior part of the left ventricle.

Right Atrium (*H*). A small part of this chamber is seen at the inferior end of the superior vena cava. This chamber is to the right of the ascending aorta.

ENCLOSING UNIT (see Fig. 10–2)

Transverse Fissure (*F*). A transversely placed slit between the superior and middle lobes of the right lung.

Oblique Fissure of the Left Lung (*P*). A slit between the superior and inferior lobes of the left lung.

Visceral Pleura (*E*). A thin sheet of membranous tissue that encases the lungs and follows surfaces of the fissures.

Parietal (Costal) Pleura (*D*). A thin sheet of membranous tissue that lines the inner wall of the thoracic cavity.

Parietal (Mediastinal) Pleura (*C*). A thin sheet of membranous tissue vertically placed adjacent to the lateral wall of the superior vena cava and the visceral pleura of the right superior lobe of the lung.

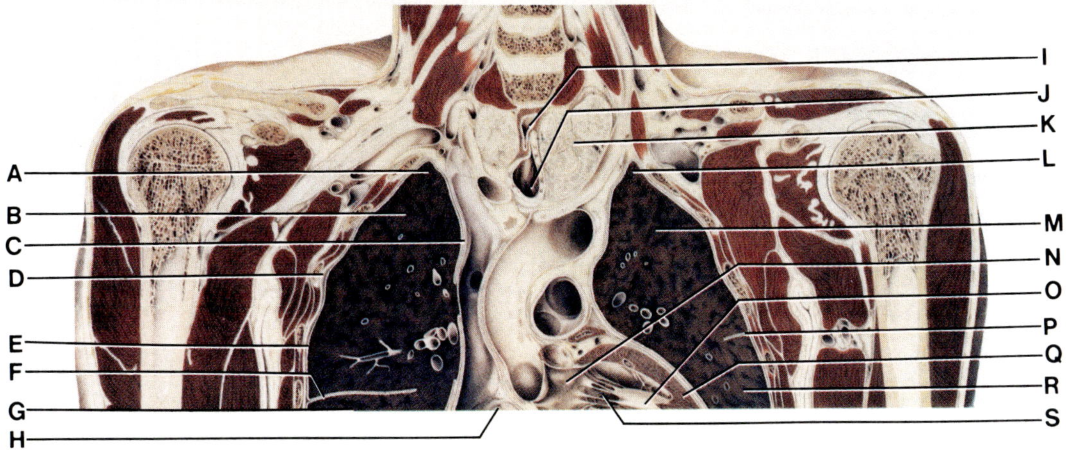

A. Apex of right lung
B. Superior lobe of right lung
C. Parietal (mediastinal) pleura
D. Parietal (costal) pleura
E. Visceral pleura
F. Transverse fissure
G. Middle lobe of right lung
H. Right atrium
I. Esophagus
J. Trachea
K. Thyroid gland
L. Apex of left lung
M. Superior lobe of left lung
N. Bicuspid (mitral) valve
O. Papillary muscles of left ventricle
P. Oblique fissure of left lung
Q. Left ventricle
R. Inferior lobe of left lung
S. Chordae tendineae of left ventricle

Figure 10–2
Midcoronal section of the thorax, highlighting both visceral and enclosing unit structures.

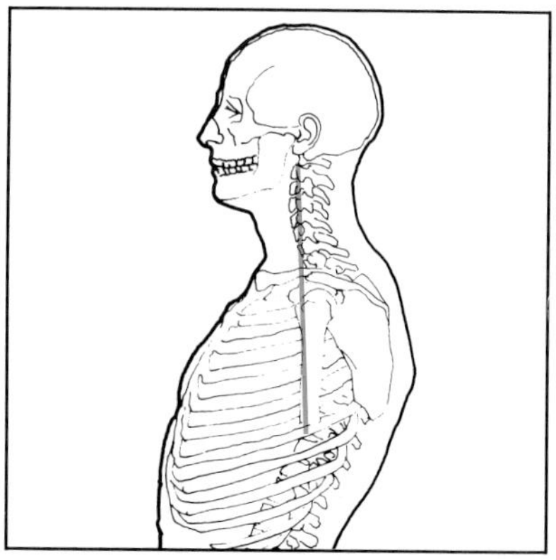

The musculoskeletal, neurovascular, and visceral unit structures are emphasized in Figure 10–3. This figure shows an MRI coronal slice of the thorax.

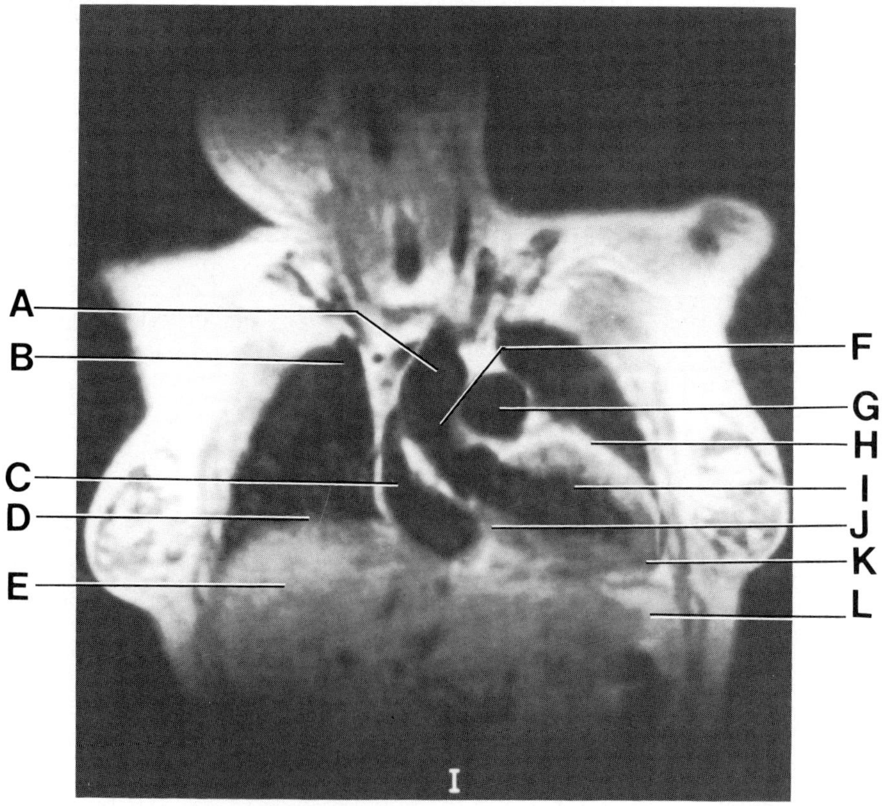

A. Ascending aorta
B. Apex of right lung
C. Right ventricle
D. Base of right lung
E. Hepatic dome of diaphragm
F. Ascending aorta
G. Pulmonary trunk
H. Cardiac notch
I. Left ventricle
J. Interventricular septum
K. Apex of heart
L. Gastric dome of diaphragm

Figure 10–3
MRI scan approximately in the midcoronal section of the thorax, highlighting musculoskeletal, neurovascular, and visceral unit structures.

CHAPTER 11

Sagittal Cadaver Sections and Images of the Thorax

One of these cadaver sagittal sections of the thorax shows the musculoskeletal and neurovascular units, and the other shows the visceral and enclosing units. In this view of the thorax, the longitudinal extent of the structures is shown, as well as the anterior/posterior dimension. Slices of structures adjacent to the median are shown on these cadaver sections. In the accompanying text, each structure is described and identified by its appearance, location, relationship of adjacent structure, dimensional orientation, origin, distribution, termination, and course of some neurovascular structures.

Two different normal, clinical, magnetic resonance images (MRI) of the thorax are presented. In each, all anatomic categories are labeled.

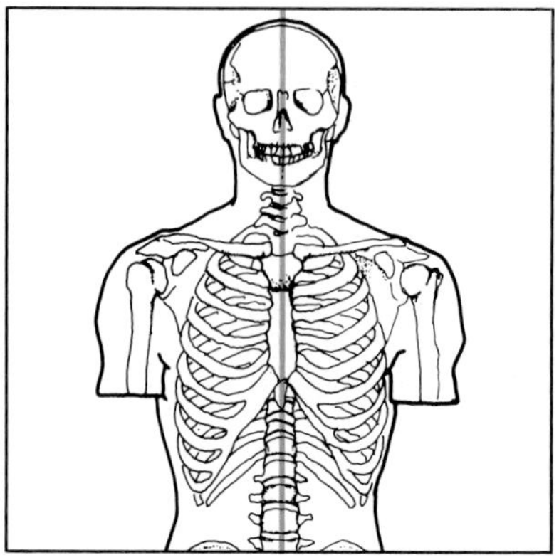

ized by a disk between bone ends and is designated an intervertebral symphysis joint type.

Spinous Process of a Thoracic Vertebra (*P*). A moderately elongated, oblique bone slice posterior to the spinal canal and anterior to the back muscles. The posterior end is slightly inferior to the anterior end.

II. Muscular Portion

Trapezius Muscle (*J*). A vertical strip of superficial muscle seen posterior to the deep back muscles in this sagittal view.

Deep Back Muscles (*N*). Several short strips of muscle sections seen deep to the trapezius muscle and posterior to the spinous processes of the vertebrae.

MODULE 1

This is a sagittal section of the head, neck, and thorax. Some of the structures seen in this plane are the ascending aorta, atria, thymus gland, esophagus, and inferior part of the trachea.

MUSCULOSKELETAL UNIT (Fig. 11–1)

I. Skeletal Portion

Manubrium of the Sternum (*B*). A vertical dimension bone slice, superior to the sternal angle and anterior to the superior mediastinum. This is a moderately wide (anterior-to-posterior) superior portion of the sternum.

Sternal Angle (Manubriosternal Joint) (*C*). The junction between the inferior part of the manubrium and the superior part of the body of the sternum. Although fibrocartilage is between these bone parts, this joint is frequently obliterated by synostosis in old age.

Body of the Sternum (*G*). A vertically placed bone slice between the sternal angle and the xiphisternal joint anterior to the mediastinum. The inferior end is slightly anterior to the superior end.

Xiphisternal Joint (*H*). The junction between the inferior part of the body of the sternum and the superior part of the xiphoid process. This cartilage usually ossifies in early adult life.

Xiphoid Process (*I*). A moderately thin, vertical bone slice inferior to the xiphisternal joint. This bone feature is the smallest of the three parts of the sternum. The bone part of the xiphoid process is covered or surrounded with hyaline cartilage. This bone gets larger as an individual gets older.

Body of a Thoracic Vertebra (*L*). A rectangular bone slice between the intervertebral disks and anterior to the spinal canal. It is posterior to the esophagus.

Intervertebral Disk (*O*). A thin, platelike, fibrocartilaginous structure between the bodies of the vertebrae. This amphiarthrodial (secondary cartilaginous) joint, a feature of the vertebral column, is character-

NEUROVASCULAR UNIT (see Fig. 11–1)

I. Arteries

Ascending Aorta (*E*). A vertical vascular structure superior to the left ventricle and anterior to the left atrium. Although the left ventricle is not seen in this view, its location in relationship to the structure seen indicates its position.

Arch of the Aorta (*D*). A curved, vertical, hollow vascular structure anterior to the right pulmonary artery and superior to the ascending aorta.

Brachiocephalic Artery (*K*). An oval, hollow vascular structure posterior to the left brachiocephalic vein and superior to the arch of the aorta. It is anterior to the inferior part of the trachea.

Right Coronary Artery (*F*). A small, circular, hollow vascular structure superior to the right ventricle and inferior to the right auricle of the heart.

Origin. Anterior part of the ascending aorta.

Distribution. This artery shares blood with the following branches: (1) marginal branch, which supplies blood to the anterior and posterior surfaces of the right ventricle, (2) posterior interventricular branch, which disperses blood to the right and left ventricles and the interventricular septum before anastomosing at the apex with the anterior interventricular artery of the left coronary artery, (3) twigs, which apportion blood to the ascending aorta and the pulmonary trunk, and (4) right coronary artery branches, which supply blood to the front surface of the right atrium into the superior part of the anterior right ventricle.

Right Pulmonary Artery (*M*). A rounded, moderately large, hollow vascular structure posterior to the arch of the aorta and superior to the left atrium. It is superior and posterior to the ascending aorta.

II. Veins

Left Brachiocephalic Vein (*A*). A rather large, circular, hollow venous structure superior to the arch of the aorta and anterior to the right pulmonary artery.

A. Brachiocephalic vein
B. Manubrium of sternum
C. Sternal angle (manubriosternal joint)
D. Arch of aorta
E. Ascending aorta
F. Right coronary artery
G. Body of sternum
H. Xiphisternal joint
I. Xiphoid process
J. Trapezius muscle
K. Brachiocephalic artery
L. Body of thoracic vertebra
M. Right pulmonary artery
N. Deep back muscles
O. Intervertebral disk
P. Spinous process of thoracic vertebra

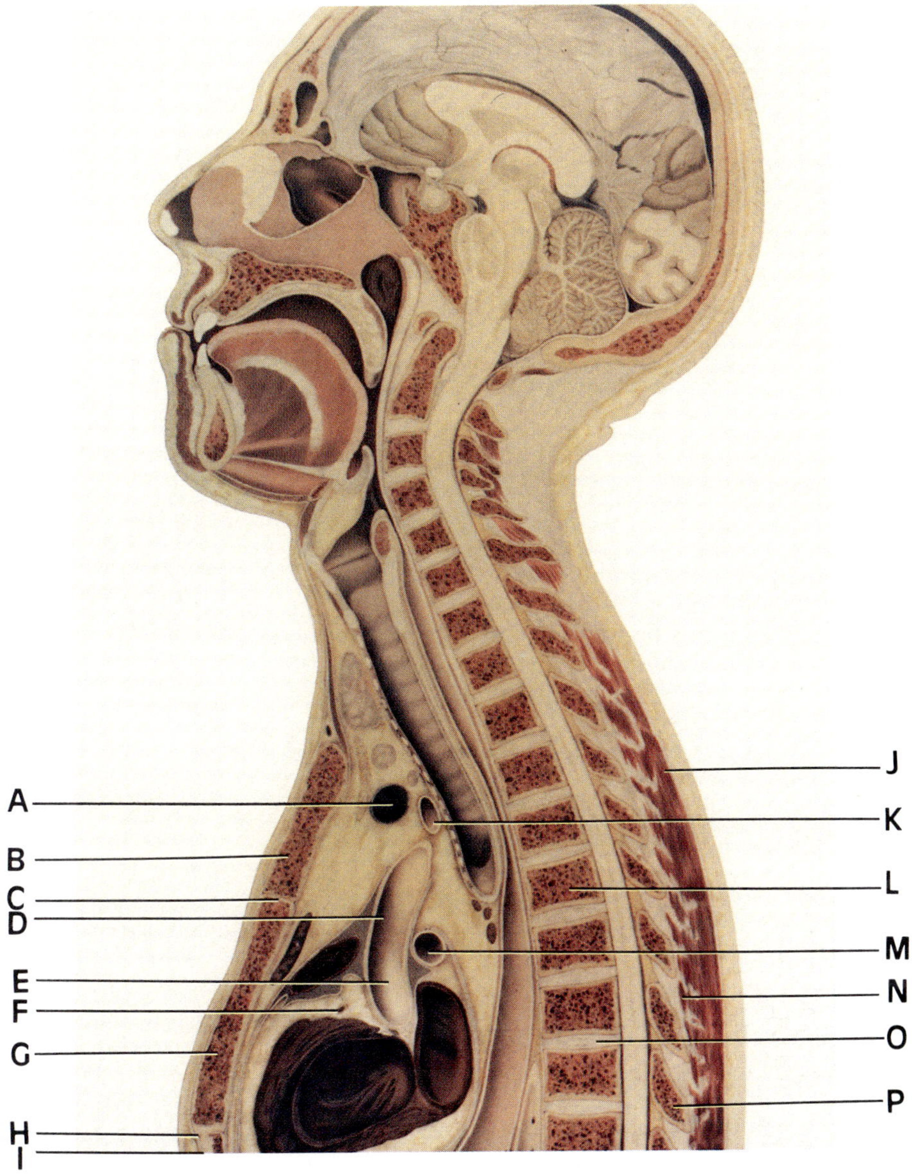

Figure 11-1
An approximately midsagittal section of the thorax, highlighting both musculoskeletal and neurovascular unit structures.

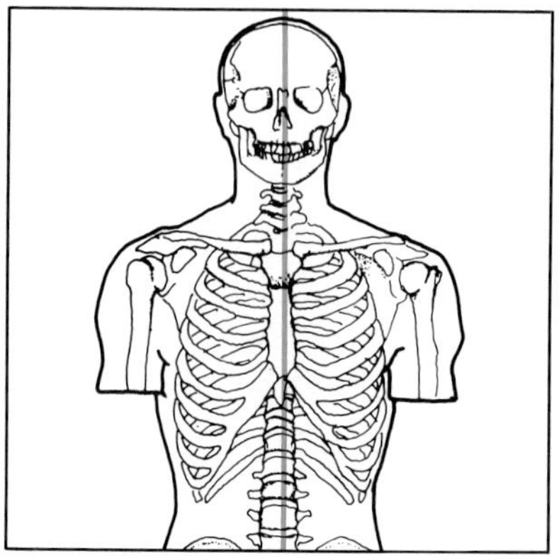

VISCERAL UNIT (Fig. 11–2)

Thymus Gland (*C*). A vertical strip of glandular tissue posterior to the superior part of the body of the sternum and anterior to the right auricle of the heart. This structure is in the inferior part of the anterior part of the superior mediastinum. The thymus gland is much larger in the newborn infant than in the adult and is more extensive.

Right Auricle (*D*). A vertical, oblique slice of the right auricle anterior to the arch of the aorta and superior to the right ventricle. The anterior end of this cardiac structure is more anterior than the superior end.

Right Atrium (*E*). A large, rounded cardiac structure anterior to the left atrium and superior to the right ventricle.

Left Atrium (*N*). An oblong (superior-to-inferior) cardiac structure posterior to the right atrium and anterior to the esophagus. It is inferior to the right pulmonary artery.

Right Ventricle (*G*). A rather large, V-shaped cardiac structure inferior and anterior to the right atrium. This structure is between the anterior and posterior mediastinal areas.

Esophagus (*L*). An elongated (superior-to-inferior), hollow digestive structure seen anterior to the bodies of the vertebrae. The portion of the esophagus posterior to the area for the bifurcation of the trachea is narrower than the portion below this point.

Trachea (*H*). An elongated (superior-to-inferior), hollow respiratory structure anterior to the esophagus and inferior to the cricoid cartilage of the larynx. Several cartilaginous rings form the anterior and lateral walls of this structure.

Right Primary Bronchus (*I*). A rounded, proximal portion of this respiratory structure is located at the inferior end of the trachea and superior to the bronchial lymph nodes.

Left Primary Bronchus (*K*). A U-shaped respiratory structure at the inferior end of the trachea and superior to the bronchial lymph nodes.

Carina of the Trachea (*J*). A ridge on the internal trachea where it bifurcates to form the right and left primary bronchi.

Spinal Cord (*M*). An elongated (superior-to-inferior) aggregate of nerve tissue located within the spinal canal.

Lymph Nodes (*B*). Oval structures seen inferior to the primary bronchi and anterior to the esophagus.

ENCLOSING UNIT (see Fig. 11–2)

Superior Mediastinum (*A*). The upper area of the thorax that contains the trachea, brachiocephalic veins, brachiocephalic artery, thymus gland in the adult, superior vena cava, and other structures.

Anterior Mediastinum (*F*). The front area between the heart and the sternum. This is a part of the inferior mediastinum.

Posterior Mediastinum (*O*). The back area between the heart and the vertebral column. This is a part of the inferior mediastinum.

Middle Mediastinum. The area where the heart is located. The heart and the origin/termination of eight great vessels are located within this mediastinal area.

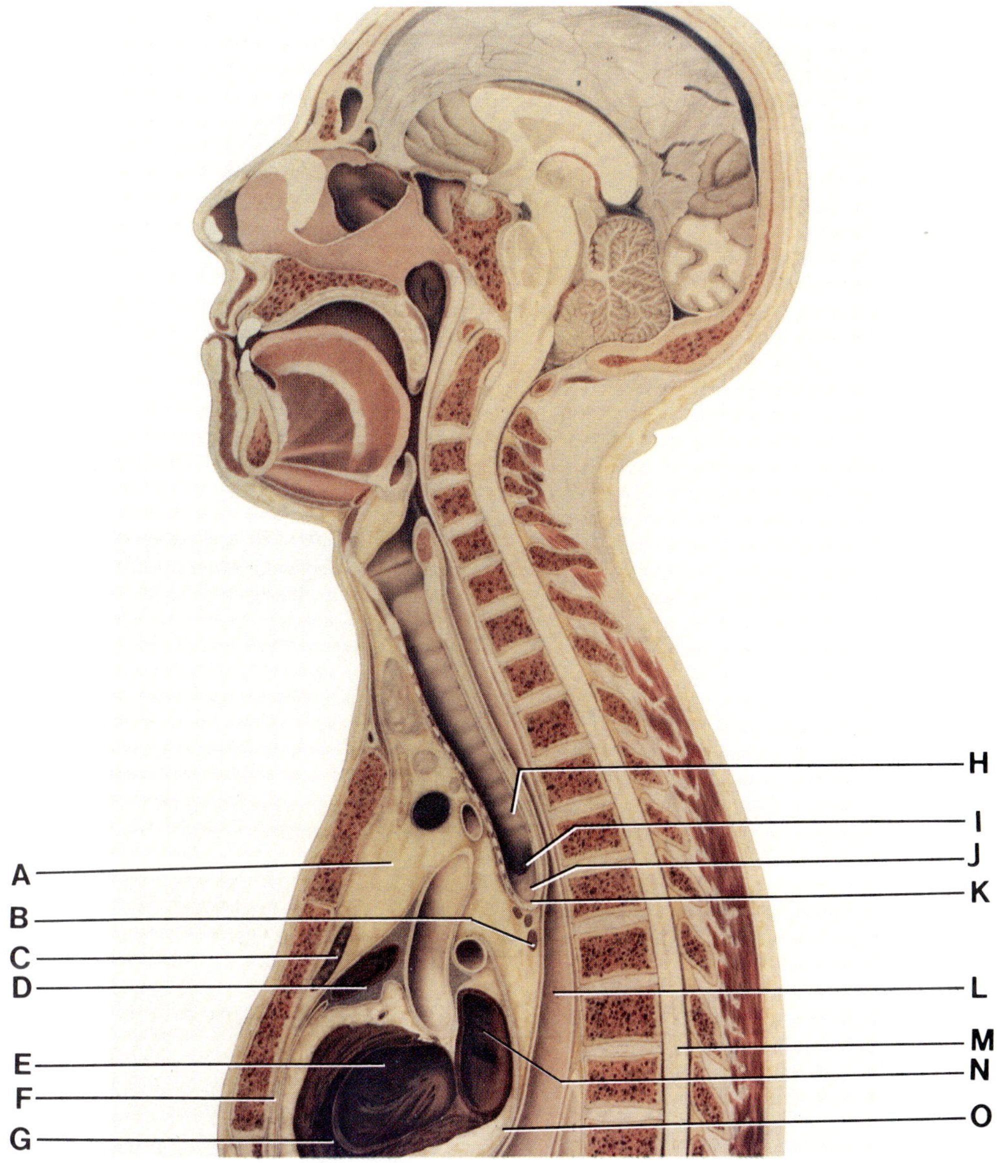

A. Superior mediastinum
B. Lymph nodes
C. Thymus gland
D. Right auricle
E. Right atrium
F. Anterior mediastinum
G. Right ventricle
H. Trachea
I. Right primary bronchus
J. Carina of trachea
K. Left primary bronchus
L. Esophagus
M. Spinal cord
N. Left atrium
O. Posterior mediastinum

Figure 11–2
Sagittal section of the thorax, highlighting both visceral and enclosing unit structures.

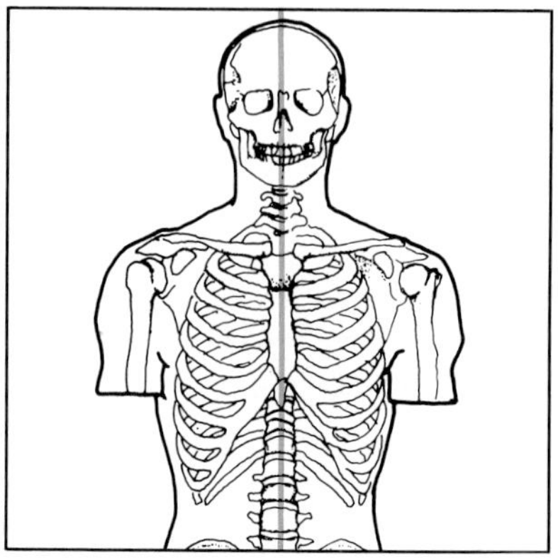

Figure 11–3 shows an MRI sagittal section of the thorax. This figure emphasizes the neurovascular, visceral, and enclosing unit structures.

Another MRI sagittal section of the thorax is seen in Figure 11–4. This figure highlights the neurovascular, visceral, and enclosing unit structures.

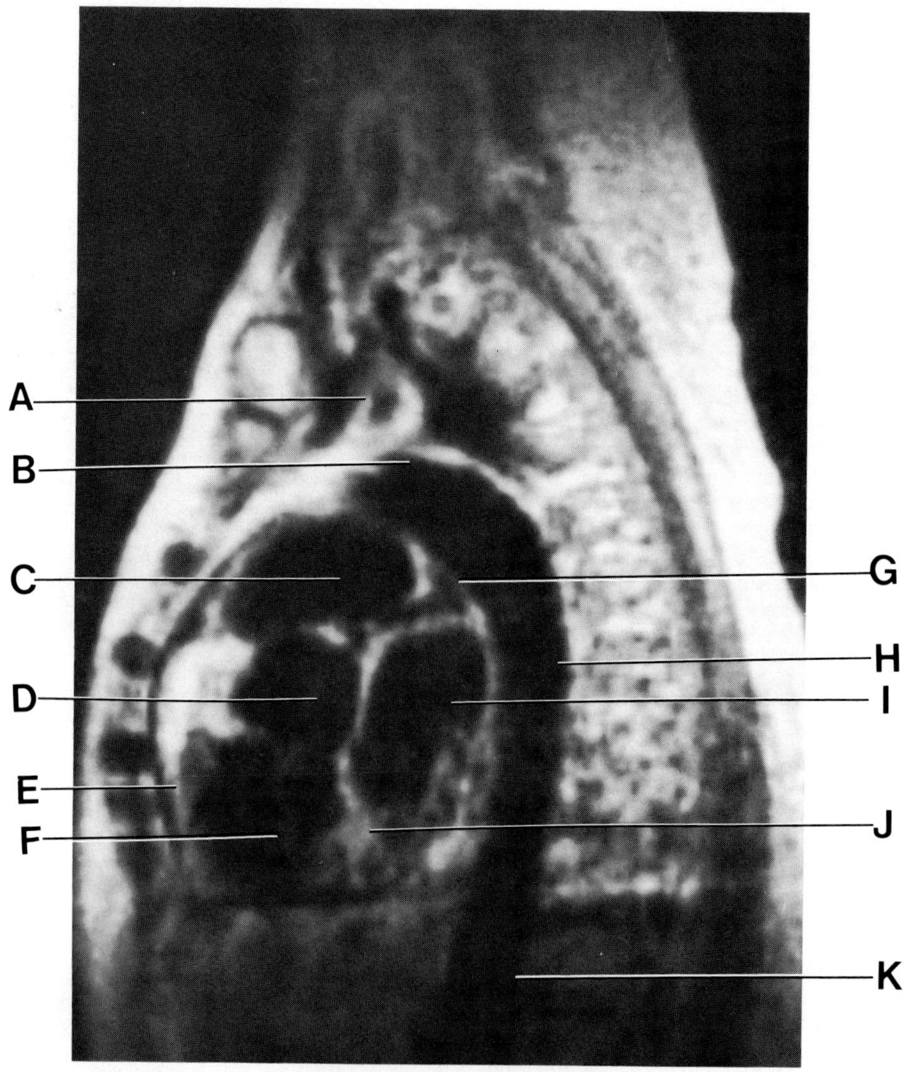

A. Superior mediastinum
B. Arch of aorta
C. Pulmonary trunk
D. Ascending aorta
E. Anterior mediastinum
F. Right ventricle
G. Left pulmonary bronchus
H. Descending thoracic aorta
I. Left atrium
J. Interventricular septum
K. Descending abdominal aorta

Figure 11–3
An approximately midsagittal MRI section of the thorax, highlighting neurovascular, visceral, and enclosing unit structures.

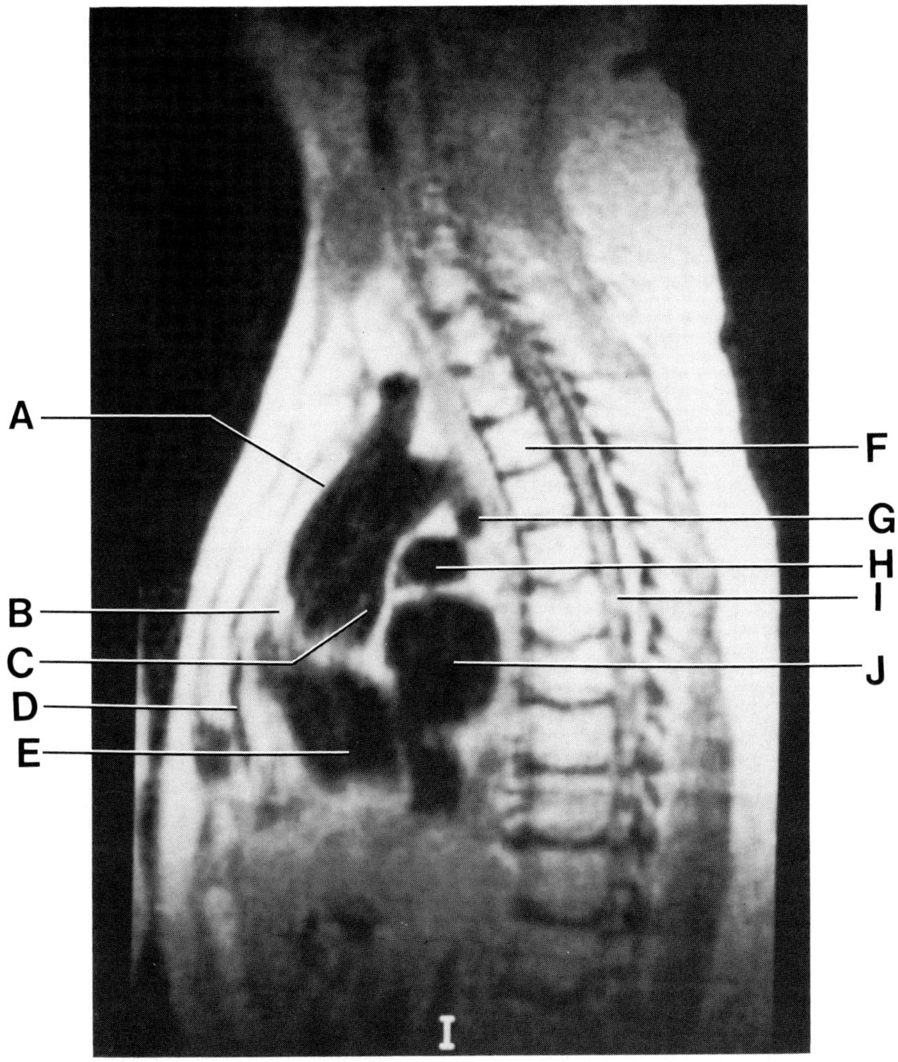

A. Ascending aorta
B. Anterior mediastinal area
C. Aortic root
D. Anterior mediastinum
E. Right ventricle
F. Body of thoracic vertebra
G. Carina
H. Pulmonary artery
I. Spinal cord
J. Left atrium

Figure 11–4
An MRI sagittal section of the thorax, highlighting neurovascular, visceral, and enclosing unit structures.

Appendix to Part IV: Clinical Correlation for the Thoracic Region

CASE ONE (Figs. AIV–1 and AIV–2)

History. The patient has lung carcinoma. There is an elevated right hemidiaphragm. Liver metastasis is being clinically considered. Because of the marked elevation of the right hemidiaphragm, two different sequences of high-resolution, thin-section T1- and T2-weighted axial images and T1-weighted coronal images were obtained from approximately the level of the carina up to the midabdominal region to cover the region of the liver.

Findings. No discrete lesions involving the liver are visualized.

No retroperitoneal adenopathy or discrete lesions involving the spleen, adrenals, pancreas, or kidneys were visualized in this study.

Incidentally, adenopathy is visualized in the carinal region as well as in the right pulmonary hilum. A large mass in the periphery of the right upper lobe is also visualized, compatible with clinically known carcinoma of the lung. This lesion is not fully covered in the current study.

Impression. Right upper lobe carcinoma, with associated mediastinal adenopathy as described. A markedly elevated right hemidiaphragm.

No evidence of metastatic disease involving the liver.

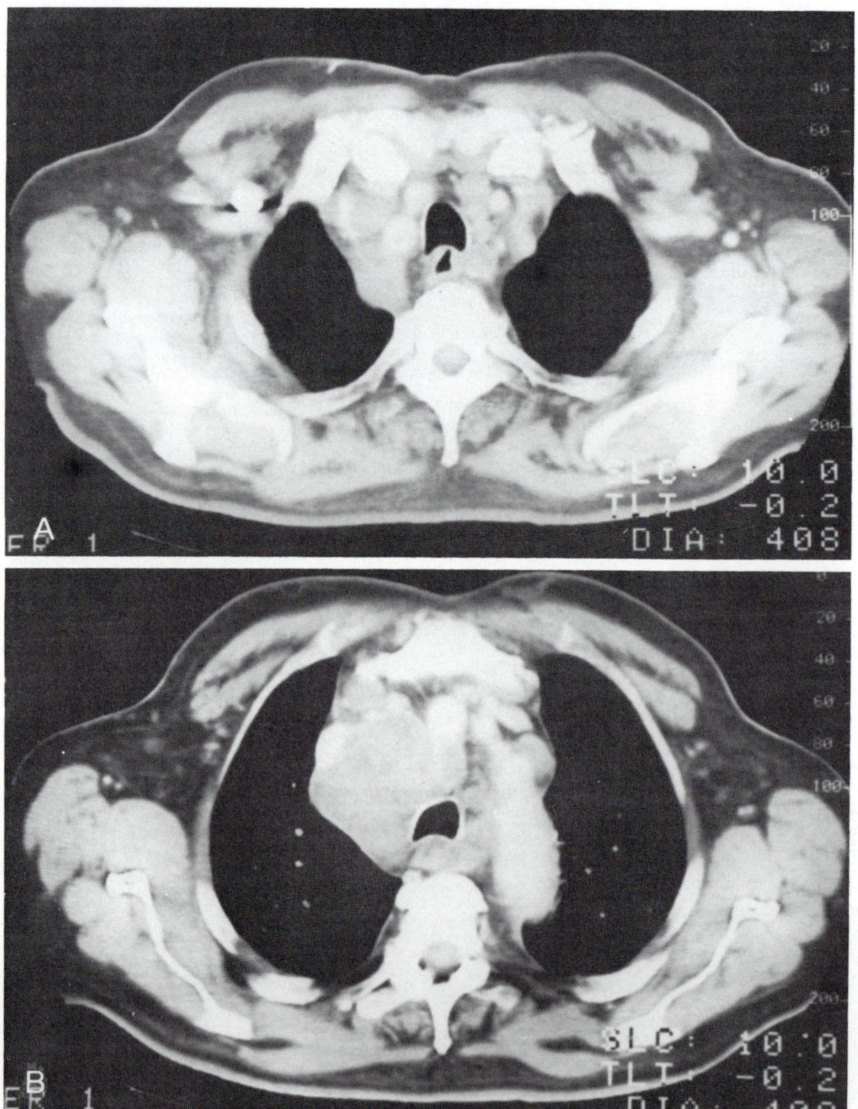

Figure AIV–1
A and *B,* CT transverse images of the thorax, showing a primary cancer of the lung.

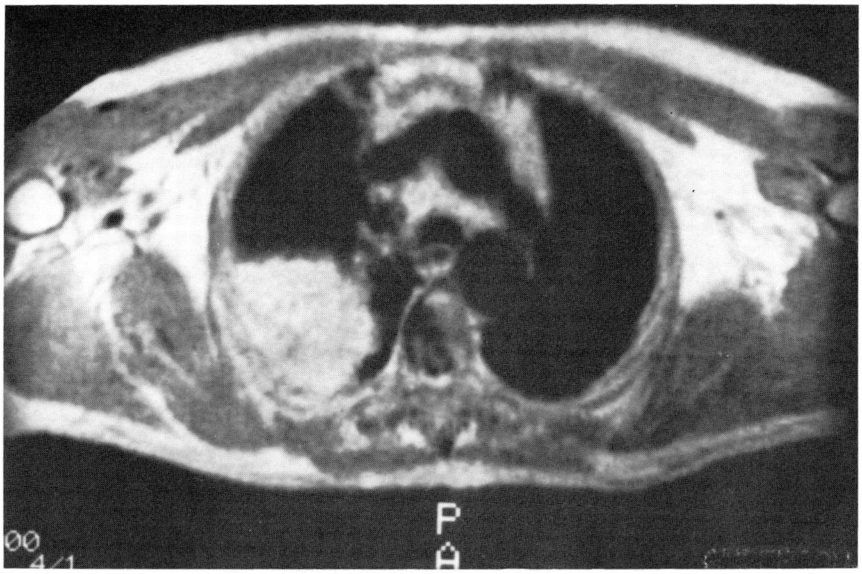

Figure AIV–2
An MRI transverse section of the thorax, showing carcinoma of the lung.

PART V

THE ABDOMEN

INTRODUCTION

The abdomen starts at the diaphragm and ends at the transtubercular plane, which is at the level of the highest point on the iliac crest. It is divided into compartments based on the location of the transverse colon. The area above the transverse colon constitutes the *supracolic compartment,* and the area below is known as the *infracolic compartment.* Each compartment is subdivided into smaller areas. For example, the area of the supracolic compartment anterior to the right lobe of the liver is known as the right subphrenic space. The space posterior to the right of the liver is the subdivision called the right subhepatic space, or Morrison's pouch. The left subphrenic space is the area of the supracolic compartment anterior to the left lobe of the liver and left of the falciform ligament.

The infracolic compartment has four smaller subdivisions: (1) the right (paracolic) gutter, (2) the gutter to the right of the mesentery, (3) the gutter to the left of the mesentery, and (4) the left (paracolic) gutter.

The subdivisions of both the supra- and infracolic compartments comprise the *greater sac.*

The bones and bony structures associated with the abdomen include the five lumbar vertebrae. The subcostal zone of the abdomen is surrounded by the lower portion of the rib cage. The subcostal zone consists of three abdominal regions: (1) the right hypochondriac region, (2) the epigastric region, and (3) the left hypochondriac region. The umbilical zone is immediately inferior to the subcostal zone and consists of three abdominal regions: (1) the right lumbar region, (2) the umbilical region, and (3) the left lumbar region. The umbilical zone is immediately superior to the hypogastric zone, which consists of pelvic regions.

The muscles of the abdomen are summarized in the tables that follow.

Several normal clinical images are included in Part V, as well as some pathologic conditions that appear in the Appendix to this part.

Table V-1 Anterior and Anterolateral Muscles of Abdomen

Muscle	Origin	Insertion	Innervation	Vasculature	Action
Rectus abdominis	5th and 6th costal cartilages and xiphoid process	Ridge (crest) of pubic bone	Branches from the iliohypogastric and ilioinguinal, and the anterior primary rami of the lower 6 intercostal nerves	Superior and inferior epigastric (muscular branches)	Vertebral column flexion and some compression of abdominal wall
External oblique abdominal (obliquus abdominis externus)	Slips of muscles interdigitate with outer surface of lower 8 ribs	Anterior half of lateral lip of iliac crest, and wide aponeuroses of abdominal wall for remainder of this muscle	Front branches of lumbar nerves 1 and 2, the anterior primary rami of thoracic nerves 7–12	Superior and inferior epigastric (muscular branches)	Contraction in forced expiration, abdominal support, compression of abdomen, trunk rotation
Internal oblique abdominal (obliquus abdominis internus)	Thoracolumbar fascia, anterior iliac crest, and lateral two thirds of inguinal ligament	Superior fibers insert into the cartilage of ribs 10–12 and into the aponeurosis, which extends from the 10th rib to the pubic bone	Front branches of lumbar nerves 1–4, the anterior primary rami of thoracic 7–12	Deep circumflex iliac, superior, and epigastric (muscular branches)	Contraction in forced expiration, abdominal support, compression of abdomen, trunk rotation
Transverse abdominal (transversus abdominis)	Deep surface of inferior costal cartilages 7–12, lumbodorsal fascia, iliac crest, and inguinal ligament	Linea alba	Anterior primary rami of intercostal nerves 7–12, ilioinguinal and iliohypogastric nerves	Inferior epigastric and deep iliac circumflex arteries	Abdominal compression, abdominal visceral support, increases abdominal pressure to aid expiration

Table V-2 Deep Back Muscles (Erector Spinal Group of Thorax and Abdomen)

Muscle	Origin	Insertion	Innervation	Vasculature	Action
Spinalis thoracis	Spinous processes of 1st and 2nd lumbar, 12th thoracic vertebrae	Spinous processes of superior vertebrae 1–4 and sometimes as far as thoracic 8 vertebrae	Posterior primary rami of spinal nerve	Posterior rami of intercostal arteries	Extension, abduction, and axial rotation of spinal column
Longissimus m. of thorax (longissimus thoracis)	Lumbodorsal fascia and transverse processes of lumbar vertebrae	Proximal to angles of ribs 3–12	Posterior primary rami of spinal nerve	Posterior rami of intercostal arteries	Extension, abduction, and rotation of spinal column
Iliocostal m. of thorax (iliocostalis thoracis)	Angles of ribs 7–12	Superior margins of rib angles from ribs 1–6	Posterior primary rami of spinal nerve	Posterior rami of intercostal arteries	Extension, abduction, and rotation of spinal column
Iliocostal m. of loins (iliocostalis lumborum)	Aggregate of erector spinae in lumbar region	Inferior margins of rib angles from ribs 7–12	Posterior primary rami of spinal nerves	Posterior rami of lumbar arteries	Extension, abduction, and rotation of spinal column

Table V-3 Deep Back Muscles (Transversospinalis Group of Thorax and Abdomen)

Muscle	Origin	Insertion	Innervation	Vasculature	Action
Semispinal m. of thorax (semispinalis thoracis)	Transverse processes of thoracic vertebrae 7–12	Spinous processes of thoracic 1–4 and cervical 6–7 vertebrae	Posterior primary rami of thoracic nerves 1–6	Posterior intercostal (muscular branches) arteries	Extension and abduction of vertebral column
Multifidus	Spinous process to transverse process adjacent to lamina; this muscle column extends from sacrum to axis	Spinous processes of vertebrae immediately superior to vertebra of origin	Posterior primary rami of all spinal nerves	Posterior intercostal and lumbar arteries	Extension, abduction, and axial rotation of vertebral column
Long and short (longus, brevis) rotator	Transverse processes of thoracic vertebrae	Long: inferior and lateral border of lamina of 2nd vertebra above. Short: inferior and lateral border of vertebra next above its origin	Posterior primary rami of spinal nerves	Posterior intercostal (muscular branches) arteries	Synergist for axial rotation of vertebral column

Table V-4 Other Deep Back Muscles (See Text)

Muscle	Origin	Insertion	Innervation	Vasculature	Action
Interspinal	Spinous process below (lateral to interspinous ligament); these are paired muscles	Spinous process above one of origin below	Posterior primary rami of spinal nerves	Posterior intercostal (muscular branches) and lumbar arteries. *Note:* in neck region by costocervical trunk (deep cervical branch)	Synergist for extension of vertebral column
Inter-transverse	Transverse processes below the insertion	Transverse processes above one of origin below	Posterior primary rami of spinal nerves except lateral intertransverse in lumbar and inferior thoracic regions, front and back intertransverse in cervical region, which are supplied by anterior primary rami of spinal nerves	Posterior intercostal (muscular branches) and lumbar arteries. *Note:* in neck region by costocervical trunk (deep cervical branch)	Synergist for abduction (lateral flexion) of vertebral column

Table V-5 Posterior Back Muscles*

Muscle	Origin	Insertion	Innervation	Vasculature	Action
Latissimus dorsi	Iliac crest, lumbodorsal fascia, ribs 9–12, and spinous processes of thoracic vertebrae 6–12	Deep surface of intertubercular sulcus of humerus	Thoracodorsal	Transverse cervical artery	Extension, adduction, and medial rotation of humerus
Psoas major	Front surface of transverse processes of all lumbar vertebrae, inferior portion of thoracic vertebra 12, and bodies of lumbar vertebrae	Lesser trochanter of femur along with iliac (iliopsoas)	Lumbar nerves 2 and 3	Iliolumbar (lumbar branch)	Flexion of thigh and of vertebral column on pelvis when legs are fixed
Psoas minor	Vertebral borders of 12th thoracic and 1st lumbar vertebrae and disk between them	Pectineal line of pelvic bone and iliopectineal eminence	Lumbar nerves 1 or 2	Iliolumbar (lumbar branch)	Synergist to psoas major when present (absent in 40% of population)

*These muscles are associated with the abdomen but function in other areas.

CHAPTER 12

Transverse (Axial) Cadaver Sections and Images of the Abdomen

Abdominal cadaver sections derived from levels of thoracic vertebrae 10 and 12 and lumbar vertebrae 2 and 3 are included in this chapter. Several transverse clinical images are illustrated. The lumbar vertebra 3 cadaver section is a sheet plastinated human slice, 2.5 to 4mm thick. The other cadaver sections are approximately 20 mm thick. Both the cadaver sections and the clinical images are depicted from the inferior surface.

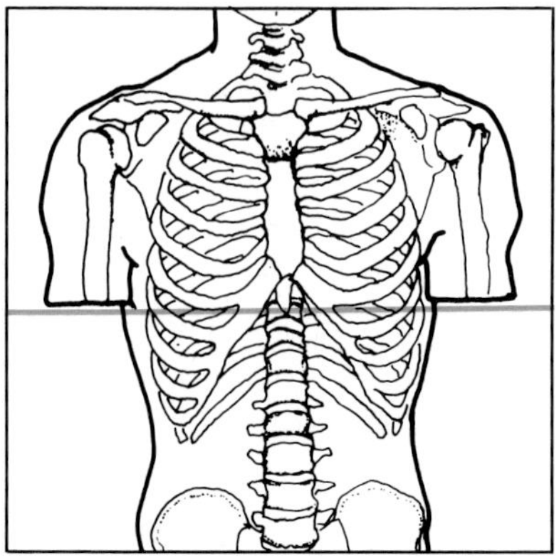

MODULE 1

This is a transverse (axial) section at the level of thoracic vertebra 10.

MUSCULOSKELETAL UNIT (Fig. 12–1)

I. Skeletal Portion

Body of a Thoracic Vertebra (*J*). An aggregate, oval bone slice that is the most anterior part of a vertebra. This feature is located in the posterior part of this body section.

Transverse Process of a Thoracic Vertebra (*E*). A lateral, oblique, bony, bilateral projection slice lateral and anterior to the deep back muscles. The lateral end is more posterior than the medial end.

Lamina of a Thoracic Vertebra (*K*). An oblique, bilateral bone section anterior and medial to the deep back muscles. The posterior end is more medial than the anterior end.

Spinous Process of a Thoracic Vertebra (*G*). A posterior median projection between the right and left groups of deep back muscles. Its anterior end is connected to the posterior ends of the laminae.

Costal Cartilage (*A*). A bilateral, hyaline cartilage slice located between the anterior end of the body of the rib and the sternum.

Costochondral Joint (*B*). A bilateral joint between the anterior part of the rib body and the costal cartilage. There are 10 pairs of these joints in a human body.

Body (Shaft) of a Rib (*C*). Rather elongated (anterior-to-posterior) bone slices that represent sections of some ribs. The anterior and posterior ends of these bone features are narrower than the tissue between the ends.

II. Muscular Portion

Erector Spinae (Sacrospinalis) Muscle (*L*). This bilateral, aggregate muscle group represents the first group of deep back muscles. It is located lateral to the spinous process and consists of three columns of muscle: lateral, intermediate, and medial.

The lateral column consists of:

The *iliocostalis lumborum muscle,* which arises from the crest of the ilium and ends on the inferior six or seven ribs.

The *iliocostalis thoracis muscle,* which originates from the inferior six ribs to end on the upper six ribs.

The *iliocostalis cervicis muscle,* which arises from the first four or five ribs and terminates on the transverse process of cervical vertebrae 4, 5, and 6.

The intermediate column consists of:

The *longissimus thoracis muscle,* which arises from the transverse processes of all five lumbar vertebrae and inserts on all thoracic vertebrae (transverse processes), superior lumbar vertebrae (transverse processes), and the ninth and tenth ribs proximal to their angles.

The *longissimus cervicis muscle* originates from the transverse processes of the superior six thoracic vertebrae and ends on the transverse processes of the cervical vertebrae (second, third, fourth, fifth, and sixth).

The *longissimus capitis muscle,* which arises from the transverse processes of the superior fifth or sixth thoracic vertebra and ends on the mastoid process.

The medial column consists of:

The *spinalis thoracis muscle,* which arises from the spinous processes of superior lumbar vertebrae 1 and 2 and the spinous processes of inferior thoracic vertebrae 11 and 12 and inserts on the spinous processes of the superior thoracic vertebrae (fourth, fifth, sixth, seventh, and eighth).

The *spinalis cervicis muscle,* which originates from the spinous processes of thoracic vertebrae 1, 2, and 7, as well as the inferior part of the nuchal ligament, and ends on the spinous process of cervical vertebra 2.

The *spinalis capitis muscle,* which originates from the transverse processes of superior thoracic vertebrae 6, 5, 4, 3, 2, and 1, as well as the articular processes of cervical vertebrae 4, 5, and 6, and ends between the superior and inferior nuchal lines of the occipital bone.

Latissimus Dorsi Muscle (*F*). An elongated, curved, bilateral muscle slice on the posterior lateral part of this body section. The concave surface is more medial than the convex surface.

Intercostal Muscles (*D, I*). Bilateral muscle sections between the rib body (shaft) sections. Typically, there are three layers of muscle. The outermost layer between the ribs is the external intercostal, the intermediate layer is the internal intercostal, and the innermost layer is the innermost intercostal.

Rectus Abdominis Muscle (*H*). A transversely placed, anterior, bilateral muscle section immediately lateral to the linea alba. The rectus abdominis is the most anterior muscle at this level.

Oblique Abdominal (Obliquus Abdominis) Muscle. An elongated (anterior-to-posterior), curved,

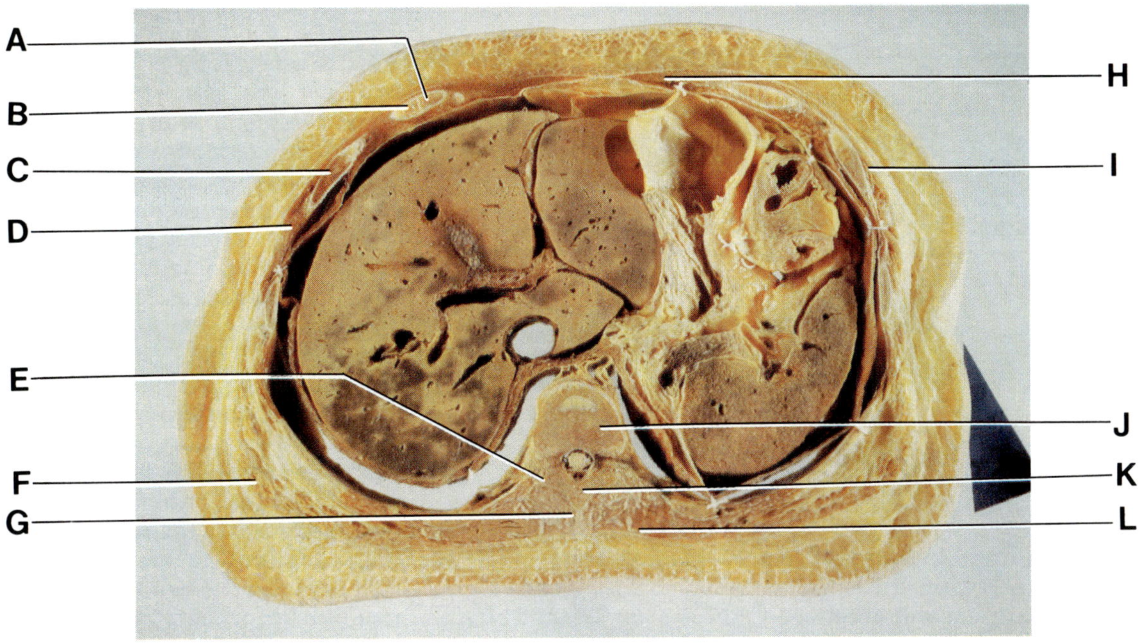

A. Costal cartilage
B. Costochondral joint
C. Body (shaft) of rib
D. Intercostal muscles (right)
E. Transverse process of thoracic vertebra
F. Latissimus dorsi muscle
G. Spinous process of thoracic vertebra
H. Rectus abdominis muscle
I. Intercostal muscles (left)
J. Body of thoracic vertebra
K. Lamina of thoracic vertebra
L. Erector spinae (sacrospinalis) muscle

Figure 12–1
Transverse (axial) section of the abdomen at the level of thoracic vertebra 10, highlighting musculoskeletal unit structures.

bilateral muscle slice lateral to the thoracic wall. The concave surface is medial to the convex surface. The anterior end is adjacent to the rectus abdominis muscle, and the posterior end is adjacent to the midaxillary line. Typically, there are three layers of muscle on the anterior and lateral walls of the abdomen. The outer muscle is known as the oblique external abdominal muscle (obliquus abdominis externus), the intermediate one is the oblique internal abdominal muscle (obliquus abdominis internus), and the innermost layer is called the transverse abdominal muscle (transversus abdominis).

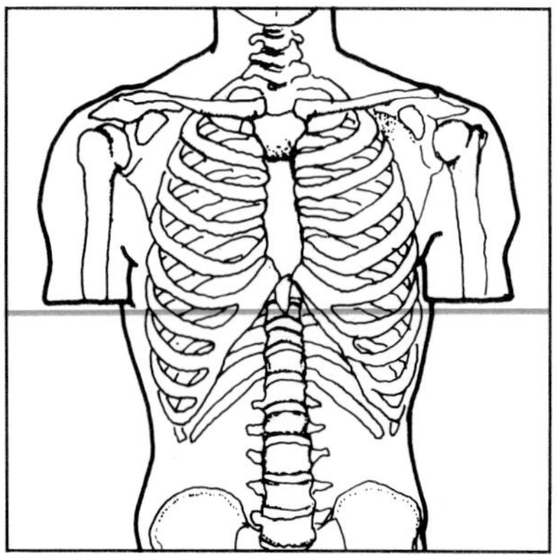

NEUROVASCULAR UNIT (Fig. 12–2)

I. Arteries

Descending Abdominal Aorta (*A*). A hollow, partially collapsed, thick-walled vessel to the left of the thoracic duct and anterior to the left front part of the vertebral body.

Origin. A continuation of the descending thoracic aorta.

Distribution. The descending abdominal aorta is the main systemic artery of the abdomen and distributes blood to the following vessels: (1) right and left inferior phrenic arteries, (2) celiac trunk (axis), (3) superior mesenteric artery, (4) right and left middle adrenal arteries, (5) right and left renal arteries, (6) right and left gonadal arteries, (7) inferior mesenteric artery, and (8) right and left common iliac arteries.

Course. It passes downward anterior to the left part of the lumbar vertebrae, to the left of the inferior vena cava, and terminates at about the level of lumbar vertebra 4.

Splenic Vessel (*G*). A rather oblique, hollow, elongated vascular structure. Its posterior end is more lateral than its anterior end and is adjacent to the hilus of the spleen.

Origin (for splenic artery). Celiac axis (trunk).

Distribution (for splenic artery). Spleen, pancreas, and fundus of the stomach.

Course (for splenic artery). The splenic artery takes the horizontal path to the spleen. In its course it passes adjacent to the pancreas and the inferior surface of the stomach to reach the spleen.

Origin (for splenic vein). Spleen.

Termination (for splenic vein). Hepatic portal vein.

Course (for splenic vein). It courses adjacent to the splenic artery and forms a groove in the superior and posterior parts of the pancreas inferior to the splenic artery. It terminates posterior to the neck of the pancreas as it unites with the superior mesenteric vein to form the portal vein.

Left Gastroepiploic Vessel (*I*). A rounded, hollow vascular structure medial to the spleen and posterior to the pancreas.

Origin (for left gastroepiploic artery). Splenic artery.

Distribution (for left gastroepiploic artery). Stomach and greater omentum.

Course (for left gastroepiploic artery). A terminal branch of the splenic artery passes between the layers of the gastrosplenic ligament and adjacent to the inferior left part of the greater curvature of the stomach. It continues toward the right yielding branches to the anterior posterior parts of the stomach and the greater omentum prior to its anastomosis with the right gastroepiploic artery.

Origin (for right gastroepiploic artery). Gastroduodenal artery.

Distribution (for right gastroepiploic artery). Greater curvature of the stomach and anterior/posterior portions of the inferior part of the stomach. This vessel also distributes blood to the greater omentum.

Course (for right gastroepiploic artery). It takes a horizontal path from the right toward the left side of the body adjacent to the greater curvature of the stomach between the two anterior layers of the greater omentum prior to its anastomosis with the left gastroepiploic artery.

Origin (for left gastroepiploic vein). Branches from the greater omentum and the stomach.

Termination (for left gastroepiploic vein). Splenic vein.

Course (for left gastroepiploic vein). It takes a leftward path adjacent to the greater curvature of the stomach and between the anterior two layers of the greater omentum. It passes within the gastrosplenic ligament to approach the hilum of the spleen, where it terminates in the splenic vein.

Origin (for right gastroepiploic vein). Receives blood from both the anterior and posterior parts of the stomach.

Termination (for right gastroepiploic vein). Superior mesenteric vein.

Course (for right gastroepiploic vein). It takes a horizontal path from left to right adjacent to the greater curvature of the stomach and between the front two layers of the greater omentum to reach the pylorus of the stomach, where it terminates in the superior mesenteric vein.

II. Veins

Azygos Vein (*F*). This is seen on the right side of this body section as a small, hollow vascular structure right of the thoracic duct and anterior to the body of a thoracic vertebra.

Inferior Vena Cava (*D*). A large, hollow, spherical vessel between the right and left caudate lobes of the liver.

Right Branch of the Hepatic Portal Vein (*C*). A hollow, elongated vascular structure located within the right lobe of the liver.

Origin. Main hepatic portal vein.

Termination. Right lobe of the liver.

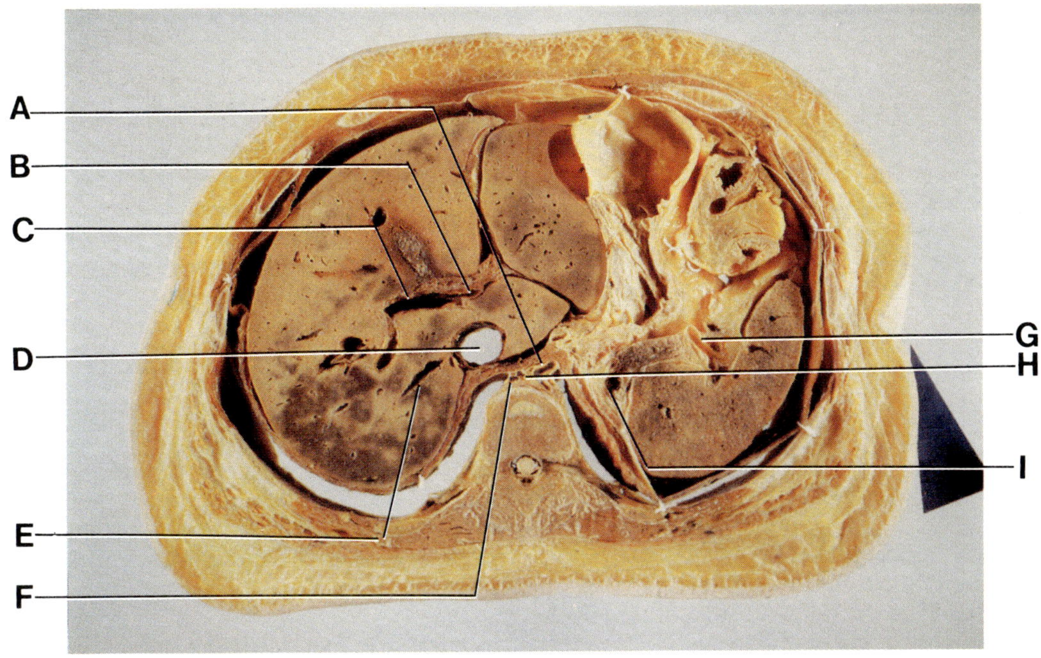

A. Descending abdominal aorta
B. Main hepatic portal vein
C. Right branch of hepatic portal vein
D. Inferior vena cava
E. Hepatic vein
F. Azygos vein
G. Splenic vessel
H. Thoracic duct
I. Left gastroepiploic vessel

Figure 12–2
Transverse section of the abdomen at the level of thoracic vertebra 10, highlighting neurovascular unit structures.

Main Hepatic Portal Vein (*B*). A transverse, hollow vascular structure seen at this level anterior to the caudate lobe of the liver and posterior to the medial part of the right lobe of the liver.

Origin. Superior mesenteric and splenic veins.

Termination. Right and left branches of the hepatic portal vein.

Course. From its origin, the hepatic portal vein takes a vertical path between the layers of the lesser omentum to arrive at the porta hepatis. This vessel splits into two branches (right hepatic portal and left hepatic portal veins).

Hepatic Vein (*E*). A transversely oblique, hollow vascular structure that runs toward the inferior vena cava on the right side of this body section. The medial end is more anterior than the lateral end. There are three hepatic veins (right hepatic, left hepatic, and middle hepatic).

Origin. Sublobular veins within the liver tissues.

Termination. Inferior vena cava.

Course. These veins exit from the posterior surface of the liver and terminate immediately into the inferior vena cava.

III. Lymphatic Structures

Thoracic Duct (*H*). A small, hollow tube anterior to the body of a thoracic vertebra between the azygos vein and abdominal aorta. The thoracic duct is the largest lymph vessel of the body and runs posterior to the esophagus while in the thorax.

Origin. Cisterna chyli at the level of lumbar vertebra 2 to the right of the abdominal aorta.

Termination. Left subclavian vein near the distal end of the internal jugular vein.

Course. This structure takes a vertical path from its origin and passes through the diaphragm within the aortic hiatus. This structure continues its vertical path in the thorax adjacent to the vertebral column between the descending thoracic aorta and azygos vein. At about thoracic vertebra 5, this large lymph vessel curves to the left and continues its vertical course adjacent to the left edge of the esophagus. It then takes a lateral horizontal course between the left carotid sheath and left vertebral artery to reach the left subclavian vein near the distal end of the internal jugular vein.

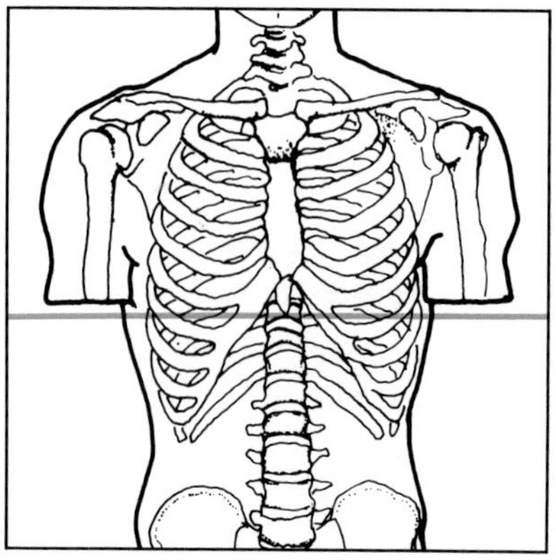

VISCERAL UNIT (Fig. 12–3)

Left Lobe of the Liver (*A*). An oblong structure medial to the right lobe of the liver and anterior to the caudate lobe of the liver. It is located in the epigastrium. Its posterior end is located between the stomach and the caudate lobe.

Right Lobe of the Liver (*B*). A large, aggregate, oval liver section that occupies nearly the entire right half of the peritoneal cavity. It is posterior to the right subphrenic space and anterior to the right subhepatic space (Morrison's pouch).

Caudate Lobe of the Liver (*C*). A triangular liver part left of the right lobe of the liver and medial to the inferior vena cava. The apex of this structure is posterior to the left lobe of the liver.

Spleen (*L*). An oblong, curved slice of splenic tissue within the posterior left part of the peritoneal cavity. The concave surface is deep to its convex surface. The spleen occupies a position in the left hypochondriac region corresponding to the vertebral level from approximately thoracic vertebra 10 to the upper part of lumbar vertebra 2. It is between the posterior surface of the stomach and the diaphragm. The spleen's average length is 5 inches; its average width is 3 inches; and its approximate weight is 7 ounces.

Hilum of Spleen (*K*). An anterior, medial indentation of the spleen, the site of entrance and exit of the splenic vessel as well as its nerve supply.

Stomach (*E*). An elongated (anterior-to-posterior), irregular, hollow structure with some stellate projections within its lumen. It is located in the left hypochondrium anterior to the midaxillary line between the left lobe of the liver and the splenic flexture. The stomach lies in a corresponding position from approximately thoracic vertebra 10 to upper lumbar vertebra 2.

Greater Curvature of the Stomach (*F*). An irregular, curved portion of the left wall of the stomach. The anterior part of the curve shows a convex surface toward the left side of the section, whereas the posterior part shows a concave surface toward the left side.

Lesser Curvature of the Stomach (*J*). A slightly curved portion of the medial wall of the stomach adjacent to the left lobe of the liver. This slight curve shows a convex surface toward the left side of this body section.

Transverse Colon (*G*). A circular part of the transverse colon lateral to the anterior part of the stomach. This part represents the distal end of the transverse colon and is adjacent to the splenic flexture.

Course. The transverse colon is approximately 20 inches (50 cm) in length and begins at the hepatic flexure in the upper right lumbar region. It is transversely and obliquely positioned across the abdomen to reach the left hypochondriac region. The left end is slightly superior to the right end. It terminates in the descending colon after taking part in the formation of the splenic flexure.

Splenic Flexure (Left Colic Flexure) (*H*). A portion of the colon between the distal end of the transverse colon and the proximal end of the descending colon. This feature is located on the anterior lateral part of the peritoneal cavity.

Descending Colon (*I*). A hollow feature on the left side of this body section, with a thick wall anterior to the spleen and psoterior to the transverse colon. The descending colon is 10 inches long in an adult human body. It connects to the sigmoid colon.

Course. It starts in the left hypochondria and upper left lumbar regions and courses inferiorly adjacent to the lateral edge of the left kidney and the angle between the psoas major and quadratus lumborum muscles, to reach the crest of the ilium. Then it takes a more medial route anterior to the iliac and psoas major muscles. It terminates in the sigmoid colon at the level of the superior part of the lesser (true) pelvis.

Spinal Cord (*D*). A spherical nerve structure within the spinal foramen. Note the dura mater lining the inner wall of the spinal foramen.

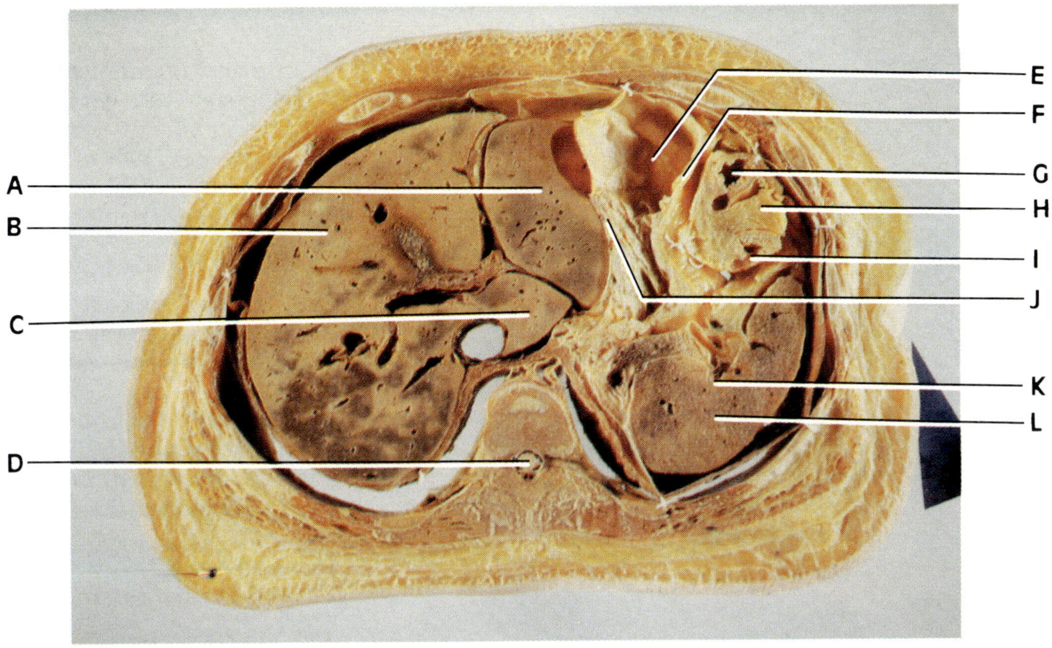

A. Left lobe of liver
B. Right lobe of liver
C. Caudate lobe of liver
D. Spinal cord
E. Stomach
F. Greater curvature of stomach
G. Transverse colon
H. Splenic flexure (left colic flexure)
I. Descending colon
J. Lesser curvature of stomach
K. Hilum of spleen
L. Spleen

Figure 12–3
Transverse section of the abdomen at the level of thoracic vertebra 10, highlighting visceral unit structures.

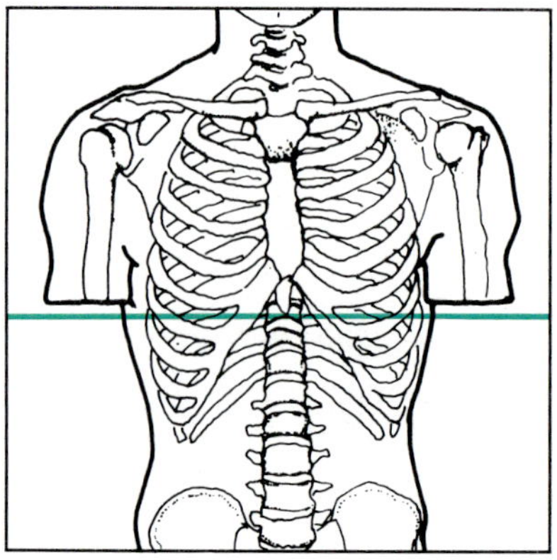

ENCLOSING UNIT (Fig. 12–4)

Falciform Ligament of Liver (*C*). A thick band of tissue between the right and left lobes of the liver. The falciform ligament extends from the umbilicus on the anterior median wall of the peritoneal cavity to the diaphragm and attaches to it.

Right Subhepatic Space (Morrison's Pouch) (*B*). A curved, hollow posterior space lateral to the body of thoracic vertebra 10 and posterior to the right lobe of the liver. The concave surface is adjacent to the posterior surface of the right lobe of the liver.

Right Subphrenic Space (*A*). A narrow, slightly circular space located primarily anterior to the right lobe of the liver. The concave surface is adjacent to the anterior part of the right lobe of the liver.

Parietal Peritoneum (*D*). A sheet of tissue (membrane) adjacent to the inner wall of the peritoneal cavity.

Dura Mater (*E*). A membrane that lines the inner wall of the spinal foramen.

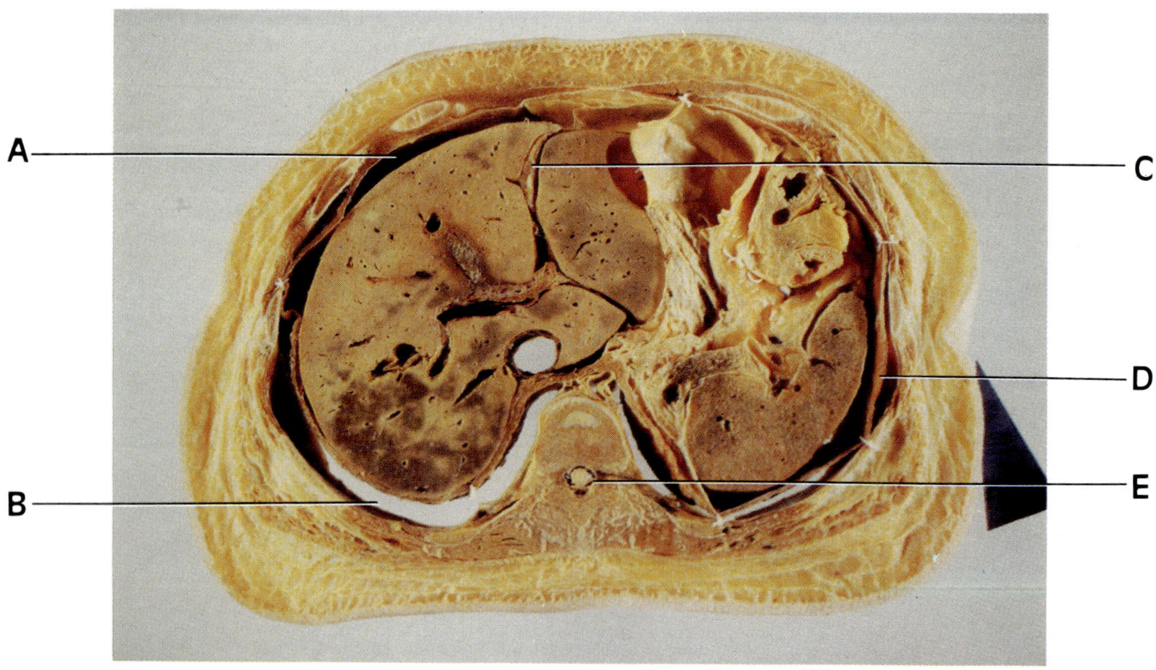

A. Right subphrenic space
B. Right subhepatic space (Morrison's pouch)
C. Falciform ligament of liver
D. Parietal peritoneum
E. Dura mater

Figure 12–4
Transverse section of the thorax at the level of thoracic vertebra 10, highlighting enclosing unit structures.

An MRI scan (Fig. 12–5) shows structures located within the upper abdomen. This part of the abdomen is referred to as the subcostal zone. The subcostal zone is subdivided into three abdominal regions: (1) right hypochondriac, (2) epigastric, and (3) left hypochondriac.

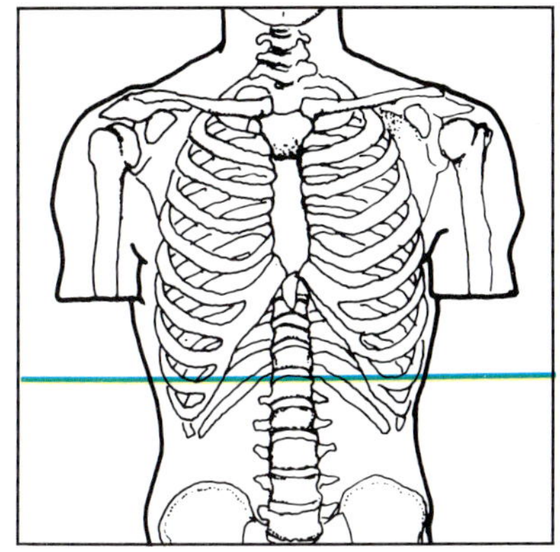

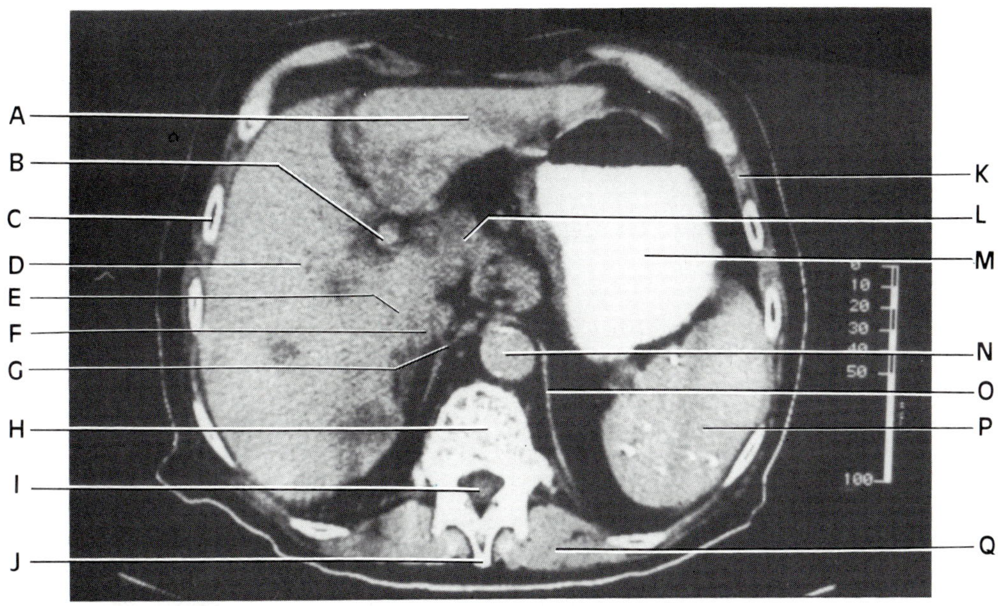

A. Left lobe of liver
B. Main portal vein
C. Rib
D. Right lobe of liver
E. Caudate lobe of liver
F. Inferior vena cava
G. Right crus of diaphragm
H. Body of thoracic vertebra
I. Spinal foramen
J. Spinous process of thoracic vertebra
K. Intercostal muscles
L. Quadrate lobe of liver
M. Stomach
N. Descending abdominal aorta
O. Left crus of diaphragm
P. Spleen
Q. Erector spinae muscle

Figure 12–5
MRI transverse image of the abdomen, highlighting musculoskeletal, neurovascular, and visceral unit structures.

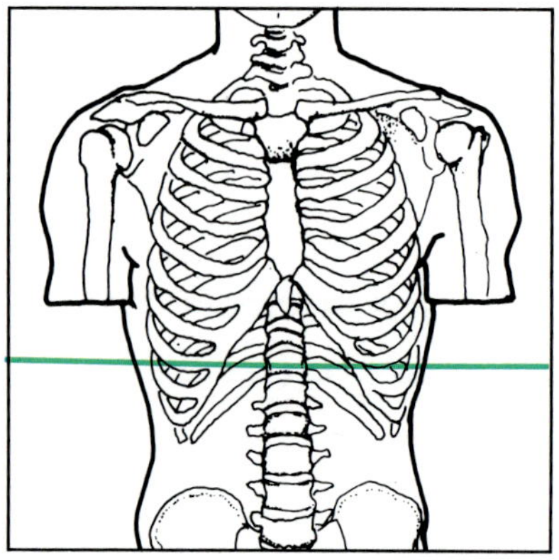

MODULE 2

This is a transverse (axial) section at the level of thoracic vertebra 12.

MUSCULOSKELETAL UNIT (Fig. 12–6)

I. Skeletal Portion

Body of a Thoracic Vertebra (*B*). An oval bone slice anterior to the pedicles and the spinal cord. The body is the most anterior feature of a thoracic vertebra.

Transverse Process of a Thoracic Vertebra (*C*). A lateral, oblique, bilateral bone projection lateral to the spinal foramen and anterior to the lamina. The lateral end is more posterior than the medial end.

Lamina of a Thoracic Vertebra (*D*). An oblique, bilateral bone section posterior and medial to the transverse process. The posterior end of this feature is more medial than the anterior end.

Spinous Process of a Thoracic Vertebra (*E*). A median, rather sharp posterior bone section between the erector spinae muscles. The proximal end of this feature starts at the distal end of the laminae.

Body (Shaft) of a Rib (*A*). Submarine-shaped, bilateral bone slices that represent sections of some ribs (body parts). These bone slices are on the peripheral edge of the peritoneal cavity.

II. Muscular Portion

Erector Spinae Muscle (*H*). Lateral, oblong, bilateral muscle aggregate lateral to the spinous process. The part adjacent to the spinous process is thicker than the lateral portion of this muscle group.

Latissimus Dorsi Muscle (*G*). This appears bilaterally as an elongated (anterior-to-posterior), curved section of muscle tissue. The concave surface is more medial than the convex surface.

Rectus Abdominis Muscle (*F*). This appears bilaterally as a thin, elongated (anterior-to-posterior) muscle section lateral to the linea alba. It is the most anterior superficial muscle seen at this level.

A. Body (shaft) of rib
B. Body of thoracic vertebra
C. Transverse process of thoracic vertebra
D. Lamina of thoracic vertebra
E. Spinous process of thoracic vertebra
F. Rectus abdominis muscle
G. Latissimus dorsi muscle
H. Erector spinae muscle

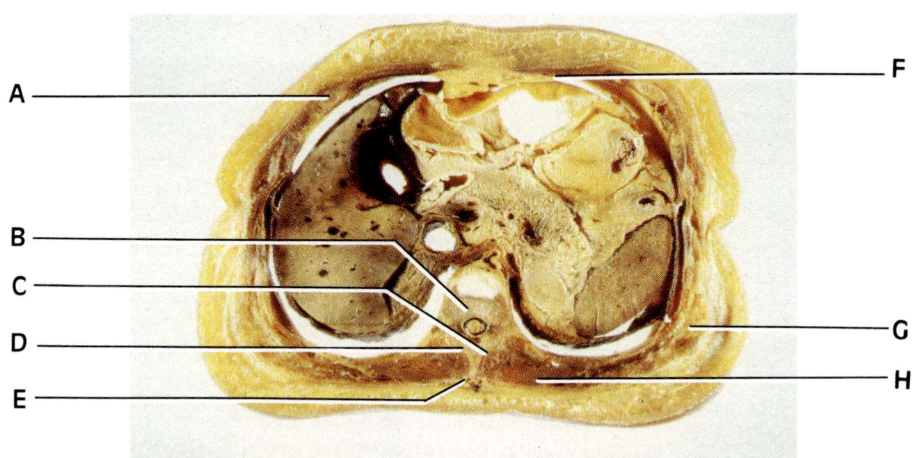

Figure 12–6
Transverse section of the abdomen at the level of thoracic vertebra 12, highlighting musculoskeletal unit structures.

NEUROVASCULAR UNIT (Fig. 12–7)

I. Arteries

Descending Abdominal Aorta (*B*). A partially collapsed, hollow vascular structure anterior to the anterior left part of a thoracic vertebra and posterior to the pancreas.

Gastroduodenal Artery (*C*). This is seen here on the left side of the median as a small, spherical vascular structure posterior to the body of the pancreas and left of the hepatic portal vein.

Origin. Common hepatic trunk.

Distribution. Stomach, duodenum, pancreas, and greater omentum. This artery disperses blood to three branches: (1) the right gastroepiploic, (2) the superior pancreaticoduodenal, and (3) the retroduodenal arteries that commonly arise from this vessel.

Course. This vessel arises posterior to the duodenum and takes a downward path between the duodenum and the neck of the pancreas. In its course, this artery is usually on the left side of the bile duct. At the inferior posterior part of the duodenum, this vessel divides to form the superior pancreaticoduodenal artery and the right gastroepiploic artery. The superior pancreaticoduodenal artery bifurcates to form the anterior superior pancreaticoduodenal artery, which passes anterior to the neck of the pancreas, and the posterior superior pancreaticoduodenal artery, which passes posterior to the neck of the pancreas. Each anastomoses with its respective anterior and posterior inferior pancreaticoduodenal arteries. The inferior pancreaticoduodenal artery arises from the superior mesenteric artery and yields both the anterior and posterior inferior pancreaticoduodenal arteries.

II. Veins

Inferior Vena Cava (*A*). This appears on the right side of the median as a large, hollow, spherical venous structure posterior to the hepatic portal vein and medial to the right lobe of the liver.

Hepatic Portal Vein (*D*). This is seen here on the right side of the median as a hollow, oblong, vascular venous vessel anterior to the inferior vena cava and posterior to the head of the pancreas.

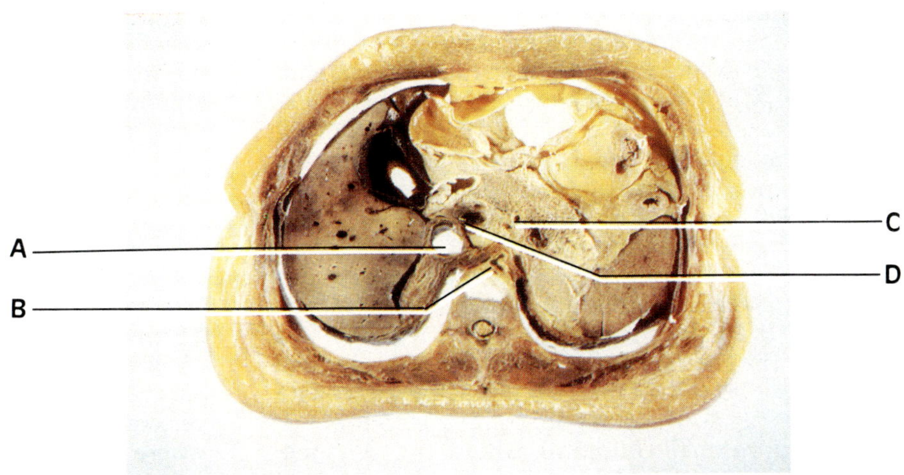

A. Inferior vena cava
B. Descending abdominal aorta
C. Gastroduodenal artery
D. Hepatic portal vein

Figure 12–7
Transverse section of the abdomen at the level of thoracic vertebra 12, highlighting neurovascular unit structures.

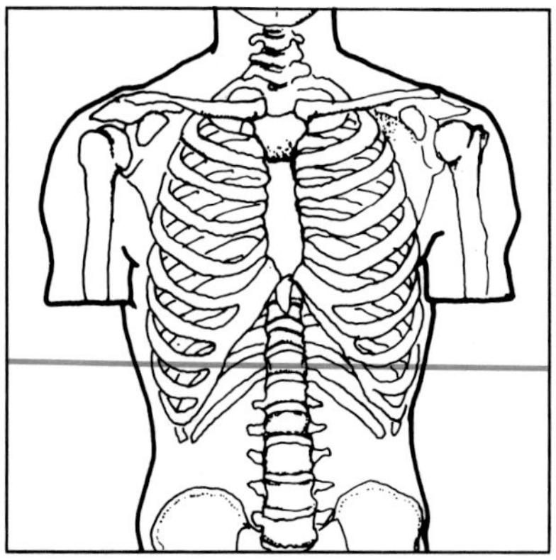

VISCERAL UNIT (Fig. 12–8)

Gallbladder (*B*). An oval, oblong, saccular slice located on the anterior medial surface of the right lobe of the liver at this level. This structure is found to the right of the median and anterior to the inferior vena cava. The gallbladder is housed in a fossa on the posterior surface of the right lobe of the liver.

Right Lobe of the Liver (*D*). A large, irregular, sickle-shaped aggregate of liver tissue lateral to the inferior vena cava and the gallbladder. The anterior end of this liver slice is thinner than the posterior end.

Spleen (*M*). A large, oblong, oblique splenic section in the posterior lateral part of the peritoneal cavity. The anterior end is more lateral than the posterior end. The posterior end is immediately lateral to the left kidney slice seen at this level.

Left Kidney (*N*). An oval slice of kidney tissue medial to the posterior part of the spleen seen at this level and posterior to the left suprarenal gland section. The left kidney is located on the left side of a vertebra. The cranial edge of the left kidney is located at the upper portion of the 12th thoracic vertebra and extends as far down as the 3rd lumbar vertebra.

Left Suprarenal (Adrenal) Gland (*K*). A small, triangular slice of glandular tissue anterior to the left kidney and posterior to the body of the pancreas. This structure is on the left of the descending abdominal aorta. This gland lies in a position corresponding from approximately the lower half of thoracic vertebra 11 to the lower edge of thoracic vertebra 12.

Head of the Pancreas (*G*). The most medial portion of the pancreas, located in the concave portion of the duodenum. Its medial end, as seen at this level, shows a concave appearance on the right side of the pancreas.

Body of the Pancreas (*H*). A transversely oblique, glandular slice anterior to the descending abdominal aorta and posterior to the descending colon. It is to the left of the duodenum. The posterior end is more lateral than the anterior end.

Tail of the Pancreas (*L*). A tapered lateral part of the pancreas seen adjacent to the posterior medial part of the spleen at this level. It is lateral to the left kidney and medial to the descending colon.

Stomach (*A*). A triangular section at this level anterior to the head of the pancreas and to the left of the anterior part of the right lobe of the liver. It is anterior to the descending duodenum.

Descending Duodenum (Second Segment of the Duodenum) (*C*). This is seen here immediately right of the median, anterior to the inferior vena cava and posterior to the stomach. This is an oval, thick-walled, hollow structure. *Note:* The duodenum is the first segment of the small intestine.

Course (for the first part of the duodenum). The first (superior) part of the duodenum is approximately 5 cm in length and is attached to the pylorus of the stomach. This part of the duodenum is at a level corresponding to the first lumbar vertebra to the right of the median. The quadrate lobe of the liver and the gallbladder are anterior to this part of the duodenum. The following structures are posterior to this part of the duodenum: (1) the proximal half of the lesser sac, (2) the gastroduodenal artery, (3) the bile duct, (4) the hepatic portal vein, and (5) the inferior vena cava.

Course (for the second part of the duodenum). The second (descending) part of the duodenum is approximately 7.5 cm in length. This structure is anterior to the renal arteries/veins and the medial parts of the kidneys. It terminates at the level of the third lumbar vertebra, anterior to the right psoas major muscle.

Course (for the third part of the duodenum). The third (horizontal) part of the duodenum is about 10 cm in length and is transversely positioned anterior to the inferior vena cava and aorta.

Course (for the fourth part of the duodenum). The fourth (ascending) part of the duodenum is approximately 2.5 cm in length and takes an upward, oblique course on the left side of the aorta and the head of the pancreas. It terminates opposite the second lumbar vertebra anterior to the left psoas major muscle, where it connects with the jejunum.

Descending Colon (*I, J*). A hollow, circular, thick-walled digestive structure immediately anterior to the lateral part of the spleen at this level. This structure is adjacent to the inner wall of the left part of the peritoneal cavity.

Right Crus of the Diaphragm (*E*). A strip of tissue that connects the diaphragm to the right lateral parts of lumbar vertebral bodies 1, 2, and 3, as well as the lower part of thoracic vertebra 12.

Left Crus of the Diaphragm (*O*). A strip of tendinous tissue that attaches the diaphragm to the left lateral parts of the vertebral bodies of lower thoracic vertebra 12 and lumbar vertebrae 1 and 2.

Spinal Cord (*F*). This is seen here as spherical aggregate nerve tissue located within the spinal foramen.

A. Stomach
B. Gallbladder
C. Descending duodenum (second segment of duodenum)
D. Right lobe of liver
E. Right crus of diaphragm
F. Spinal cord
G. Head of pancreas
H. Body of pancreas
I. Colon
J. Colon
K. Left suprarenal (adrenal) gland
L. Tail of pancreas
M. Spleen
N. Left kidney
O. Left crus of diaphragm

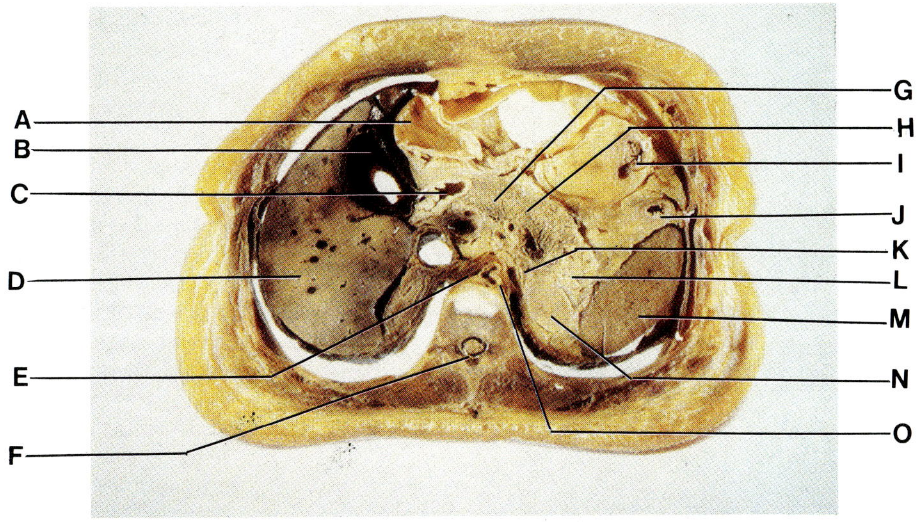

Figure 12-8
Transverse section of the abdomen at the level of thoracic vertebra 12, highlighting visceral unit structures.

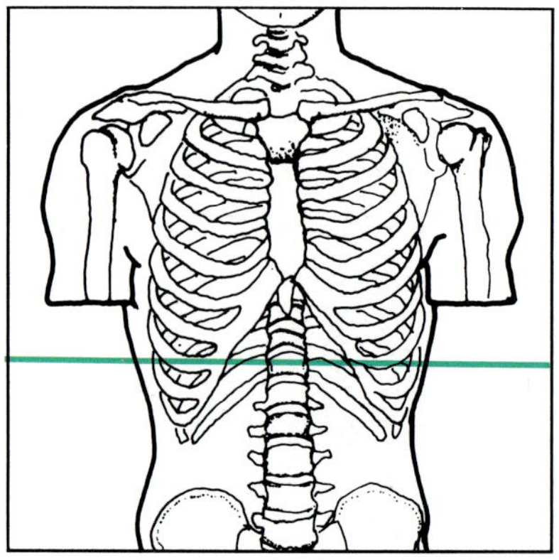

ENCLOSING UNIT (Fig. 12–9)

Right Subhepatic Space (Morrison's Pouch) (*C*). A C-shaped, hollow area posterior to the right lobe of the liver. The concave part is adjacent to the posterior part of the right lobe of the liver. The medial end of this cavity is posterior to the inferior vena cava and to the right of a thoracic vertebral body.

Right Subphrenic Space (*B*). An arched cavity anterior to the right lobe of the liver. The concave surface is adjacent to the anterior part of the right lobe of the liver.

Renal (Gerota's) Fascia (*G*). A fascia that surrounds the kidneys outside the renal capsule.

Transversalis Fascia (*A*). A layer of tissue within the wall of the abdominal cavity, deep to the transversus abdominis muscle and superficial to the subserous fatty layer in the wall of the abdominal cavity.

Greater Sac (*E*). An anterior potential space in the abdominal cavity, seen at this level as a rather narrow, curved, empty space. The concave surface is adjacent to the anterior parts of the abdominal viscera.

Linea Alba (*D*). An anterior median plane section between the left and right rectus abdominis muscles.

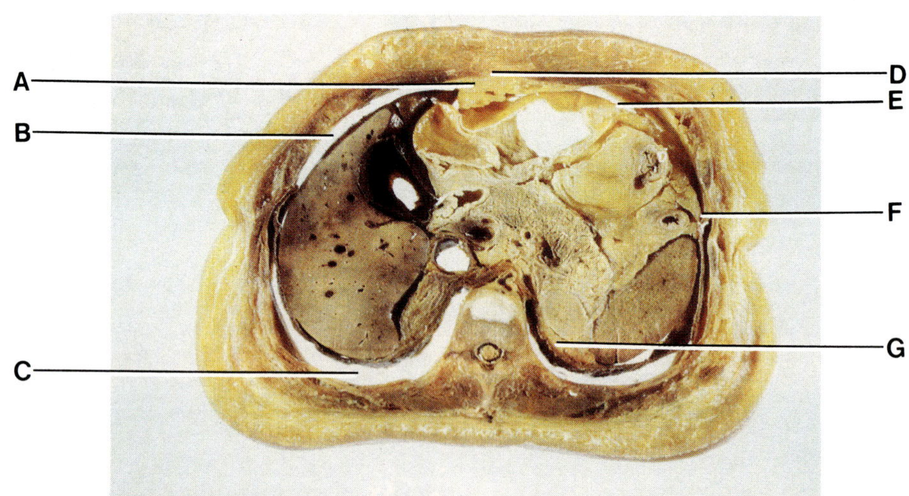

A. Transversalis fascia
B. Right subphrenic space
C. Right subhepatic space (Morrison's pouch)
D. Linea alba
E. Greater sac
F. Left colic gutter
G. Renal (Gerota's) fascia

Figure 12–9
Transverse section of the abdomen at the level of thoracic vertebra 12, highlighting enclosing unit structures.

Figure 12–10 is an MRI scan showing structures of the abdomen in a transverse plane. Most of the structures depicted are located in the umbilical zone of the abdomen. The three regions of the umbilical zone are (1) the right lumbar, (2) the umbilical, and (3) the left lumbar. Some structures, such as the spleen, stomach, and lobes of the liver, usually are seen more extensively in the subcostal zone.

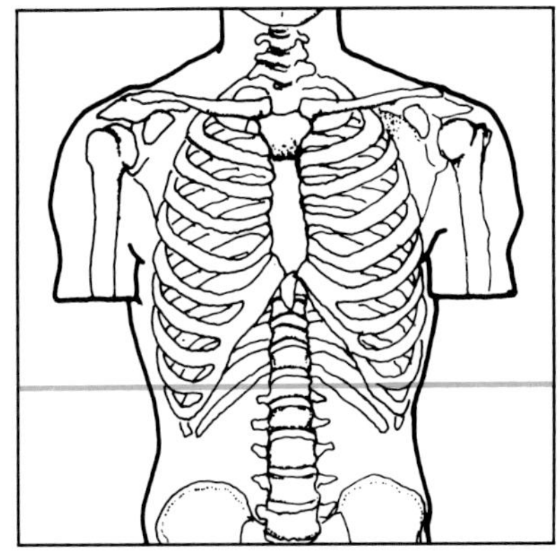

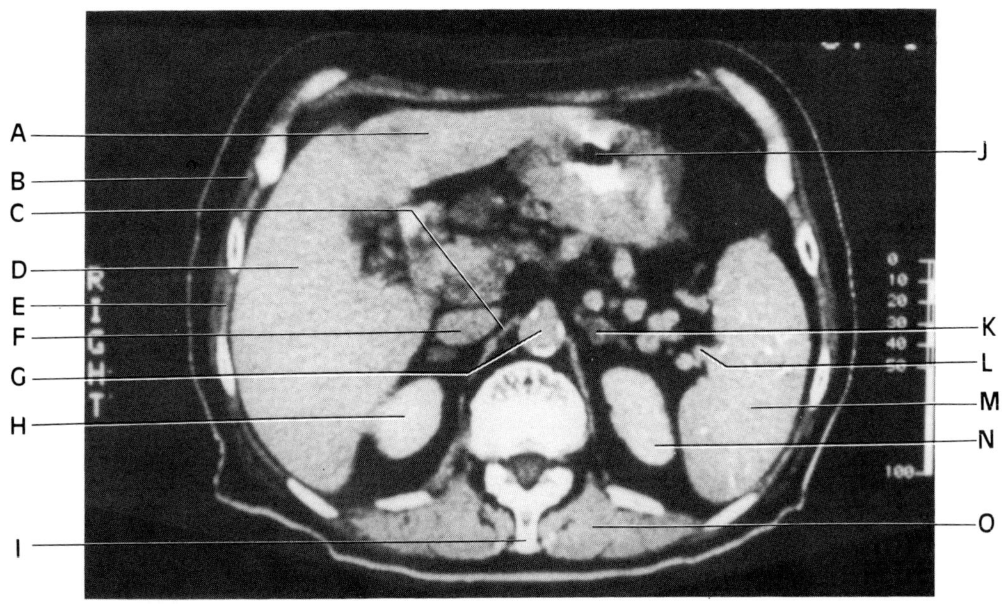

A. Left lobe of liver
B. Oblique muscle of the abdomen (obliquus abdominis muscle)
C. Right crus of diaphragm
D. Right lobe of liver
E. Intercostal muscles
F. Inferior vena cava
G. Descending abdominal aorta
H. Right kidney
I. Spinous process of vertebra
J. Stomach
K. Left suprarenal (adrenal) gland
L. Renal vessel
M. Spleen
N. Left kidney
O. Erector spinae muscle

Figure 12–10
MRI transverse image of the abdomen, highlighting musculoskeletal, neurovascular, and visceral unit structures.

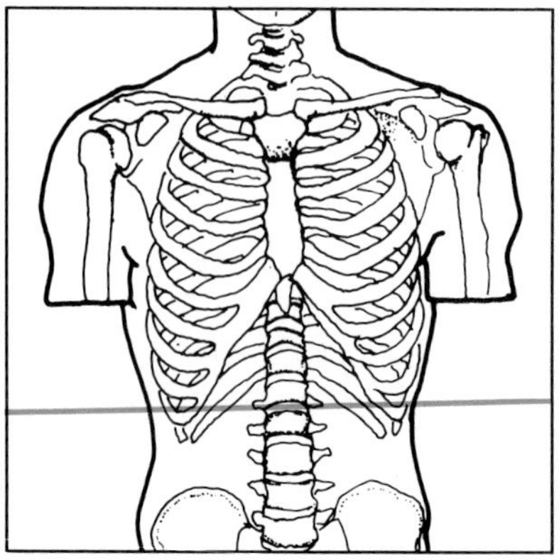

MODULE 3

This is a transverse (axial) section at the level of lumbar vertebra 2.

MUSCULOSKELETAL UNIT (Fig. 12–11)

I. Skeletal Portion

Body of a Lumbar Vertebra (*A*). An oval aggregate of bone tissue with a posterior indentation anterior to the spinal foramen. The posterior lateral parts of this feature are between the right and left psoas major muscles.

Pedicle of a Lumbar Vertebra (*C*). A short, narrow strip of bilateral bone tissue immediately posterior to the lateral part of the vertebral body. This feature is immediately anterior to the transverse process and lateral to the spinal foramen.

Transverse Process of a Lumbar Vertebra (*D*). A short, narrow, bilateral projection anterior to the multifidus muscle and posterior to the vertebral pedicle.

Lamina of a Lumbar Vertebra (*B*). An oblique, bilateral bone section posterior and lateral to the spinal foramen. The posterior end is more medial than the anterior end. It joins with the spinous process in the posterior median.

Spinous Process of a Lumbar Vertebra (*E*). A posterior, median, narrow bone slice immediately between the deep back muscles. Its proximal end is attached to the posterior ends of the lamina.

II. Muscular Portion

Erector Spinae (Sacrospinalis) Muscle (*M*). An oval, bilateral muscle aggregate lateral to the multifidus muscle and the spinous process. This muscle is posterior to the psoas major muscle.

Multifidus Muscle. A rather small, oblong, oblique, bilateral muscle section adjacent to the lamina and part of the spinous process. The posterior end is slightly more medial than the anterior end.

Psoas Major Muscle (*L*). A triangular (anterior-to-posterior), bilateral muscle slice. The base of the triangular muscle section is adjacent to the transverse process of a vertebra and lateral to the pedicle.

Latissimus Dorsi Muscle (*F*). A rather narrow strip of bilateral muscle located on the posterior lateral part of this transverse body section. The medial end is immediately posterior to the erector spinae muscle, and the anterior end is superficial but adjacent to the external oblique muscle.

Serratus Posterior Inferior Muscle (*G*). A rather small, triangular, bilateral muscle slice seen at this level immediately lateral to the erector spinae muscle and anterior to the latissimus dorsi muscle. The medial part is wider than the lateral part.

External Oblique (Obliquus Abdominis Externus) Muscle (*K*). A curved strip of bilateral muscle tissue superficial to the internal oblique and located on the lateral part of this transverse body section. The concave surface is on its medial part. This muscle extends from the posterior part of this body section to the linea semilunaris (a connective tissue layer between the external oblique and rectus abdominis muscles).

Internal Oblique (Obliquus Abdominis Internus) Muscle (*J*). A curved strip of bilateral muscle tissue immediately deep to the external oblique muscle. The medial surface shows its concave part.

Transversus Abdominis Muscle (*I*). A curved, internal, deep layer of muscle tissue located on the anterior and lateral abdominal wall at this level. This muscle is immediately deep to the internal oblique muscle and superficial to the transversalis fascia.

Rectus Abdominis Muscle (*H*). An anterior, superficial, transversely (medial-to-lateral) placed, bilateral muscle strip, with each end narrower than the other parts of the muscle. This muscle is located immediately lateral to the linea alba and medial to the linea semilunaris.

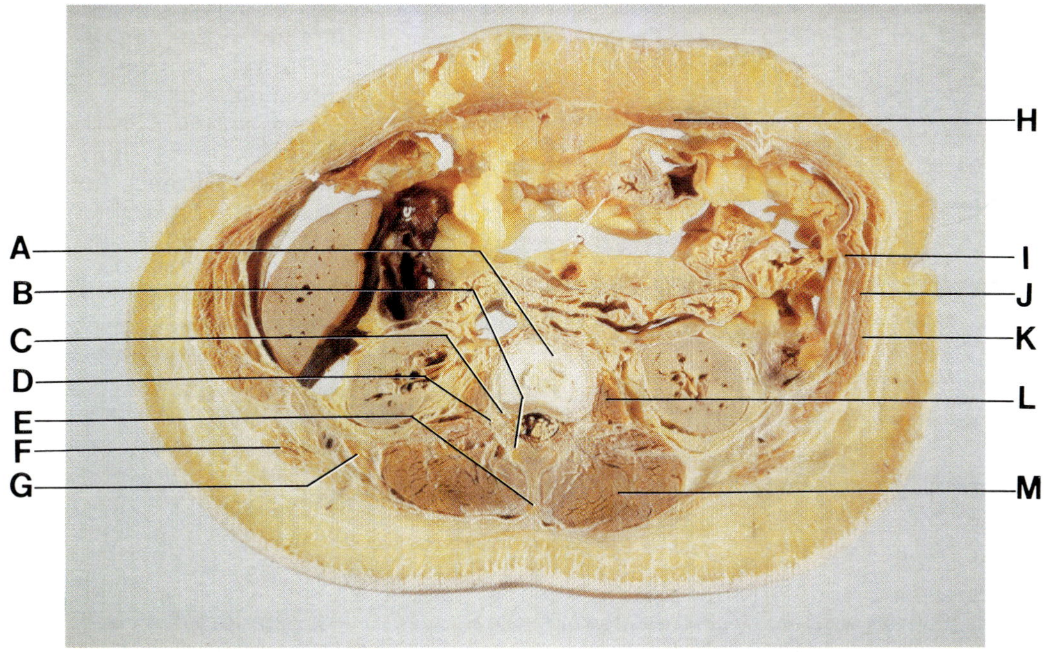

A. Body of lumbar vertebra
B. Lamina of lumbar vertebra
C. Pedicle of lumbar vertebra
D. Transverse process of lumbar vertebra
E. Spinous process of lumbar vertebra
F. Latissimus dorsi muscle
G. Serratus posterior inferior muscle
H. Rectus abdominis muscle
I. Transversus abdominis muscle
J. Internal oblique (obliquus abdominis internus) muscle
K. External oblique (obliquus abdominis externus) muscle
L. Psoas major muscle
M. Erector spinae (sacrospinalis) muscle

Figure 12–11
Transverse (axial) section of the abdomen at the level of lumbar vertebra 2, highlighting musculoskeletal unit structures.

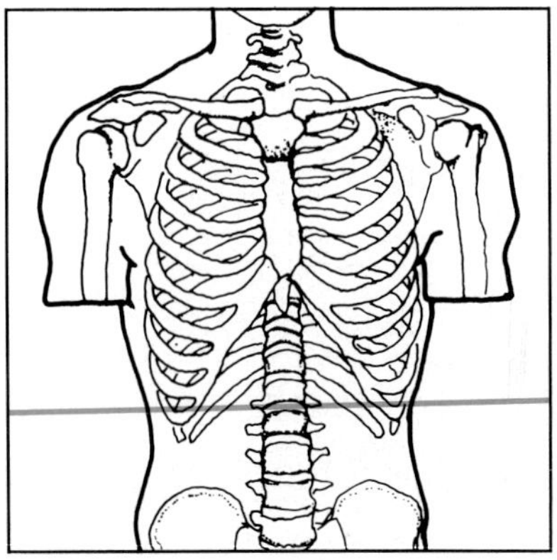

NEUROVASCULAR UNIT (Fig. 12–12)

I. Arteries

Descending Abdominal Aorta (*B*). A hollow, partially collapsed vascular structure anterior to the body of the lumbar vertebra and to the left of the inferior vena cava.

Superior Mesenteric Artery (*A*). A rather large, oval, hollow vascular structure located in the median anterior to the descending abdominal aorta and the descending (horizontal) duodenum.

Origin. Descending abdominal aorta.

Distribution. The superior mesenteric artery is the main vessel to vascularize the following: (1) cecum, (2) right lateral portion of the transverse colon, (3) ascending colon, (4) vermiform appendix, and (5) small intestine. The branches of this major artery and the structures it supplies are (1) the intestinal arteries supply blood to the small intestine, (2) the ileocolic artery supplies blood to the distal end of the ileum, cecum, and vermiform appendix, (3) the right colic artery supplies blood to the ascending colon, (4) the middle colic artery supplies blood to the transverse colon, and (5) the inferior pancreaticoduodenal artery supplies blood to the superior part of the duodenum and the medial end of the pancreas.

Course. This artery arises from the anterior part of the aorta immediately inferior to the celiac trunk (axis). In its path it runs posterior to the splenic vein and the body of the pancreas, as well as anterior to the renal vein. As it continues to course downward and forward, this vessel passes anterior to the uncinate process of the pancreas, the horizontal part of the duodenum, and between the layers of the mesentery to reach the right iliac fossa.

Right Renal Artery (C). This is a transverse (medial-to-lateral) vascular structure immediately medial to the right kidney and lateral to the inferior vena cava.

Origin (for the right and left renal arteries). Descending abdominal aorta.

Distribution (for the right and left renal arteries). Right and left kidneys, right and left suprarenal (adrenal) glands by a branch (inferior adrenal) from the right renal artery. The right renal artery also supplies blood to the right and left ureters.

Course (for the right renal artery). This artery arises from the descending abdominal aorta at the approximate level of lumbar vertebra 2 and courses posterior to the inferior vena cava to enter the hilus of the right kidney.

Course (for the left renal artery). This artery arises from the descending abdominal aorta at the approximate level of lumbar vertebra 1 and courses posterior to the body of the pancreas and the renal and splenic veins to reach the left kidney.

II. Veins

Right Renal Vein (*E*). A small, hollow vascular feature immediately lateral to the inferior vena cava.

Origin (for the right and left renal veins). Right and left kidneys.

Termination (for the right and left renal veins). Inferior vena cava.

Course (for the right renal vein). This vein courses posterior to the descending part of the duodenum to reach the inferior vena cava, where it terminates.

Course (for the left renal vein). This vein is three times longer than the right renal vein and courses anterior to the left renal artery as well as posterior to the splenic vein to reach the inferior vena cava, where it terminates.

III. Nerves

Left Sympathetic Nerve Trunk (Sympathetic Chain) (*F*). An oval nerve structure anterior to the psoas major muscle, medial to the left kidney, and lateral to the body of a lumbar vertebra. The sympathetic trunk (also referred to as vertebral chain ganglia) is a series of ganglia that lie in a vertical row on the left side of the vertebral column, extending from the base of the skull to the coccyx.

Right Sympathetic Nerve Trunk (*G*). This is located lateral to the body of a lumbar vertebra and anterior to the right psoas major muscle. This nerve trunk (chain) lies on the right side of the vertebral column, extending from the base of the skull to the coccyx. The fusion of the right and left sympathetic nerve trunks anterior to the coccyx forms a structure known as the ganglion impar (no pair).

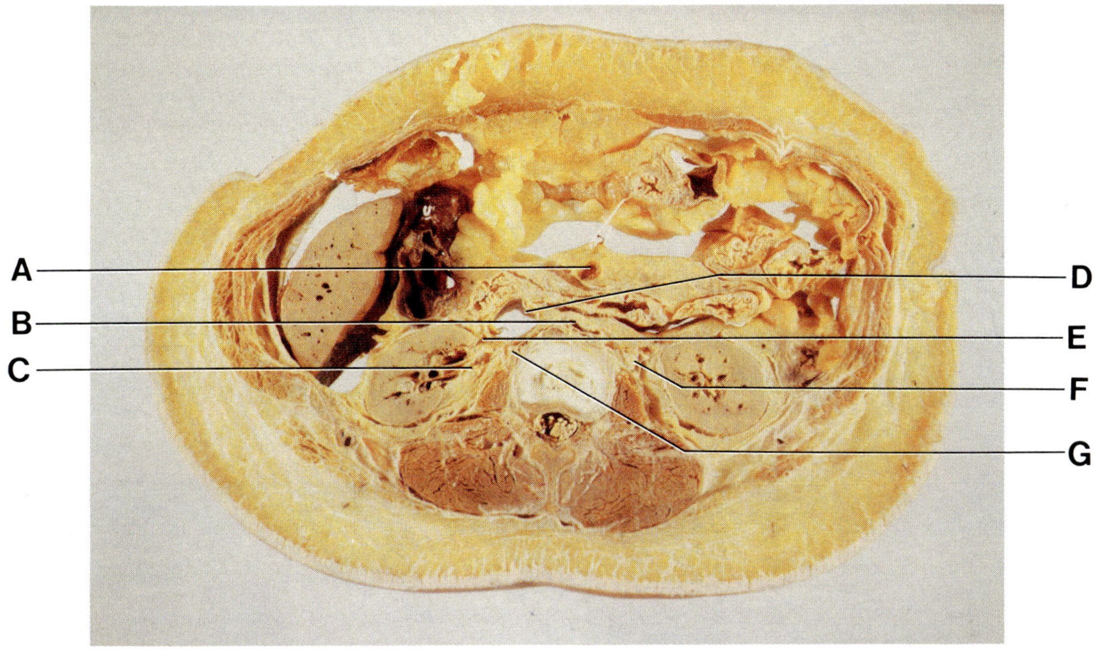

A. Superior mesenteric artery
B. Descending abdominal aorta
C. Right renal artery
D. Inferior vena cava
E. Right renal vein
F. Left sympathetic nerve trunk (sympathetic chain)
G. Right sympathetic nerve trunk

Figure 12–12
Transverse section of the abdomen at the level of lumbar vertebra 2, showing neurovascular unit structures.

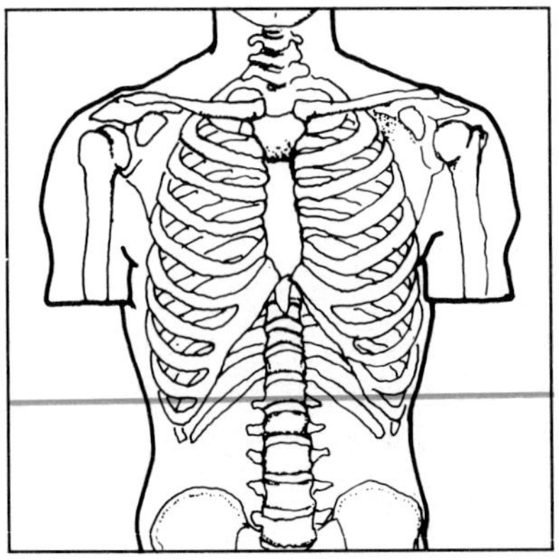

A CT scan with contrast (Fig. 12–13) depicts musculoskeletal and neurovascular structures of the abdomen in the transverse plane. All structures shown are typically found in the subcostal zone. Many of these structures extend throughout the abdominal area.

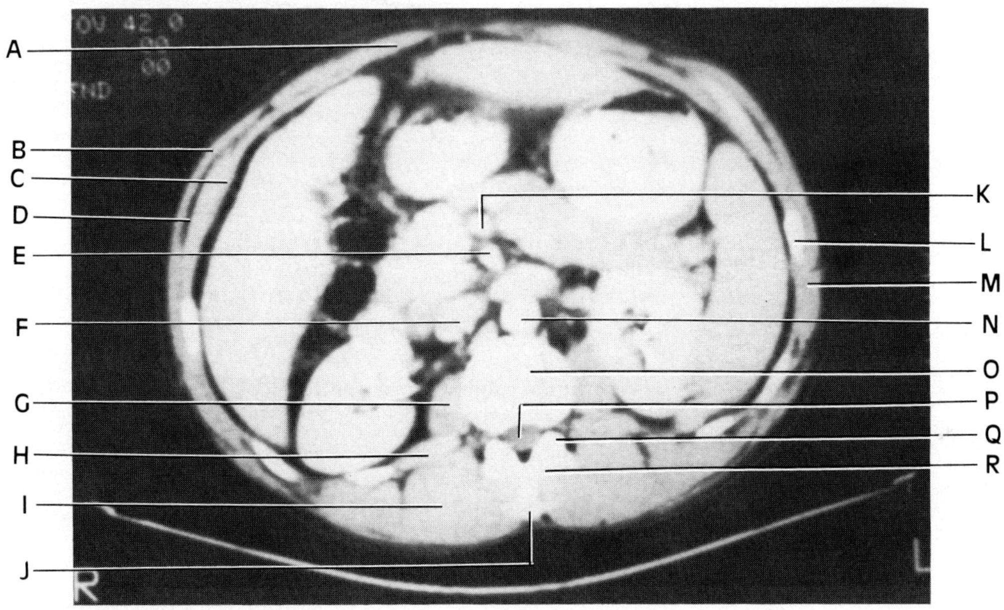

A. Rectus abdominis muscle
B. External oblique (obliquus abdominis externus) muscle
C. Transverse abdominal muscle
D. Internal oblique (obliquus abdominis internus) muscle
E. Superior mesenteric artery
F. Inferior vena cava
G. Psoas major muscle
H. Quadratus lumborum muscle
I. Erector spinae (sacrospinalis) muscle group
J. Spinous process of lumbar vertebra
K. Superior mesenteric vein
L. Rib
M. Intercostal muscles
N. Descending abdominal aorta
O. Body of lumbar vertebra
P. Spinal foramen
Q. Transverse process of lumbar vertebra
R. Lamina of lumbar vertebra

Figure 12–13
CT (with contrast) transverse section of the abdomen, highlighting both musculoskeletal and neurovascular unit structures.

VISCERAL UNIT (Fig. 12–14)

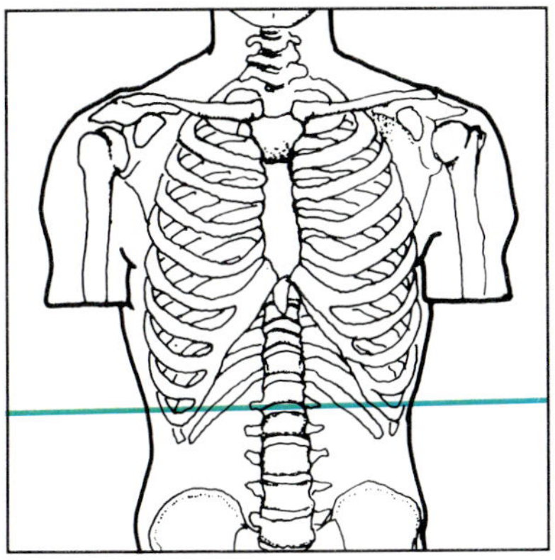

Structures of the kidney seen at this level:

(1) **Left Renal Pelvis** (*L*). An oblong trough formed by the proximal end of the left ureter. It is located lateral to the left psoas major muscle.

(2) **Cortex of the Left Kidney** (*M*). The outer layer of the left kidney, with a granular appearance. It contains the glomeruli.

(3) **Right Renal Pelvis** (*E*). The proximal end of the right kidney seen lateral to the right psoas major muscle.

(4) **Cortex of the Right Kidney** (*F*). The outer layer of the right kidney.

Right Lobe of the Liver (*D*). A submarine-shaped liver slice placed obliquely in the right part of the peritoneal cavity. The anterior end is more medial than the posterior end. The ascending colon is immediately medial to the anterior part of this liver slice.

Descending Duodenum (*B*). A rather thick-walled, hollow digestive structure anterior to the inferior vena cava and medial to the ascending colon. This second part of the duodenum is anterior and medial to the right kidney.

Horizontal Duodenum (*H*). A transversely oblique, hollow digestive structure anterior to the descending abdominal aorta. This third part of the duodenum extends from the anterior part of the inferior vena cava to the anterior part of the left

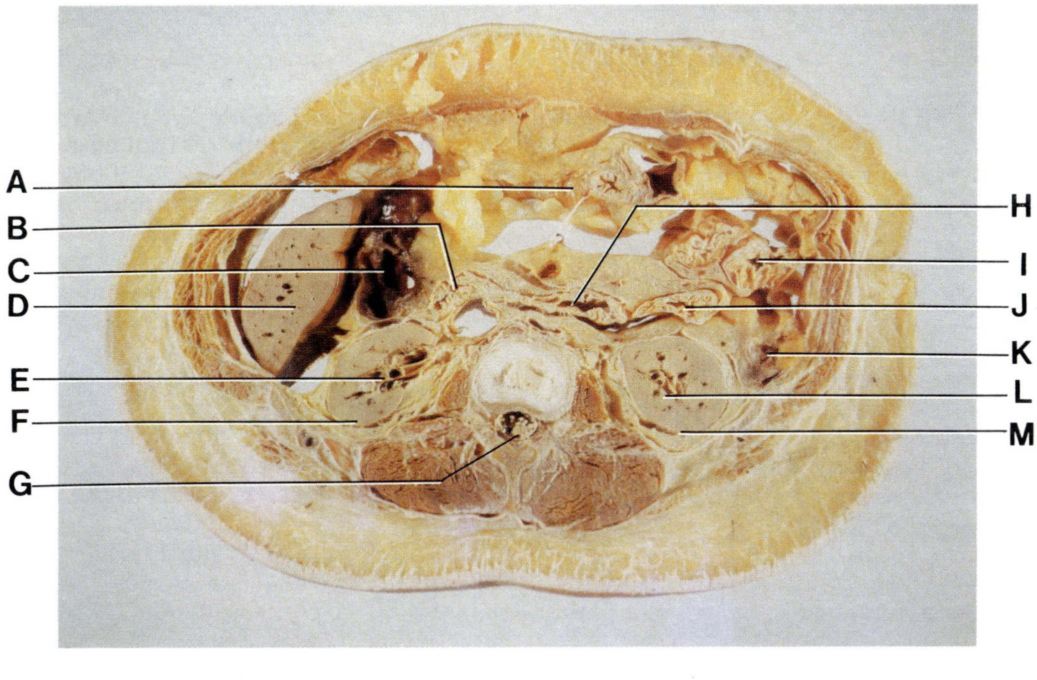

A. Transverse colon
B. Descending duodenum
C. Ascending colon
D. Right lobe of liver
E. Right renal pelvis
F. Cortex of right kidney
G. Cauda equina
H. Horizontal duodenum
I. Jejunum
J. Ascending duodenum
K. Descending colon
L. Left renal pelvis
M. Cortex of left kidney

Figure 12–14
Transverse section of the abdomen at lumbar vertebra 2, highlighting visceral unit structures.

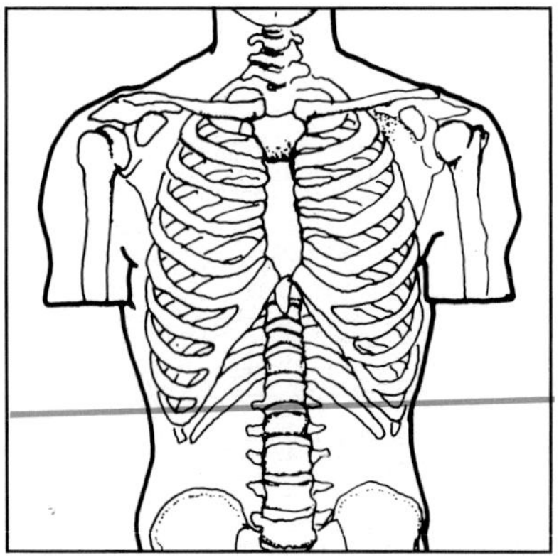

kidney. The left end is slightly posterior to the right end.

Ascending Duodenum (*J*). A hollow digestive structure anterior to the left kidney and posteriorly medial to a part of the jejunum. This fourth and terminal part of the duodenum connects with the jejunum.

Jejunum (*I*). An oval slice of the hollow digestive structure seen anterior and lateral to the ascending duodenum. It is located anterior to a section of the descending colon. The jejunum is two fifths of the small intestine; it extends from the duodenum to the ileum.

Course. This structure connects with the ascending part of the duodenum in a recess between the anterior surface of the left part and the left part of the transverse mesocolon. It descends inferiorly and, with its many coils, lies in the umbilical region deep to the greater omentum.

Ascending Colon (*C*). This is seen on the right side of this body section as a hollow, oblong feature medial to the right lobe of the liver and lateral to the descending duodenum. The ascending colon is a segment of the large intestine (colon).

Course. It starts in the lower right posterior part of the peritoneal cavity and takes a vertical superior path to reach the caudal surface of the right lobe of the liver. It takes part in the formation of the hepatic flexure with the transverse colon. This structure is retroperitoneal.

Transverse Colon (*A*). A transversely placed part of the colon located in the anterior part of this body. The transverse colon is a segment of the large intestine (colon).

Descending Colon (*K*). A rather large, oval structure immediately lateral to the left kidney and posterior to a section of the jejunum seen at this level. The descending colon is a segment of the large intestine (colon).

Cauda Equina. Several small, rounded nerves within the spinal foramen inside of lumbar vertebra 2.

Course. Several nerve roots course from the inferior part of the spinal cord within the spinal foramen surrounded by meninges. This cluster of nerves is located around the filum terminale. Each nerve pair exits through its appropriate intervertebral foramen.

ENCLOSING UNIT (Fig. 12–15)

Renal (Gerota's) Fascia (*J*). A rather thick, fatty, circular tissue layer that surrounds the posterior and lateral parts of the right and left kidneys.

Linea Alba (*F*). A narrow connective tissue structure in the median between the right and left rectus abdominis muscles.

Left Linea Semilunaris (*G*). A bilateral connective tissue structure immediately lateral to the rectus abdominis muscle.

Transversalis Fascia (*C*). A rather thick layer of connective tissue immediately deep to the transversus abdominis muscle and superficial to the subserous layer of the abdominal wall.

Parietal Peritoneum (*A*). A sheet of tissue that lines the inner wall of the abdomen.

Right Subphrenic Space (*B*). An area within the greater sac anterior to the right lobe of the liver.

Right Subhepatic Space (Morrison's Pouch) (*E*). An area posterior to the right lobe of the liver, lateral to the right kidney, and posterior to the ascending colon as seen at this level.

Epiploic Appendages (*H*). Globular fat bodies attached to the outer surface of the colon, seen most clearly in this transverse body section adjacent to the transverse colon.

Posterior Part of the Mesocolon (*I*). A flat sheet of mesentery somewhat irregularly shaped. It is seen at this level with a thicker left part than a right part. The medial end is closely associated with the superior mesenteric artery.

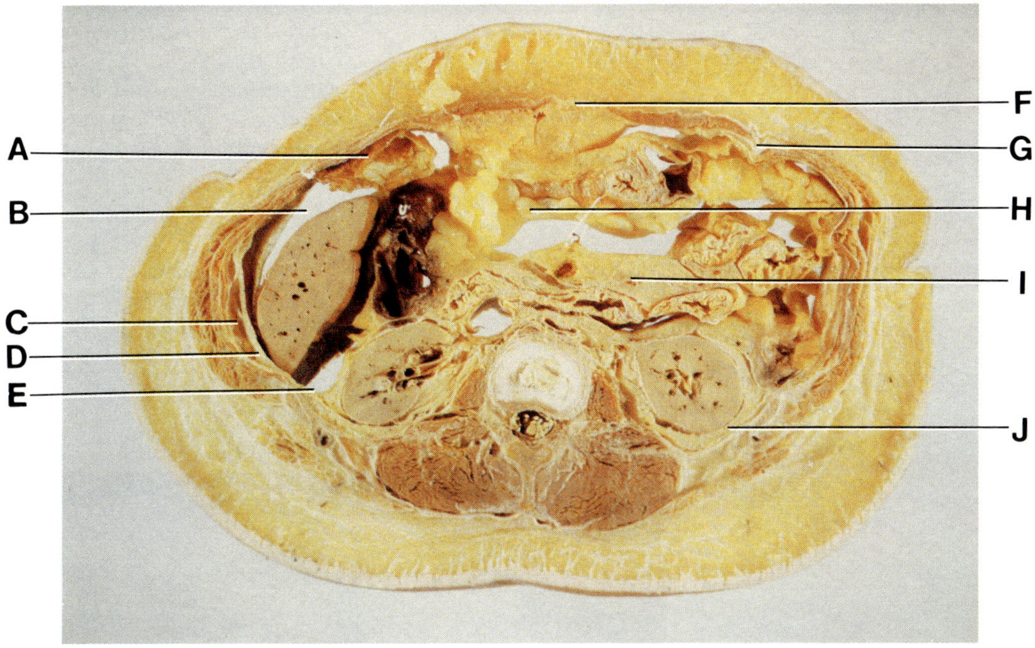

A. Parietal peritoneum (translocated)
B. Right subphrenic space
C. Transversalis fascia
D. Parietal peritoneum
E. Right subhepatic space (Morrison's pouch)
F. Linea alba
G. Left linea semilunaris
H. Epiploic appendages
I. Posterior part of mesocolon
J. Renal (Gerota's) fascia

Figure 12–15
Transverse section of the abdomen at the level of lumbar vertebra 2, highlighting enclosing unit structures.

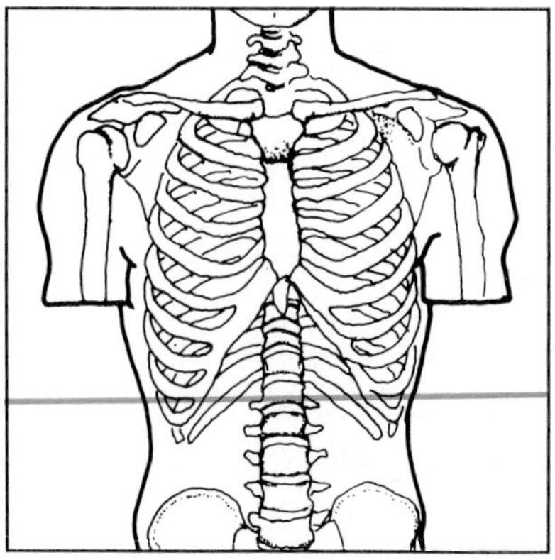

Many visceral and enclosing unit structures of the subcostal zone are depicted in a transverse CT scan (with contrast) in Figure 12–16. Figure 12–17 is a picture of an MRI transverse slice of the abdomen. Sections of several structures typically found in the upper part of the umbilical zone are noted.

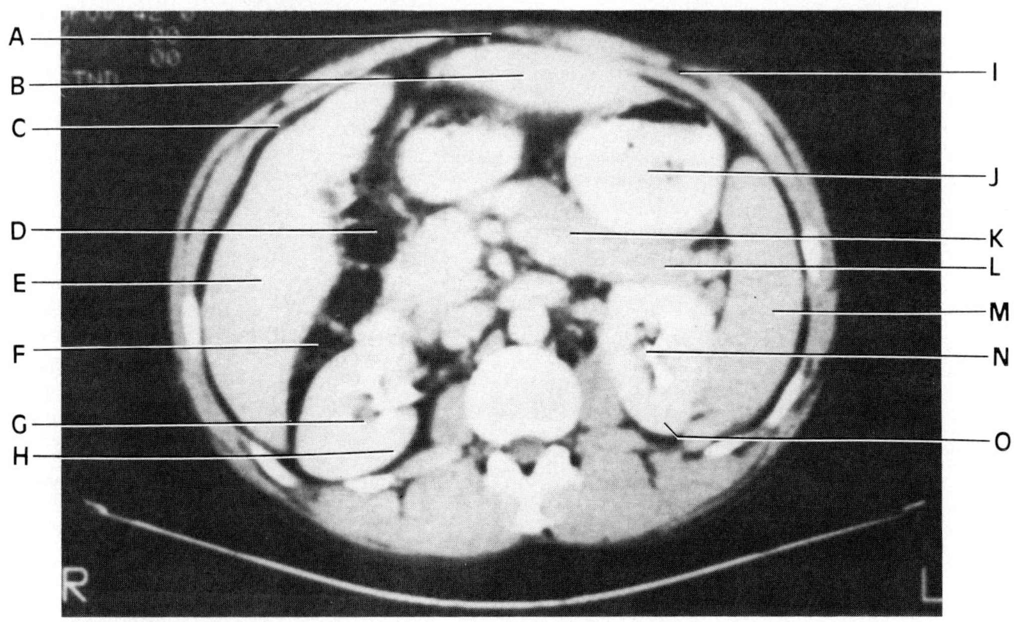

A. Linea alba
B. Left lobe of liver
C. Right subphrenic space
D. Colon
E. Right lobe of liver
F. Right subhepatic space (Morrison's pouch)
G. Pelvis of right kidney
H. Cortex of right kidney
I. Linea semilunaris
J. Stomach
K. Body of pancreas
L. Tail of pancreas
M. Spleen
N. Pelvis of left kidney
O. Cortex of left kidney

Figure 12–16
CT (with contrast) transverse image of the abdomen, highlighting both visceral and enclosing unit structures.

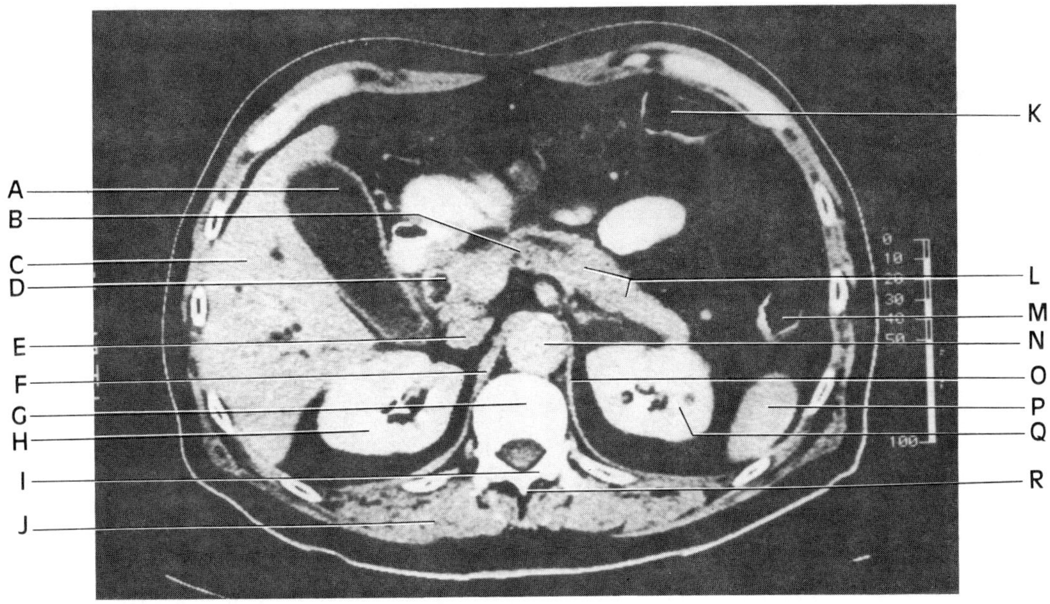

A. Ascending colon
B. Head of pancreas
C. Right lobe of liver
D. Duodenum
E. Inferior vena cava
F. Right crus of diaphragm
G. Body of vertebra
H. Right kidney
I. Lamina of vertebra
J. Erector spinae muscles
K. Transverse colon
L. Body of pancreas
M. Descending colon
N. Descending abdominal aorta
O. Left crus of diaphragm
P. Spleen
Q. Left kidney
R. Spinous process of vertebra

Figure 12–17
MRI transverse section of the abdomen, highlighting musculoskeletal, neurovascular, and visceral unit structures.

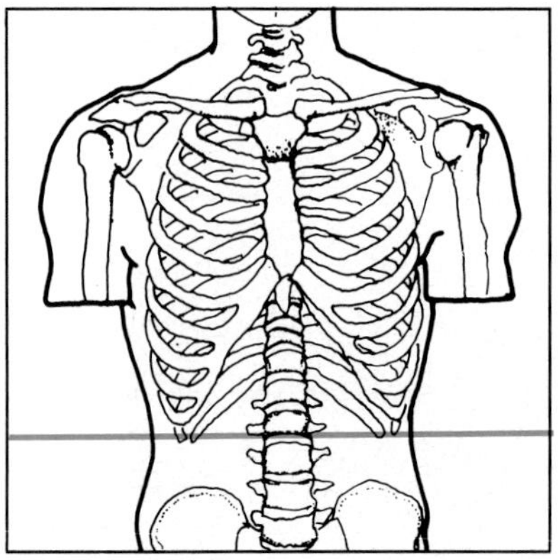

MODULE 4

This is a sheet plastinated (2.5 mm thick) transverse section at the level of lumbar vertebra 3.

MUSCULOSKELETAL UNIT (Fig. 12–18)

I. Skeletal Portion

Body of a Lumbar Vertebra (*A*). An oval aggregate bone slice. The lateral parts are immediately anterior to the pedicles. This feature is between the right and left psoas major muscles and anterior to the spinal foramen.

Pedicle of a Lumbar Vertebra (*B*). A rather wide, short, bilateral bone section immediately posterior to the lateral part of the vertebral body and anterior to the medial end of the transverse process.

Transverse Process of a Lumbar Vertebra (*C*). A transversely oblique, lateral projection immediately anterior to the erector spinae and posterior to the psoas major muscles. The medial part of this bilateral bone section is more anterior than the lateral part.

Lamina of a Lumbar Vertebra (*J*). An oblique, rather thin, bilateral bone slice posterior to the spinal foramen. The posterior end is more medial than the anterior end and is attached to the proximal end of the spinous process of a vertebra.

Spinous Process of a Lumbar Vertebra (*D*). A median, posterior, rather thin bone section immediately lateral to the multifidus muscle and posterior to the distal ends of the laminae.

II. Muscular Portion

Multifidus Muscle (*M*). An oblong (anterior-to-posterior), bilateral muscle slice immediately lateral to the spinous process and posterior to the lamina. It is medial to the erector spinae muscle.

Erector Spinae (Sacrospinalis) Muscle Group (*L*). A rather triangular, aggregate, bilateral muscle tissue immediately lateral to the multifidus muscle. The medial part is wider than the lateral part and is immediately posterior to the lumbar vertebra as seen at this level. The lateral part is posterior to the medial part of the quadratus lumborum muscle.

Quadratus Lumborum Muscle (*K*). A transversely placed strip of bilateral muscle tissue immediately lateral to the transverse process of a lumbar vertebra, as seen in this transverse body section. The medial end is immediately anterior to the lateral part of the erector spinae muscle.

Psoas Major Muscle (*I*). A rather triangular, bilateral muscle slice immediately lateral to the body and pedicle of a lumbar vertebra. The posterior part is wider than the anterior part and immediately anterior to the transverse process.

Oblique External Muscle of the Abdomen (Obliquus Abdominis Externus Muscle) (*H*). A rather narrow, curved superficial strip of bilateral muscle tissue that extends from the quadratus lumborum muscle to the lateral edge of the rectus abdominis muscle. The concave surface is medial to the convex surface.

Oblique Internal Muscle of the Abdomen (Obliquus Abdominis Internus Muscle) (*G*). A narrow strip of bilateral muscle tissue immediately deep to the oblique external muscle of the abdomen and superficial to the transversus abdominis muscle. Each end is narrower than the other parts of the muscle.

Transverse Muscle of the Abdomen (Transversus Abdominis Muscle) (*F*). A narrow strip of curved, bilateral muscle tissue deep to the oblique internal muscle of the abdomen. The medial surface shows a concave part.

Rectus Abdominis Muscle (*E*). A transverse (medial-to-lateral), bilateral muscle slice immediately lateral to the linea alba and medial to the linea semilunaris.

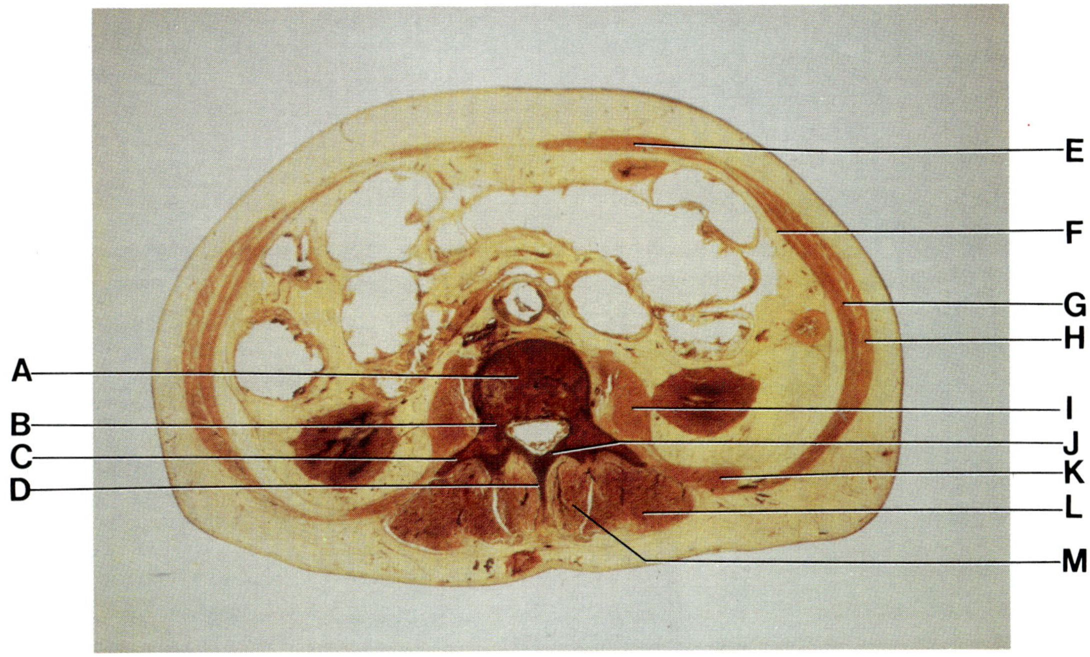

A. Body of lumbar vertebra
B. Pedicle of lumbar vertebra
C. Transverse process of lumbar vertebra
D. Spinous process of lumbar vertebra
E. Rectus abdominis muscle
F. Transverse muscle of the abdomen (transversus abdominis muscle)
G. Internal oblique abdominal (obliquus abdominis internus) muscle
H. External oblique abdominal (obliquus abdominis externus) muscle
I. Psoas major muscle
J. Lamina of lumbar vertebra
K. Quadratus lumborum muscle
L. Erector spinae (sacrospinalis) muscle group
M. Multifidus muscle

Figure 12-18
Transverse (axial) section of the abdomen at the level of lumbar vertebra 3, highlighting musculoskeletal unit structures.

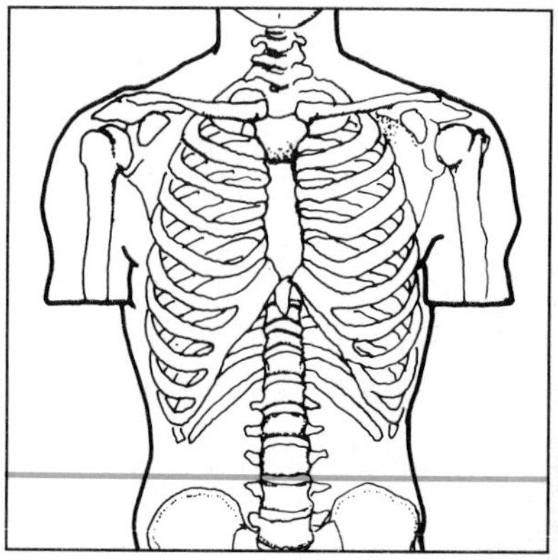

Musculoskeletal structures are shown on a CT transverse scan (with contrast) of the abdomen (Fig. 12–19). *These structures are common to each abdominal region.*

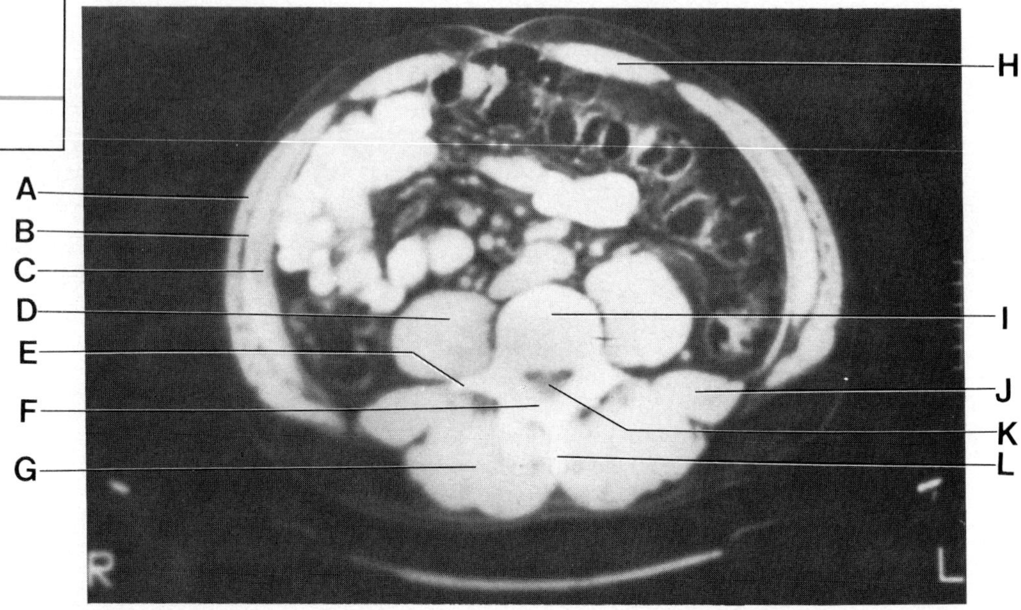

A. External oblique (obliquus abdominis externus) muscle
B. Internal oblique (obliquus abdominis internus) muscle
C. Transverse abdominal (transversus abdominis) muscle
D. Psoas major muscle
E. Transverse process of lumbar vertebra
F. Lamina of lumbar vertebra
G. Erector spinae (sacrospinalis) muscle
H. Rectus abdominis muscle
I. Body of lumbar vertebra
J. Quadratus lumborum muscle
K. Spinal foramen
L. Spinous process of lumbar vertebra

Figure 12–19
CT (without contrast) transverse image of the abdomen, highlighting musculoskeletal unit structures.

NEUROVASCULAR UNIT (Fig. 12–20)

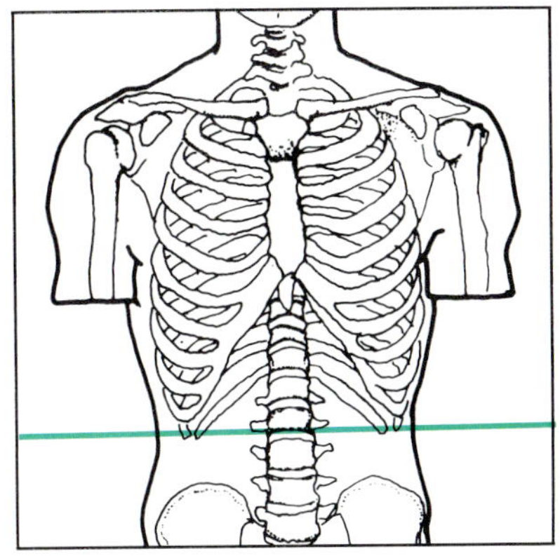

I. Arteries

Descending Abdominal Aorta (*B*). A rounded, hollow vascular structure immediately anterior to the middle part of the body of lumbar vertebra 3.

Superior Mesenteric Artery (*E*). An oval, hollow vascular structure to the left of the anterior part of the descending abdominal aorta and the superior mesenteric vein.

II. Veins

Superior Mesenteric Vein (*A*). A rather transversely placed, hollow, vascular venous structure immediately to the right of the superior mesenteric artery and anterior to the descending abdominal aorta.

Origin. This vein starts by the fusion of tributaries from the distal ends of the ileum, cecum, vermiform appendix, small intestine, ascending colon, and transverse colon.

Termination. Hepatic portal vein.

Course. This vein takes a path within the mesentery to the right of the superior mesenteric artery and travels superiorly. It passes anterior to the right ureter, uncinate process of the pancreas, horizontal part of the duodenum, and inferior vena cava. It terminates into the hepatic portal vein.

Inferior Vena Cava (*C*). A collapsed venous structure to the right of the posterior part of the descending abdominal aorta and anterior to the right part of lumbar vertebra 3.

III. Nerves

Right Sympathetic Trunk (Chain) (*D*). A small, rounded nerve structure immediately lateral to the anterior part of lumbar vertebra 3 and medial to the anterior part of the psoas major muscle as seen at this level.

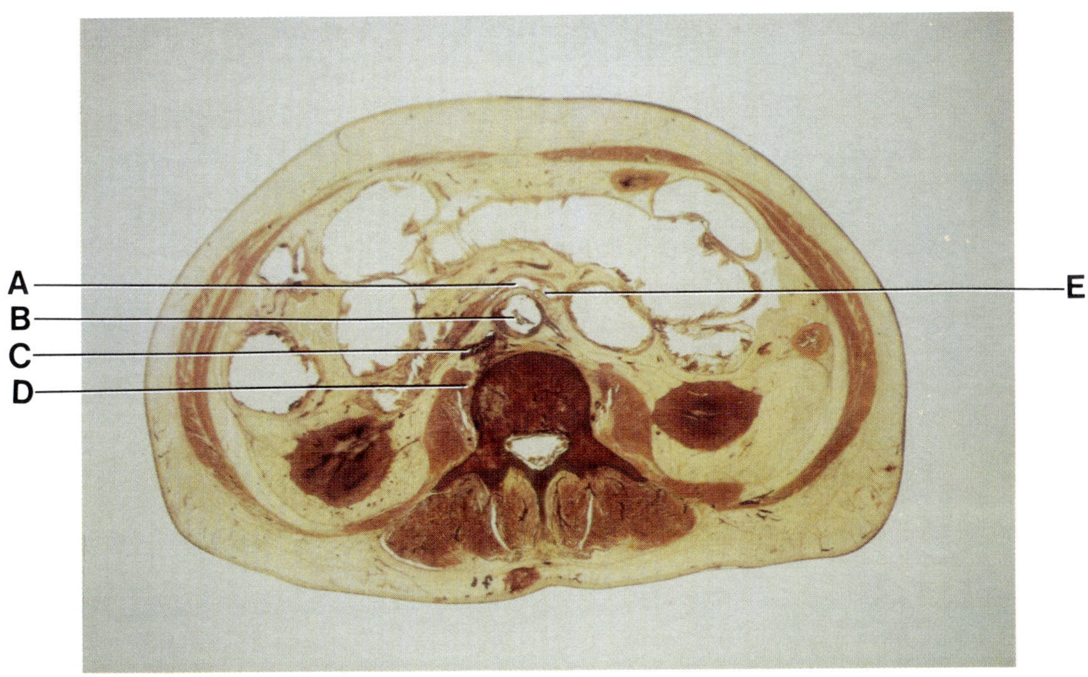

A. Superior mesenteric vein
B. Descending abdominal aorta
C. Inferior vena cava
D. Right sympathetic trunk (chain)
E. Superior mesenteric artery

Figure 12–20
Transverse section of the abdomen at the level of lumbar vertebra 3, highlighting neurovascular unit structures.

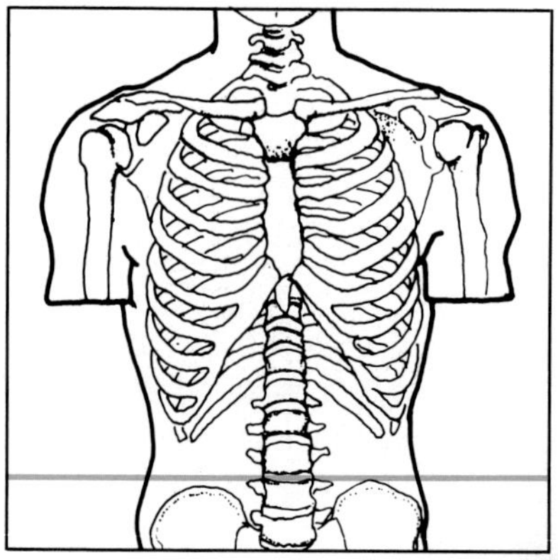

Figure 12–21 is a transverse CT scan (with contrast) that highlights neurovascular structures. The slice is from the lower abdomen.

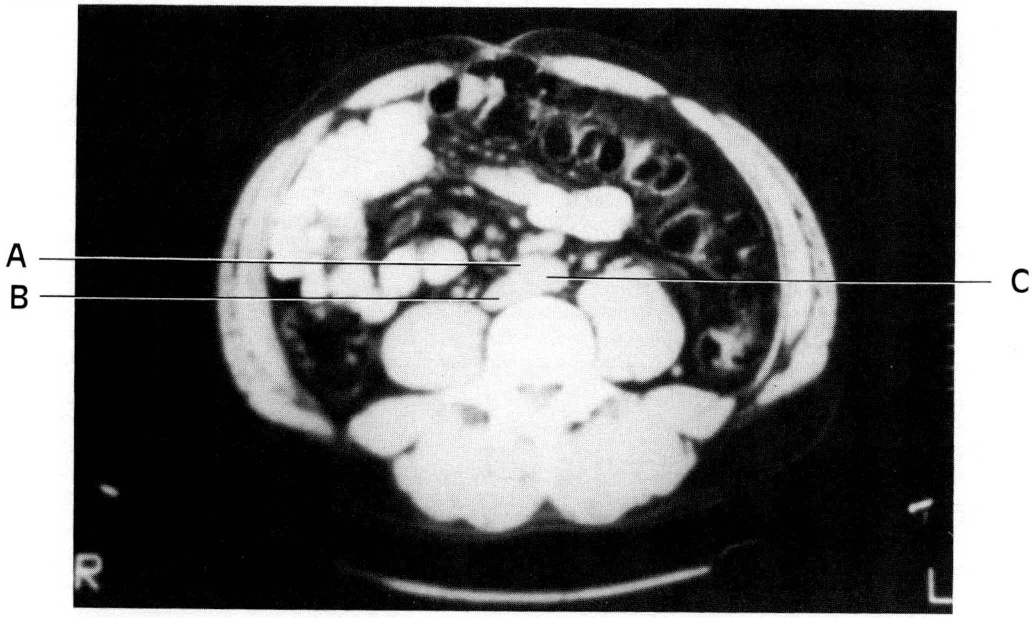

A. Area of right common iliac artery
B. Inferior vena cava
C. Left common iliac artery

Figure 12–21
CT (without contrast) transverse image of the abdomen, highlighting neurovascular unit structures.

VISCERAL UNIT (Fig. 12–22)

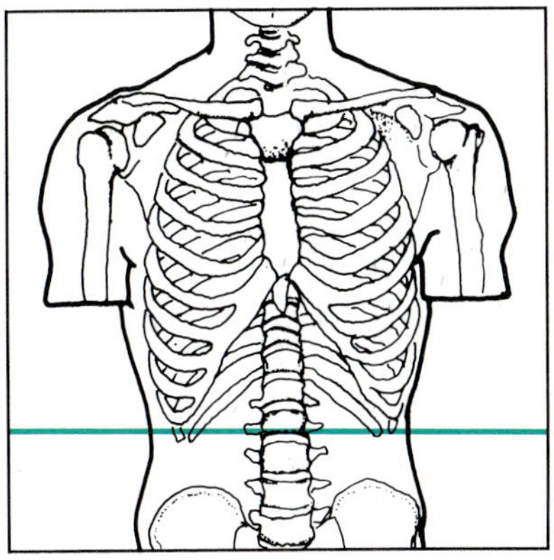

Right Kidney (*D*). An oval slice of renal tissue immediately posterior to the ascending colon and lateral to the psoas major muscle. The right kidney is located on the right side of a vertebra. The cranial edge of the right kidney is located at the lower portion of the 12th thoracic vertebra and extends as far down as the lower part of the 3rd lumbar vertebra.

Left Kidney (*F*). An oval slice of renal tissue immediately lateral to the psoas major muscle and anterior to the quadratus lumborum muscle. Unlike the right kidney, the left kidney starts at a slightly higher position.

Ascending Colon (*B*). A large, round, hollow digestive organ immediately anterior to the lateral part of the right kidney and lateral to the portion of the transverse colon.

Transverse Colon (*A*). A transversely placed, elongated, hollow structure located in the anterior part of the peritoneal cavity. In addition to this transversely placed transverse colon slice, there are some oval parts of this structure.

Descending Colon (*E*). A rounded, hollow digestive structure located on the left part of the peritoneal cavity and lateral to the left part of the transverse colon.

Descending Duodenum (*C*). An obliquely placed, hollow digestive segment medial to the anterior part of the right kidney at this level. The posterior end is more lateral than the anterior end. This part of the duodenum is posterior to a right part of the transverse colon.

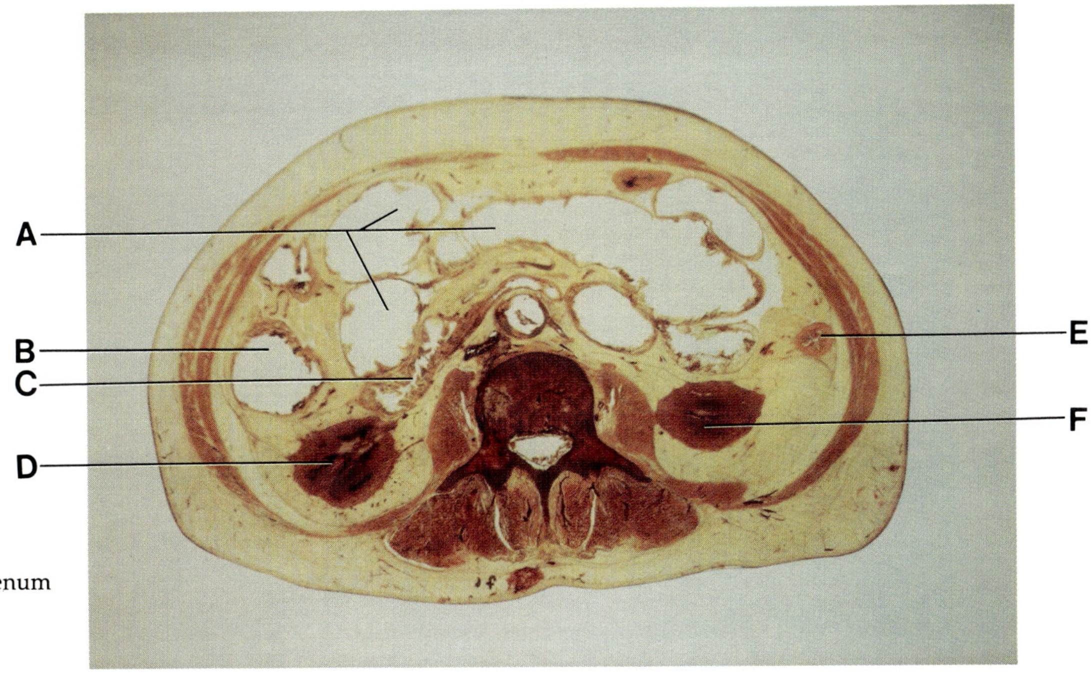

A. Transverse colon
B. Ascending colon
C. Descending duodenum
D. Right kidney
E. Descending colon
F. Left kidney

Figure 12–22
Transverse section of the abdomen at the level of lumbar vertebra 3, showing visceral unit structures.

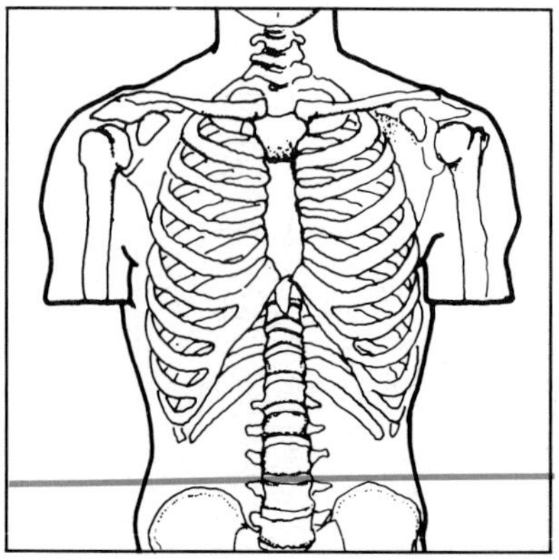

Visceral structures of the lower abdomen are emphasized in Figure 12–23. This is a picture of a CT scan (with contrast) in the transverse plane.

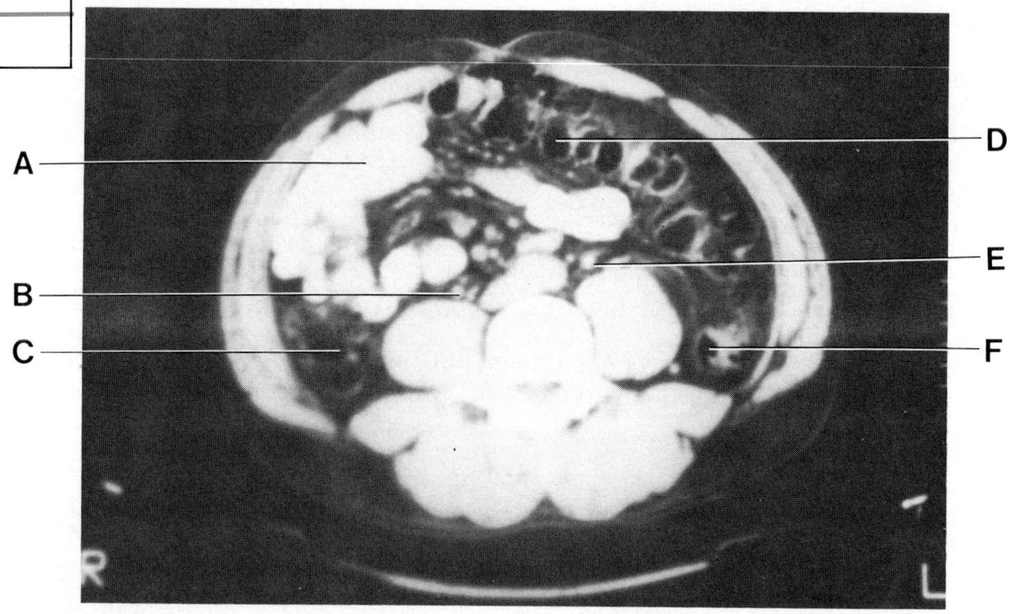

A. Small bowel
B. Right ureter
C. Cecum
D. Transverse colon
E. Left ureter
F. Descending colon

Figure 12–23
CT (without contrast) transverse image of the abdomen, highlighting visceral unit structures.

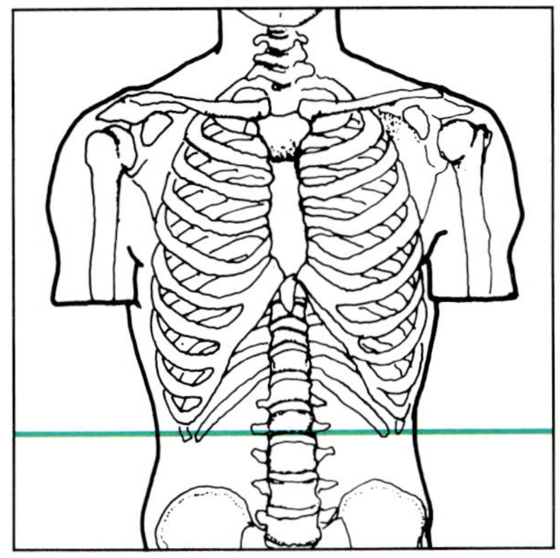

ENCLOSING UNIT (Fig. 12–24)

Linea Alba (*A*). A narrow strip of connective tissue between the rectus abdominis muscles.

Linea Semilunaris (*C*). A bilateral connective tissue feature between the rectus abdominis and the obliquus abdominis muscles.

Renal (Gerota's) Fascia (*E*). A rather thick, bilateral slice of fatty tissue that surrounds the kidney. The posterior and lateral portions are thicker than the anterior and medial portions.

Parietal Peritoneum (*B*). A thin sheet of tissue that lines the inner wall of the peritoneal cavity.

Mesocolon (*D*). A part of the mesentery seen at this level posterior to the transversely placed part of the transverse colon. This sheet of tissue contains both the superior mesenteric artery and mesenteric vein. It is attached to the posterior abdominal wall and crosses anterior to the descending abdominal aorta as well as the inferior vena cava.

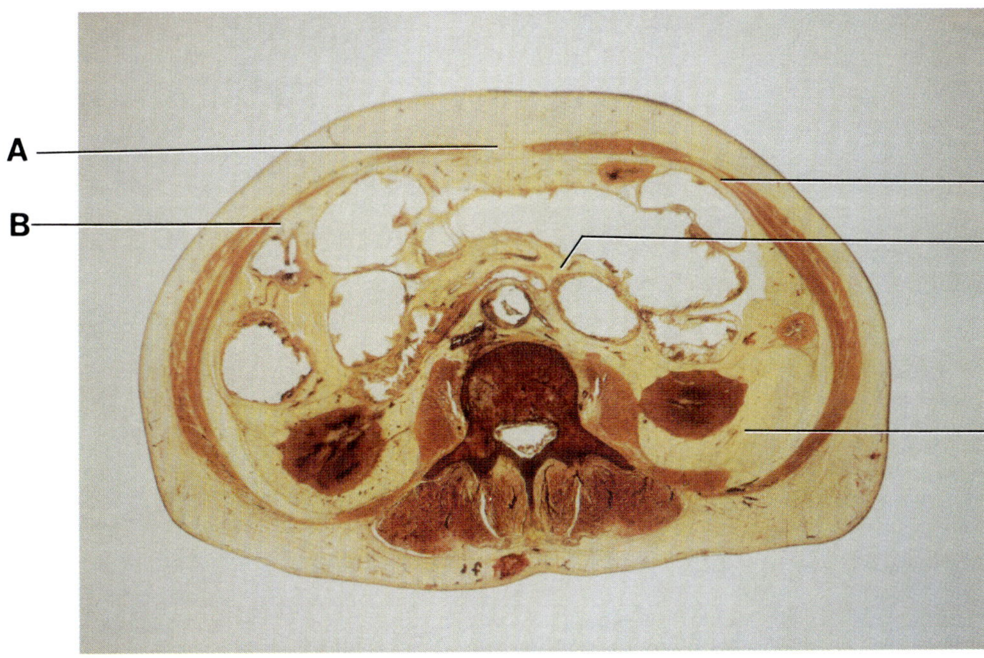

A. Linea alba
B. Parietal peritoneum
C. Linea semilunaris
D. Mesocolon
E. Renal (Gerota's) fascia

Figure 12–24
Transverse section of the abdomen at the level of lumbar vertebra 3, highlighting enclosing unit structures.

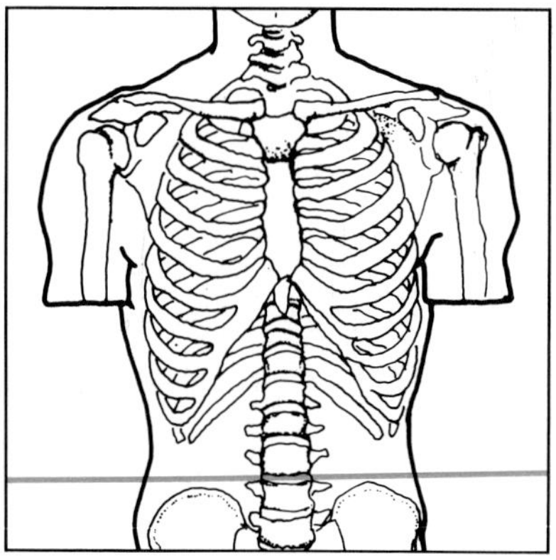

Figure 12–25 highlights enclosing unit structures of the abdomen. This is a picture of a CT scan (without contrast) in the transverse plane.

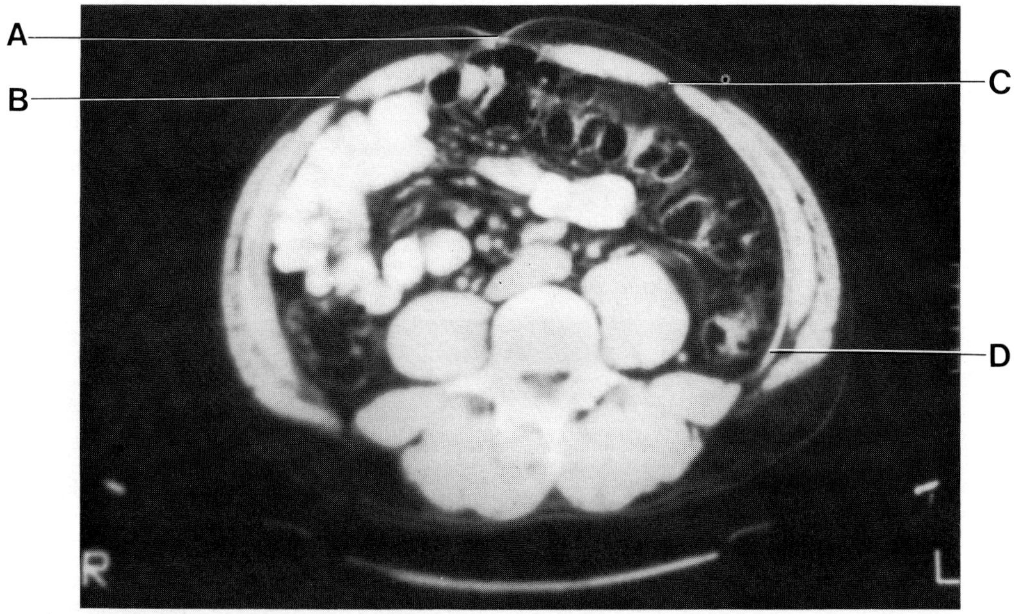

A. Linea alba
B. Right linea semilunaris
C. Left linea semilunaris
D. Parietal peritoneum

Figure 12–25
CT (without contrast) transverse image of the abdomen, highlighting enclosing unit structures.

CHAPTER 13

Coronal (Frontal) Cadaver Sections and Images of the Abdomen

This chapter consists of coronal (frontal) cadaver pictures showing the major areas of the greater sac and the supra- and infracolic compartments. Located within the greater sac are several abdominal viscera. Another cadaver illustration highlights the major blood vessels for the digestive organs within the abdomen. Three more cadaver pictures from the same part of the body separate and emphasize the musculoskeletal, visceral, and enclosing units. Another cadaver picture from a more posterior part of the abdomen focuses on the neurovascular unit.

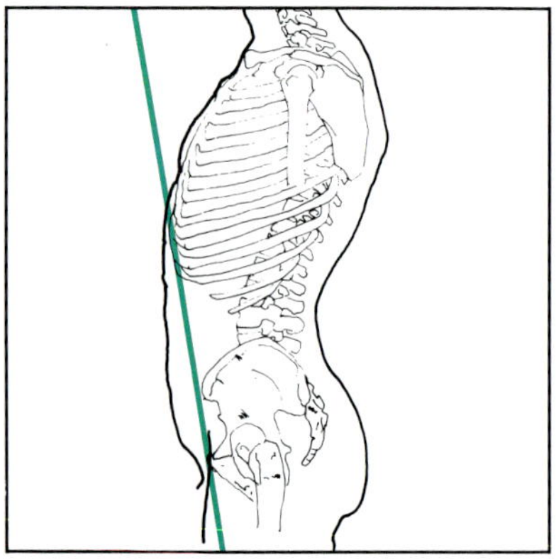

MODULE 1

This module contains two coronal (frontal) sections of the abdomen. One illustration shows three anatomic categories (musculoskeletal, visceral, and enclosing units), and the other depicts the neurovascular unit. The typical order of these sections to illustrate the anatomic categories is followed in this module.

MUSCULOSKELETAL UNIT (Fig. 13–1)

I. Skeletal Portion

Xiphoid Process (*C*). A vertical, rather broad plate of skeletal tissue located in the median inferior chest. This sword-shaped feature is the most inferior portion of the sternum.

Infrasternal Angle (*B*). The junction formed by the seventh costal cartilage and the superior part of the xiphoid process.

Seventh Costal Cartilage (*A*). A vertically elongated, oblique hyaline cartilage structure. The inferior end is more lateral than the superior end.

Anterior Superior Iliac Spine (ASIS). A rather moderately sharp projection on the anterior part of the iliac crest.

Costal Arch (*D*). A curved portion of the inferior rib cage, which includes the seventh costal cartilage and the remaining anterior portions of the ribs.

Iliac Crest. A superior ridge on the ilium. This crest extends from the ASIS to the posterior superior iliac spine (PSIS). The most superior point of the crest is usually at the level of lumbar vertebra 4.

Iliac Tubercle. A lateral elevated swelling on the ilium at the level of lumbar vertebra 5 in the midaxillary line.

II. Muscular Portion

External Oblique Abdominal (Obliquus Abdominis Externus) Muscle (*F*). A vertical, oblique, bilateral muscle slice seen in this section at the anterior lateral pelvis. The inferior end of this muscle section is more medial than the superior end. This muscle is adjacent to the inguinal ligament.

Internal Oblique Abdominal (Obliquus Abdominis Internus) Muscle (*E*). A vertical, oblique, bilateral muscle slice, seen in this section between the external oblique and transversus abdominis muscles. The inferior end is more medial than the superior end.

Transversus Abdominis Muscle (*G*). This vertical muscle slice is seen deep to the internal oblique abdominal muscle and superficial to the parietal peritoneum. This muscle section is seen in an oblique position, with its inferior end more medial than the superior end.

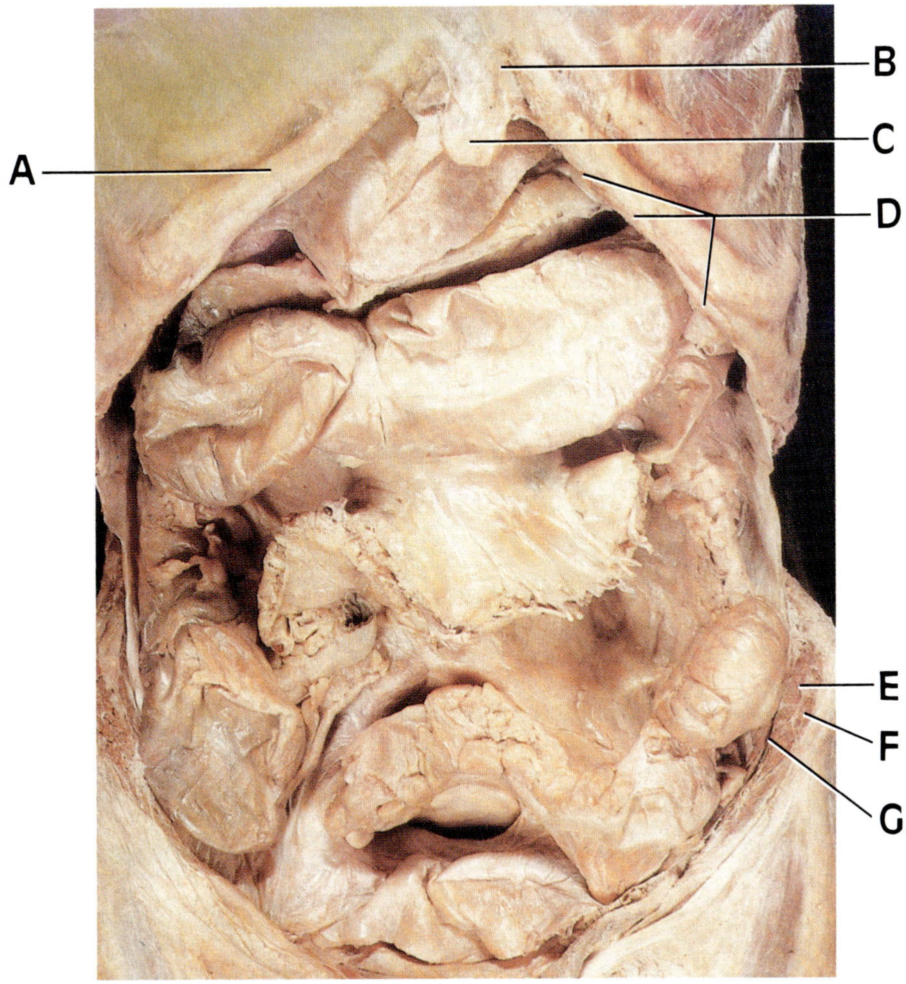

A. Seventh costal cartilage
B. Infrasternal angle
C. Xiphoid process
D. Costal arch
E. Internal oblique abdominal (obliquus abdominis internus) muscle
F. External oblique abdominal (obliquus abdominis externus) muscle
G. Transversus abdominis muscle

Figure 13-1
Coronal view of the abdomen, highlighting musculoskeletal unit structures.

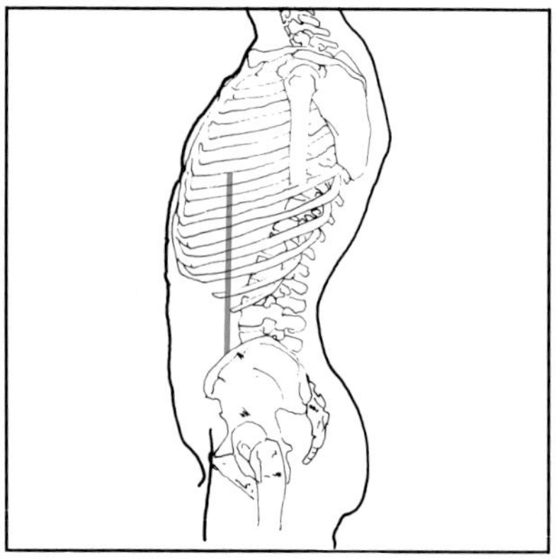

NEUROVASCULAR UNIT (Fig. 13–2)

I. Arteries

Superior Mesenteric Artery (*F*). A rather large, nearly vertical vascular structure showing the following arterial branches: (1) middle colic, (2) right colic (in this case it is seen that the right colic arises from a common trunk with the middle colic), (3) ileocolic, (4) jejunal, and (5) ileal arteries. *Note:* Usually the jejunal and the ileal arteries are known as the intestinal arteries.

Middle Colic Artery (*A*). This vascular structure is seen in a vertical, slightly oblique, median position of the abdomen. The superior end is at a more lateral (right) place than the inferior end.

Origin. Arises from the superior mesenteric artery, in this case as a common trunk with the right colic artery.

Distribution. Middle colic.

Course. This vessel courses behind the mesocolon, and as it passes toward the middle colic it bifurcates into right and left branches. The right colic artery and the right branch of the middle colic artery anastomose. The left colic artery and the left branch of the middle colic artery anastomose.

Right Colic Artery (*B*). A transverse vessel that extends from a common trunk with the middle colic artery, in this case to the ascending colon on the right of the abdomen.

Origin. Usually arises directly from the superior mesenteric artery, but, as mentioned earlier, in this case it arises from a common trunk with the middle colic artery.

Distribution. Ascending colon.

Course. Travels retroperitoneally toward the right of the body on the posterior abdominal wall. At its terminal end adjacent to the ascending colon, it divides into ascending and descending branches. The descending branch and the ascending branch of the ileocolic artery anastomose. The ascending branch of the right colic and the middle colic arteries anastomose.

Ileocolic Artery (*D*). A vertical vascular structure obliquely positioned in the lower abdomen. The inferior end is more lateral than the superior end.

Origin. Right side of the superior mesenteric artery; it is the most inferior branch of this artery.

Distribution. Inferior portion of the ascending colon and cecum and yields an appendicular artery as well as an ileal branch. The appendicular artery supplies blood to the vermiform appendix, while the ileal branch supplies blood to the distal portion of the ileum.

Course. This artery passes inferiorly from the superior mesenteric artery to the right iliac fossa. At this location it bifurcates to form superior and inferior branches. The superior branch anastomoses with the right colic artery, and the inferior branch gives rise to the appendicular artery and the ileal branch.

Jejunal Arteries (*G*). Several rather small, vascular, oblique structures. The inferior parts of these vessels are placed in a more lateral (right) position than the superior ends.

Origin. The left side of the superior mesenteric artery.

Distribution. Jejunum.

Course. These arteries pass toward the left of the abdomen to reach the jejunum and form a series of arcades before reaching their destination. The vasa recta (small straight vessels) continue from the arcades to reach the jejunum.

Ileal Arteries (*H*). Several rather moderately sized vessels in a vertical, oblique position. The inferior ends are more lateral (left) than the superior ends.

Origin. Superior mesenteric artery inferior to the jejunal arteries.

Distribution. Ileum.

Course. These arteries pass inferiorly and slightly leftward to reach the ileum. A series of arcades is formed before they reach the ileum by way of the vasa recta.

II. Veins

Superior Mesenteric Vein. A rather large, venous vascular structure seen to the right of the superior mesenteric artery. Note that the tributaries of this vein correspond to the branches of the superior mesenteric artery.

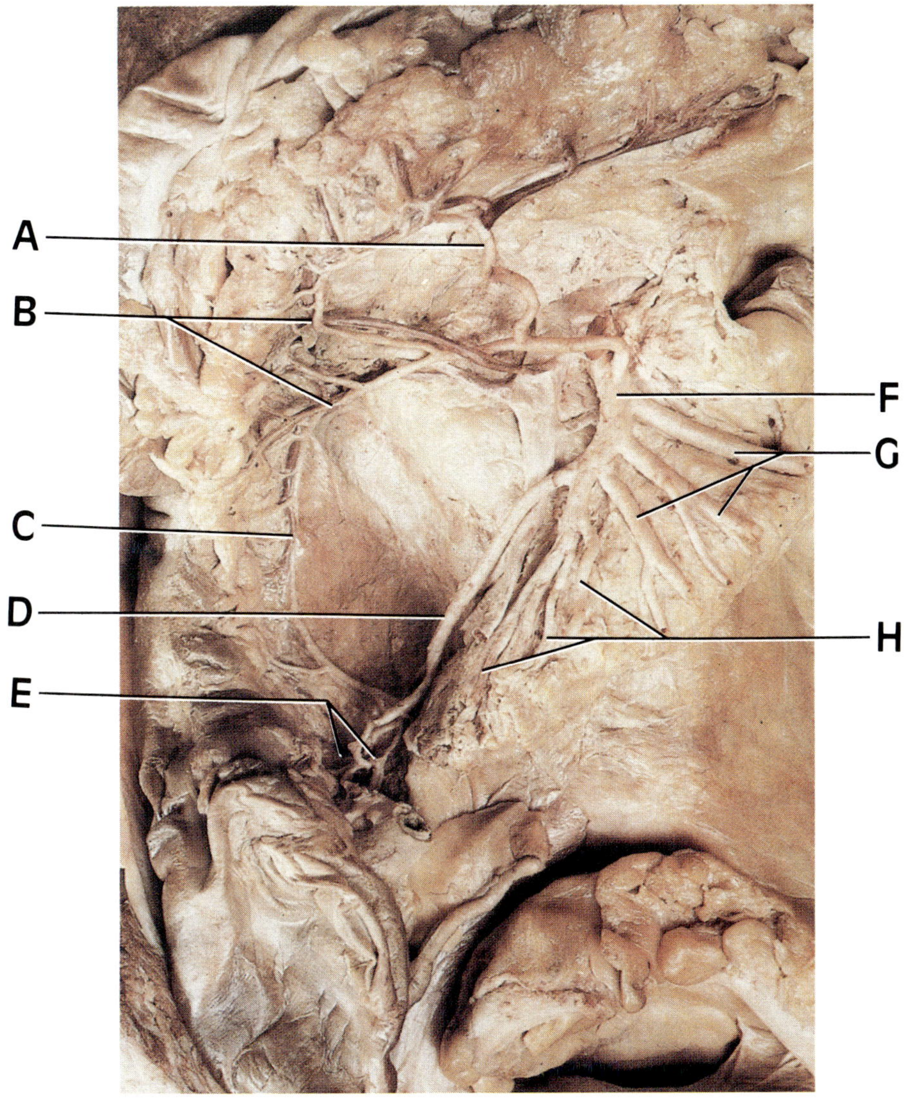

A. Middle colic artery
B. Right colic artery
C. Marginal artery
D. Ileocolic artery
E. Cecal artery
F. Superior mesenteric artery
G. Jejunal arteries
H. Ileal arteries

Figure 13–2
Coronal view of the abdomen, highlighting neurovascular unit structures.

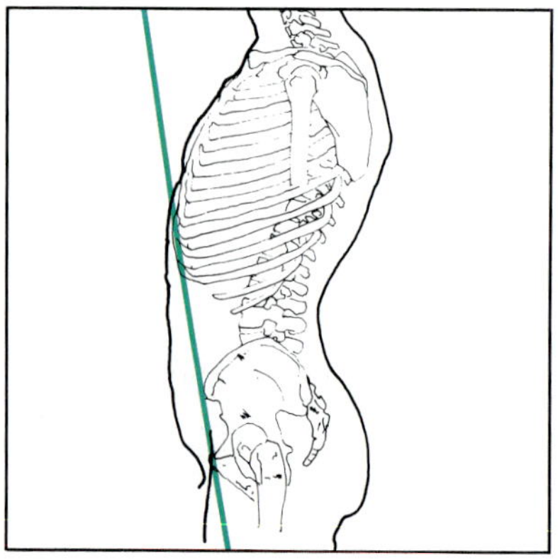

VISCERAL UNIT (Fig. 13–3)

Cecum (*D*). An oval portion of the colon seen in the right iliac fossa area. This is the first segment of the colon, and it receives materials from the distal end of the ileum.

Vermiform Appendix (*C*). A rather thin, elongated, worm-shaped, oblique portion of the colon. The inferior end is attached to the cecum and is more lateral (right) than its terminal end.

Ascending Colon (*A*). A vertical, elongated segment of the colon on the right side of the abdomen between the right colic gutter and the gutter to the right of the mesentery. This structure extends from the cecum to the hepatic flexure (right colic flexure).

Transverse Colon (*F*). A transverse part of the colon in an oblique position. The left end is at a more superior location than the right end. This structure extends from the hepatic flexure to the splenic flexure (left colic flexure).

Descending Colon (*G*). A vertical, elongated part of the colon, seen on the left side of the abdomen between the left colic gutter and the gutter to the left of the mesentery. This structure extends from the splenic flexure to the sigmoid colon.

Hepatic Flexure (Right Colic Flexure). A bend of the colon on the right side of the upper abdomen adjacent to the liver. This flexure is formed by the ascending and right end of the transverse colon.

Splenic Flexure (Left Colic Flexure) (*E*). A bend of the colon on the left side of the upper abdomen adjacent to the spleen. This flexure is at a higher position than the hepatic flexure and is formed by the left end of the transverse colon and the superior end of the descending colon.

Sigmoid Colon (*H*). A rather S-shaped part of the colon, seen in the lower part of the abdomen extending from the left iliac fossa area to the right of the median. This portion of the colon extends from the inferior end of the descending colon to the superior part of the rectum.

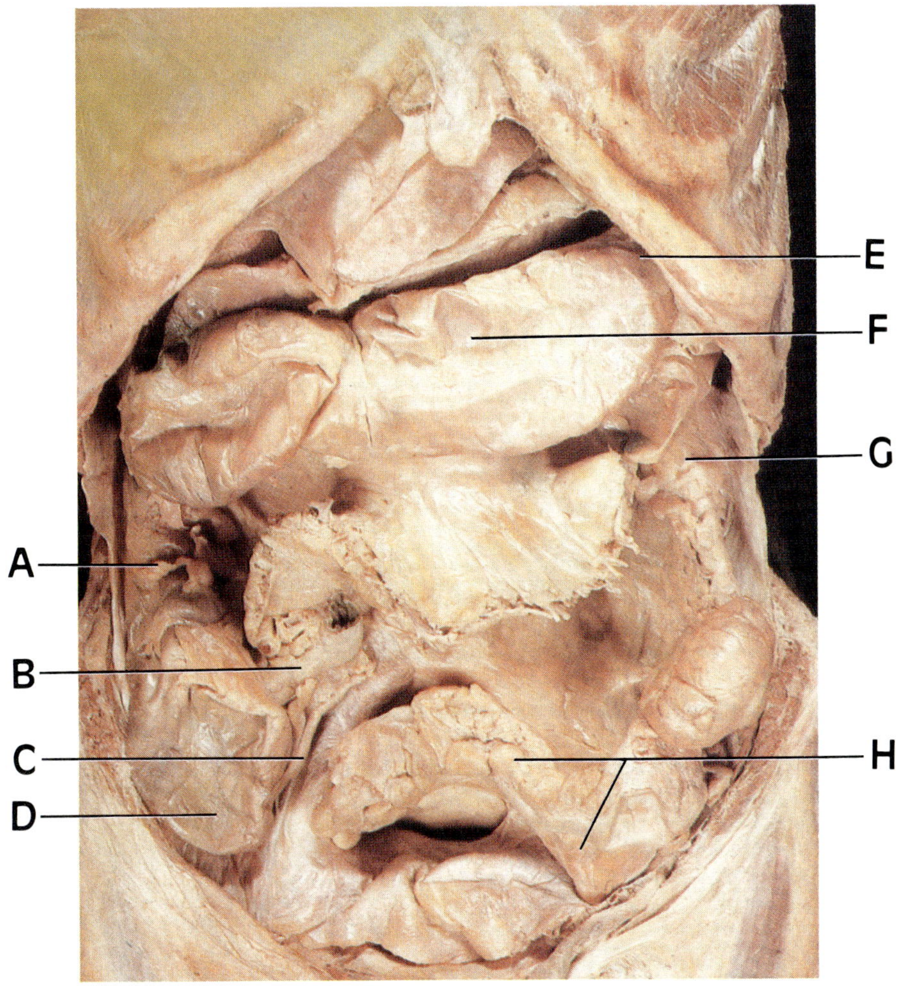

A. Ascending colon
B. Ileum
C. Vermiform appendix
D. Cecum
E. Splenic flexure (left colic flexure)
F. Transverse colon
G. Descending colon
H. Sigmoid colon

Figure 13–3
Coronal view of the abdomen, highlighting visceral unit structures.

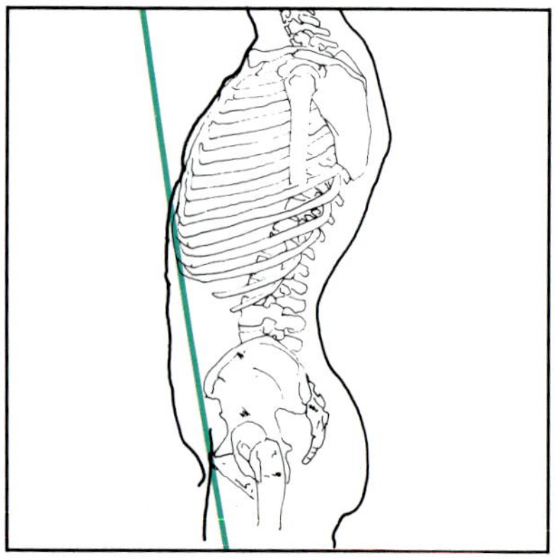

ENCLOSING UNIT (Fig. 13–4)

Supracolic Compartment. An area of the abdomen above the transverse colon that is a part of the greater sac and consists of the following:

(1) **Right Subphrenic Space** (*A*). An area anterior to the right lobe of the liver and to the right of the falciform ligament. This space extends from the diaphragm to the level of the transverse colon.

(2) **Left Subphrenic Space** (*E*). A region anterior to the left lobe of the liver and to the left of the falciform ligament. This area extends from the diaphragm to the level of the transverse colon.

(3) **Right Subhepatic Space (Morrison's Pouch).** This is not seen in this coronal view because it is posterior to the right lobe of the liver, which covers this space in this section.

Falciform Ligament (*B*). A vertical, oblique ligament found between the right and left lobes of the liver. The inferior end is more to the right than the superior end in this view. This ligament connects the liver to the inferior surface of the diaphragm and to the back area of the anterior abdominal wall. The anterior portion of this ligament is found in the median and extends from the diaphragm to the umbilicus.

Infracolic Compartment. This area of the abdomen is below the level of the transverse colon and is divided into four smaller areas:

(1) **Right Lateral (Paracolic) Gutter** (*C*). A vertical, elongated potential space between the ascending colon and the right wall of the abdomen. This space extends from the cecum to the hepatic flexure.

(2) **Gutter to the Right of the Mesentery** (*D*). An area bounded by the mesentery, the ascending colon, and the right half of the transverse colon. This gutter is closed at both the superior and inferior ends.

(3) **Gutter to the Left of the Mesentery** (*H*). An area bounded by the mesentery, the descending colon, and the left portion of the transverse colon. This gutter is open at its inferior end.

(4) **Left Lateral (Paracolic) Gutter** (*F*). A vertical potential space between the descending colon and the left wall of the abdomen. This gutter extends from the splenic flexure to the sigmoid colon.

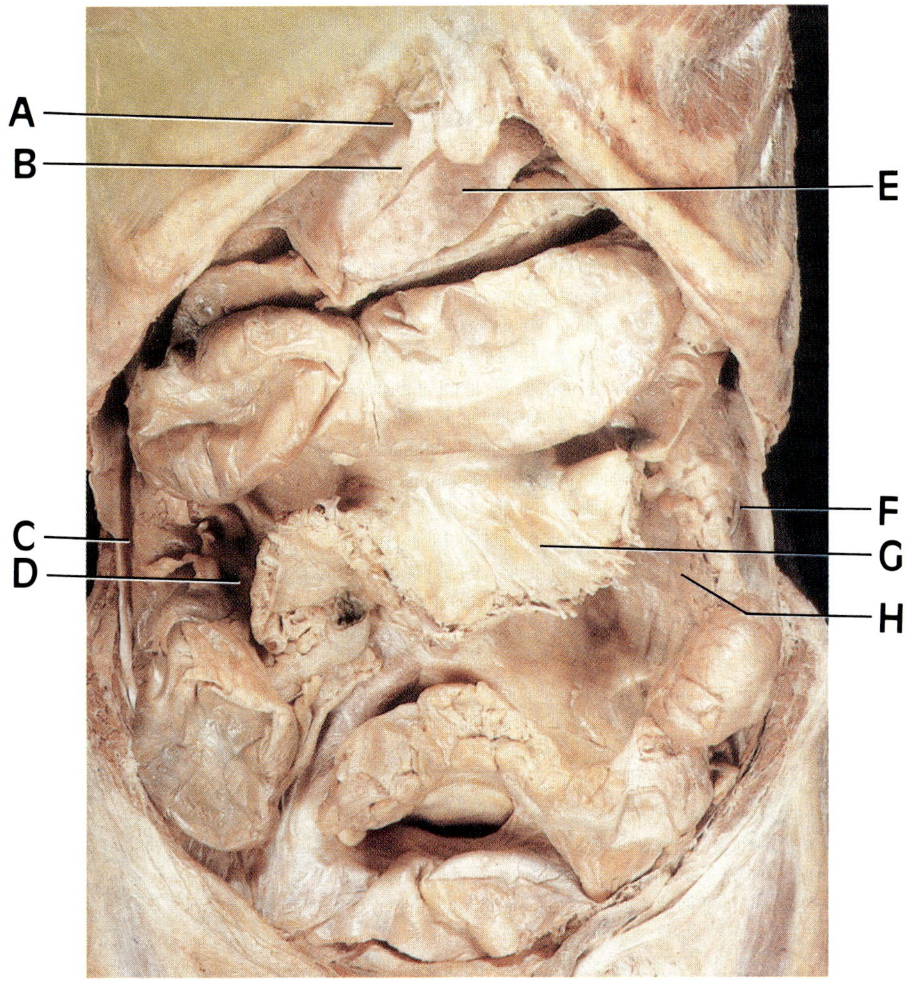

A. Right subphrenic space
B. Falciform ligament
C. Right lateral (paracolic) gutter
D. Gutter to right of mesentery
E. Left subphrenic space
F. Left lateral (paracolic) gutter
G. Mesentery
H. Gutter to left of mesentery

Figure 13–4
Coronal view of the abdomen, highlighting enclosing unit structures.

CHAPTER 14

Sagittal Cadaver Section and Image of the Abdomen

The cadaver section of this chapter is derived from the median plane of the body. At this part of the body several visceral and neurovascular units may be observed clearly in the sagittal plane. Superior and inferior relationships of structures are compared, as well as anterior and posterior relationships.

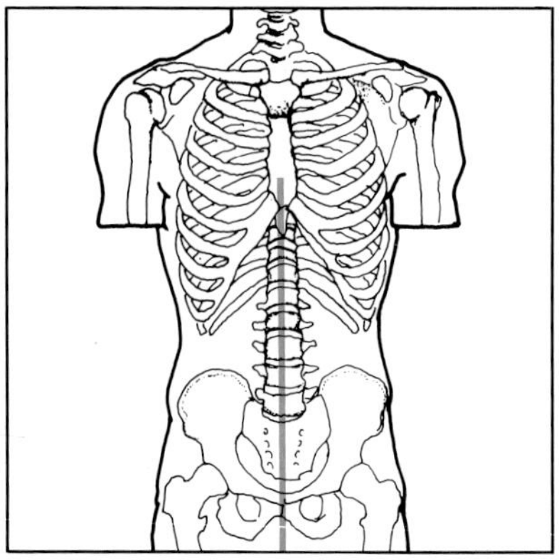

MODULE 1

This is a sagittal section of the male abdomen and pelvis. The emphasis in this module is on the abdomen. Some of the abdominal structures seen are the liver, stomach, duodenum, pancreas, and superior mesenteric artery.

MUSCULOSKELETAL UNIT (Fig. 14–1)

I. Skeletal Portion

Lumbar Vertebrae (*Y*). A series of bone slices with an intervertebral disk between them. These bone slices have rectangular bodies and posterior bony projections. These bony posterior projections are posterior to the spinal cord and cauda equina. The bodies and disks are anterior to the spinal cord.

Intervertebral Disks (*T*). A series of platelike fibrocartilaginous structures between the bodies of the vertebrae. These disks are part of the anterior wall of the spinal canal.

II. Muscular Portion

Rectus Abdominis Muscle (*N*). A vertical strip of anterior, superficial muscle tissue. Its inferior end is slightly inferior to the superior end and is attached to the pubis.

NEUROVASCULAR UNIT (Fig. 14–1)

I. Arteries

Descending Abdominal Aorta (*P*). A vertical, elongated, hollow vascular structure anterior to the bodies of lower thoracic and lumbar vertebrae. This vessel shows many proximal openings of other vessels along its course.

Superior Mesenteric Artery (*H*). A rather short, vertical, oblique vessel inferior to the posterior part of the stomach and anterior to the superior mesenteric vein. In this plane the inferior end is anterior to the superior end.

Branch of the Superior Mesenteric Artery (*K*). A vascular structure seen extending from the superior mesenteric artery.

Common Hepatic Artery (*C*). A rather small, hollow structure posterior to the anterior part of the pancreas, as seen in this sagittal view.

Origin. Celiac axis (trunk).

Distribution. This artery disperses blood to three major branches: (1) gastroduodenal, (2) right gastric, and (3) proper hepatic.

Course. It travels to the right and slightly inferior from its origin to reach the superior part of the first part of the duodenum. It passes inferior to the medial part of the epiploic foramen to reach the anterior part of the hepatic portal vein. This artery climbs between layers of lesser omentum and anterior to the epiploic foramen to approach the porta hepatis of the liver. It bifurcates to form right and left hepatic arteries.

Pancreaticoduodenal Artery (*U*). A small, oval vascular structure anterior to the inferior part of the pancreas and posterior to the superior mesenteric vein. There are two pancreaticoduodenal arteries, but only one is seen and labeled in this view. The one that originates from the gastroduodenal artery is the superior pancreaticoduodenal artery, and the one that arises from the superior mesenteric artery is the inferior pancreaticoduodenal artery.

Origin (for superior pancreaticoduodenal). Gastroduodenal artery.

Origin (for inferior pancreaticoduodenal). Superior mesenteric artery.

Distribution (for superior pancreaticoduodenal). It has two major branches: (1) the anterior superior pancreaticoduodenal artery and (2) the posterior superior pancreaticoduodenal artery.

Distribution (for inferior pancreaticoduodenal): It has two major branches: (1) the anterior inferior pancreaticoduodenal artery and (2) the posterior inferior pancreaticoduodenal artery.

Course (for superior pancreaticoduodenal). It is a continuation of the gastroduodenal artery adjacent to the medial surface of the upper part of the descending duodenum.

Course (for inferior pancreaticoduodenal). It is a branch of the superior mesenteric artery and courses adjacent to the superior part of the horizontal portion of the duodenum. It terminates by dividing into anterior and posterior inferior pancreaticoduodenal arteries. These arteries anastomose with the anterior superior pancreaticoduodenal and posterior superior pancreaticoduodenal arteries, which are branches of the superior pancreaticoduodenal artery.

Right Renal Artery (*R*). An oval, proximal end of a hollow vascular organ adjacent to the right side of the descending abdominal aorta and anterior to lumbar vertebra 1, as seen in this sagittal plane.

Right Common Iliac Artery (*L*). An oval, proximal end of a hollow vascular structure, seen as a branch of the descending abdominal aorta. This proximal portion of the right common iliac artery is seen adjacent to the anterior superior part of lumbar vertebra 4.

Origin. The inferior part of the descending abdominal aorta.

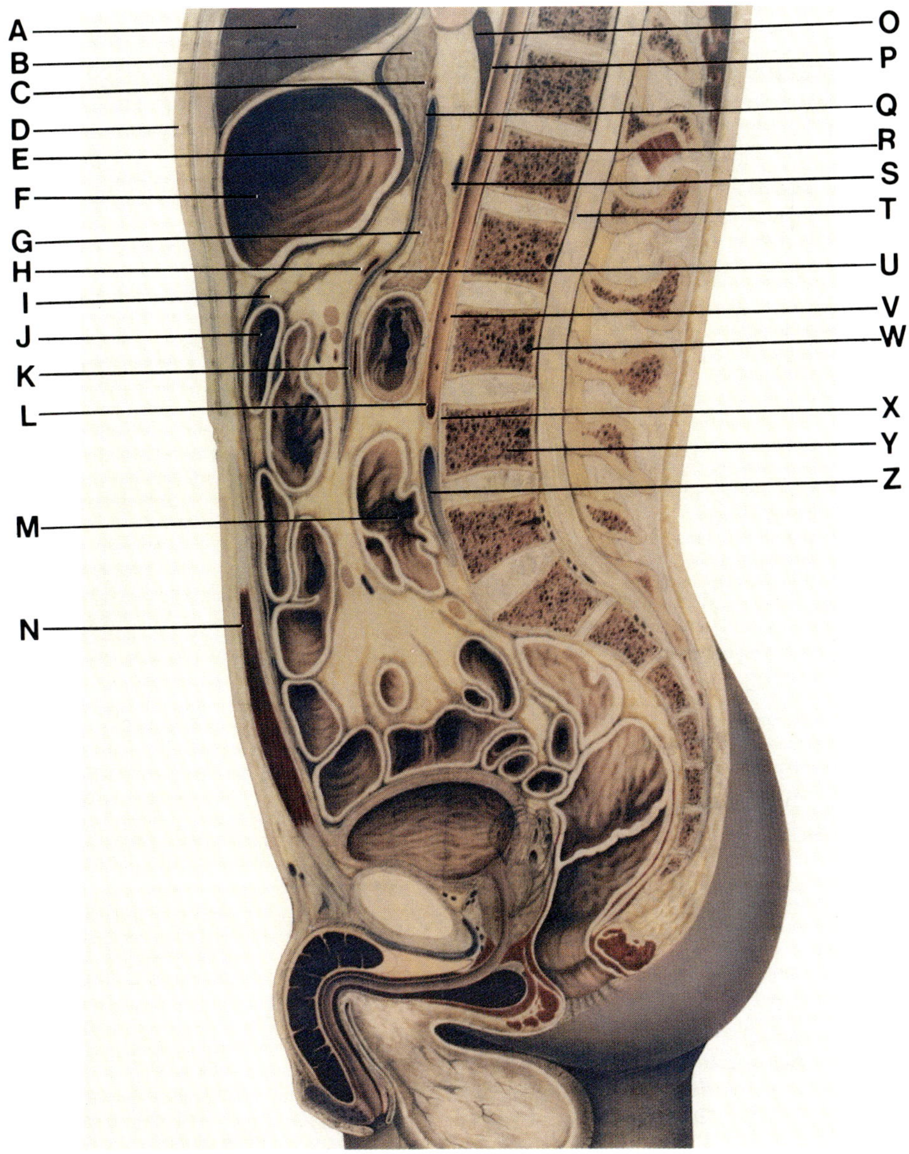

A. Liver
B. Pancreas
C. Common hepatic artery
D. Linea alba
E. Lesser sac (omental bursa)
F. Stomach
G. Pancreas
H. Superior mesenteric vessel
I. Mesocolon
J. Transverse colon
K. Branch of superior mesenteric artery
L. Right common iliac artery
M. Sigmoid colon
N. Rectus abdominus muscle
O. Diaphragm
P. Descending abdominal aorta
Q. Superior mesenteric vein
R. Right renal artery
S. Right renal vein
T. Intervertebral disk
U. Pancreaticoduodenal artery
V. Diaphragm (crura)
W. Posterior longitudinal ligament
X. Anterior longitudinal ligament
Y. Lumbar vertebra
Z. Left common iliac vein

Figure 14–1
Male section, approximately midsagittal, highlighting some structures of the abdomen.

Distribution. There are two terminal branches: (1) the right internal and (2) the right external iliac arteries.

Course. It courses obliquely (inferior and lateral) anterior to the proximal end of the inferior vena cava, sympathetic trunk, and bodies of lumbar vertebrae 4 and 5. This vessel is retroperitoneally located and therefore posterior to the intestines.

II. Veins

Superior Mesenteric Vein (*Q*). A vertically elongated, curved, hollow vascular vessel. Its superior end is seen posterior to a part of the pancreas, and its course is anterior to another part of the pancreas and anterior to the duodenum, as well as the branch of the superior mesenteric artery. It terminates near the anterior superior part of the sigmoid colon.

Right Renal Vein (*S*). A rather short, vertical, slitlike structure seen posterior to the pancreas and anterior to the inferior part of lumbar vertebra 1.

Left Common Iliac Vein (*Z*). A vertical, hollow venous structure anterior to the inferior part of lumbar vertebra 4 and the superior part of lumbar vertebra 5, as well as the disk between these two vertebral bodies.

Origin. The left internal and external iliac veins unite to form the left common iliac vein anterior to the sacroiliac joint.

Termination. Inferior part of the inferior vena cava.

Course. It courses medial to the left common iliac artery before coursing posteriorly to the right common iliac artery. On its course it travels posterior to the root of the sigmoid mesocolon.

VISCERAL UNIT (see Fig. 14–1)

Diaphragm (*O*). A vertical, slightly oblique diaphragmatic tissue seen anterior to thoracic vertebrae 11 and 12 and the superior anterior part of lumbar vertebra 1. The inferior end is narrower than and slightly posterior to the superior end.

Liver (*A*). A large, triangular slice of hepatic tissue. The base of this triangular section is superior and the apex is inferior, as well as anterior to the superior part of the stomach, as seen in this sagittal section.

Pancreas (*B*). A vertical aggregate of glandular tissues posterior to the stomach and superior to the horizontal duodenum, as seen in this sagittal body section. It is seen as two portions; the superior portion is immediately posterior to the lesser sac, and the inferior part (*G*) is posterior to the superior mesenteric vein.

Stomach (*F*). A rounded, hollow, large digestive organ inferior to the bulk of the liver and anterior to the lesser sac. It is anterior to the pancreas and superior to the transverse colon.

Horizontal Segment of Duodenum. A large, oval, hollow structure inferior to the pancreas and superior to the sigmoid colon. It is posterior to the superior mesenteric vein.

Small Bowel Loops. Several irregularly oval and oblong small intestines that represent the jejunum and ileal segments. These bowel loops are anterior within the peritoneal cavity and inferior to the transverse colon. Some of these intestinal slices are immediately superior to the urinary bladder and pouch of Douglas.

Transverse Colon (*J*). An oblong (superior-to-inferior), hollow, rather large digestive segment inferior to the stomach and anterior to the mesenteric vein, as well as anterior to the horizontal part of the duodenum, as seen in this sagittal plane.

Sigmoid (Pelvic) Colon (*M*). A rather oblong, large digestive segment of the large intestine, seen anterior to lumbar vertebrae 4 and 5.

Course. The three parts of the sigmoid colon take the following path as they descend to join the proximal part of the rectum. The first part runs downward adjacent to the left pelvic wall. The second part runs transversely in the pelvic cavity between the rectum and uterus in women but between the bladder and rectum in men. This transverse part frequently reaches the right pelvic wall before becoming the third part. The third part of the sigmoid colon curves posteriorly and medially to reach the median plane, where it bends inferiorly to connect with the proximal part of the rectum.

ENCLOSING UNIT (see Fig. 14–1)

Lesser Sac (Omental Bursa) (*E*). A curved, vertical space posterior to the stomach and anterior to the pancreas. The concave part of this curved structure is adjacent to the posterior surface of the stomach. The upper recess is seen posterior to the liver, and the inferior recess is posterior to the stomach and greater omentum.

Transverse Mesocolon (*I*). A vertical, oblique, moderately narrow strip of mesentery seen inferior to the stomach, and the anterior end is closely associated with the superior part of the transverse colon. The inferior end is anterior to the posterior end.

Mesentery Proper. An irregular, vertical sheet of tissue attached to the small bowel loops anteriorly and inferiorly. It is attached to the posterior abdominal wall and extends 15 cm obliquely and inferiorly from the left side of lumbar vertebra 2 to the right superior part of the sacral iliac joint. In its course it passes anterior to the horizontal portion of the duodenum, the inferior vena cava, the right ureter, the right psoas major muscle, and the descending abdominal aorta.

Anterior Longitudinal Ligament (*X*). A vertical strip of regular connective tissue seen anterior to the bodies of the vertebrae and the disks between them. It is posterior to the descending abdominal aorta.

Posterior Longitudinal Ligament (*W*). A vertical strip posterior to the bodies of the vertebrae and the disks between them. It is also anterior to the spinal cord and cauda equina.

Linea Alba (*D*). A vertical strip of tendinous tissue on the anterior abdominal wall. It is seen in this sagittal section superior to a portion of the rectus abdominis muscle.

Appendix to Part V: Clinical Correlations for the Abdominal Region

CASE ONE (Fig. AV-1)

History. The patient has lung carcinoma. The right hemidiaphragm is elevated. A possible liver metastasis is being clinically considered. Because of the marked elevation of the right hemidiaphragm, two different sequences of high-resolution, thin-section T1- and T2-weighted axial images and T1-weighted coronal images were obtained from approximately the level of the carina up to the midabdominal region to cover the region of the liver.

Findings. No discrete lesions involving the liver are visualized.

Adenopathy is visualized in the carinal region, as well as in the right pulmonary hilum. A large mass in the periphery of the right upper lobe is also visualized, compatible with clinically known carcinoma of the lung. This lesion is not fully covered in the current study.

Impression. Right upper lobe carcinoma with associated mediastinal adenopathy as described. Marked elevated right hemidiaphragm. No evidence of metastatic disease involving the liver.

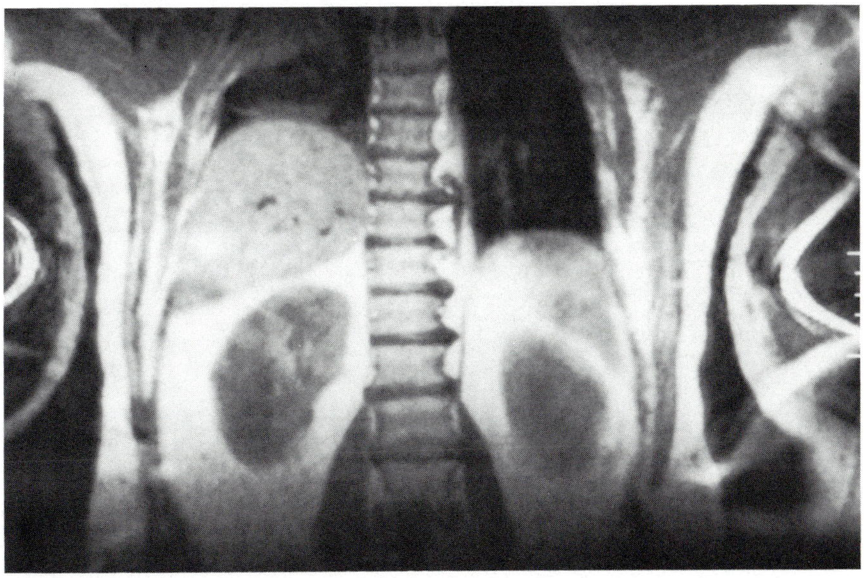

Figure AV-1.
MRI coronal image of the abdomen, showing right upper lobe carcinoma.

CASE TWO (Fig. AV-2)

History. The patient was examined to determine the status following back trauma. Low back pain and painful bilateral lower extremities were reported.

Findings. A borderline central bulging of the L4-L5 disk is seen. Minimal encroachment of this bulging disk on the anterior aspect of the thecal sac is visualized. The remainder of the lumbar disks shows no definite abnormality. No spinal stenosis or significant degree of facet arthropathy is visualized. The conus medullaris is visualized at approximately the T12 level.

Impression. Mild central bulge of the L4-L5 disk, as described.

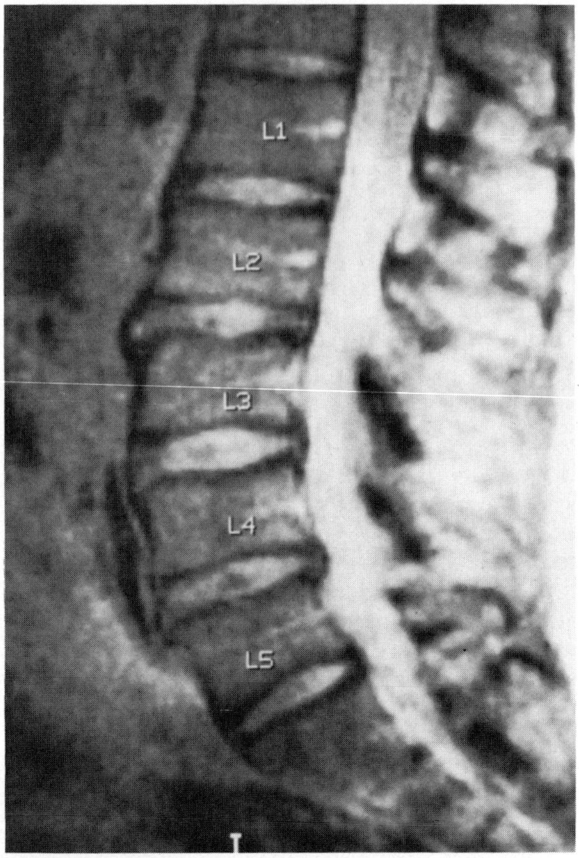

Figure AV-2.
Magnetic resonance sagittal image of the lumbar vertebrae, showing a mild central bulge of the L4-L5 disk.

PART VI

THE PELVIS

INTRODUCTION

The bones of the pelvis are the right and left os coxae (hip bones or pelvic bones), sacrum, and coccyx. The pelvis is divided into the greater and lesser pelvis by the arcuate line (pelvic inlet). The features that indicate the pelvic inlet from posterior to anterior include the sacral promontory, the ala of the sacrum, the arcuate line of the ilium, the pectineal line of the pubis, and the crest of the pubis as far as the pubic symphysis. The muscles of the pelvis are presented in Table VI–1.

The pelvis is divided into three regions: (1) right iliac, (2) hypogastric, and (3) left iliac. These three regions are subdivisions of the hypogastric zones. The viscera of the pelvis can be located using these three regions as guides. For example, some structures in the right iliac region include the cecum and the vermiform appendix. In the hypogastric region, structures such as the distal end of the ileum, the urinary bladder, the uterus (nonpregnant), and the prostate gland are found. The sigmoid colon, left ovary, and distal end of the left ureter are located in the left iliac region.

Table VI–1 Muscles of Pelvis and Perineum

Muscle	Origin	Insertion	Innervation	Vasculature	Action
Levator ani	Frontward from pelvic area of pubis and back inner area of ischial spine	Anterior and lateral parts of coccyx and median raphe (anococcygeal)	Pudendal plexus (muscular branches)	Internal pudendal (muscular branches), inferior gluteal, inferior rectal	Arranges floor of pelvic cavity, constricts inferior terminal of vagina and rectum, helps maintain viscera in position
Coccygeal (ischiococcygeal)	Medial area of ischial spine and sacrospinous ligament	Edge of coccygeus and lateral inferior segment of sacrum	Pudendal plexus (muscular branches)	Internal pudendal (muscular branches), inferior gluteal	Protects and pulls coccyx anteriorly
Superficial transverse m. of perineum (transversus perinei superficialis)	Tuberosity (front part)	Perineal body that is the central tendinous place of perineum	Pudendal (perineal branches)	Internal pudendal (perineal branches)	Adjusts central tendinous place of perineum
Bulbospongiosus (male)	Anterior raphe and central tendinous place of perineum	Inferior surface of perineal membrane, back area of corpus spongiosum and fascia (deep) on penile dorsum	Pudendal (perineal branches)	Internal pudendal (perineal branches)	Penile erection; near termination of micturition empties the urethra
Ischiocavernous	Pelvic surface of ischial tuberosity, ischiopubic ramus	Lateral parts of crus of penis or clitoris into corpus cavernosum and edge of pubic arch	Pudendal (perineal branches)	Internal pudendal (perineal branches)	Erection of penis and clitoris. When the crus of penis is compressed it impedes the return of blood
External sphincter m. of anus (sphincter ani externus)	Tip of coccyx and surrounding fascia	Tendinous center of perineum	Pudendal (perineal and inferior rectal branches)	Transverse perineal, inferior rectal	Shuts the orifice of anus
Deep transverse m. of perineum (transversus perinei profundus) (male)	Medial area of bottom part of ischial rami	Perineal body and inner tendinous raphe	Pudendal (perineal branches)	Internal pudendal (perineal branches)	Protects the prostate gland, fixes perineal body, enhances action of sphincter
Deep transverse perineal (female)	Medial area of bottom part of ischial rami	Lateral areas of vagina	Pudendal (perineal branches)	Internal pudendal (perineal branches)	Protects vagina, helps fix perineal body, enhances action of sphincter
Bulbospongiosus (vaginal sphincter)	Fibers around vaginal opening bulb of vestibule and central tendinous place of perineum	Lateral parts of pubic arch and corpora cavernosa of clitoris	Pudendal (perineal branches)	Internal pudendal (perineal branches)	Helps in clitoral erection, constricts vaginal orifice
Sphincter m. of urethra (sphincter urethrae)	Ischiopubic ramus (external fibers) and area around membranous urethra (internal fibers)	The raphe anterior and posterior to urethra (male); in female, it encloses urethra and attaches to lateral part of vagina	Pudendal (perineal branches)	Internal pudendal (perineal branches)	Functions in compressing urethra upon completion of micturition

CHAPTER 15

Transverse (Axial) Cadaver Sections and Images of the Pelvis

These transverse cadaver sections are derived from a female, at the level of S1 and S5, and from a male, showing the coccyx, greater trochanter, and pubic tubercle. The male transverse section is a sheet plastinated specimen, 2.5 to 4.0 mm thick as are others in this text.

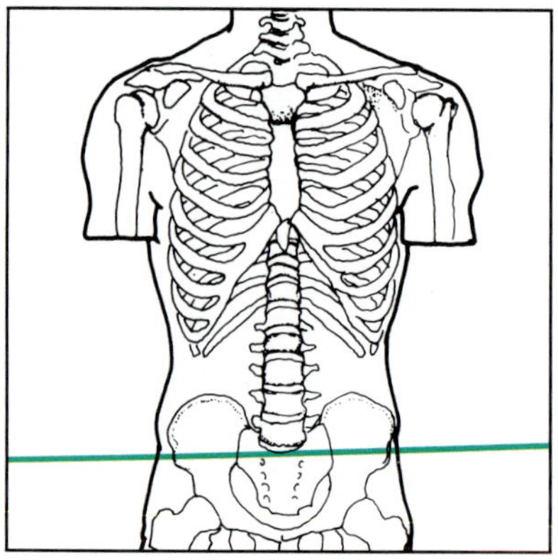

MODULE 1

This is a transverse (axial) section of the female body at the vertebral level of sacral vertebra 1, showing the sacroiliac joint, cecum, iliac and rectus abdominis muscles, and common iliac artery and vein.

MUSCULOSKELETAL UNIT (Fig. 15–1)

I. Skeletal Portion

Anterior Superior Iliac Spine (*E*). An oblique, anterior, bilateral projection of the ilial section. The anterior end is slightly lateral to the posterior part.

Ilium (*G*). An elongated (anterior-to-posterior), oblique, rather thin bone slice. The posterior end of this bilateral bone is thicker and more medial than the anterior end. Its posterior medial surface is adjacent to the lateral part of the sacrum and takes part in the formation of the sacroiliac joint.

Sacroiliac Joint (*I*). A rather elongated (anterior-to-posterior), oblique junction between the posterior medial part of the ilium and the lateral part of the sacrum. The posterior end of this bilateral joint is more medial than the anterior end.

Sacral Promontory (*F*). An oval, anterior projection of the upper segment of the first sacral vertebra between the common iliac vessels and anterior to the main body of the sacral vertebra.

Superior Sacral Tubercle (*Q*). A posterior median projection between the multifidus muscles and posterior to the sacral canal.

Lamina of Sacral Vertebra 1 (*J*). A rather short (anterior-to-posterior), oblique, bilateral bone slice between the superior sacral tubercle and the superior articular process. The posterior end is more medial than the anterior end.

Superior Articular Process (*L*). A lateral projection of the sacrum adjacent to the posterior medial part of the ilium and posterior to the psoas major muscle. This structure is bilateral.

Sacral Canal (*P*). A triangular space posterior to the body of the sacral vertebra anterior to the superior sacral tubercle and medial to the laminae of the sacrum. The base of the triangular feature is adjacent to the body of the sacral vertebra.

Body of Sacral Vertebra (*H*). A rather large bone slice anterior to the sacral canal; its anterior part contains the sacral promontory.

II. Muscular Portion

Rectus Abdominis Muscle (*A*). A transversely placed, bilateral muscle slice immediately lateral to the linea alba and medial to the oblique muscles of the abdomen.

External Oblique Abdominal (Obliquus Abdominis Externus) Muscle (*D*). An elongated (anterior-to-posterior), oblique bilateral muscle strip superficial to the internal oblique abdominal muscle and immediately deep to Scarpa's fascia. The anterior end is more medial than the posterior end.

Internal Oblique Abdominal (Obliquus Abdominis Internus) Muscle (*C*). An elongated (anterior-to-posterior), oblique bilateral muscle strip between the external oblique abdominal and transverse abdominal muscles. The anterior end is more medial than the posterior end.

Transverse Abdominal (Transversus Abdominis) Muscle (*B*). An elongated (anterior-to-posterior), oblique, bilateral muscle section deep to the internal oblique abdominal muscle. The anterior end is more medial than the posterior end.

Psoas Major Muscle (*M*). A rounded, bilateral muscle aggregate lateral to the common iliac vessels and anterior to the medial part of the iliac muscle. The lumbar plexus is seen adjacent to its posterior surface.

Iliac Muscle (*N*). An elongated (anterior-to-posterior), oblique, bilateral muscle slice adjacent to the iliac fossa, which is anterior to the ilium. Its anterior end is more lateral than its posterior end. The posterior part of this muscle is posterior to the psoas major muscle.

Gluteus Medius Muscle (*O*). An elongated (anterior-to-posterior), oblique, bilateral muscle slice immediately posterior to the ilium. The anterior end is more lateral than the posterior end.

Multifidus Muscle (*R*). A rather short (anterior-to-posterior), oblique, bilateral muscle that extends from the superior sacral tubercle to the anterior part of the lamina of the sacrum. It is deep to the sacrospinalis (erector spinae) muscle. The anterior end is more lateral than the posterior end.

Sacrospinalis (Erector Spinae) Muscle (*K*). A rather short (anterior-to-posterior), thick, oblique, bilateral muscle aggregate lateral to the posterior part of the ilium and posterior to the multifidus muscle. Its anterior end is more lateral than its posterior end.

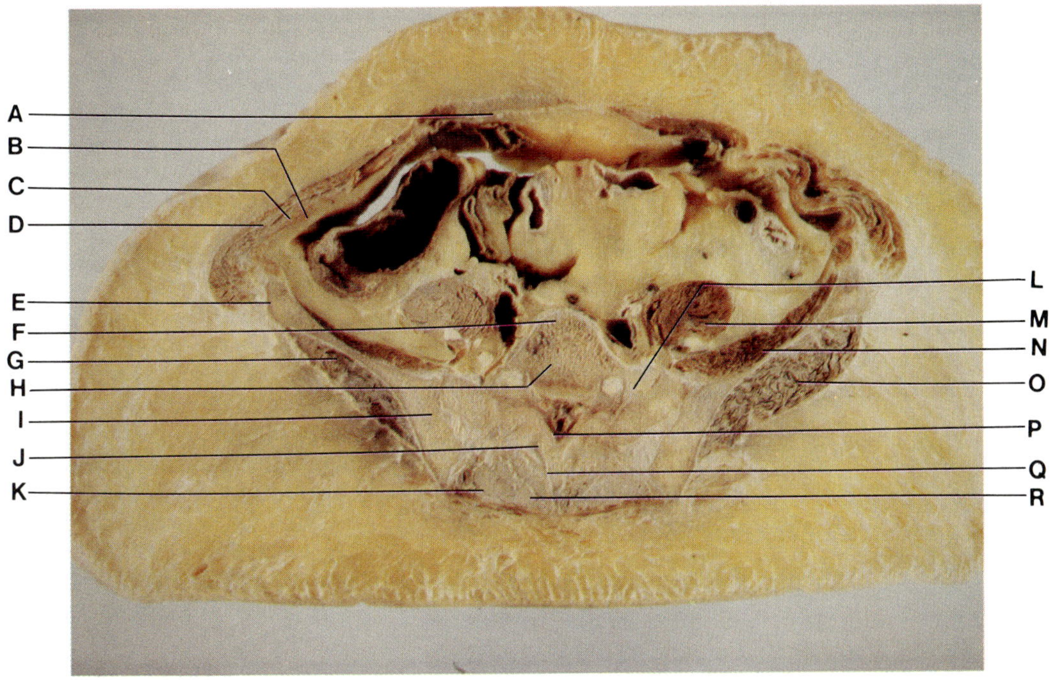

A. Rectus abdominis muscle
B. Transverse abdominal muscle
C. Internal oblique abdominal (obliquus abdominis internus) muscle
D. External oblique abdominal (obliquus abdominis externus) muscle
E. Anterior superior iliac spine
F. Sacral promontory
G. Ilium
H. Body of sacral vertebra
I. Sacroiliac joint
J. Lamina of sacral vertebra 1
K. Sacrospinalis (erector spinae) muscle
L. Superior articular process
M. Psoas major muscle
N. Iliac muscle
O. Gluteus medius muscle
P. Sacral canal
Q. Superior sacral tubercle
R. Multifidus muscle

Figure 15–1
A transverse section of the female pelvis at the level of sacral vertebra 1, highlighting musculoskeletal unit structures.

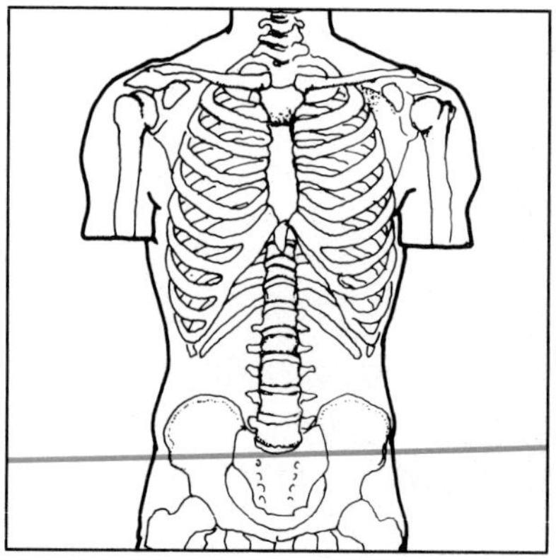

NEUROVASCULAR UNIT (Fig. 15–2)

Common Iliac Artery (*E*). A partially collapsed, bilateral vascular structure immediately anterior to the common iliac vein and medial to the anterior part of the psoas major muscle.

Origin. Descending abdominal aorta as its terminal branches.

Distribution. External and internal iliac arteries are the terminal branches. The other branches are (1) to the peritoneum, (2) to the ureter, (3) to the areolar tissue in the adjacent area, and (4) to the parietal peritoneum in the adjacent area.

Course (for right common iliac artery). This 5-cm (2-inch) long, oblique (the inferior end is more lateral than the superior end) artery begins on the left inferior part of lumbar vertebra 4 and terminates anterior to the right sacroiliac joint at approximately the level of the lumbosacral intervertebral disk. On its course it passes posterior to the sympathetic fibers of the superior hypogastric plexus and the ureter at its distal end. It is in the retroperitoneal space throughout its entire course and crosses anterior to the sympathetic trunk.

Course (for left common iliac artery). This 4-cm (1.6-inch), oblique (the inferior end is more lateral than the superior end) vessel starts at the inferior lateral surface of the fourth lumbar vertebra and terminates anterior to the left sacroiliac joint at the level of the lumbosacral intervertebral disk. This vessel is retroperitoneal throughout its course and passes posterior to the ilium, superior rectal artery, and sympathetic fibers for the superior hypogastric plexus.

II. Veins

Common Iliac Vein (*F*). A rather large, oval, bilateral venous structure immediately lateral to the anterior part of the sacrum and medial to the psoas major muscle. It is posterior to the common iliac artery.

Origin. External and internal iliac veins.

Termination. Inferior vena cava.

Course (for right common iliac vein). This rather short vessel, in a nearly vertical dimension, climbs posterior to the right common iliac artery for a short distance and then continues to climb lateral to the right common iliac artery to reach the inferior vena cava.

Course (for left common iliac vein). This rather long vessel is obliquely positioned and ascends for a short distance medial to the left common iliac artery and then continues to ascend posterior to the right common iliac artery.

III. Nerves

Lumbar Plexus (*G*). An obliquely (anterior-to-posterior) arranged, mixed nerve structure posterior to the psoas major muscle and lateral to the articular process of the sacrum. The anterior nerve structure is more lateral than the posterior nerve structure.

Origin. Anterior primary rami of lumbar nerves 1 through 4.

Distribution. It has six branches: (1) iliohypogastric, (2) femoral, (3) obturator, (4) genitofemoral, (5) ilioinguinal, and (6) lateral femoral cutaneous.

A Nerve Within the Sacral Foramen (*K*). A rounded, bilateral nerve slice inside a sacral foramen. This sacral foramen is medial to the articular process of the sacrum and anterior to the lamina.

VISCERAL UNIT (Fig. 15–2)

Jejunum (*I*). A circular, hollow digestive organ that shows many foldlike projections adjacent to the lumen. This structure is seen in the left peritoneal cavity near its anterior part.

Ileum (*H*). A rather large, hollow digestive structure seen in this body transverse section medial to the cecum and posterior to the right rectus abdominis muscle. The ileum is the last segment of the small bowel.

Cecum (*C*). A large, elongated (anterior-to-posterior), oblique, hollow digestive structure located in the right lateral peritoneal cavity. Its anterior end is more medial in this transverse body section than its posterior end.

ENCLOSING UNIT (Fig. 15–2)

Parietal Peritoneum (*D*). A thin sheet of tissue seen on the wall of the peritoneal cavity.

Mesentery (*J*). A rather broad sheet of tissue seen attached to the small bowel and posterior structures within the peritoneal cavity.

Right Paracolic Gutter (*B*). A space in an oblique dimension (the anterior part is more medial than the posterior part) between the cecum and the wall adjacent to it. The right paracolic gutter extends as far up as the distal end of the ascending colon between it and the wall adjacent to it.

Paracolic Gutter to the Right of the Mesentery. A space medial to the cecum and lateral to the mesentery.

Linea Alba (*A*). A median connective tissue structure between the right and left rectus abdominis muscles.

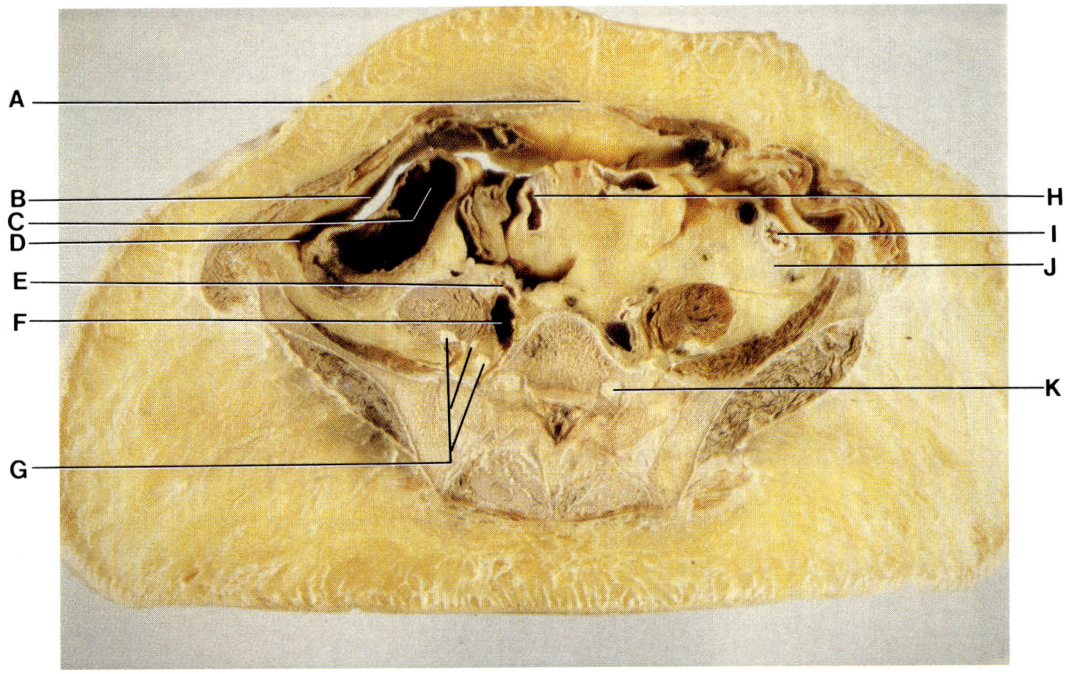

A. Linea alba
B. Right paracolic gutter
C. Cecum
D. Parietal peritoneum
E. Right common iliac artery
F. Right common iliac vein
G. Lumbar plexus
H. Ileum
I. Jejunum
J. Mesentery
K. Nerve within sacral foramen

Figure 15–2
A transverse section of a female pelvis at the level of sacral vertebra 1, highlighting neurovascular, visceral, and enclosing unit structures.

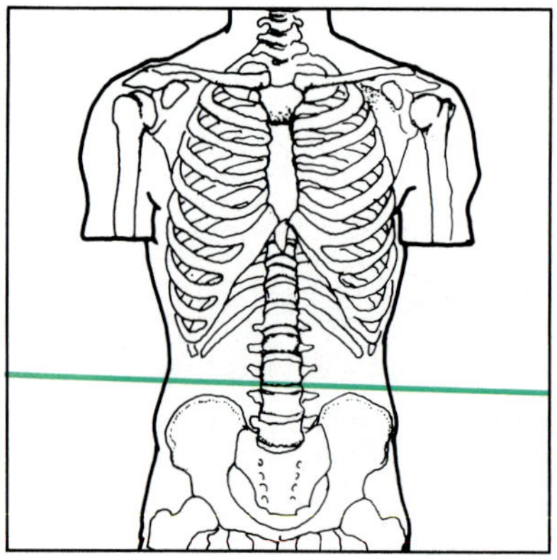

Figure 15–3 is a CT (with contrast) transverse image of the abdomen. This slice is derived from the upper area of the umbilical zone and highlights musculoskeletal, visceral, and neurovascular unit structures.

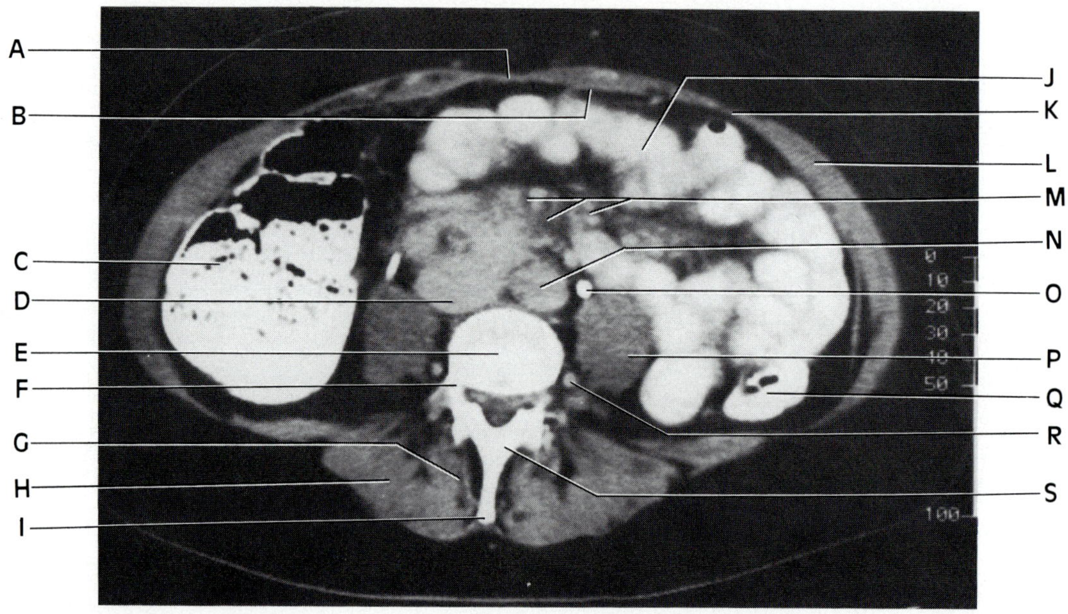

A. Linea alba
B. Rectus abdominis muscle
C. Ascending colon
D. Inferior vena cava
E. Body of lumbar vertebra
F. Pedicle of lumbar vertebra
G. Multifidus muscle
H. Erector spinae
I. Spinous process of lumbar vertebra
J. Transverse colon
K. Linea semilunaris (left)
L. External oblique abdominal (obliquus abdominis externus) muscle
M. Mesenteric fat and vessels of superior mesenteric artery and vein
N. Descending abdominal aorta
O. Left ureter
P. Psoas major muscle
Q. Descending colon
R. Sympathetic trunk (left)
S. Lamina of lumbar vertebra

Figure 15–3
CT (with contrast) transverse image of the abdomen, highlighting some structures of the pelvis.

Structures of the pelvis are seen in Figure 15–4. This is a picture of a CT (with contrast) transverse section.

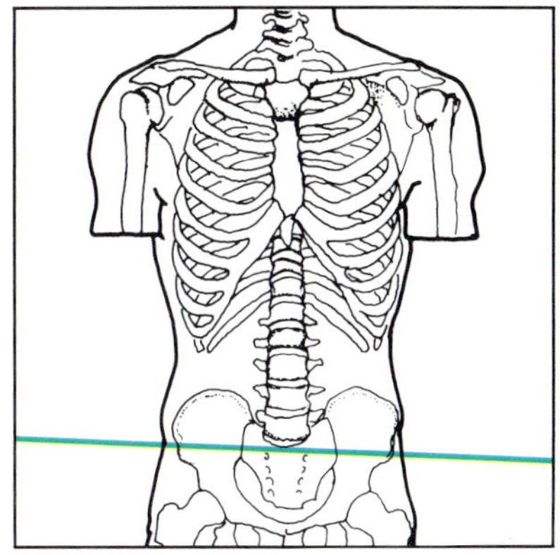

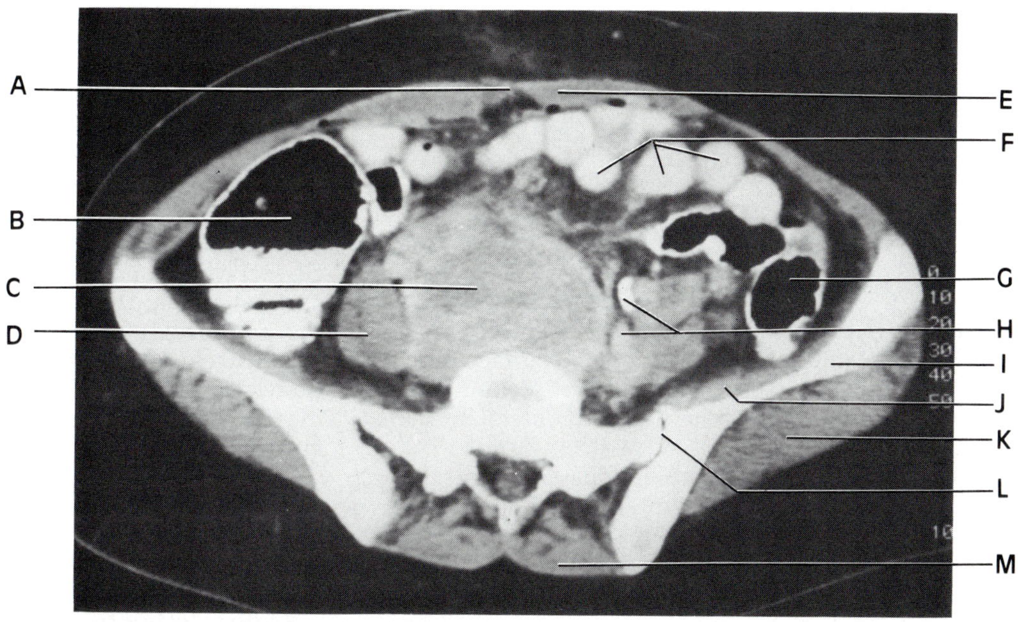

A. Linea alba
B. Cecum
C. Urinary bladder
D. Psoas major muscle
E. Rectus abdominis muscle
F. Small bowel (intestine)
G. Sigmoid colon
H. Obturator vessels
I. Ilium
J. Iliacus muscle
K. Gluteus medius muscle
L. Sacroiliac joint
M. Erector spinae muscle

Figure 15–4
CT (with contrast) transverse image of the abdomen, highlighting some structures of the pelvis.

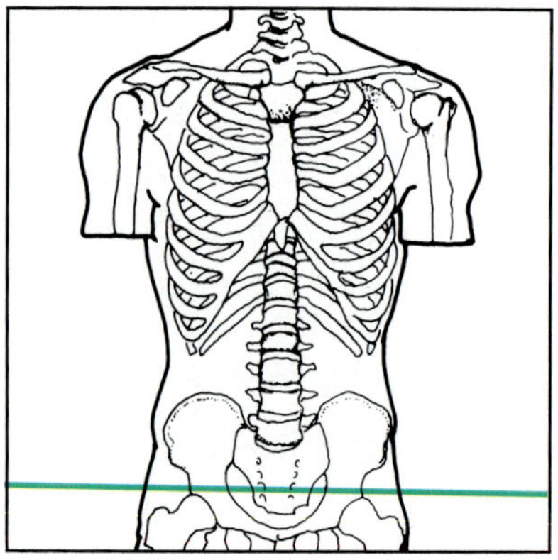

MODULE 2

This is a transverse (axial) section of a woman at the level of sacral vertebra 5.

MUSCULOSKELETAL UNIT (Fig. 15–5)

I. Skeletal Portion

Sacrum (*H*). An oblong, transversely placed bone slice with an anterior, slightly indented surface in the posterior median part of this body section. The gluteal muscles are seen immediately lateral to this bone.

Ilium (*D*). An oblique, oblong, bilateral bone slice between the iliopsoas and gluteus medius muscles. The anterior end is more lateral than the posterior end.

II. Muscular Portion

Gluteus Maximus Muscle (*G*). A large, transversely elongated, oblong, bilateral muscle slice located immediately lateral to the sacrum and posterior to the gluteus medius muscle. The medial end is adjacent to the sacrum and is more tapered than the lateral end.

Gluteus Medius Muscle (*F*). A transversely elongated, sausage-shaped, bilateral muscle section. The lateral end is much wider than the medial end. This muscle is between the gluteus maximus and gluteus minimus muscles.

Gluteus Minimus Muscle (*E*). An oblong (anterior-to-posterior), oblique, bilateral muscle slice. The anterior end is more lateral than the posterior end. The medial surface is adjacent to the iliac bone, and the lateral surface is adjacent to the lateral part of the gluteus medius muscle.

Iliopsoas Muscle (*C*). An oblong (anterior-to-posterior), oblique, bilateral muscle slice deep to the ilium bone section seen at this level. The posterior end is more medial than the anterior end.

Tensor Muscle of the Fascia Lata (Tensor Fasciae Latae Muscle) (*K*). A rather small, thin, triangular, bilateral muscle section obliquely placed lateral to the gluteus medius muscle. The anterior end is slightly wider than the posterior end and is more medial.

Sartorius Muscle (*J*). A small, oval, bilateral muscle section immediately lateral to the iliopsoas muscle and anterior to the tensor muscle of the fascia lata.

Aponeurosis of the External Oblique Abdominal Muscle (*I*). A thin, bilateral strip of tendinous tissue immediately lateral to the rectus abdominis muscle and medial to the lateral part of the iliopsoas muscle.

Internal Oblique Abdominal Muscle (*B*). A rather narrow strip of bilateral muscle tissue immediately anterior to the lateral part of the iliopsoas muscle and posterior to the aponeurosis of the external oblique abdominal muscle.

Rectus Abdominis Muscle (*A*). A transverse, bilateral muscle section immediately lateral to the linea alba and medial to the oblique abdominal muscles.

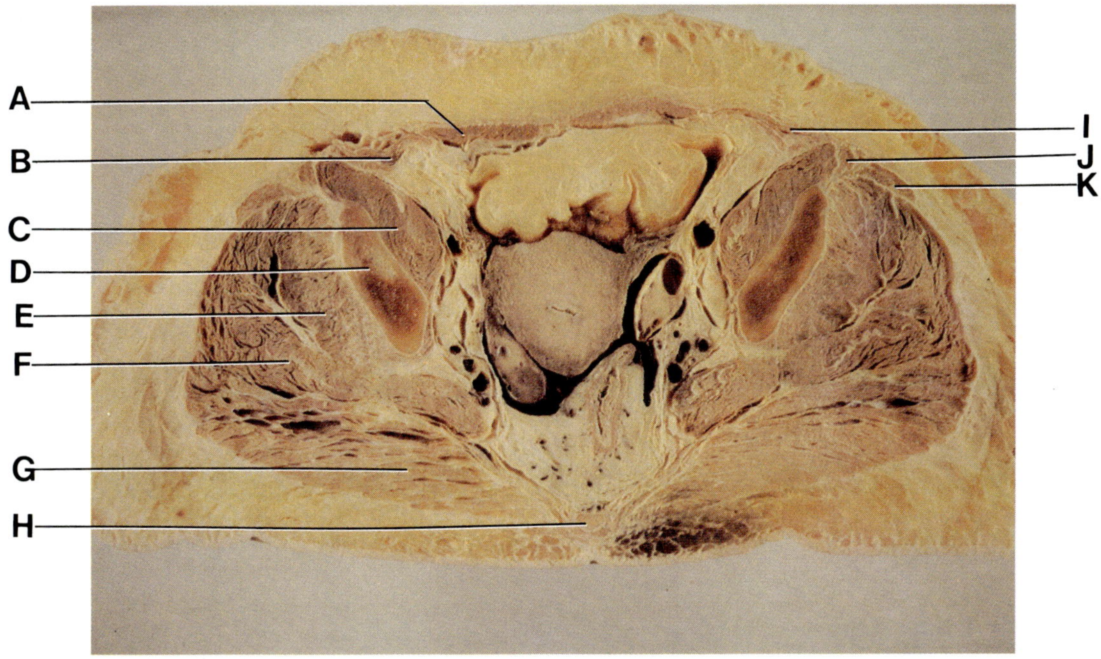

A. Rectus abdominis muscle
B. Internal oblique abdominal (obliquus abdominis internus) muscle
C. Iliopsoas muscle
D. Ilium
E. Gluteus minimus muscle
F. Gluteus medius muscle
G. Gluteus maximus muscle
H. Sacrum (inferior portion)
I. Aponeurosis of external oblique abdominal muscle
J. Sartorius muscle
K. Tensor muscle of fascia lata

Figure 15–5
Transverse section of a female pelvis at the level of sacral vertebra 5, highlighting musculoskeletal unit structures.

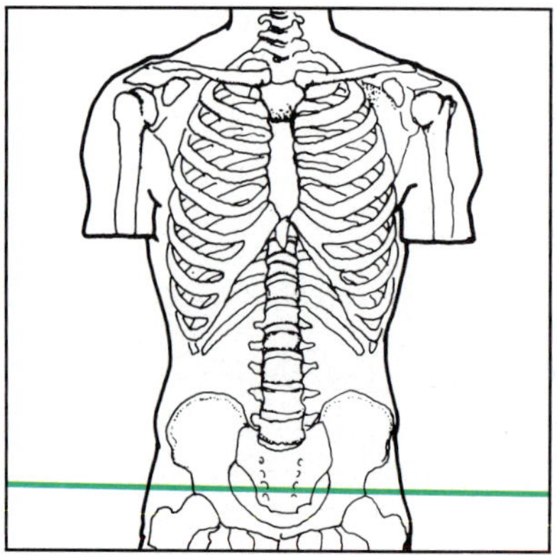

NEUROVASCULAR UNIT (Fig. 15–6)

I. Arteries

External Iliac Artery (*D*). This is seen medial to the iliopsoas muscle and lateral to the uterovesical space. This bilateral artery is a rounded, hollow vascular structure.

Origin. Common iliac artery.

Distribution. This artery distributes blood to the following: (1) a branch known as the inferior epigastric artery, (2) the deep circumflex iliac artery, and (3) the continuation of this artery beyond the inguinal ligament, the femoral artery.

Course. This artery descends obliquely along the medial border of the psoas major muscle, which is on the lateral wall of the pelvis. This artery passes deep to the inguinal ligament, equidistant from the pubic symphysis and the anterior superior iliac spine, where it becomes the femoral artery.

Internal Iliac Artery (*E*). A rounded, hollow vascular structure lateral to the ureter and anterior to the internal iliac vein.

Origin. Common iliac artery.

Distribution. This artery distributes blood to nine branches. It divides into two divisions, anterior and posterior. The branches of the anterior division are (1) the umbilical, continued as the superior vesical, (2) inferior gluteal, (3) obturator, (4) uterine and vaginal (female); inferior vesical (male), (5) middle sacral, and (6) internal pudendal. The branches of the posterior division are (1) superior gluteal, (2) iliolumbar, and (3) lateral sacral.

Course. It descends within the pelvic cavity in a path that passes posterior to the ureter and in front of its corresponding internal iliac vein.

Inferior Gluteal Artery (*C*). A rounded, hollow, bilateral vascular structure posterior to the internal iliac vein and medial to the posterior part of the gluteus medius muscle.

Origin. Arises from the anterior division of the internal iliac artery.

Distribution. Gluteus maximus muscle.

Course. This artery courses inferiorly anterior to the sacral plexus and piriformis. It passes posterior to the internal pudendal artery and exits from the pelvis through the greater sciatic notch (foramen) inferior to the sciatic nerve. It passes between the greater trochanter and ischial tuberosity to reach the gluteus maximus muscle.

II. Veins

External Iliac Vein (*B*). A partially collapsed, bilateral venous structure immediately posterior to the external iliac artery and medial to the posterior part of the iliopsoas muscle, as seen at this level.

Origin. It starts posterior to the inguinal ligament, midway between the anterior superior iliac spine and the pubic symphysis.

Termination. Anterior to the sacroiliac joint by fusion with the opposite external iliac vein to form the common iliac vein.

Course. The right external iliac vein courses medial to the external iliac vein initially and posterior as it ascends to join the common iliac vein. The left external iliac vein courses medial to the external iliac artery.

Internal Iliac (Hypogastric) Vein (*F*). A rounded, hollow, bilateral vascular structure posterior to the internal iliac artery and medial to the posterior part of the gluteus medius muscle.

Origin. This artery arises from tributaries that correspond with internal iliac artery branches, such as the internal pudendal, obturator, gluteal, lateral sacral, vesical, uterine, middle rectal, and vaginal veins.

Termination. It terminates into the common iliac vein anterior to the sacroiliac joint.

Course. It starts adjacent to the superior part of the greater sciatic notch (foramen) and travels superiorly to the medial surface of the internal iliac artery to join the external iliac vein. The internal iliac and external iliac veins form the common iliac vein at the level of the pelvic inlet anterior to the sacroiliac joint.

III. Nerves

Femoral Nerve (*A*). An oval, bilateral motor and sensory nerve tissue slice on the medial surface of the iliopsoas muscle embedded within its fascia.

Origin. Lumbar plexus (lumbar nerves 2, 3, and 4).

Distribution. Quadriceps muscles: (1) rectus femoris, (2) vastus medialis, (3) vastus lateralis, and (4) vastus intermedius. It also supplies nerve innervation for iliac, sartorius, and pectinus muscles.

Course. This nerve descends between the iliac and psoas major muscles and enters the thigh area posterior to the inguinal ligament, halfway between the anterior superior iliac spine and the pubic symphysis. It penetrates the femoral triangle and yields many branches.

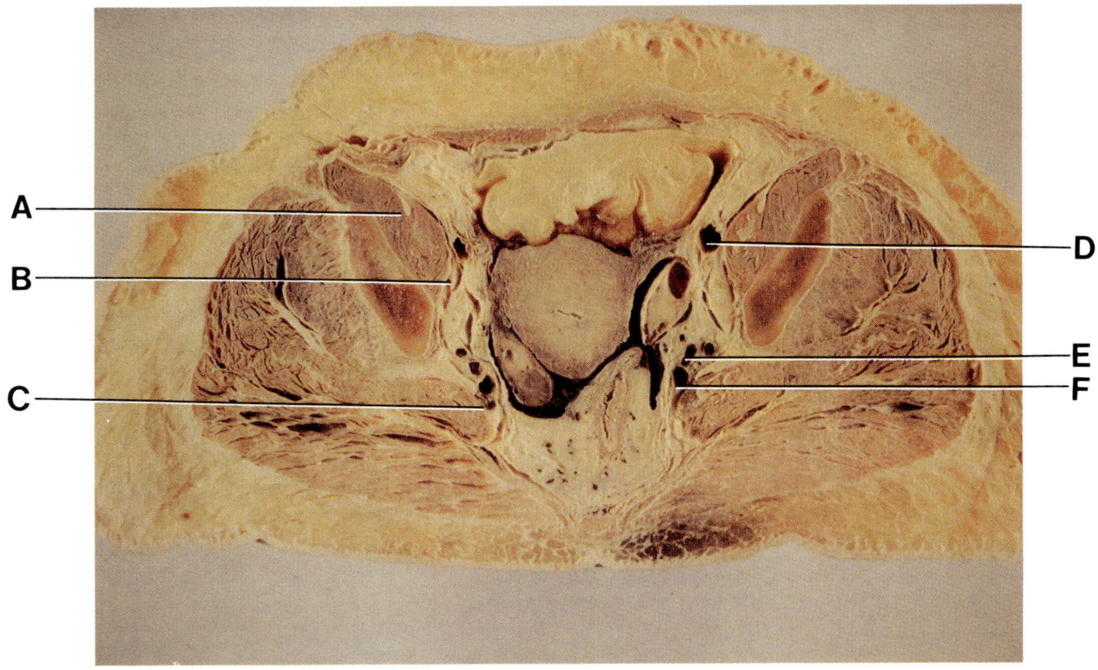

A. Femoral nerve
B. External iliac vein
C. Inferior gluteal artery
D. External iliac artery
E. Internal iliac artery
F. Internal iliac (hypogastric) vein

Figure 15–6
Transverse section of a female pelvis at the level of sacral vertebra 5, highlighting neurovascular unit structures.

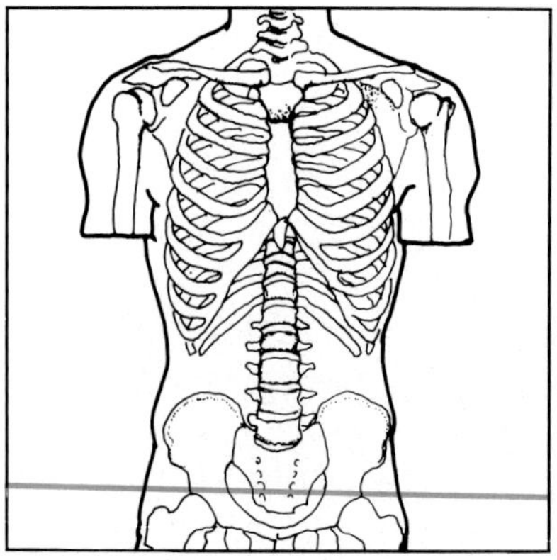

VISCERAL UNIT (Fig. 15–7)

Rectum (*I*). A rather elongated (anterior-to-posterior) slit posterior to the rectouterine pouch and anterior to the sacrum. The rectum is about 12.5 cm in length.

Course. It descends in the median of the upper sacrum and along its anterior curve, as well as the curve of the coccyx as it passes through the pelvis. In men, it takes its route posterior to the pouch of Douglas; in women, its route is posterior to the rectouterine pouch.

Ovary (*B*). An oblong (anterior-to-posterior), bilateral organ posterior to the broad ligament and lateral to the body of the uterus. It is posterior and medial to the external iliac artery. The posterior part is adjacent to the lateral part of the rectouterine pouch. The ovary is attached to the posterior part of the broad ligament by the mesovarian ligament.

Uterus (Womb) (*A*). A large, rounded, thick-walled structure with a transverse, slightly oblique slit in the center of the organ. The left end of the slit is slightly more posterior than the right end. The nonpregnant uterus is located within the true pelvis deep to the hypogastric region between the urinary bladder and rectum. The long axis of the uterus extends from the vagina to the pelvic inlet and forms a right angle with the vagina by anteflexion of the fundus. The length of the uterus is about 7.5 cm (3 inches), and it is 5 cm (2 inches) at its widest point. Its thickness is approximately 2.5 cm (1 inch).

Ureter (*G*). A small, rounded, hollow, bilateral organ medial to the internal iliac artery and posterior to the ovary. It is anterior and lateral to the rectum.

Origin. Renal pelvis of the kidney.

Termination. Base of the urinary bladder.

Course. The ureter starts as a continuation of the renal pelvis and takes a path about 5 cm lateral to the median. This structure is located behind the peritoneum (retroperitoneum) and courses slightly medial as it enters the pelvic cavity. As it runs in the pelvic cavity, it follows along the lateral pelvic wall, still deep to the peritoneum and anterior to the internal iliac artery and vein. It is medial to the psoas major and internal obturator muscles.

ENCLOSING UNIT (Fig. 15–7)

Rectouterine Pouch (*C*). A transversely oblique, hollow space posterior to the uterus and anterior to the rectum. The left end of the pouch is more anterior than the right end in this transverse body section.

Uterovesical (Vesicouterine) Pouch (*F*). A transverse, hollow space posterior to the urinary bladder and anterior to the uterus. The lateral part of this space is anterior to the broad ligaments.

Retropubic Space (*D*). A transversely placed area anterior to the urinary bladder and posterior to the distal end of the rectus abdominis muscle as well as the pubic bones.

Pararectal Fossa (*H*). A bilateral slit (anterior-to-posterior) immediately lateral to the rectum and medial to the internal iliac vessels.

Paravesical Fossa (*E*). A bilateral slit (anterior-to-posterior) lateral to the urinary bladder and medial to the external iliac vessels.

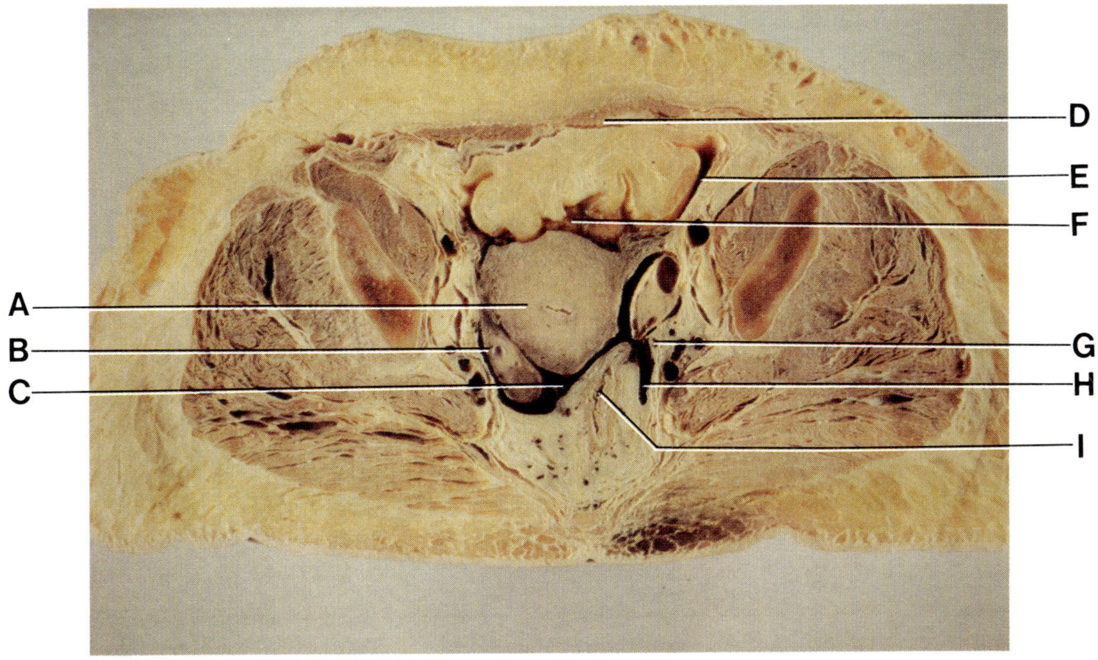

A. Uterus (womb)
B. Ovary
C. Rectouterine pouch
D. Retropubic space
E. Paravesical fossa
F. Uterovesical (vesicouterine) pouch
G. Ureter
H. Pararectal fossa
I. Rectum

Figure 15–7
Transverse section of a female pelvis at the level of sacral vertebra 5, highlighting visceral and enclosing unit structures.

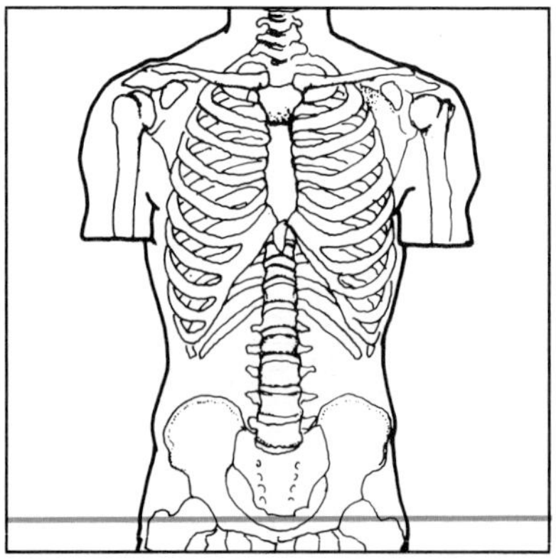

MODULE 3

This is a sheet plastinated (2.5 mm thick) transverse (axial) section of a man, showing the coccyx and greater trochanter as well as the pubic tubercles.

MUSCULOSKELETAL UNIT (Fig. 15–8)

I. Skeletal Portion

Coccyx (*H*). A rounded slice of bone tissue posterior to the rectum and medial to the gluteus maximus muscle.

Pubic Symphysis (*A*). An amphiarthroidal (secondary cartilaginous) joint between the right and left pubic bones.

Pubic Bone (*K*). An oblique, rather thick, bilateral bone slice anterior to the internal obturator muscle and lateral to the pubic symphysis. The anterior end is more medial than the posterior end.

Pubic Tubercle (*B*). An anterior, bilateral projection immediately medial to the pectinus muscle and lateral to the pubic crest.

Pubic Crest (*J*). A transversely placed, rather wide, bilateral bone slice between the pubic tubercle and the pubic symphysis.

Acetabulum (*F*). A medial, bilateral, rather deep, invaginated bone feature of the pelvic bone seen medial to the head of the femur and anterior to a part of the ischium.

Ischium (*R*). An oblong, bilateral bone slice posterior to the acetabulum and lateral to the posterior part of the internal obturator muscle.

Head of the Femur (*O*). A rounded, bilateral bone slice with its medial end adjacent to the acetabulum and lateral to the pectinus muscle.

Neck of the Femur (*P*). A transverse, bilateral bone slice between the head of the femur and the greater trochanter.

Greater Trochanter (*Q*). A large, lateral bone projection medial to the anterior part of the gluteus maximus muscle and lateral to the neck of the femur as seen in this transverse body section. This bone slice is bilateral.

II. Muscular Portion

Pectineus Muscle (*C*). A transversely oblique, bilateral muscle slice lateral to the pubic tubercle and anterior to the body of the pubis. The lateral part is more posterior than the anterior part and is posterior to the femoral vessels.

Tensor Muscle of the Fascia Lata (*D*). An oval, bilateral muscle slice posterior and slightly lateral to the rectus femoris muscle. The anterior part of this muscle slice is lateral to the rectus femoris.

Sartorius Muscle (*L*). A transverse, rather short, oblong, bilateral muscle slice anterior and medial to the rectus femoris muscle.

Iliopsoas Muscle (*N*). An irregularly oval, bilateral muscle section immediately anterior to the head of the femur and medial to the rectus femoris muscle, as seen at this level.

Gluteus Maximus Muscle (*I*). A transverse, thick, elongated, bilateral muscle slice with a tapering lateral end. The medial end is adjacent to the coccyx.

Rectus Femoris Muscle (*M*). An oval, bilateral muscle slice posterior to the sartorius muscle and lateral to the iliopsoas muscle.

Internal Obturator Muscle (*E*). An elongated (anterior-to-posterior) strip of bilateral muscle tissue that courses medial to the pelvic bone as seen at this level. It is lateral to the pelvic viscera.

Iliococcygeus Muscle (*G*). An elongated (anterior-to-posterior), narrow strip of oblique, bilateral muscle medial to the internal obturator muscle; the posterior end is adjacent to the coccyx. The medial side shows as a concave surface.

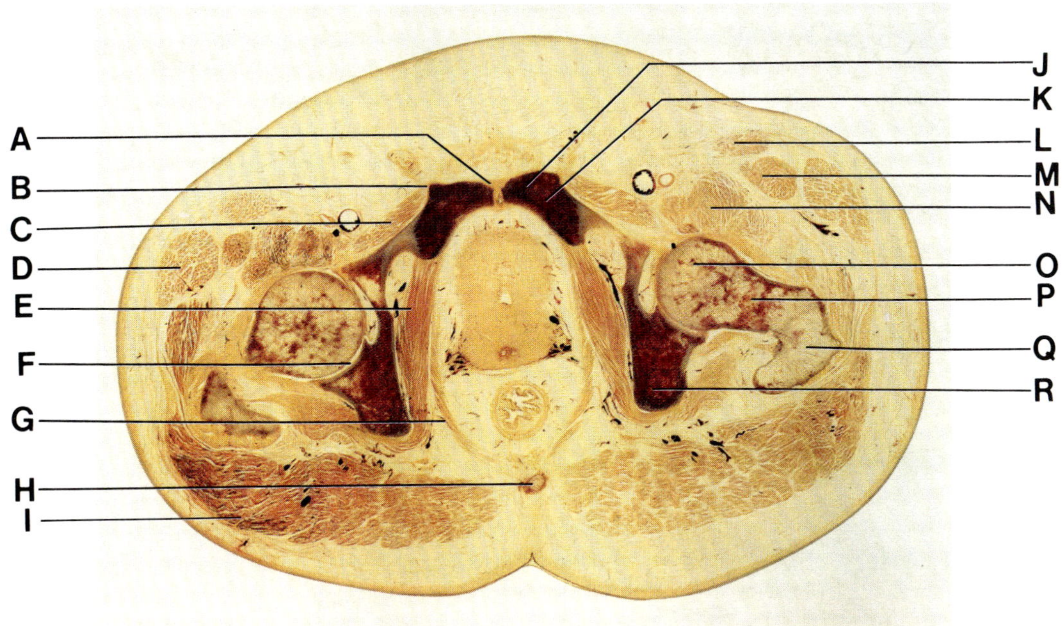

A. Pubic symphysis
B. Pubic tubercle
C. Pectineus muscle
D. Tensor muscle of fascia lata
E. Internal obturator muscle
F. Acetabulum
G. Iliococcygeus muscle
H. Coccyx
I. Gluteus maximus muscle
J. Pubic crest
K. Pubic bone
L. Sartorius muscle
M. Rectus femoris muscle
N. Iliopsoas muscle
O. Head of femur
P. Neck of femur
Q. Greater trochanter
R. Ischium

Figure 15–8
Plastinated transverse section of a male pelvis, highlighting musculoskeletal unit structures.

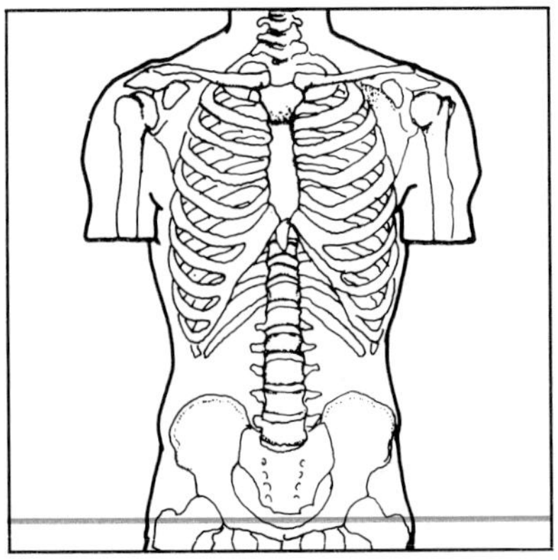

NEUROVASCULAR UNIT (Fig. 15–9)

I. Arteries

Femoral Artery (*E*). A small, rounded, hollow, bilateral vascular structure immediately lateral to the femoral vein and medial to the iliopsoas muscle.

Origin. Continuation of the external iliac artery.

Distribution. It distributes blood to seven branches: (1) superficial epigastric, (2) descending genicular, (3) superficial external pudendal, (4) deep external pudendal, (5) profunda femoris, (6) superficial circumflex iliac, and (7) muscular arteries.

Course. The femoral artery starts deep to the inguinal ligament midway between the anterior iliac spine and the pubic symphysis. This artery takes a course within the femoral sheath in its lateral compartment, which is in the femoral triangle. It enters the adductor (Hunter's) canal at the distal end of the femoral triangle and finally pierces a tendon of a great adductor (adductor magnus) muscle to reach the popliteal space, where it becomes the popliteal artery.

Spermatic Cord (*D*). An oval, bilateral structure anterior to the femoral vessels and medial to the sartorius muscle. This structure contains three arteries: (1) the artery of the ductus deferens (vas deferens) originates from the superior vesical artery and anastomoses with the testicular artery, (2) the cremasteric arises from the inferior epigastric artery and is the blood supply for the cremaster muscle, the covering of the spermatic cord, and anastomoses with the testicular artery, and (3) the testicular artery originates from the descending abdominal aorta and distributes blood to the testis, including the mediastinal testis. The fat around the kidney, the iliac lymph nodes, the cremaster muscle, and the ureter also receive some blood from this vessel. This bilateral artery takes an oblique course posterior to the peritoneum and enters the internal inguinal ring. It traverses the inguinal canal and exits from the external inguinal ring to reach the testis.

The major venous structure is the pampiniform plexus. This venous plexus arises from the posterior part of the testis and is the chief component of the spermatic cord. It passes upward anterior to the ductus deferens and terminates into three veins inferior to the superficial inguinal ring. These three or four veins traverse the inguinal ring to enter the abdomen, where they combine into two veins. These two veins ascend anterior to the psoas major muscle and the ureter and eventually unite to form the testicular vein. The right testicular vein drains into the inferior vena cava below the testicular vein, and the left testicular vein drains into the left renal vein.

Testicular lymphatic vessels are included in the spermatic cord. The fluid from these vessels drains into the lumbar lymph nodes.

The autonomic nerves are associated with both the spermatic and pelvic plexuses.

The vas deferens is included in the spermatic cord and is discussed in the visceral unit.

II. Veins

Femoral Vein (*A*). A rather large, hollow, bilateral venous structure immediately medial to the femoral artery and anterior to the lateral part of the pectinus muscle.

Origin. It begins as a continuation of the popliteal vein.

Termination. External iliac vein.

Course. It starts in the posterior median distal lower third of the thigh at the opening in the great adductor muscle. It ascends adjacent to the femoral artery and enters the adductor (Hunter's) canal. While coursing in the adductor canal it is posterolateral to the femoral artery. As it climbs through the femoral triangle, it is medial to the femoral artery and finally terminates deep to the inguinal ligament as it changes to the external iliac vein.

III. Nerves

Femoral Nerve (*B*). A rounded slice of bilateral, mixed (motor and sensory) nerve tissue anterior to the iliopsoas muscle and medial to the sartorius muscle. It is lateral to the femoral artery at this level.

Origin. Anterior primary rami of lumbar nerves 2, 3, and 4. The main spinal nerve for this structure is lumbar nerve 4.

Distribution. Quadriceps, iliac, sartorius, and pectineus muscles, as well as the skin of the anterior and medial thigh.

Course. It starts in the muscle tissue of the psoas major muscle and exits on its lateral border immediately inferior to the iliac crest. From this point it takes a downward path in the groove between the psoas major and iliac muscles. This nerve makes its entrance into the thigh area posterior to the median inguinal ligament, which is known as the muscular compartment. In this compartment it courses lateral to the femoral vein. Upon entering the femoral triangle, several branches emerge to supply the aforementioned structures.

Sciatic Nerve (*C*). An oval slice of bilateral, mixed

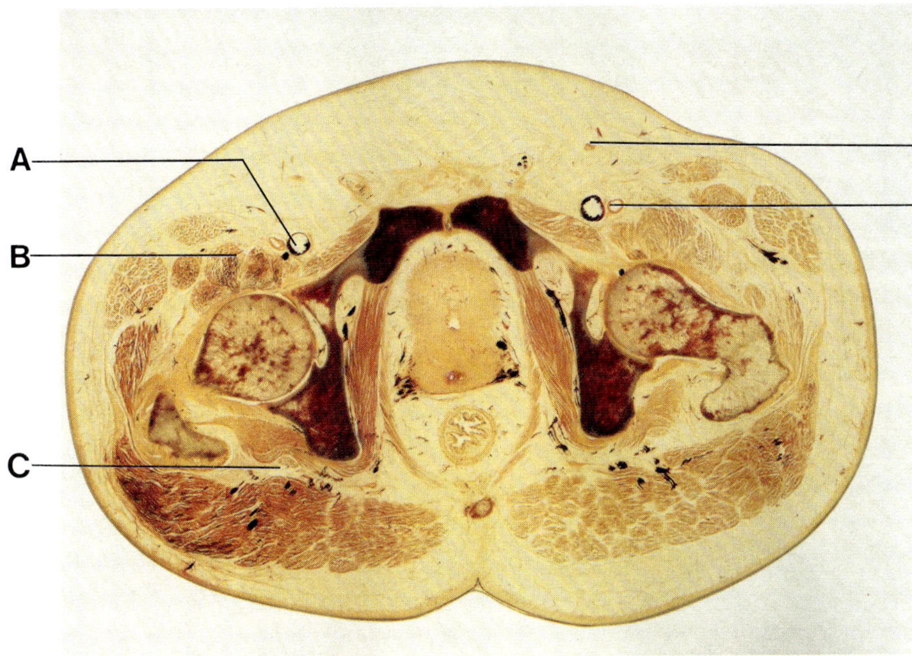

A. Femoral vein
B. Femoral nerve
C. Sciatic nerve
D. Cutaneous vessel
E. Femoral artery

Figure 15-9
Plastinated transverse section of a male pelvis, highlighting neurovascular unit structures.

(motor and sensory) nerve anterior to the middle part of the gluteus maximus muscle and midway between the posterior part of the pelvic bone and the medial part of the greater trochanter. The sciatic nerve is the largest nerve in a human body and represents two nerves (the peroneal and the tibial).

Origin. The part of the sacral plexus that arises from lumbar nerve 4 to sacral nerve 3.

Distribution. Hamstrings (biceps femoris, semitendinosus, and semimembranosus) and the head of the great adductor muscle attached to the ischial tuberosity. Finally, the sciatic nerve divides into the common peroneal and tibial nerves.

Course. It exits from the pelvis through the greater sciatic notch (foramen) immediately inferior to the piriform muscle and travels downward, midway between the ischial tuberosity and the greater trochanter, to reach the posterior thigh. It ends in the lower third of the posterior thigh to become the common peroneal and tibial nerves.

Cutaneous Vessel (*D*). Small skin vascular structure.

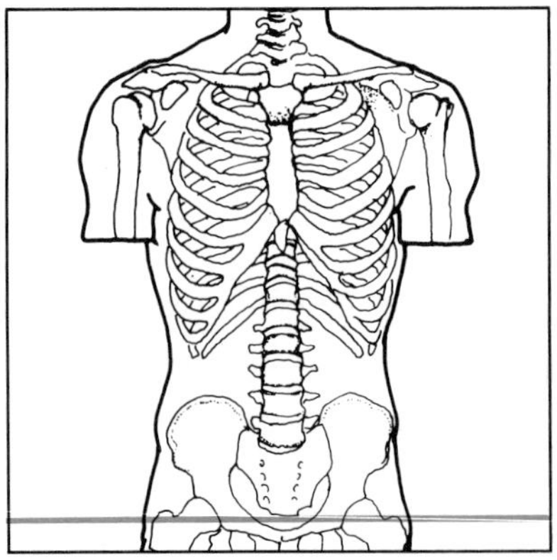

VISCERAL UNIT (Fig. 15–10)

Prostate Gland (*B*). A large, oblong (anterior-to-posterior) glandular tissue between the iliococcygeus muscles and posterior to the retropubic space. The posterior part shows two median openings (ejaculatory ducts), and in the center there is a rather large opening (prostatic urethra).

Course and Location. It surrounds the neck of the urinary bladder and the proximal portion of the urethra (prostatic urethra).

Median Lobe of the Prostate Gland (*C*). A flat portion of the prostate gland at this level, located between the urethra and ejaculatory duct. Enlargement of this lobe is responsible for the improper closing of the proximal end of the urethra. This improper closing allows urine to flow into the prostatic urethra, which makes the urge to urinate almost constant.

Prostatic Urethra (*G*). A rather large, irregularly oval, hollow structure located midway between the anterior and posterior parts of the prostate gland.

Origin. It arises at the neck of the bladder.

Termination. Membranous urethra.

Course. This 3-cm-long tube is the most dilated and widest portion that traverses the prostate gland within its center.

Vas Deferens (Ductus Deferens). A hollow structure within the spermatic cord.

Course. It starts as a continuation of the tail of the epididymis and passes upward from the posterior border of the testis adjacent to the medial surface of the epididymis to reach the superficial inguinal ring, where it enters the inguinal canal. It exits from the inguinal canal through the deep inguinal ring and starts its course inside the pelvis separate from the other structures of the spermatic cord to reach the medial surface of the seminal vesical. There it joins with its duct to form the ejaculatory duct. On its course from the deep inguinal ring to the ejaculatory duct, this structure passes adjacent to the lateral surface of the inferior epigastric artery, in front of the external iliac artery, deep to the parietal peritoneum on the lateral wall of the true pelvis, adjacent to the medial surface of the obliterated umbilical artery, adjacent to the medial surface of the obturator nerve and vessels, adjacent to the medial surface of the vesical vessels, and anterior to the ureters prior to becoming adjacent to the medial surface of this structure. Finally, the vas deferens courses adjacent to the posterior surface of the bladder to reach its base, where it joins the duct of the seminal vesical to form the ejaculatory duct.

Ejaculatory Duct (*D*). A small, oval, hollow, bilateral structure within the prostate gland near its median plane, anterior to the rectovesical pouch (pouch of Douglas).

Origin. A combination of the duct of the seminal vesicle and the distal part of the vas deferens.

Termination. It terminates within a minute opening upon the colliculus of the urethral crest, adjacent to the utricle.

Course. It starts at the base of the prostate gland and courses anteroinferiorly within the median between the right and left lobes of the prostate gland. It passes adjacent to the sides of the prostatic utricle.

Rectum (*I*). A rather large, rounded, hollow structure, with projections inside, adjacent to its lumen anterior to the posterior parts of the iliococcygeus muscles and posterior to the rectovesical pouch.

ENCLOSING UNIT (Fig. 15–10)

Retropubic Space (*A*). An area anterior to the urinary bladder/prostate gland and posterior to the medial parts of the pubic bones.

Paravesical Fossa (*F*). A bilateral area between the urinary bladder/prostate gland and the medial wall of the lesser (true) pelvis.

Rectovesical Pouch (Pouch of Douglas) (*H*). A downward, saclike area between the urinary bladder/prostate gland and the rectum.

Pararectal Fossa (*E*). A bilateral depression between the rectum and iliococcygeus muscle as seen at this level.

Ischiorectal Fossa (*J*). A triangular area between the posterior part of the internal obturator and the iliococcygeus muscles. The base of the triangular area is adjacent to the medial part of the gluteus maximus at this level.

Figure 15–11 shows an MRI transverse image of a male pelvis. Musculoskeletal, neurovascular, visceral, and enclosing unit structures are marked.

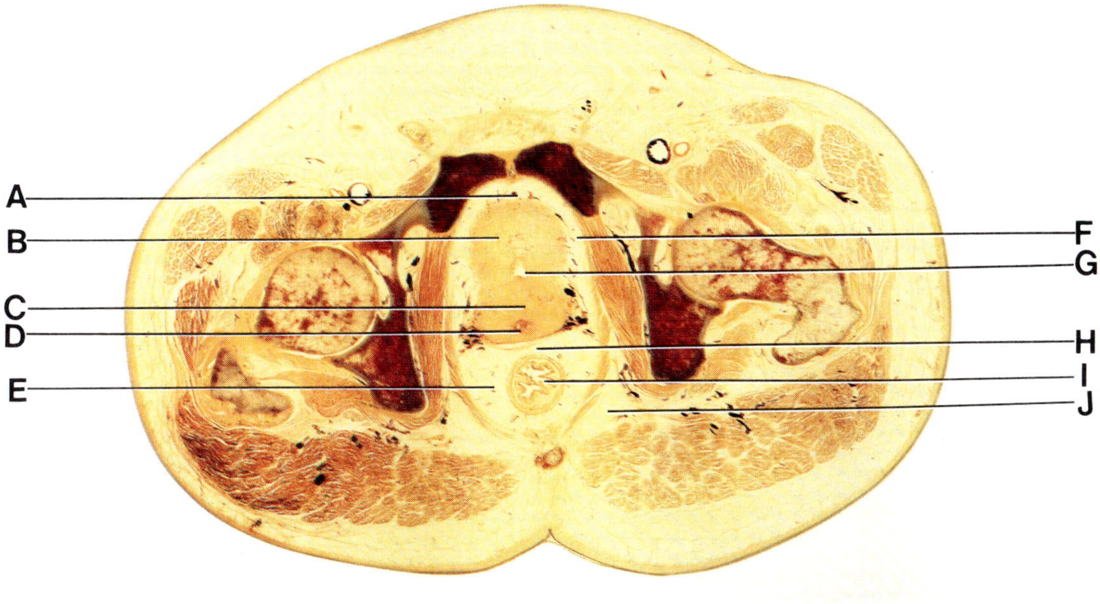

A. Retropubic space
B. Prostate gland
C. Median lobe of prostate gland
D. Ejaculatory duct
E. Pararectal fossa
F. Paravesical fossa
G. Prostatic urethra
H. Rectrovesical pouch (pouch of Douglas)
I. Rectum
J. Ischiorectal fossa

Figure 15–10
Plastinated transverse section of a male pelvis, highlighting visceral and enclosing unit structures.

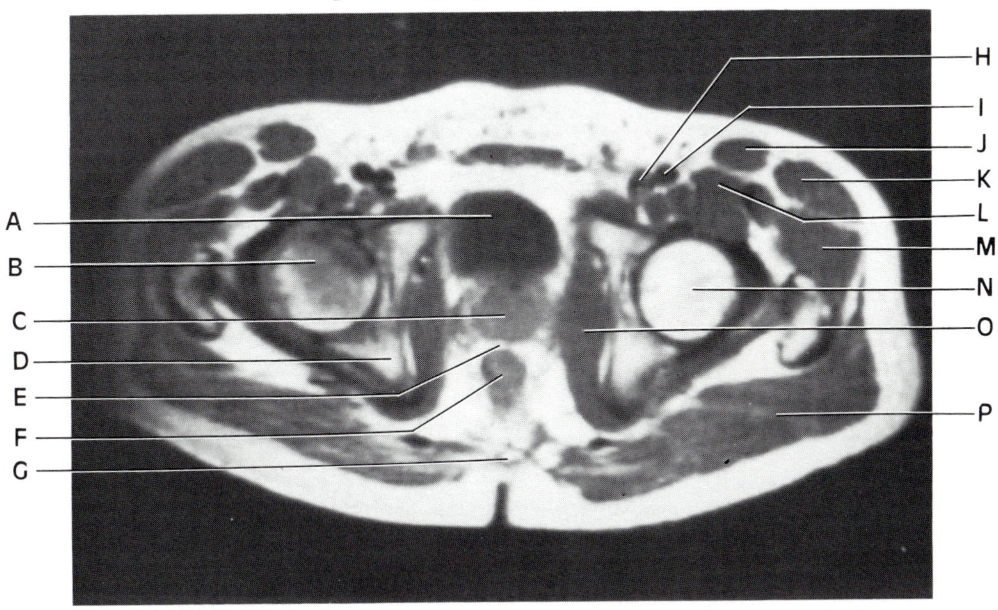

A. Urinary bladder
B. Right femoral head
C. Prostate gland
D. Ischial spine
E. Pouch of Douglas
F. Rectum
G. Coccyx
H. Femoral vein
I. Femoral artery
J. Sartorius muscle
K. Tensor muscle of fascia lata
L. Iliopsoas muscle
M. Gluteus medius muscle
N. Head of femur (left)
O. Internal obturator muscle
P. Gluteus maximus muscle

Figure 15–11
MRI transverse image of a male pelvis, highlighting some pelvic structures. Note the changes in the right femoral head (for further information see the Appendix to Part VI).

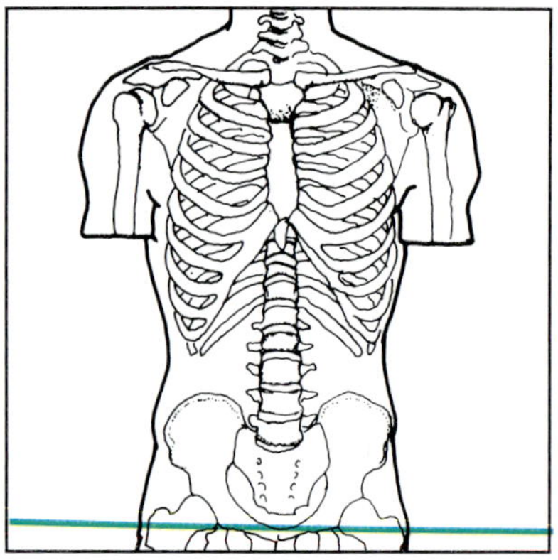

A transverse MRI image of a male pelvis (Fig. 15–12) is at a lower level than unit structures.

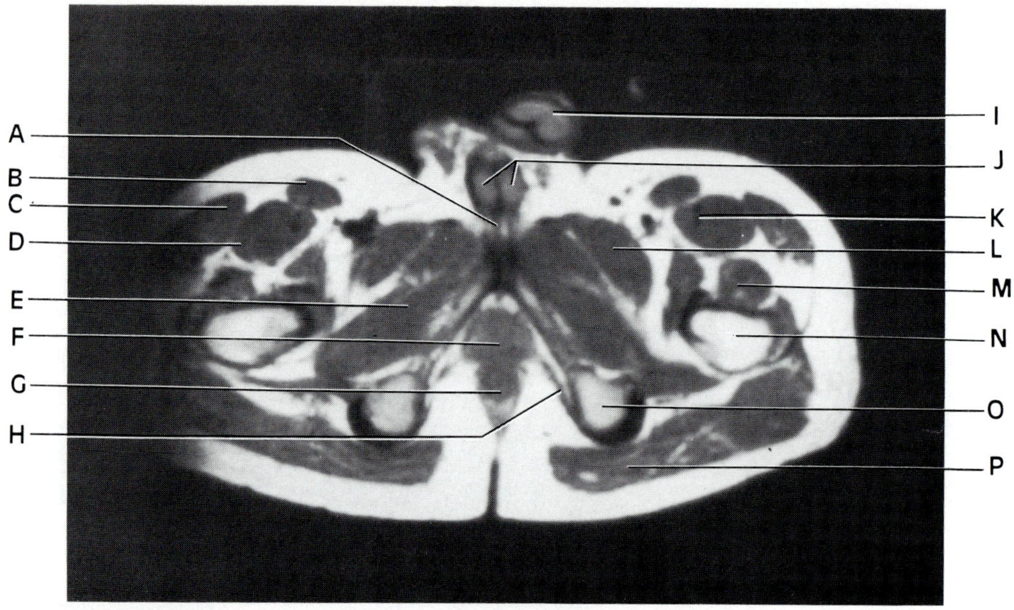

A. Bulb of penis
B. Sartorius muscle
C. Tensor muscle of fascia lata
D. Rectus femoris muscle
E. External obturator muscle
F. Prostate gland
G. Rectum
H. Internal obturator muscle
I. Glans penis
J. Corpora cavernosa of penis
K. Iliopsoas muscle
L. Pectineus muscle
M. Vastus intermedius muscle
N. Femur
O. Ischial tuberosity
P. Gluteus maximus muscle

Figure 15–12
MRI transverse image of a male, highlighting some structures of the pelvis.

A CT (with contrast) transverse image of a female pelvic region is shown in Figure 15–13. Note the ovarian cancer in this image.

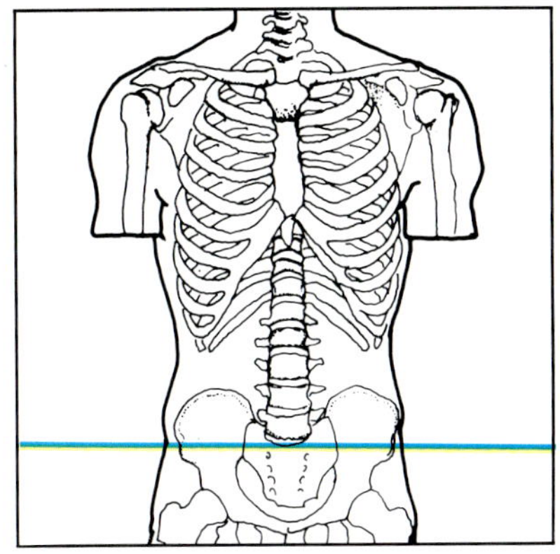

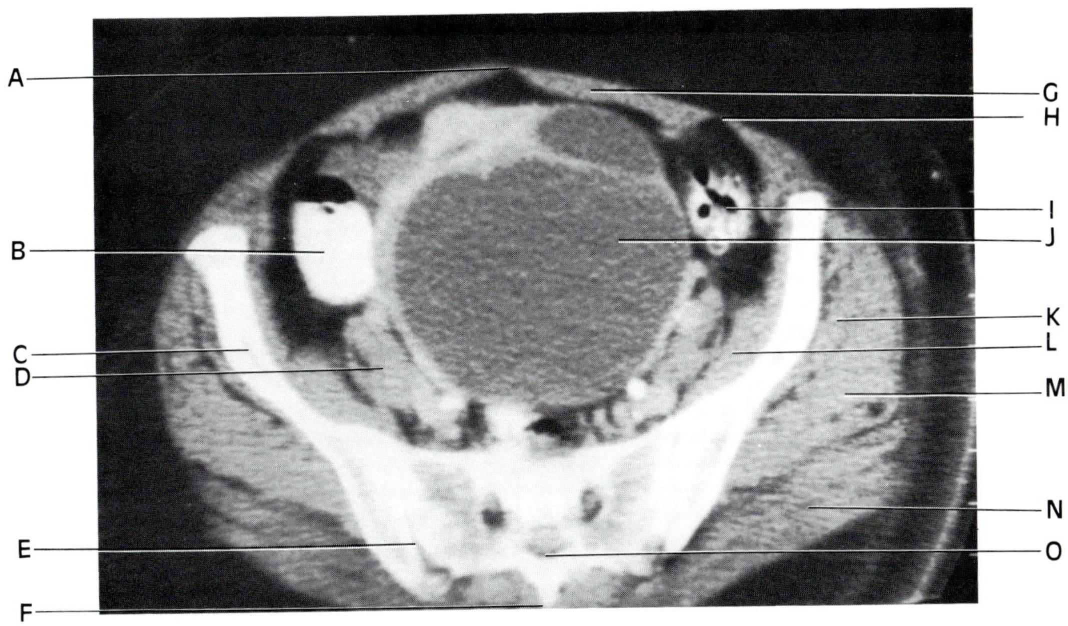

A. Linea alba
B. Cecum
C. Ilium
D. Psoas major muscle
E. Sacroiliac joint
F. Median sacral crest
G. Rectus abdominis muscle
H. Ilium
I. Sigmoid colon
J. Urinary bladder
K. Gluteus minimus muscle
L. Iliac muscle
M. Gluteus medius muscle
N. Gluteus maximus muscle
O. Sacral canal

Figure 15–13
CT (with contrast) transverse pelvic image of a female.

Figure 15–14 is an ultrasound transverse image, which shows a fetus. Note the urinary bladder and rectum in this picture.

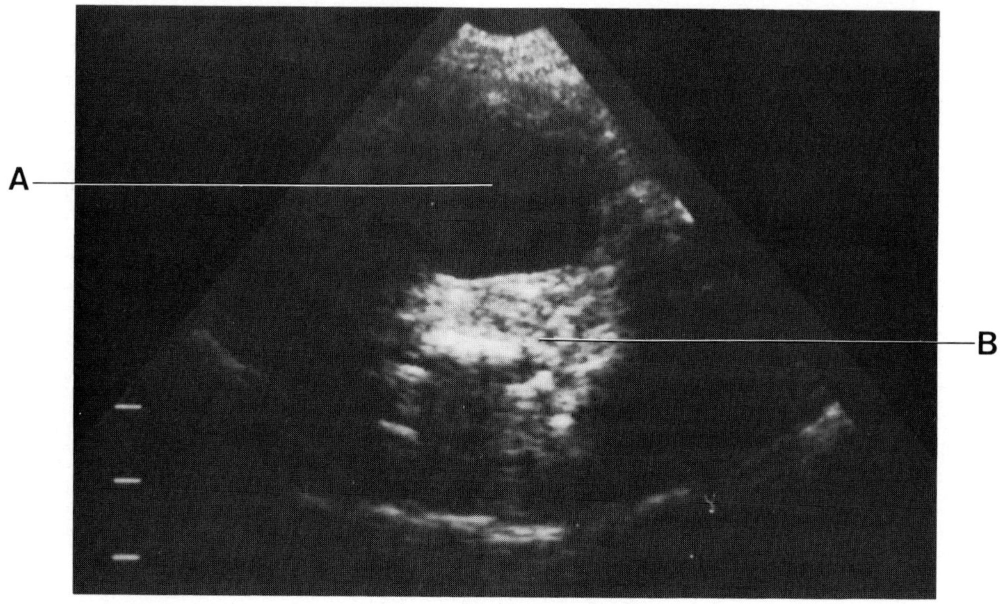

A. Urinary bladder
B. Rectum

Figure 15–14
Ultrasound (sonography) transverse image of a fetus.

CHAPTER 16

Coronal (Frontal) Cadaver Sections and Images of the Pelvis

This coronal cadaver section shows structures that are located near and at the midaxillary line of the male pelvis. In this plane, superior and inferior relationships of structures are observed, and medial and lateral relationships are also clearly seen. For example, the psoas major muscle is seen medial to the iliac muscle, and the vermiform appendix is seen superior to the psoas major muscle.

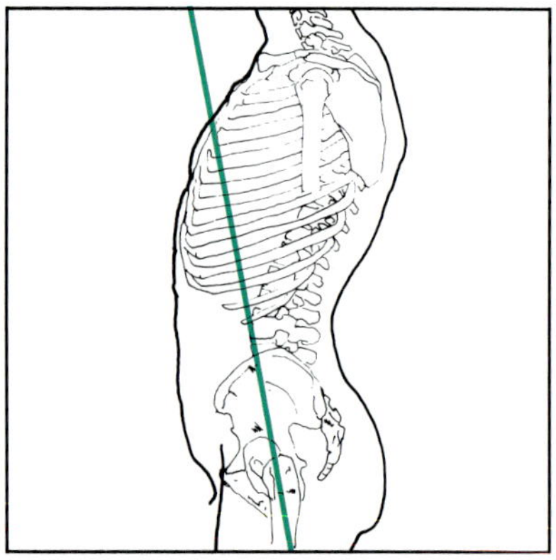

MODULE 1

This is a coronal (frontal) section of the pelvis and upper part of the lower limbs. In this module, the emphasis is on the pelvic structures, some of which are the psoas major muscle, vermiform appendix, external iliac artery and vein, urinary bladder, and levator ani muscle.

MUSCULOSKELETAL UNIT (Fig. 16–1)

I. Skeletal Portion

Iliac Crest (*S*). A rather sharp ridge at the superior part of the ilium, medial to the inferior part of the external and internal oblique abdominal muscles.

Ilium (*T*). An oblique, vertical bone slice between the iliac and gluteus minimus muscles. It is superior to the head of the femur, and the superior end is more lateral than the inferior end.

Pubic Bone (*P*). An oval, rather small, bilateral bone slice medial to the external obturator muscle and inferior to the inferior end of the internal obturator muscle.

Ischiopubic Ramus (*W*). An inferior part of the pelvic bone, rounded in appearance and inferior to the head of the femur.

II. Muscular Portion

External Oblique Abdominal (Obliquus Abdominis Externus) Muscle (*C*). An oblique, vertical, bilateral muscle strip superior to the ilium and superficial to the internal oblique abdominal muscle. The inferior end is more medial than the superior end.

Internal Oblique Abdominal (Obliquus Abdominis Internus) Muscle (*B*). An oblique, vertical, bilateral muscle strip between the external oblique abdominal and the transverse abdominal muscles. The inferior part is more medial than the superior end.

Transverse Abdominal Muscle (*A*). An oblique, vertical, bilateral muscle strip deep to the internal oblique abdominal muscle and superior to the iliac muscle.

Iliac Muscle (*I*). An oblique, bilateral muscle aggregate adjacent to the medial surface of the ilium and lateral to the psoas major muscle. The inferior end is more medial than the superior end.

Psoas Major Muscle (*G*). A vertical, bilateral muscle medial to the iliac muscle and femoral nerve. It is inferior to the vermiform appendix and cecum.

Internal Obturator Muscle (*M*). An oblique, bilateral muscle strip medial to the ischiopubic ramus and the external obturator muscle. The inferior end is more medial than the superior end and immediately lateral to the levator ani muscle.

External Obturator Muscle (*O*). A transversely oblong, bilateral muscle slice inferior to the internal obturator muscle and superior to the great adductor muscle.

Levator Ani Muscle (*X*). A rather small, oblong, oblique, bilateral muscle tissue adjacent to the lateral and inferior parts of the urinary bladder. The inferior end is more medial than the superior end.

Gluteus Medius Muscle (*L*). An elongated (superior-to-inferior), bilateral muscle strip that extends from the superior part of the ilium to the greater trochanter, superficial to the gluteus minimus muscle.

Gluteus Minimus Muscle (*K*). A vertical, bilateral muscle strip medial to the gluteus medius muscle and adjacent to the lateral part of the ilium. It extends to the greater trochanter of the femur.

NEUROVASCULAR UNIT (see Fig. 16–1)

I. Arteries

External Iliac Artery (*H*). An oval, hollow, bilateral vascular structure superior to the external iliac vein and medial to the psoas major muscle.

Internal Iliac Artery (*J*). Seen as a vertical hollow vascular structure.

II. Veins

External Iliac Vein. A vertical, oblique, bilateral venous structure inferior to the external iliac artery and medial to the inferior part of the psoas major muscle. The superior end is more medial than the inferior end.

VISCERAL UNIT (see Fig. 16–1)

Ileum (*F*). A hollow, irregularly-shaped digestive structure located to the right of the median in this coronal view. It is seen medial to the psoas major muscle.

Jejunum (*U*). A hollow, irregularly-shaped digestive structure to the left of the median. The internal surface of this structure shows numerous projections.

Vermiform Appendix (*D*). An oval, hollow slice of the digestive structure medial to the cecum and superior to the psoas major muscle.

Cecum (*E*). An irregularly-shaped proximal portion

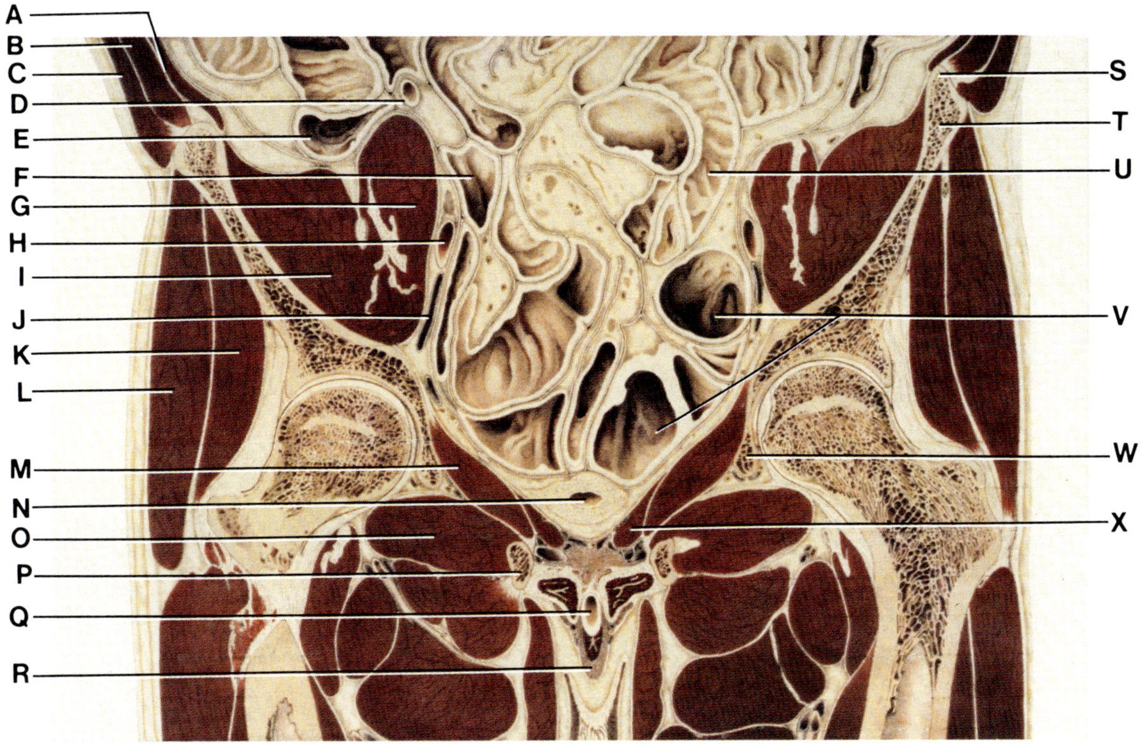

A. Transverse abdominal muscle
B. Internal oblique abdominal (obliquus abdominis internus) muscle
C. External oblique abdominal (obliquus abdominis externus) muscle
D. Vermiform appendix
E. Cecum
F. Ileum
G. Psoas major muscle
H. External iliac artery
I. Iliac muscle
J. Internal iliac artery
K. Gluteus minimus muscle
L. Gluteus medius muscle
M. Internal obturator muscle
N. Urinary bladder
O. External obturator muscle
P. Pubic bone
Q. Penile urethra
R. Bulb of penis
S. Iliac crest
T. Ilium
U. Jejunum
V. Sigmoid colon
W. Ischiopubic ramus
X. Levator ani muscle

Figure 16–1
Approximately midcoronal (frontal) section of a male, highlighting musculoskeletal, neurovascular, and visceral unit structures.

of the colon seen lateral to the vermiform appendix and superior to the groove between the psoas major and iliac muscles.

Sigmoid Colon (*V*). Two rather large, oval portions of this structure are seen inferior and slightly medial to the psoas major muscle on the left side of this coronal plane.

Urinary Bladder (*N*). A thick-walled, hollow excretory organ in the median of the inferior pelvis, seen between the sections of the levator ani.

Penile Urethra (*Q*). An oval, vertical, hollow structure inferior to the urinary bladder and superior to the bulb of the penis.

Origin. Membranous urethra as its continuation.

Termination. Orifice of the penis.

Course. It runs within the corpus spongiosum of the penis.

Bulb of Penis (*R*). An oblong structure of the proximal penis, inferior to the penile urethra and medial to the superior part of the gracilis muscle.

Corpus Cavernosum of Penis. A rather irregularly-shaped slice of bilateral penile tissue lateral to the superior part of the penile urethra. This structure forms the main back and lateral parts of the penis. It terminates as rounded projections covered by the prepuce of the penis.

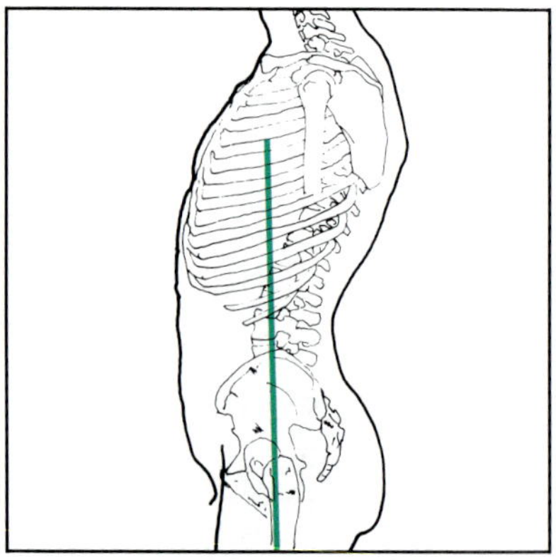

Figure 16–2 shows a magnetic resonance coronal image of a male pelvis. Musculoskeletal unit structures are highlighted.

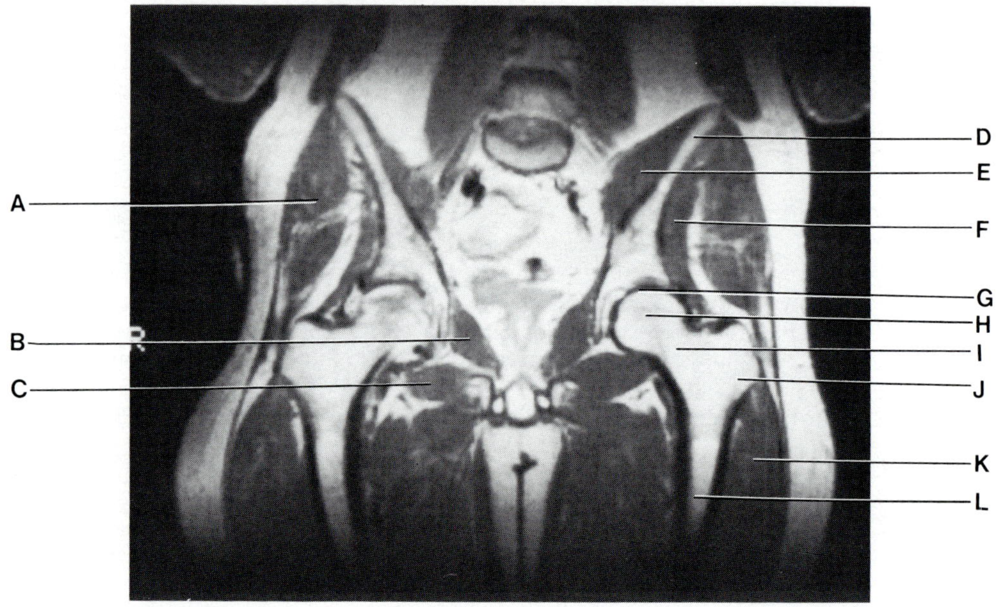

A. Gluteus medius muscle
B. Internal obturator muscle
C. External obturator muscle
D. Ilium
E. Iliac muscle
F. Gluteus minimus muscle
G. Acetabulum
H. Head of femur (left)
I. Left femoral neck
J. Greater trochanter
K. Vastus lateralis muscle
L. Femur (left)

Figure 16–2
Magnetic resonance coronal (frontal) image of a male, highlighting some pelvic structures. Note the differences between the right and left femoral heads (for more information see the Appendix to Part VI).

A more posterior MR coronal image is shown in Figure 16–3. This image emphasizes musculoskeletal and visceral unit structures.

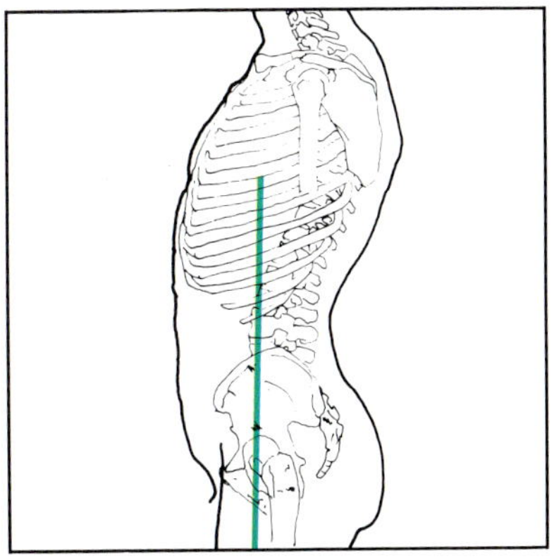

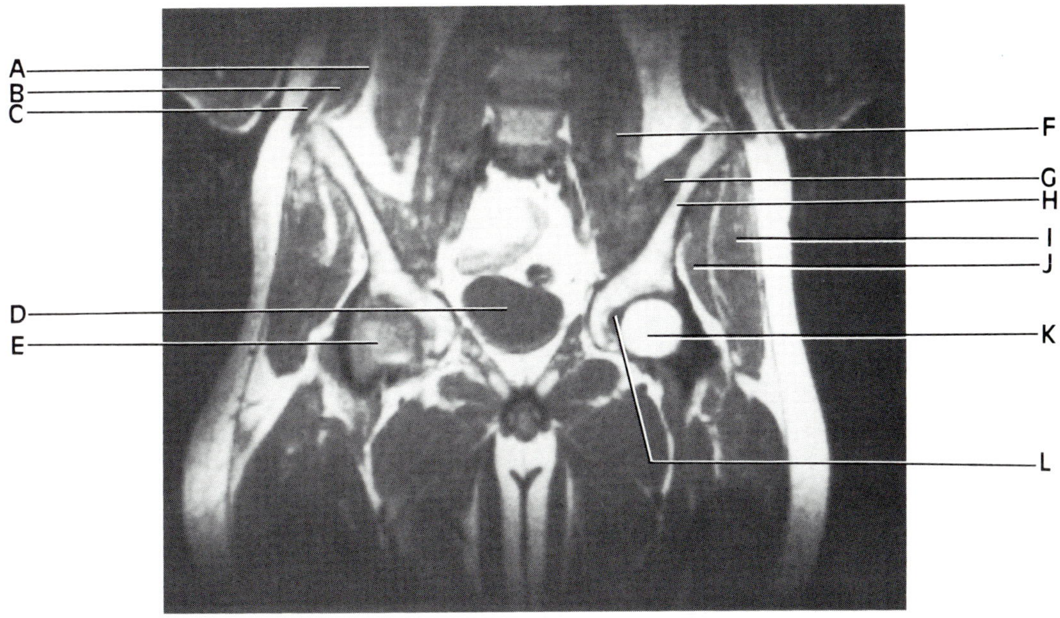

A. Transverse abdominal muscle
B. Internal oblique abdominal muscle
C. External oblique abdominal muscle
D. Urinary bladder
E. Head of femur (right)
F. Psoas major muscle
G. Iliac muscle
H. Ilium
I. Gluteus medius muscle
J. Gluteus minimus muscle
K. Head of femur (left)
L. Acetabulum

Figure 16–3
Magnetic resonance coronal (frontal) image of a male, highlighting some of the pelvic structures. Note the differences between the right and left femoral heads (for more information see the Appendix to Part VI).

CHAPTER 17

Sagittal Cadaver Sections and Images of the Pelvis

There are four different cadaver sections. One shows the male pelvic structures, two show the female pelvic structures, and the fourth shows the female neurovascular structures. However, the neurovascular unit, other than the uterine and vaginal vessels, applies to the male as well.

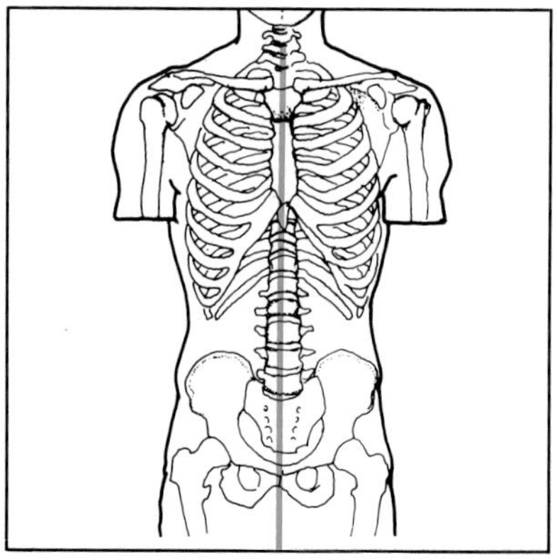

MODULE 1

This is a sagittal section of the male abdomen and pelvis. The emphasis in this module is on the pelvic structures. Some of the structures in the pelvis of this sagittal section are the prostate gland, prostatic urethra, bulb of the penis, urinary bladder, rectum, and pubic bone.

MUSCULOSKELETAL UNIT (Fig. 17–1)

I. Skeletal Portion

Pubic Bone (*E*). An oval (superior-to-inferior), oblique bone slice anterior to the urinary bladder and superior to the bulb of the penis. The inferior part of this bone section is posterior to the superior part.

Sacral Vertebra 1. A rather irregularly vertical, rectangular, oblique bone slice inferior to the fifth lumbar vertebra and anterior to the cauda equina. The inferior end is narrower and more posterior than the superior end.

Sacral Promontory. A projection of superior anterior sacral vertebra 1, superior to the proximal part of the rectum.

Coccyx. A curved, segmented end of the vertebral column, inferior to the fifth sacral vertebra and superior to the external sphincter ani muscle.

II. Muscular Portion

External Sphincter Ani Muscle (*U, W*). Two irregularly shaped aggregates of muscle tissue. One is posterior to the anal canal in this sagittal body section, and one is anterior to the anal canal. The posterior one is larger than the anterior one.

Internal Sphincter Ani Muscle (*R, T*). Two curved, vertical strips of muscle tissue. One is posterior and located anterior to the posterior portion of the external sphincter ani muscle. The internal sphincter ani muscle is posterior to the anterior portion of the external sphincter ani muscle. The concave surface of the posterior strip of the internal sphincter ani muscle is adjacent to the posterior portion of the external sphincter ani muscle. A concave surface of the anterior strip of the internal sphincter ani muscle is adjacent to the anal canal.

Deep Transverse Perineal (Transversus Perinei Profundus) Muscle (*Q*). A rather small muscle aggregate anterior to the internal sphincter ani muscle and posterior to the sphincter muscle of the urethra. This thin sheet of muscle is found in the deep perineal space, along with the sphincter muscle of the urethra.

Sphincter Muscle of the Urethra. A rather small muscle structure adjacent to the distal end of the prostatic urethra and anterior to the deep transverse perineal muscle.

A. Seminal vesicle
B. Prostatic urethra
C. Urinary bladder
D. Membranous urethra
E. Pubic bone
F. Suspensory ligament of penis
G. Dorsal vein of penis
H. Cowper's gland (bulbourethral gland)
I. Penile urethra
J. Corpus cavernosum
K. Corpus spongiosum
L. Glans penis
M. Prepuce (foreskin)
N. Proximal rectum
O. Pouch of Douglas
P. Prostate gland
Q. Deep transverse perineal (transversus perinei profundus) muscle
R. Internal sphincter ani muscle
S. Distal rectum
T. Internal sphincter ani muscle
U. External sphincter ani muscle
V. Bulb of penis
W. External sphincter ani muscle
X. Testis
Y. External urethral orifice (orifice of penis)

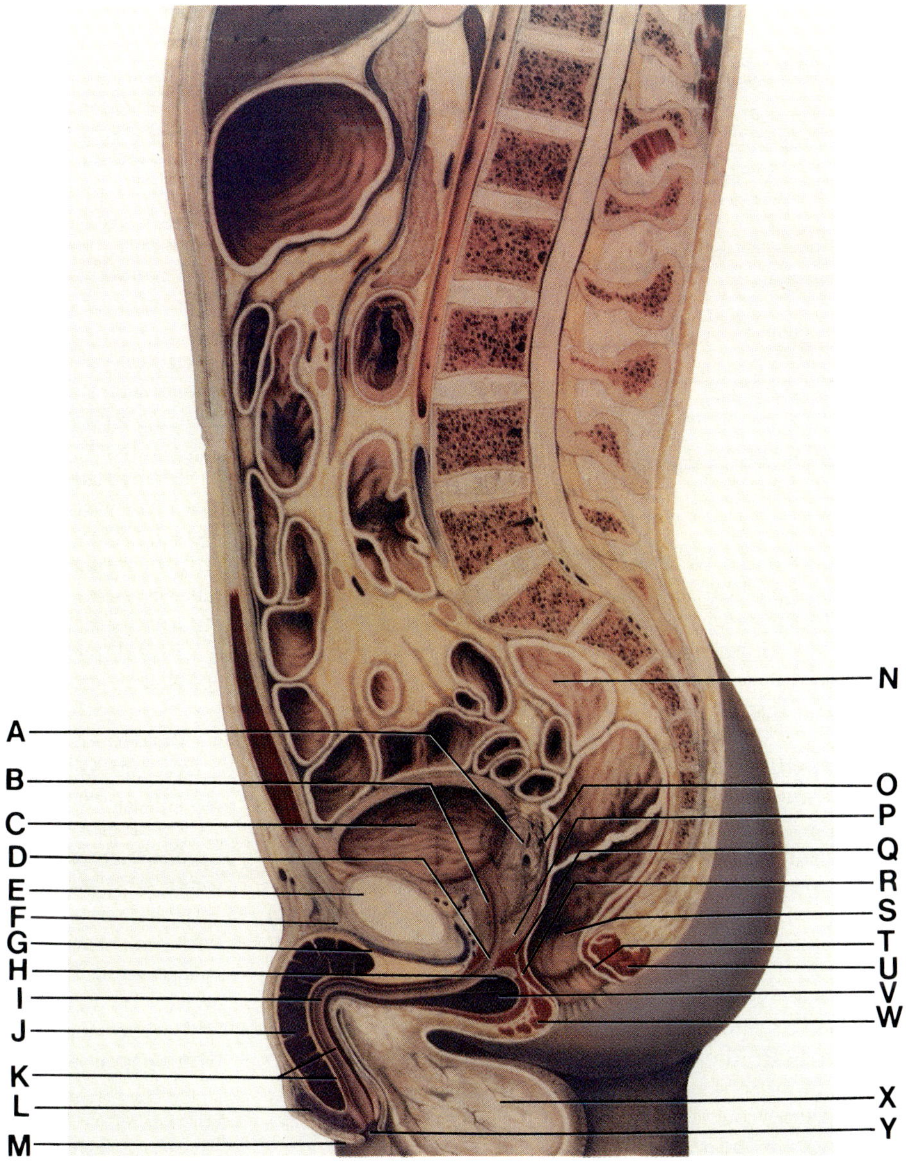

Figure 17-1
Midsagittal male section, highlighting musculoskeletal, neurovascular, visceral, and enclosing unit structures.
A. Seminal vesicle

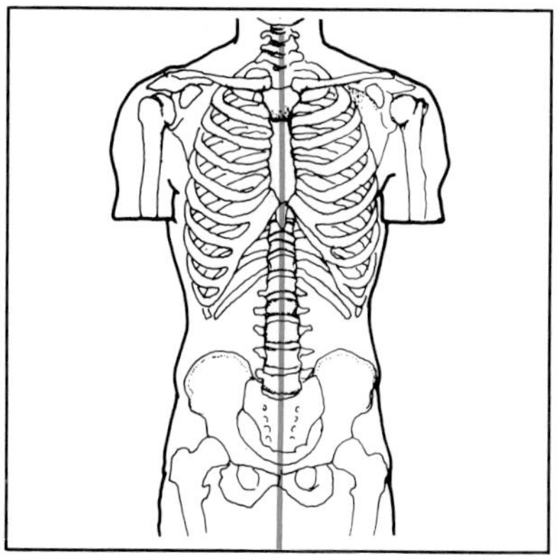

NEUROVASCULAR UNIT (see Fig. 17–1)

I. Veins

Dorsal Vein of Penis (*G*). An elongated, curved venous structure seen starting at the posterior part of the inferior pubis and continuing beneath the pubis, along the dorsal part of the penile body within the deep penile fascia.

Origin. It receives blood from the corpus cavernosa and glans penis.

Termination. The prostatic venous plexus, which then drains into the internal pudendal vein.

Course. The deep dorsal vein courses within the deep penile fascia between the right and left dorsal arteries of the penis to reach the prostatic venous plexus.

VISCERAL UNIT (see Fig. 17–1)

Proximal Rectum (*N*). A vertical, hollow, rather large, oblique slice of digestive structure anterior to the superior parts of the sacrum. The inferior part is posterior to the superior part.

Distal Rectum (*S*). A vertical, hollow, rather large, oblique slice of digestive structure anterior to the coccyx and posterior to the prostate gland. It is superior to the anal canal. The inferior part is anterior to the superior part.

Anal Canal. A hollow, vertical, rather short, oblique, terminal end of the digestive tract between the sphincter ani muscles. The inferior end is slightly posterior to the superior end.

Seminal Vesicle (*A*). A vertical glandular tissue posterior to the urinary bladder and anterior to the rectum.

Ejaculatory Duct. A vertical, hollow, oblique tube that runs through the posterior part of the prostate gland to reach the prostatic urethra. The inferior end is more anterior than the posterior end. Fusion of the duct from the seminal vesicle and the vas deferens forms this structure.

Urinary Bladder (*C*). A rather delta-shaped, hollow urinary structure, with the apex adjacent to and surrounded by the prostate gland. The wider portion is superior to the apex and superior to the pubic bone. This structure is posterior to the retropubic space and anterior to the rectum.

Prostate Gland (*P*). An irregularly oval slice of glandular tissue with two canals passing through it. The vertical canal is the prostatic urethra, and the vertical, oblique canal is the ejaculatory duct. This organ surrounds the neck of the bladder and the prostatic urethra.

Prostatic Urethra (*B*). A vertical canal that starts at the inferior part of the urinary bladder and terminates at the inferior surface of the prostate gland. This structure divides the prostate gland into anterior and posterior portions. The posterior portion is slightly larger than the anterior portion and contains the ejaculatory duct.

Membranous Urethra (*D*). A short slice of a tubular structure between the muscles of the deep perineal space. The specific structure that the membranous urethra passes through is known as the urogenital diaphragm. The structures that form the urogenital diaphragm are listed from superficial to deep: (1) the perineal membrane, (2) the muscular layer of the deep perineal space, and (3) the deep fascia on the upper surface of the muscles within the deep perineal space.

Penile Urethra (*I*). An elongated slice of a tubular structure seen within the corpus spongiosum of the penis. It extends from the distal end of the membranous urethra to the orifice of the penis.

Bulb of the Penis (*V*). A transverse, bulb-shaped slice of tissue anterior to the sphincter ani muscles and inferior to the sphincter muscle of the urethra and the deep transverse perineal muscle. It is immediately inferior to the bulbourethral (Cowper's) gland. The wider part of the bulb-shaped structure is posterior to the narrow part. The anterior part is continuous with the corpus spongiosum of the penile body.

Corpus Spongiosum (*K*). An elongated slice of tissue that surrounds a tubelike structure (spongy urethra) within the body of the penis. This tissue is ventral to the two bodies within the penis known as the corpora cavernosa. The corpus spongiosum is seen on the distal end of the penis as an enlarged portion known as the glans penis.

Corpus Cavernosum (*J*). An elongated slice of the erectile tissue of the penile body dorsal to the corpus spongiosum. It extends from the proximal penis to the glans penis. There are two structures like this within the body of the penis, but only one is seen in this sagittal view.

Glans Penis (*L*). An enlarged terminal portion of the corpus spongiosum. The orifice of the penis passes through it.

Prepuce (Foreskin) (*M*). A slice of skin covering both the anterior and posterior parts of the glans penis as seen in this sagittal plane. This is the skin commonly removed surgically in a procedure known as circumcision.

External Urethral Orifice (Orifice of Penis) (*Y*). The terminal and distal opening of the spongy urethra that exits from the glans penis.

Testis (*X*). An oval reproductive structure seen within the scrotum. There are two testes, although only one is seen in this sagittal view.

ENCLOSING UNIT (see Fig. 17–1)

Scrotum. A slice of skin, with an oval appearance, that encloses the testes.

Dartos Tunic of Scrotum. A membranous layer of the superficial fascia that contains both smooth muscles and fascia within the inner lining of the skin of the scrotum.

Pouch of Douglas (*O*). A vertical space between the rectum and the seminal vesicles, prostate gland, and urinary bladder.

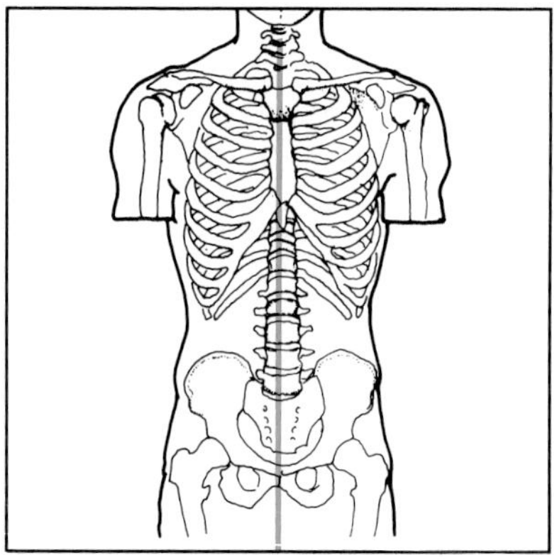

MODULE 2

This module includes a midsagittal section of a female pelvis and a sagittal section, showing branches of the internal iliac artery as well as the ureter.

MUSCULOSKELETAL UNIT

I. Skeletal Portion

Sacral Promontory. An anterior, rounded process of the first sacral vertebra near its superior part.

First Sacral Vertebra. A parallelogram-shaped bone immediately inferior to the fifth lumbar vertebra. The inferior end is narrower (anterior-to-posterior) than the superior end.

Coccyx. A curved bony segment of four bones, with the inferior end curled slightly anterior.

Sacrococcygeal Joint. An amphiarthrodial joint (secondary cartilaginous) between the fifth segment of the sacrum and the first coccygeal bone.

Lumbosacral Joint. An amphiarthrodial joint between the fifth lumbar vertebra and the first sacral bony segment.

Pubic Bone (*E*). A sickle-shaped bone slice located on the anterior pelvis. This view of the pubic bone shows some interpubic disk.

NEUROVASCULAR UNIT (Fig. 17–2)

I. Arteries

Right Internal Iliac Artery (*C*). A rather large, vertical, oblique vascular structure. The inferior end is slightly posterior to its superior end in this section.

Origin. Right common iliac artery.

Distribution. The internal iliac artery distributes blood to an anterior and a posterior division.

The anterior division has the following branches: (1) inferior gluteal, which distributes blood to the gluteus maximus, (2) obturator (*B*), which distributes blood to the medial thigh, (3) internal pudendal, which distributes blood to the perineum and external genitalia, (4) umbilical (superior vesical), which supplies the ductus deferens and upper surface of the bladder, (5) inferior vesical, which supplies the prostate gland and seminal vesical, as well as the superior surface of the levator ani and the inferior surface of the bladder, (6) middle rectal, which supplies in men the prostate and seminal vesical, in women the vagina, and the rectum in both men and women, (7) uterine, which supplies the body and fundus of the uterus, cervix, and vagina (*note:* the uterine artery sometimes arises directly from the middle rectal artery), and (8) vaginal artery (or arteries), which supplies the posterior surface of the vagina, base of the bladder, rectum, and bulb of the vestibule.

The posterior division has three branches: (1) iliolumbar, which supplies the iliac muscle, ilium, psoas major muscle, quadratus lumborum muscle, and some vasculature to the cauda equina, (2) lateral sacral, which supplies blood to the sacrum, and (3) superior gluteal, which supplies blood to the gluteus minimus and medius and some twigs to the gluteus maximus.

Course. The internal iliac artery starts at the level of the lumbosacral joint immediately anterior to the sacroiliac joint and courses downward and obliquely for about 4 cm to reach the upper of margin of the greater sciatic foramen (greater sciatic notch), where it bifurcates to form anterior and posterior divisions.

External Iliac Artery. A vertical, elongated, oblique vascular structure. The inferior end is more anterior than the superior end.

Ovarian vessels (*A*). Seen as hollow vascular structures in this sagittal view.

Uterine Artery and Vein (*E*). Seen as hollow vascular structures adjacent to the uterus in this section.

II. Veins

Internal Iliac Vein (*C*). A vertical, oblique venous structure immediately posterior to the internal iliac artery. The inferior end is slightly posterior to the superior end.

Uterine Vein (*E*). A vertical, elongated, oblique venous structure posterior to the uterine artery. In the course of this vessel a slight anterior bend is formed, but the inferior end is slightly anterior to the superior end.

Origin. Tissue of uterus.

Termination. Internal iliac vein.

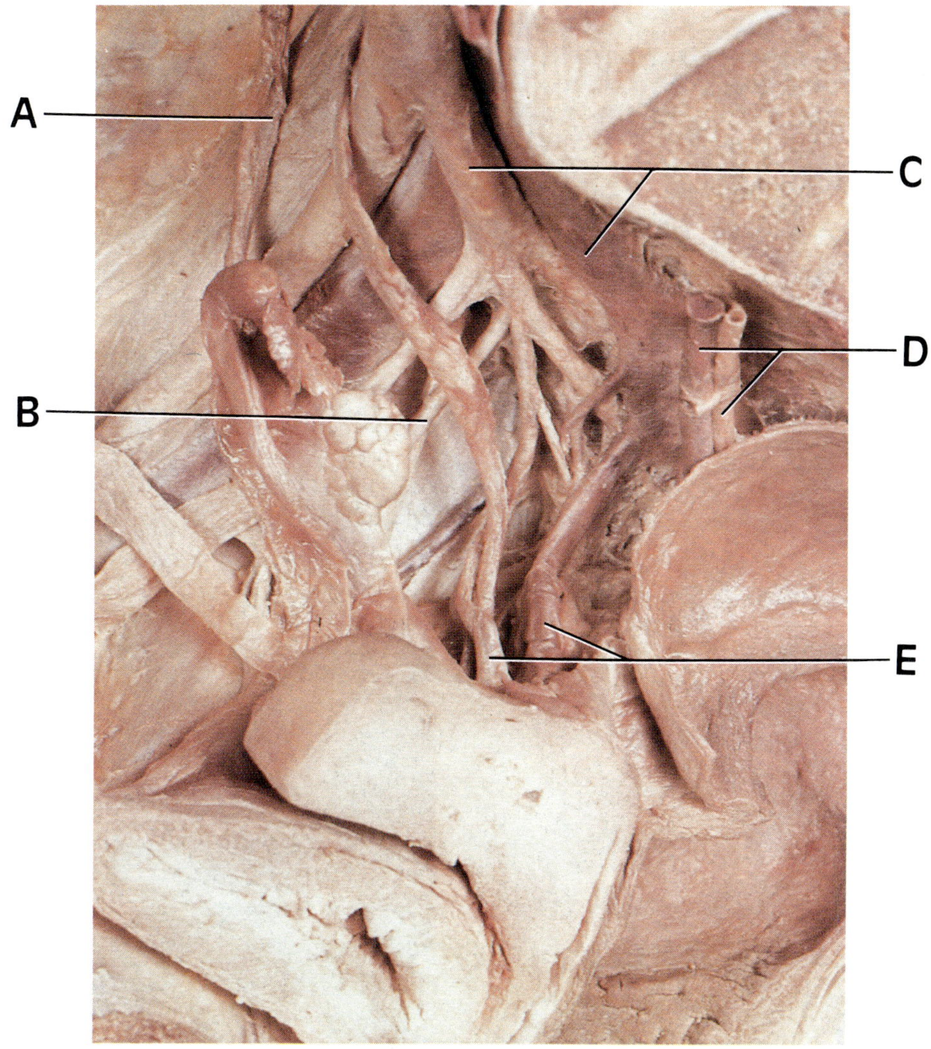

A. Ovarian vessels
B. Obturator artery
C. Internal iliac artery and vein
D. Superior rectal vessels
E. Uterine artery and vein

Figure 17–2
Sagittal view of a female, highlighting neurovascular unit structures.

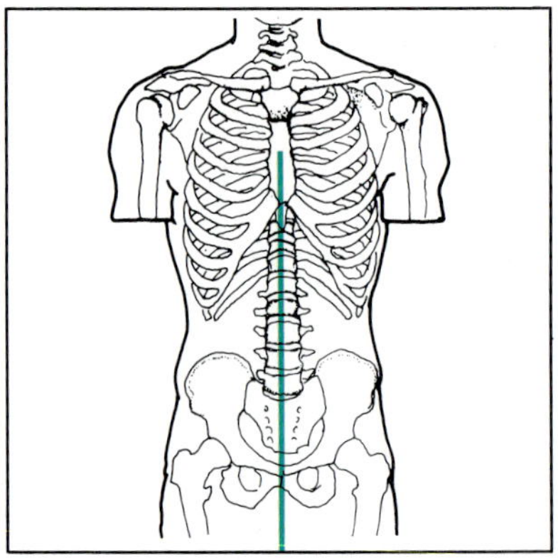

VISCERAL UNIT (Figs. 17–3 and 17–4)

Uterus (Fig. 17–3*B*, Fig. 17–4*C*). This is divided into three main parts: (1) the fundus is the upper, flattened, anteriorly tilted part of the uterus; (2) the body is a rather large vertical and oblique portion of the uterus between the fundus and the cervix—the inferior end is more posterior than the superior end; and (3) the cervix is the part of the uterus adjacent to the deep end of the vagina and inferior to the body of the uterus. The uterus is between the urinary bladder and rectum in this section.

Lumen of Urinary Bladder (Fig. 17–4*E*). A hollow, elongated structure (superior to inferior), obliquely positioned. The superior end is anterior to the inferior end, as seen in this section.

Vagina (Fig. 17–3*H*). A vertically elongated, slitlike feature, as seen in this section, between the rectum and the distal end of the bladder and associated structures.

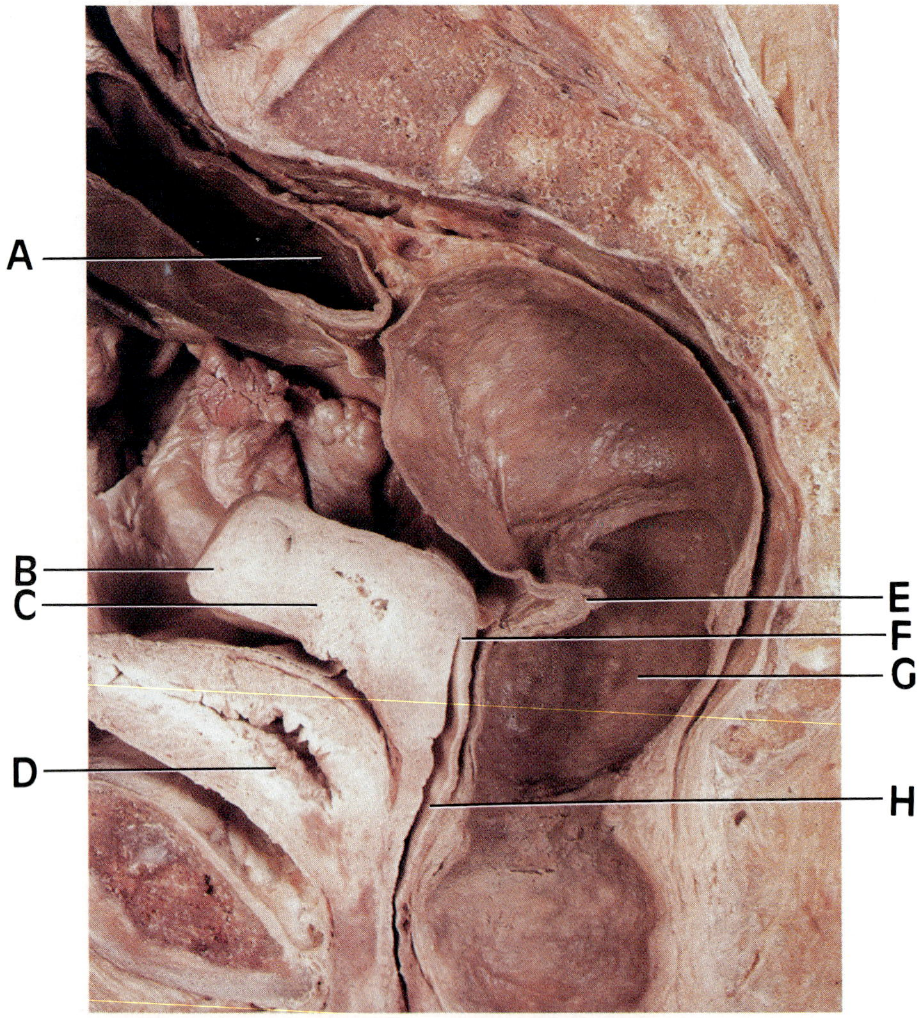

A. Sigmoid colon
B. Fundus of uterus
C. Body of uterus
D. Urinary bladder
E. Horizontal fold
F. Posterior fornix
G. Rectum
H. Vagina

Figure 17–3
Sagittal view of a female, highlighting visceral unit structures.

Posterior Fornix (Fig. 17–3*F*). A slitlike feature at the junction of the posterior vaginal wall and the posterior part of the cervix. This area is in a slightly oblique position, with its superior end slightly posterior to the inferior part.

Ovary. A rounded organ on the posterior part of the broad ligament and anterior to the superior part of the rectum. This is the female gonad. Although only one is seen in this section, there are two.

Uterine Tube. A tubular structure seen in this section superior to the fundus of the uterus and broad ligament.

Urinary Bladder (Fig. 17–3*D*, Fig. 17–4*E*). A rather thick-walled, hollow organ in a vertical and oblique position in the anterior pelvis. The superior part is anterior to the inferiorpart. The inner wall of this structure shows anirregular surface.

Urethra. A vertical slit, as seen in this section, that courses between the vagina and the pubic bone.

Sigmoid Colon (Fig. 17–3*A*). A hollow structure with a rather large lumen, as seen in this section, anterior to both the lower lumbar and upper sacral bony segments. This structure is seen superior to the rectum.

Rectum (Fig. 17–3*G*). A vertical, rather large, hollow structure posterior to the uterus and vagina and anterior to the lower sacral and coccygeal segments. As seen in this section, the rectum extends below the coccyx.

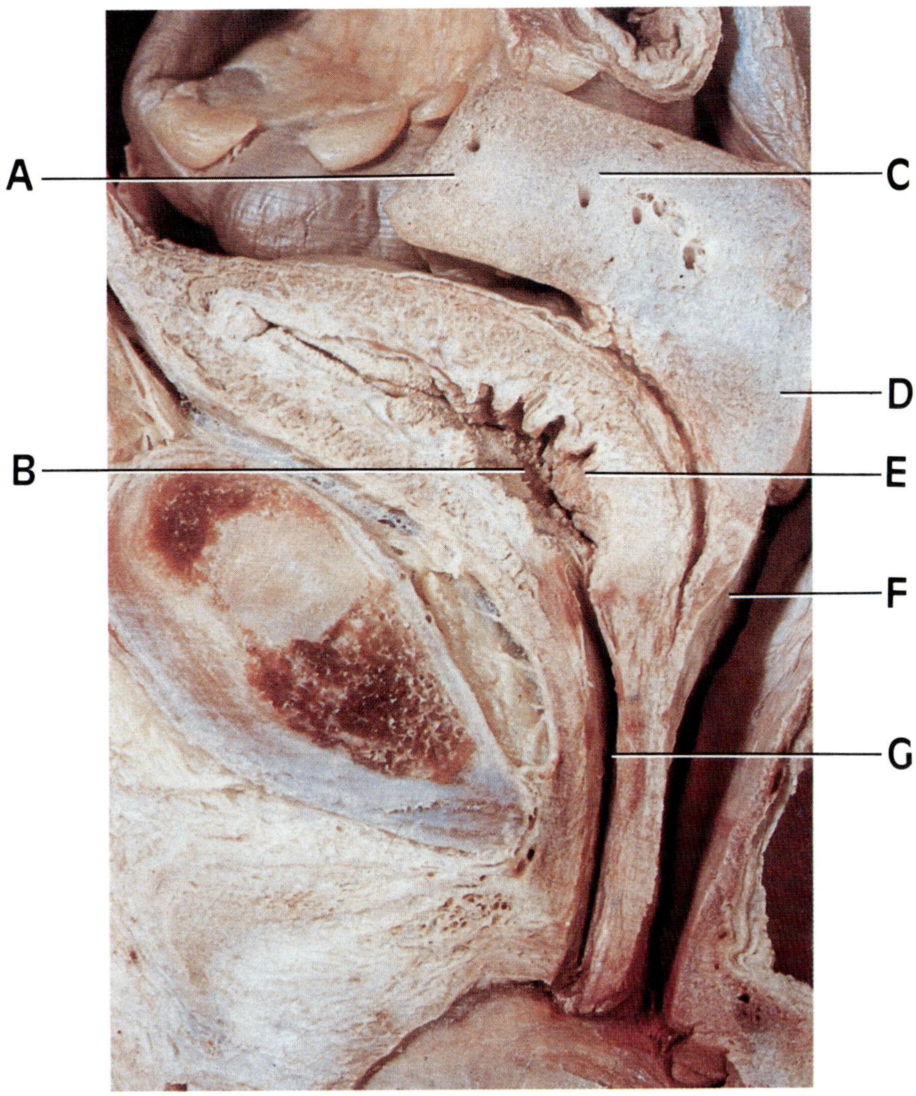

A. Fundus of uterus
B. Lumen of urinary bladder
C. Body of uterus
D. Cervix of uterus
E. Urinary bladder
F. Anterior vaginal wall
G. Ureter

Figure 17–4
Sagittal view of a female, highlighting visceral unit structures.

ENCLOSING UNIT (Fig. 17–5)

Peritoneum (*A*). A rather thin sheet of tissue that is seen draped anterior to the rectum, posterior uterus, superior fundus, anterior uterus, and posterior urinary bladder.

Vesicouterine (Uterovesical) Pouch (*B*). A deepened area between the anterior upper uterus and the posterior part of the bladder.

Rectouterine Pouch (*D*). A rather large, deepened area between the anterior upper portion of the rectum and the posterior part of the uterus.

Retropubic Space (*C*). A narrow, vertical, oblique empty space between the urinary bladder and the pubic bone.

Urachus. A superior cord attached to the superior part of the urinary bladder. This structure continues throughout life as a cord known as the median umbilical ligament.

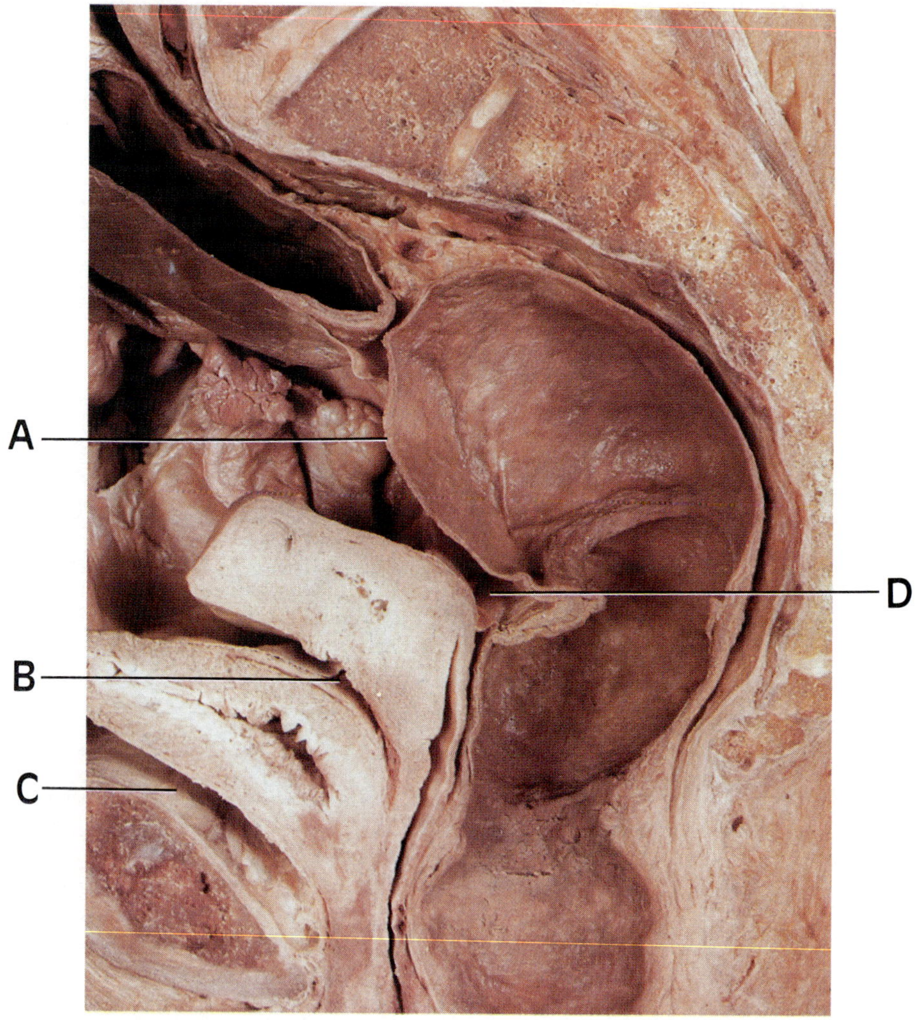

A. Peritoneum
B. Vesicouterine (uterovesical) pouch
C. Retropubic space
D. Rectouterine pouch

Figure 17–5
Sagittal view of a female pelvis, highlighting enclosing unit structures.

Appendix to Part VI: Clinical Correlation for the Pelvic Region

CASE ONE (Fig. AVI–1)

History. The patient was in a motor vehicle accident and suffered a dislocated right hip. Aseptic necrosis was to be ruled out. The patient had right hip pain.

Findings. T1-weighted images clearly show marked degrees in signal intensity of the right femoral head, compared with the left femoral head, indicating replacement of the fat. These findings, though highly typical for aseptic necrosis, are not exclusive of other types of joint disease. Further review of the plain and nuclear scan reveals abnormality in the right femoral head.

Impression. Findings consistent with aseptic necrosis of the right femoral head. Further review reveals positive uptake in this area in the nuclear scan and sudden changes in the femoral head in the plain pelvis films.

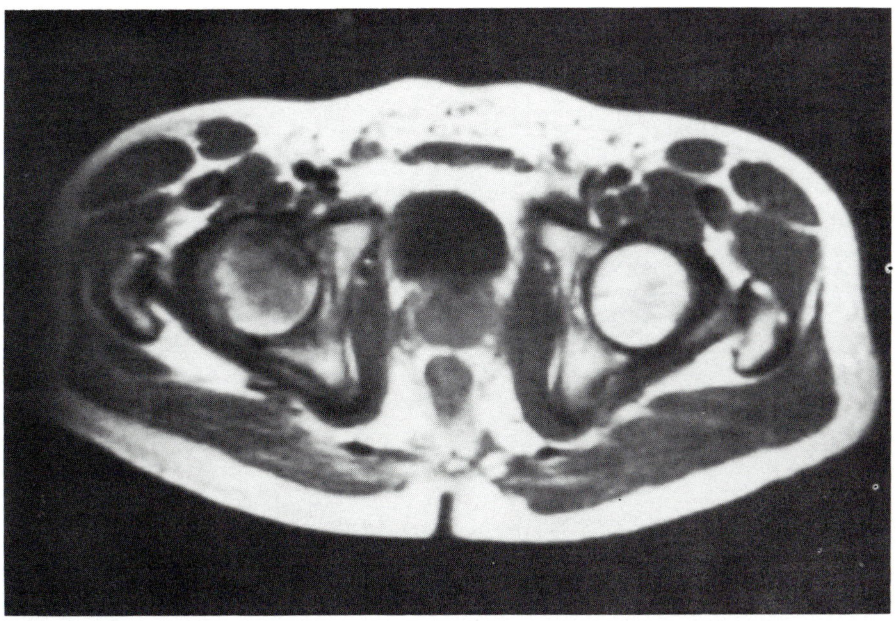

Figure AVI-1
Magnetic resonance transverse image of the pelvis, showing abnormality in the right femoral head (also see Figures 16–2 and 16–3).

PART VII

THE UPPER LIMB

INTRODUCTION

There are 32 bones in each upper limb. The bone segments from superior to inferior include the pectoral girdle (scapula and clavicle), brachium (humerus), antebrachium (radius and ulna), and manus (carpal bones, metacarpal bones, and phalanges of the digits). The joints of the upper limb consist of (1) the shoulder joint, which is a synovial ball and socket joint, (2) the humeroulnar joint (elbow joint), a synovial hinge joint, (3) the radioulnar joint, a synovial pivot joint, (4) the wrist joint, a synovial angular joint, and (5) the 25 joints of the hand, which all are members of the synovial joint class. Each hand is able to make 58 movements.

There are numerous muscles of the upper limb. For example, 10 muscles function across the shoulder joint. The shoulder joint is capable of making seven movements: (1) flexion, (2) extension, (3) abduction, (4) adduction, (5) medial rotation, (6) lateral rotation, and (7) circumduction. The muscles across the elbow joint are capable of flexion and extension at the humeroulnar joint (hinge joint). Muscles that cross the elbow joint and some deep forearm muscles are capable of medial and lateral rotations of the forearm (radioulnar hinge joint). Medial and lateral rotations of the forearm are known as pronation and supination, respectively. There are superficial and deep ventral forearm muscles, as well as superficial and deep dorsal forearm muscles. These muscles function across the wrist joint: the metacarpal, phalangeal, and interphalangeal joints.

The arterial supply to the upper limb is summarized as follows:

In the right upper limb: Left ventricle → ascending aorta → arch of aorta → brachiocephalic artery → right subclavian artery → right axillary artery → right brachial artery → radial artery → deep palmar arch → ulnar artery (*note:* the ulnar artery also arises from the brachial artery) → superficial palmar arch.

In the left upper limb: Left ventricle → ascending aorta → arch of aorta → left subclavian artery → left axillary artery → left brachial artery → radial artery → deep palmar arch → ulnar artery (*note:* the ulnar artery also arises from the brachial artery) → superficial palmar arch.

As will be seen in the tables of muscles, branches of the brachial plexus supply nerves for the muscles and other structures of the upper limbs.

Table VII–1 Muscles of the Arm

Muscle	Origin	Insertion	Innervation	Vasculature	Action
Brachioradial	Lateral supracondyloid crest of humerus	Near inferior end of radius	Radial	Radial recurrent	Flexes forearm
Biceps m. of arm (biceps brachii)	Long head: superior edge of glenoid cavity. Short head: coracoid process	Radial tuberosity	Musculocutaneous	Brachial (muscular branch)	Forearm flexion and supination, flexion of arm when forearm is fixed
Brachial	Anterior surface of inferior half of humerus	Ulnar tuberosity and coronoid process of ulna	Musculocutaneous, radial	Radial recurrent, brachial	Forearm flexion
Triceps m. of arm (triceps brachii)	Long head: scapula (axillary border). Lateral head: side and back areas of body of humerus. Medial head: back area of body of humerus inferior to lateral head	Olecranon of ulna	Radial	Deep brachial	Forearm extension
Anconeus	Posterior surface of lateral epicondyle	Back area of upper ulna and olecranon process	Radial	Deep brachial	Helps triceps m. of arm

Table VII–2 Deep Ventral Forearm Muscles

Muscle	Origin	Insertion	Innervation	Vasculature	Action
Deep flexor m. of fingers (flexor digitorum profundus)	Front and inner surface of ulnar body and inner surface of coronoid process	Four tendons at bases of distal phalanges	Ulnar, median (volar interosseous)	Ulnar artery	Flexion of medial 4 digits (terminal phalanges), helps in wrist flexion
Supinator	Lateral epicondyle of humerus (superficial head), crest on ulna, and radial ligament of elbow (deep head)	Body of radius	Radial (deep branch)	Radial recurrent, ulnar (posterior interosseous)	Forearm supination
Pronator teres	Medial epicondyle of humerus and coronoid process of ulna	Body of radius	Median	Anterior ulnar, recurrent	Pronates forearm, helps in forearm flexion
Pronator quadratus	Front surface of ulna on its lower ¼ portion	Ventral surface of radius	Volar interosseous of median	Volar interosseous of ulnar artery	Pronates forearm
Long flexor m. of thumb (flexor pollicis longus)	Anterior surface of body of radius	Thumb (base of distal phalanx)	Volar interosseous of median	Volar interosseous of ulnar artery	Flexes thumb

Table VII-3 Superficial Ventral Muscles of Wrist Joint

Muscle	Origin	Insertion	Innervation	Vasculature	Action
Radial flex or m. of wrist (flexor carpi radialis)	Fascia of antebrachium and medial epicondyle (common tendon)	Base of metacarpal bones 2 and 3	Median	Radial artery	Flexes wrist, helps in flexing forearm
Long palmar m. (palmaris longus)	Medial condyle (common tendon)	Palmar aponeurosis of hand and transverse carpal ligament	Median	Posterior ulnar recurrent artery	Flexes wrist
Ulnar flexor m. of wrist (flexor carpi ulnaris)	Humeral origin: medial epicondyle (common tendon). Ulnar origin: back surface of ulna and olecranon process	Wrist bone: pisiform and hamate. Metacarpal bone: 5th	Ulnar	Posterior ulnar recurrent artery	Flexes wrist, helps in flexing forearm
Superficial flexor m. of fingers (flexor digitorum superficialis)	Humeral: medial epicondyle (common tendon). Ulnar: coronoid process. Radial: oblique line	Middle phalanx of digit 4	Median	Ulnar and radial arteries	Flexes middle and neighboring phalanges of medial 4 digits; helps in flexing forearm and wrist

Table VII-4 Superficial Dorsal Muscles of Wrist Joint

Muscle	Origin	Insertion	Innervation	Vasculature	Action
Long radial extensor m. of wrist (extensor carpi radialis longus)	Lateral part of intermuscular septum and inferior 3rd of lateral supracondylar ridge of humerus	Back surface of base of 2nd metacarpal bone	Radial	Radial and radial recurrent arteries	Extends wrist, abducts hand
Short radial extensor m. of wrist (extensor carpi radialis brevis)	Elbow radial collateral ligament and common extensor tendon from lateral epicondyle	Back surface of base of 3rd metacarpal bone	Radial	Radial and radial recurrent arteries	Extends and abducts wrist joint
Extensor m. of fingers (extensor digitorum)	Common extensor tendon from lateral epicondyle of humerus and part of intermuscular septum	Side and back surfaces of medial 4 hand digits	Radial	Posterior interosseous of ulnar artery	Extends 4 digits (medial), helps in extending hand
Extensor m. of little finger (extensor digiti minimi)	Common extensor tendon from lateral epicondyle of humerus and part of intermuscular septum	Proximal phalanx (back portion) of 5th digit	Radial (posterior interosseous)	Posterior interosseous of ulnar artery	Extends 5th digit
Ulnar extensor m. of wrist (extensor carpi ulnaris)	Common extensor tendon from lateral epicondyle of humerus and posterior edge of ulna	5th metacarpal bone (medial portion of base)	Radial (posterior interosseous)	Posterior interosseous of ulnar artery	Extends wrist, adducts hand

Table VII-5 Deep Dorsal Muscles of Wrist Joint

Muscle	Origin	Insertion	Innervation	Vasculature	Action
Long abductor m. of thumb (abductor pollicis longus)	Back portion of ulna, interosseous membrane of forearm, middle third of back portion of radius	1st metacarpal bone on lateral side	Radial (posterior interosseous)	Posterior interosseous of ulnar artery	Abducts wrist and thumb
Long extensor m. of thumb (extensor pollicis longus)	Interosseous membrane of forearm and posterior portion of ulna (middle third)	Thumb's distal phalanx	Radial (posterior interosseous)	Posterior interosseous of ulnar artery	Extends end phalanx of thumb
Short extensor m. of thumb (extensor pollicis brevis)	Back portion of radius and interosseous of forearm	Thumb's proximal phalanx	Radial (posterior interosseous)	Posterior interosseous of ulnar artery	Extends proximal phalanx of thumb
Extensor m. of index finger (extensor indicis)	Back part of ulna and interosseous membrane of forearm	Back of proximal phalanx of 2nd digit	Radial (posterior interosseous)	Posterior interosseous of ulnar artery	Extends phalanx of index finger (proximal phalanx)

CHAPTER 18

Transverse (Axial) Cadaver Sections and Images of the Upper Limb

The three transverse cadaver sections shown here include two plastinated specimens (2.5 to 4 mm thick). One transverse section shows structures of the arm; one shows the forearm structures; and another shows a transverse section of the hand in the palmar region. The units of the upper limb are combined, but of course there is no visceral unit to be separated and emphasized.

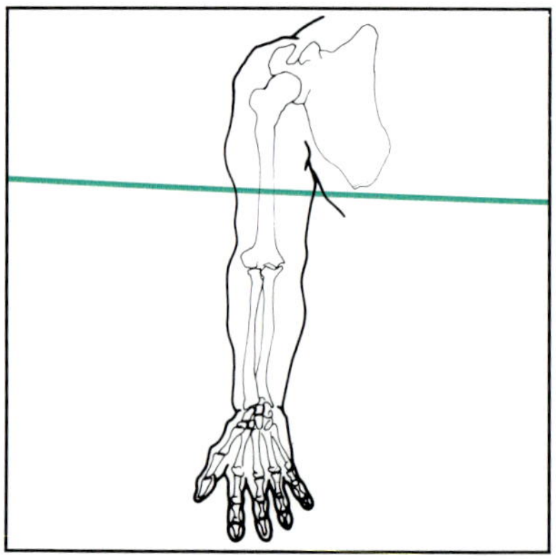

MODULE 1

This is a sheet plastinated (2.5 mm thick) transverse (axial) section of the midshaft of the arm, showing the basilar vein, brachial artery, brachialis, biceps brachii, and triceps brachii.

MUSCULOSKELETAL UNIT (Fig. 18–1)

I. Skeletal Portion

Humerus (*K*). An oval slice of bone tissue, showing a lateral projected surface. In the center of this bone section is a medullary (marrow) cavity. This bone is surrounded by the brachialis muscle anteriorly and the medial head of the triceps brachii posteriorly.

Medullary (Marrow) Cavity) (*J*). An oval, hollow space of the humerus with blood vessels noted within.

II. Muscular Portion

Biceps Brachii Muscle (*G*). A rather thick, curved, elongated (medial-to-lateral) muscle aggregate anterior to the brachialis. The concave surface is adjacent to the anterior surface of the brachial muscle.

Brachial Muscle (*H*). A rather wide, rectangular-shaped muscle slice anterior to the humerus and posterior to the biceps brachii. The widest part of this muscle is on its lateral surface.

Lateral Head of the Triceps Brachii Muscle (*L*). A moderately thick, oblong (medial-to-lateral) muscle aggregate posterior to the humerus and lateral part of the brachial muscle. It is lateral to the medial and long heads of the triceps brachii muscle.

Long Head of the Triceps Brachii Muscle (*F*). An oblong (anterior-to-posterior) muscle slice medial to the lateral head of the triceps and posterior to the medial head of the triceps brachii muscle. Its anterior part is medial to the medial head of the triceps brachii muscle, as seen in this plastinated transverse section of the midshaft of the arm.

Medial Head of the Triceps Brachii Muscle (*E*). An oval muscle section medial to the humerus and lateral to the anterior part of the long head of the triceps brachii muscle.

NEUROVASCULAR UNIT (see Fig. 18–1)

I. Arteries

Brachial Artery (*C*). An oval, hollow vascular structure medial to the medial part of the brachial muscle and anterior to the long head of the triceps brachii muscle.

Origin. Continuation of the axillary artery as it crosses the tendon of the teres major muscle.

Distribution. This vessel disperses blood to seven branches: (1) nutrient, (2) muscular, (3) profunda brachii, (4) inferior ulnar collateral, (5) superior ulnar lateral, (6) radial, and (7) ulnar.

Course. This artery starts at the inferior edge of the tendon of the teres major muscle and continues toward the elbow adjacent to the medial surface of the humerus. It winds toward the anterior arm to reach the midpoint between the median and lateral epicondyles of the humerus.

II. Veins

Brachial Vein (*A*). A partially collapsed venous structure anterior to the median nerve and medial to the anterior part of the medial surface of the brachial muscle.

Origin. Tributaries of the arm (venae comitantes).

Termination. This vein joins the axillary vein at the inferior edge of the tendon of the teres major muscle.

Course. On both the lateral and medial sides of the brachial artery, tributaries of the brachial vein are found. The medial tributary commonly unites with the basilic vein to enter the axillary vein adjacent to the inferior edge of the subscapularis muscle.

Basilic Vein (*D*). An oval, hollow, superficial venous structure adjacent to the medial posterior part of the brachial muscle and anterior to the long head of the triceps brachii muscle.

Origin. Medial continuation of the dorsal venous arch.

Termination. Commonly joins the medial tributary of the brachial vein to form the axillary vein.

Course. This superficial vein travels upward on the back surface of the ulnar side of the antebrachium. It spirals from the dorsal part of the forearm to the medial area and continues to ascend medial to the medial epicondyle. In the brachial region, this vein courses medial to the biceps before it pierces the deep fascia. It then runs adjacent to the medial part of the brachial artery before it joins the axillary vein. As it passes through the anterior medial cubital area, it is united by the median cubital vein.

III. Nerves

Median Nerve (*B*). An oval slice of mixed nerve tissue anterior to the brachial artery and posterior to the brachial vein. It is medial to the medial anterior part of the brachial muscle.

Origin. Medial and lateral cords of the brachial plexus.

Distribution. The pronator teres, flexor carpi radialis,

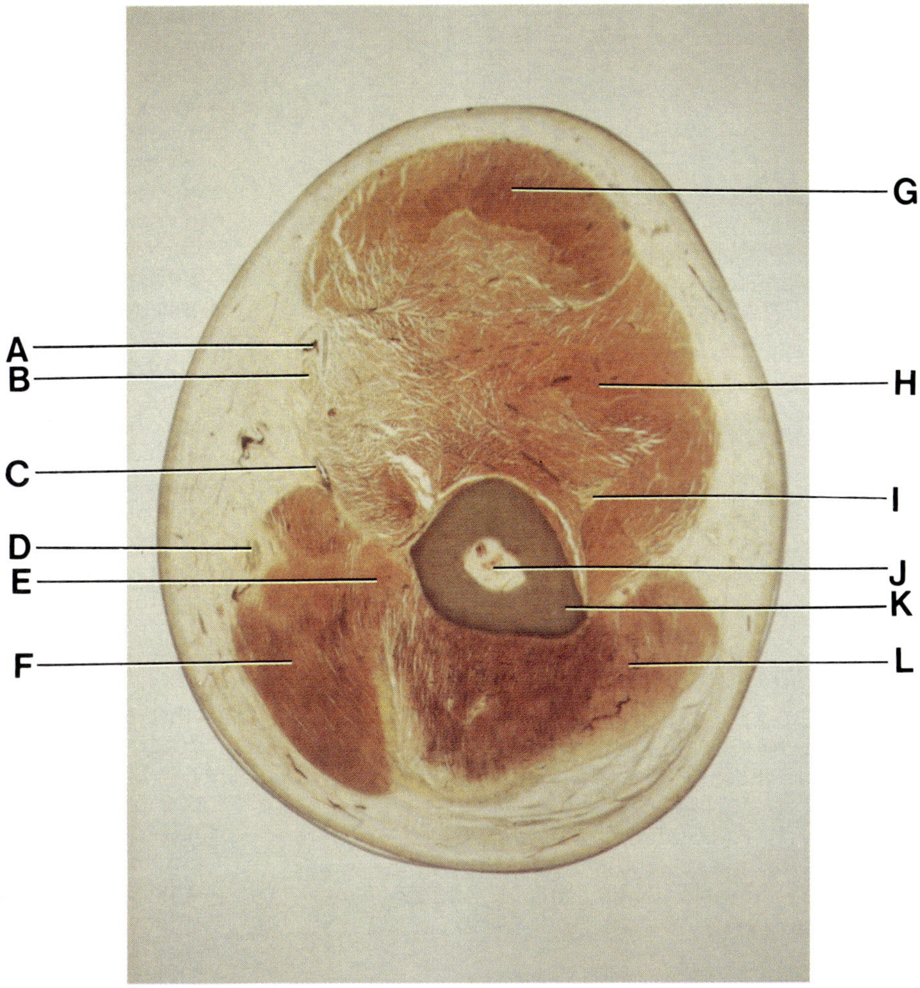

A. Brachial vein
B. Median nerve
C. Brachial artery
D. Basilic vein
E. Medial head of triceps brachii muscle
F. Long head of triceps brachii muscle
G. Biceps brachii muscle
H. Brachial muscle
I. Radial nerve
J. Medullary (marrow) cavity
K. Humerus
L. Lateral head of triceps brachii muscle

Figure 18-1
Transverse (Axial) section of arm through midshaft highlighting labels of musculoskeletal and neurovascular unit structures.

palmaris longus, flexor digitorum superficialis, flexor pollicis longus, pronator quadratus, abductor pollicis brevis, opponens pollicis, flexor pollicis brevis, lumbricales (the two lateral ones), and the lateral part of the flexor digitorum profundus.

Course. This nerve runs adjacent to the lateral surface of the axillary and brachial arteries until it reaches the midshaft. It then continues in front of the brachial artery to reach the medial surface of this artery, where it will continue its descent to reach the anterior forearm. In the forearm, it runs deep to the palmaris longus muscle and enters the carpal tunnel deep to the flexor retinaculum.

Radial Nerve (*I*). An oval slice of mixed nerve tissue embedded within the posterior lateral portion of the brachial muscle.

Origin. Posterior cord of the brachial plexus.

Distribution. Supinator, triceps brachii, anconeus, abductor pollicis longus, extensor pollicis brevis, extensor pollicis longus, extensor indicis, part of the brachial, brachioradialis, extensor carpi radialis longus, extensor carpi radialis brevis, extensor digitorum, extensor digiti minimi, and extensor carpi ulnaris.

Course. It starts as a continuation of the posterior cord posterior to the axillary artery and anterior to the subscapularis muscle. It continues posterior to the proximal portion of the brachial artery anterior to the tendons of the latissimus dorsi and teres major muscles. It spirals laterally and posterior to the humerus to reach the surface deep to the triceps brachii muscle and proceeds deep to the brachioradial muscle anterior to the lateral epicondyle, where it divides into its terminal branches.

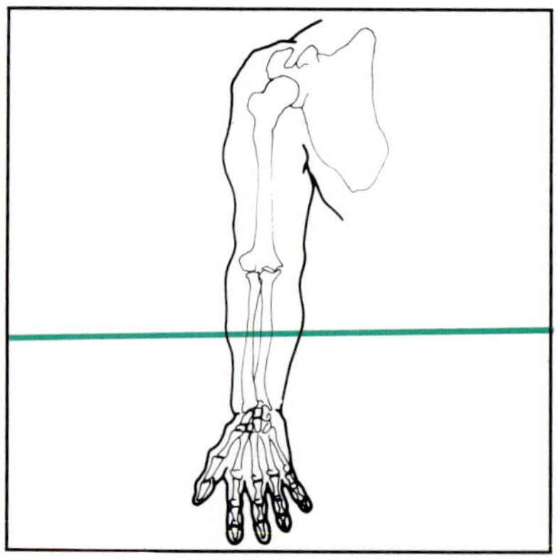

MODULE 2

This is a transverse (axial) section of the proximal third of the forearm.

MUSCULOSKELETAL UNIT (Fig. 18–2)

I. Skeletal Portion

Ulna (*D*). A roughly triangular bone slice. The apex is more posterior than the base. In the center is a rounded circular feature that represents the marrow cavity.

Radius (*I*). A rounded slice of bone tissue lateral to the ulna which is smaller in this section. In the center of the slice there is a smaller, circular, whitish feature, the marrow cavity.

II. Muscular Portion

Ulnar Extensor Wrist (Extensor Carpi Ulnaris) Muscle (*E*). A short transversely positioned muscle slice posterior to the extensor pollicis longus, medial to the extensor digiti minimi and lateral to the posterior part of the ulna.

Long Extensor Thumb (Extensor Pollicis Longus) Muscle (*K*). A rather small, rectangular muscle slice adjacent to the lateral surface of the ulna.

Extensor Little Finger (Extensor Digiti Minimi) Muscle (*L*). A small, oval muscle slice lateral to the extensor carpi ulnaris muscle and posterior to the abductor pollicis longus muscle.

Extensor Finger (Extensor Digitorum) Muscle (*J*). A rather large muscle slice lateral to the extensor pollicis longus muscle and medial to the posterior part of the radius.

Long and Short Radial Extensor Wrist (Extensor Carpi Radialis Longus and Brevis) Muscle (*H*). A rounded muscle slice lateral and adjacent to the radius. This muscle is seen posterior to the brachioradial muscle.

Brachioradial Muscle (*G*). A rather elongated oblong muscle section anterior to the radial extensor wrist muscle.

Round Pronator (Pronator Teres) Muscle (*F*). A rather elongated (anterior-to-posterior) muscle slice seen medial to the brachioradial muscle and anterior to the radial extensor wrist muscle.

Radial Flexor Wrist (Flexor Carpi Radialis) Muscle (*A*). An oval muscle slice of the anterior lateral forearm lateral to the superficial flexor muscles of the fingers.

Superficial Flexor Finger (Flexor Digitorum Superficialis) Muscles (*C*). A poorly defined muscle slice in the anterior medial forearm anterior to the deep flexor (flexor profundus) muscle.

Deep Flexor (Flexor Profundus) Muscle (*B*). A rather large muscle aggregate extending anterior to the ulna and medial to the anterior part of the radius.

A. Radial flexor wrist (flexor carpi radialis) muscle
B. Deep flexor (flexor profundus) muscle
C. Flexor finger (flexor digitorum) muscles
D. Ulna
E. Ulnar extensor wrist (extensor carpi ulnaris) muscle
F. Round pronator (pronator teres) muscle
G. Brachioradial muscle
H. Radial extensor wrist (extensor carpi radialis) muscle
I. Radius
J. Extensor finger (extensor digitorum) muscle
K. Long extensor thumb (extensor pollicis longus) muscle
L. Extensor little finger (extensor digiti minimi) muscle

Figure 18–2
Transverse (Axial) section of the forearm at the level of proximal 3rd of the forearm highlighting musculoskeletal unit structure.

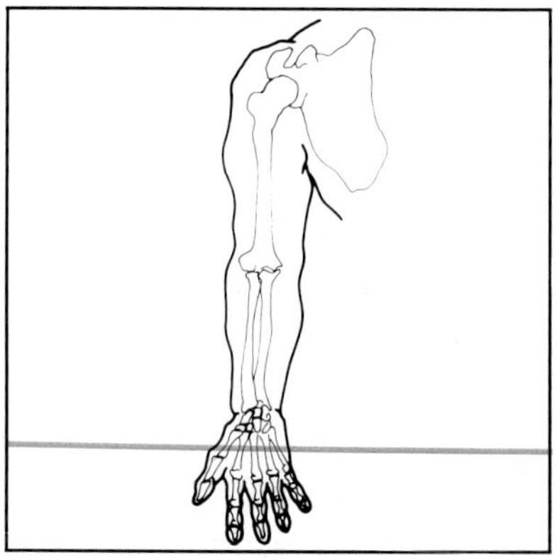

MODULE 3

This is a sheet plastinated (2.5 mm thick) transverse (axial) section of the hand through the metacarpal bones, showing the tendons of the superficial flexor muscles of the fingers, the tendons of the deep flexor muscles of the fingers, the adductor muscle of the thumb, the ulnar artery, and the median nerve.

MUSCULOSKELETAL UNIT (Fig. 18–3)

I. Skeletal Portion

Metacarpal Bones (*F, H, J, P, S*). Five oval bone sections, each with a marrow cavity on the dorsal part of this transverse section of the hand.

II. Muscular Portion

Adductor Thumb (Adductor Pollicis) Muscle (*Q*). A moderately wide, elongated (medial-to-lateral) muscle slice medial to the first metacarpal bone and extending anterior to the second and third metacarpal bones. Its medial end is posterior to the tendons of the deep flexor muscles of the fingers.

Dorsal Interosseous Muscle (*R*). An aggregate of muscle tissue between the first and second metacarpal bones, posterior to the long adductor muscle of the thumb (adductor pollicis longus muscle).

Extensor Tendons of the Fingers (Extensor Digitorum Tendons) (*I*). Rather small, oval tendinous tissues posterior to the third and fourth metacarpal bones.

Opposing Muscle of the Little Finger (Opponens Digiti Minimi Muscle) (*D*). An elongated (medial-to-lateral) strip of muscle tissue extending from the flexor tendons of the fingers to the anterior portion of the fifth metacarpal bone.

Flexor Muscle of the Little Finger (Flexor Digiti Minimi Muscle) (*C*). A moderately wide sheet of muscle tissue lateral to the abductor muscle of the little finger and anterior to the opposing muscle of the little finger. Its lateral end is anterior to the medial superficial flexor tendon of the fingers.

Abductor Muscle of the Little Finger (Abductor Digiti Minimi Muscle) (*B*). An oval aggregate of muscle tissue medial to the flexor muscle of the little finger and anterior to the medial part of the opposing muscle of the little finger.

Long Flexor Thumb (Flexor Pollicis Longus) Muscle (*O*). An elongated (medial-to-lateral) strip of muscle tissue medial to the first metacarpal bone and lateral to the median nerve. This muscle is between the opposing and adductor muscles of the thumb.

Opposing Thumb (Opponens Pollicis) Muscle (*M*). An elongated (medial-to-lateral), moderately thick strip of muscle tissue deep to the short adductor muscle of the thumb and anterior to the long flexor muscle of the thumb.

Short Abductor Thumb (Abductor Pollicis Brevis) Muscle (*K*). An elongated (medial-to-lateral), sickle-shaped muscle tissue anterior to the opposing muscle of the thumb. This is the most superficial muscle of the thenar eminence.

Superficial Flexor Finger (Flexor Digitorum Superficialis) Tendons (*L*). Oval tendinous tissues superficial to the deep flexor tendons of the fingers, deep to the lateral part of the opposing muscle of the little finger and medial part of the opposing muscle of the thumb.

Deep Flexor Finger (Flexor Digitorum Profundus) Tendons (*G*). Oval tendinous tissues deep to the superficial flexor tendons of the fingers.

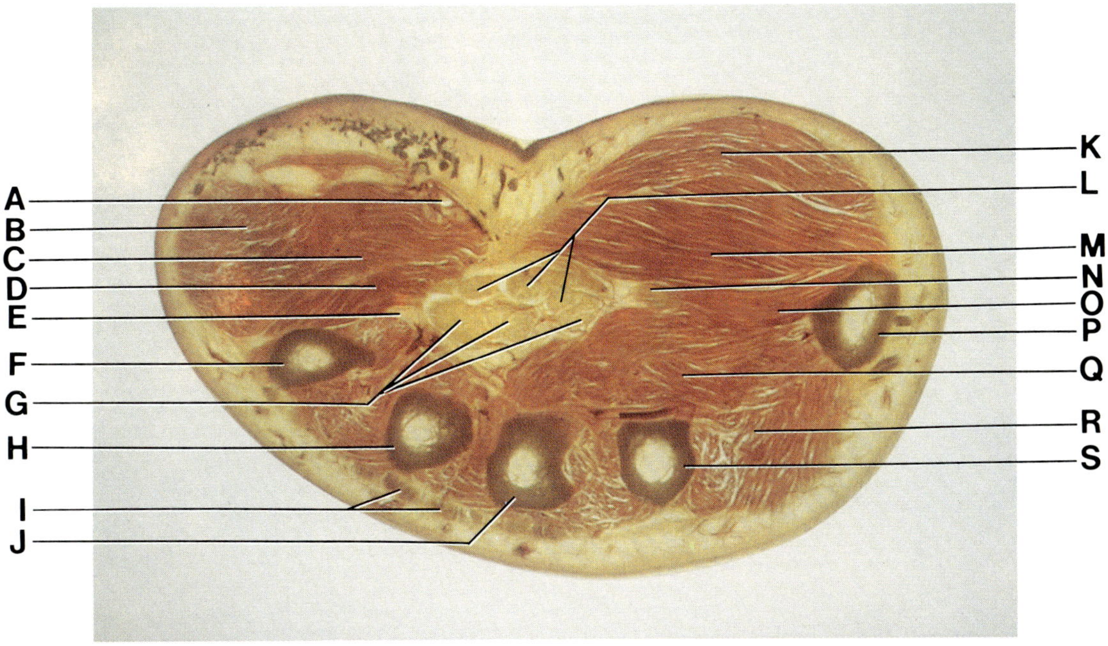

A. Ulnar artery
B. Abductor muscle of little finger (abductor digiti minimi muscle)
C. Flexor muscle of little finger (flexor digiti minimi muscle)
D. Opposing muscle of little finger (opponens digiti minimi muscle)
E. Ulnar bursa
F. Fifth metacarpal bone
G. Deep flexor finger (flexor digitorum profundus) tendons
H. Fourth metacarpal bone
I. Extensor tendons of fingers (extensor digitorum tendons)
J. Third metacarpal bone
K. Short abductor thumb (abductor pollicis brevis) muscle
L. Superficial flexor finger (flexor digitorum superficialis) tendons
M. Opposing thumb (opponens pollicis) muscle
N. Median nerve
O. Long flexor thumb (flexor pollicis longus) muscle
P. First metacarpal bone
Q. Adductor thumb (adductor pollicis) muscle
R. Dorsal interosseous muscle
S. Second metacarpal bone

Figure 18–3
Transverse (Axial) section of the hand through the metacarpal bones.

NEUROVASCULAR UNIT (see Fig. 18–3)

I. Arteries

Ulnar Artery (*A*). An oval, hollow vascular structure located on the anterior lateral part of the flexor muscle of the little finger.

Origin. It is a terminal branch of the brachial artery.

Distribution. This vessel has four major branches: (1) dorsal carpal, (2) palmar carpal, (3) deep palmar arch, and (4) superficial palmar arch. The main terminal branch of this artery is the superficial palmar arch.

Course. This artery starts in the forearm adjacent to the anterior part of the neck of the radius and descends on the medial area of the forearm lateral to the ulnar flexor muscle of the wrist and anterior to the ulnar nerve. It enters the hand area anterior to the flexor retinaculum adjacent to the lateral surface of the pisiform and is adjacent to the medial surface of the ulnar nerve.

II. Nerves

Median Nerve (*N*). An oval slice of mixed nerve tissue medial to the long flexor thumb muscle and lateral to the flexor tendons of the fingers.

ENCLOSING UNIT (see Fig. 18–3)

Ulnar Bursa (*E*). A dilation on the medial side of the midpalmar compartment, immediately lateral to the tendons of the flexor finger muscles.

CHAPTER 19

Coronal (Frontal) Cadaver Sections and Images of the Upper Limb

This chapter contains an unusually clear picture of a coronal view of the upper limb. The relationships of the structures in the chest region to the upper limb are vividly shown. It is interesting to note the difference in the appearance of structures compared with the transverse section. For further information on appearances of structures in different planes, see the description of structures for the figures in Chapters 18, 19, and 20.

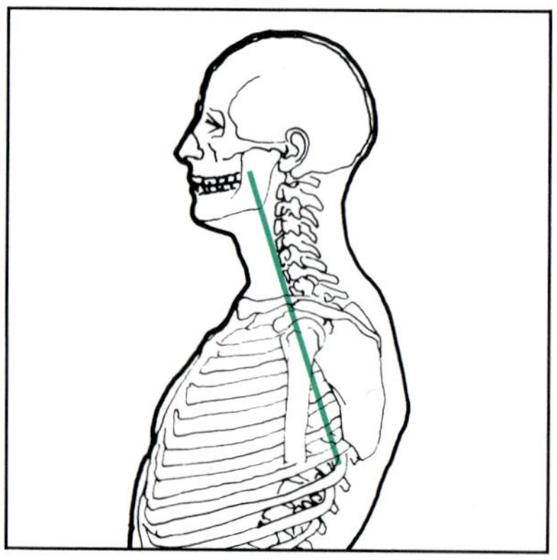

MODULE 1

This coronal (frontal) view of the thorax and upper limbs will depict and emphasize the upper limb structures. Some of the structures visualized are the head of the humerus, the deltoid muscle, the pectoralis major and minor muscles, the serratus anterior muscle, and the axillary vessels.

MUSCULOSKELETAL UNIT (Fig. 19–1)

I. Skeletal Portion

Head of the Humerus (*G*). A bulb-shaped bone slice of the proximal humerus.

Anatomic Neck of the Humerus (*O*). A slight constriction immediately below the head of the humerus. This area shows cancellous (spongy) bone and is superior to the surgical neck.

Surgical Neck of the Humerus (*Q*). A constriction immediately below the proximal epiphysis and above the proximal end of the diaphysis.

Diaphysis (Shaft) of the Humerus (*T*). A vertical bone slice seen with a vertical marrow cavity between two layers of compact bone.

Marrow (Medullary) Cavity (*U*). A vertical cavity within the diaphysis of the humerus, extending as far up as the surgical neck.

Epiphyseal Line (*F*). A transversely curved, thin strip of bone adjacent to the head and anatomic neck of the humerus. The inferior part shows a concave surface.

Coracoid Process of the Scapula. An oval bone slice inferior to the clavicle and medial to the head of the humerus. It is superior to the subscapularis muscle as seen in this coronal plane.

Acromion Process of the Scapula (*C*). A transversely oblong bone slice superior to the head of the humerus and medial to the deltoid muscle. The medial end of this slice is narrower than the lateral end.

II. Muscular Portion

Deltoid Muscle. A transversely placed bilateral muscle strip seen clearly on the left side of this coronal section. It is superior to the coracoid process and the head of the humerus.

Lateral Head of the Triceps Brachii Muscle (*S*). A vertical, elongated, bilateral muscle slice immediately lateral to the humerus. Its superior and inferior ends show constriction compared with the rest of the muscle.

Subscapularis Muscle. An irregularly rectangular, bilateral muscle medial to the head and neck of the humerus as well as inferior to the coracoid process of the scapula.

Coracobrachialis Muscle (*K*). A narrow vertical strip of bilateral muscle tissue adjacent to the medial part of the anatomic neck and upper diaphysis of the humerus.

Medial Head of the Triceps Brachii Muscle (*V*). A vertical, rather wide band of muscle tissue inferior to the axillary artery and medial to the diaphysis of the humerus.

Tendon of the Long Head of the Biceps Brachii Muscle (*L*). A transverse, short tendinous tissue located in the median of the head of the humerus between the articular capsule and the head of the humerus.

Serratus Anterior Muscle (*I*). Moderately small strips of bilateral muscle tissue adjacent to the lateral surface of the ribs and inferior to the pectoral muscles. The superior parts of some of these muscle strips are deep to the inferior parts of the pectoral muscles.

Pectoralis Major Muscle (*P*). A vertical, rather thin, bilateral muscle strip superficial to the inferior part of the pectoralis minor and the superior part of the serratus anterior muscles.

Pectoralis Minor Muscle (*H*). Rather thick, vertical, bilateral muscle tissue adjacent to the lateral parts of the upper ribs. The inferior part of this muscle slice is narrower than the upper part and deep to the pectoralis major muscle.

Omohyoid Muscle (*J*). A transverse, bilateral muscle superior to the axillary artery and medial to the deltoid muscle.

Subclavian Muscle (*D*). A transverse, rather small, bilateral muscle inferior to the clavicle. The lateral end is superior to the coracoid process of the scapula.

NEUROVASCULAR UNIT (see Fig. 19–1)

I. Arteries

Right Axillary Artery. A transverse, curved, hollow vascular structure lateral to the first rib. The lateral end is more inferior than the medial end.

Origin. Brachiocephalic artery.

Distribution. This artery disperses blood to six major branches: (1) the supreme thoracic from the first part, which is between the subclavian and the upper edge of the pectoralis minor muscles, (2) the lateral thoracic from the second part, which is deep to the pectoralis minor muscle, (3) the thoracoacromial from the second part, which is deep to the pectoralis minor muscle, and the next three branches arise from the third part of this vessel, which is between the inferior

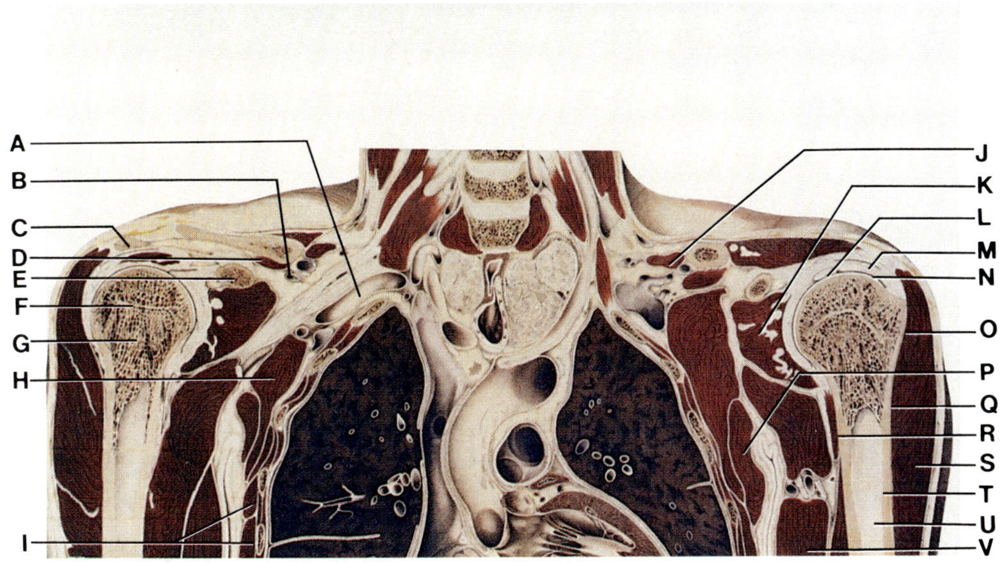

A. Right subclavian artery
B. Suprascapular (transverse scapular) artery
C. Acromion process of scapula
D. Subclavian muscle
E. Brachial plexus
F. Epiphyseal line
G. Head of humerus
H. Pectoralis minor muscle
I. Serratus anterior muscle
J. Omohyoid muscle
K. Coracobrachialis muscle
L. Tendon of long head of biceps brachii muscle
M. Articular capsule
N. Joint cavity
O. Anatomic neck of humerus
P. Pectoralis major muscle
Q. Surgical neck of humerus
R. Part of medial head of triceps brachii muscle
S. Lateral head of triceps brachii muscle
T. Diaphysis (shaft) of humerus
U. Marrow (medullary) cavity
V. Medial head of triceps brachii muscle

Figure 19–1
Approximately midcoronal section of the thorax and upper limb, highlighting some structures of the upper limb.

edge of the pectoralis minor and the tendon of the teres major, (4) the anterior humeral circumflex, (5) the posterior humeral circumflex, and (6) the subscapular.

Course. It is a continuation of the subclavian artery and runs lateral to the subclavian vein within the axilla to reach the tendon of the teres major muscle.

Right Suprascapular (Transverse Scapular) Artery (*B*). An oval, hollow vascular structure inferior to the medial part of the subclavian muscle.

Origin. Thyrocervical trunk.

Distribution. This artery disperses blood to the following arterial vessels: (1) articular twigs to the sternoclavicular, acromioclavicular, and glenohumeral joints, (2) the acromial branch, which is a cutaneous innervation to the skin covering the acromial area, and (3) the suprasternal branch, which is a cutaneous branch to the skin covering the anterior part of the chest.

Course. It runs toward the side from its origin and then turns posteriorly across the root of the neck to reach the deep surface of the sternocleidomastoid muscle. Further, this artery becomes deep to the clavicle and trapezius muscle. It also passes adjacent to the front surface of the scalenus anterior muscle, the brachial plexus, the third part of the subclavian artery, and the phrenic nerve to reach the supraspinatus fossa. To reach the supraspinatus fossa, this artery crosses the transverse scapular ligament and yields branches to the supraspinatus muscle before entering the infraspinatus fossa through the spinoglenoid notch. The infraspinatus muscle receives blood supply from this vessel. Finally, this artery anastomoses with a circumflex scapular branch of the subscapular artery as well as a deep branch of the transverse cervical artery.

ENCLOSING UNIT (see Fig. 19–1)

Articular Capsule (*M*). An enclosure of the head of the humerus by a rather thick sheet of connective tissue, attached to both the medial and lateral parts of the proximal epiphysis of the humerus in this coronal section.

Joint Cavity (*N*). A circular space between the articular capsule and articular cartilage covering the head of the humerus.

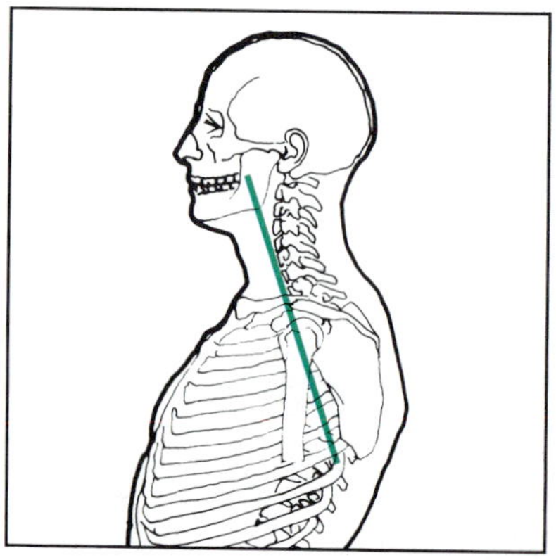

Figure 19–2 shows a magnetic resonance coronal image of the upper limb that emphasizes musculoskeletal unit structures.

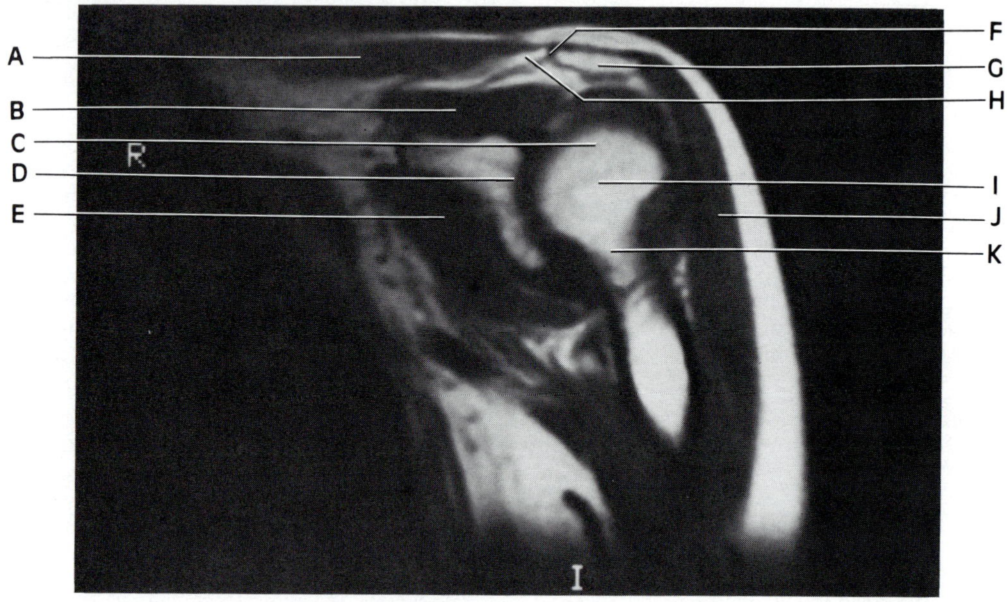

A. Trapezius muscle
B. Supraspinatus muscle
C. Head of humerus
D. Glenoid fossa
E. Subscapularis muscle
F. Acromioclavicular joint
G. Acromion
H. Clavicle
I. Anatomic neck of humerus
J. Deltoid muscle
K. Surgical neck of humerus

Figure 19–2
Magnetic resonance coronal image of the upper limb, showing some of its structures.

CHAPTER 20

Sagittal Cadaver Sections and Images of the Upper Limb

Two sagittal cadaver sections are included in one figure. One illustration focuses on structures of the sagittal view in the area of the elbow and the structures adjacent to it. The other emphasizes the structures of the sagittal view in the hand and the structures of the distal forearm.

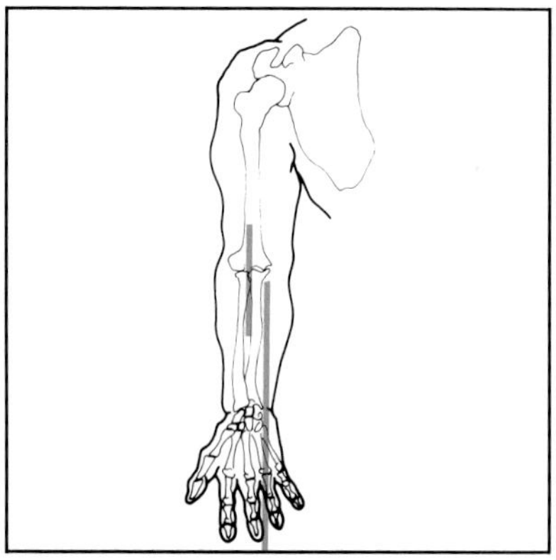

MODULE 1

This module includes two sagittal sections of the upper limb. One shows the elbow joint structures, such as the olecranon process, head of the radius, trochlear notch of the ulnar, brachial muscle, and distal ends of the triceps brachii and biceps brachii muscles. The other sagittal section shows the muscles of the forearm, the tendons of the forearm and hand, and the wrist joint.

MUSCULOSKELETAL UNIT (Fig. 20–1)

I. Skeletal Portion

Humerus (*D*). A vertical, oblique bone between the brachial and triceps brachii muscles. The inferior end is slightly posterior to the superior end in this sagittal view.

Olecranon Process (*F*). A proximal, curved bone slice of the ulna posterior to the distal end of the humerus. The concave surface, known as the trochlear notch, is adjacent to the distal end of the humerus.

Trochlear Notch (*G*). An anterior indentation of the olecranon process, seen adjacent to the distal end of the humerus in this sagittal plane.

Coronoid Process (*H*). An anterior projection of the ulna immediately inferior to the trochlear notch and the distal end of the humerus. It is seen posterior to the head of the radius in this sagittal view.

Ulna (*K*). A vertically positioned bone in this view, inferior to the humerus and posterior to the radius.

Head of the Radius (*I*). A rounded, proximal, flat, superior surface inferior to the distal end of the humerus and anterior to the coronoid process.

Neck of the Radius (*J*). A constriction near the proximal end of the radius, immediately inferior to the head of the radius and superior to the radial tuberosity.

Radial Tuberosity (*L*). An enlargement of the radius immediately inferior to the neck of the radius and posterior to the superior part of the supinator muscle.

Radius (*O, R*). This is seen in both sagittal views as a vertical bone. The sagittal view of the elbow joint shows the radius in a vertical, oblique position. The inferior part is anterior to the superior part. In the sagittal view of the wrist joint, the radius is in a vertical position with an enlarged distal portion.

Wrist Joint (*X*). A narrow, transverse space between the distal end of the radius and the proximal surface of the lunate carpal bone.

Lunate Bone (*Y*). A rounded bone slice between the distal end of the radius and the capitate carpal bone.

Capitate Bone (*Z*). A short, vertical, oblique bone inferior to the lunate and superior to the metacarpal bone. The inferior end is slightly posterior to the superior end.

Third Metacarpal Bone (*b*). A vertical bone, slightly elongated, between the capitate bone and the proximal phalanx.

Proximal Phalanx (*d*). A vertical, oblique bone slice inferior to the third metacarpal bone. This is one of the bones for digit number 3.

Third Metacarpophalangeal Joint (*c*). A transverse, narrow space between the third metacarpal bone and the third proximal phalanx.

II. Muscular Portion

Triceps Muscle of Arms (Brachii) (*B*). A vertical and oblique muscle slice posterior to the humerus. The inferior end is narrower and slightly posterior to the superior end.

Brachial Muscle (*C*). A vertical, slightly oblique muscle tissue posterior to the biceps brachii muscle and anterior to the humerus. The inferior end is slightly wider and more posterior than the superior end. The inferior end of this muscle crosses the front part of the elbow joint

Biceps Muscle of Arm (Brachii) (*A*). A vertical strip of skeletal muscle between the brachial and coracobrachial muscles.

Coracobrachial Muscle (*E*). A vertical strip of muscle tissue anterior to the biceps brachii muscle.

Ulnar Extensor Wrist (Extensor Carpi Ulnaris) Muscle (*N*). A vertical strip of muscle tissue posterior to the ulna. The superior end is narrow and adjacent to the olecranon process.

Supinator Muscle (*M*). An oblong muscle slice with an inferior narrow end inferior to the biceps brachii and brachioradial muscles. It is anterior to the long radial extensor muscle of the wrist in this sagittal view.

Long Radial Flexor Wrist (Flexor Carpi Radialis Longus) Muscle (*P*). A vertical strip of muscle tissue seen in this view posterior to the brachial muscle. The inferior part is adjacent to the anterior part of the radius.

Superficial Flexor Muscle of the Fingers (Flexor Digitorum Superficialis Muscle) (*T*). A vertical muscle superficial to the deep flexor muscle of the fingers. The tendon of this muscle extends as far down as the proximal part of the second phalanx of the third digit, as revealed in this sagittal view.

Deep Flexor Muscle of the Fingers (Flexor Digitorum Profundus Muscle) (*S*). A vertical muscle

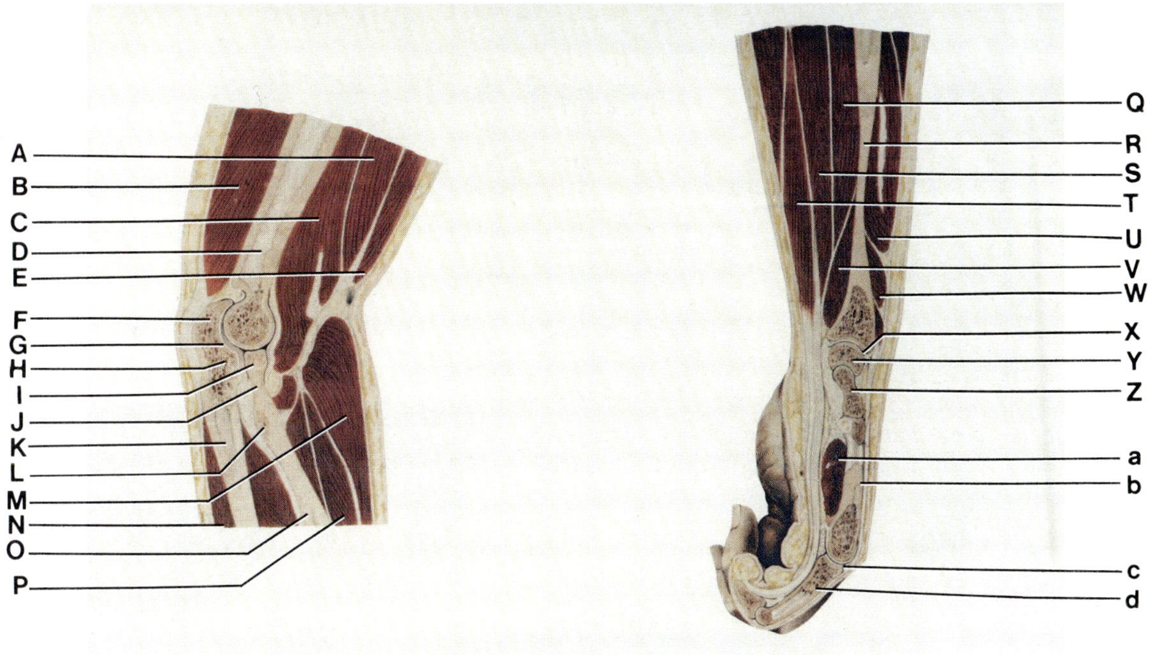

A. Biceps muscle of arm (brachii)
B. Triceps muscle of arm (brachii)
C. Brachial muscle
D. Humerus
E. Coracobrachial muscle
F. Olecranon process
G. Trochlear notch
H. Coronoid process
I. Head of radius
J. Neck of radius
K. Ulna
L. Radial tuberosity
M. Supinator muscle
N. Ulnar extensor wrist (extensor carpi ulnaris) muscle
O. Radius
P. Long radial flexor wrist (flexor carpi radialis longus) muscle
Q. Long flexor thumb (flexor pollicis longus) muscle
R. Radius
S. Deep flexor muscle of fingers (flexor digitorum profundus muscle)
T. Superficial flexor muscle of fingers (flexor digitorum superficialis muscle)
U. Short extensor thumb (extensor pollicis brevis) muscle
V. Pronator quadratus muscle
W. Extensor muscle of forearm
X. Wrist joint
Y. Lunate bone
Z. Capitate bone
a. Lumbricals
b. Third metacarpal bone
c. Third metacarpophalangeal joint
d. Proximal phalanx

Figure 20–1
Sagittal section of the arm, forearm, and hand, highlighting some of their structures.

deep to the superficial flexor muscle of the fingers. The tendon of this muscle extends as far down as the distal phalanx of the third digit.

Long Flexor Thumb (Flexor Pollicis Longus) Muscle (*Q*). A vertical strip of muscle posterior to the deep flexor muscle of the fingers and superior to the pronator quadratus. The inferior end is narrower in this view than the superior end.

Pronator Quadratus Muscle (*V*). An oblong, oblique muscle slice deep to the inferior part of the deep flexor muscle of the fingers and anterior to the distal part of the radius. The superior part is narrower than and slightly posterior to the inferior part.

Short Extensor Thumb (Extensor Pollicis Brevis) Muscle (*U*). A vertical muscle strip adjacent to the posterior part of the radius and superior to the long extensor muscle of the thumb.

Long Extensor Thumb (Extensor Pollicis Longus) Muscle. A vertical, rather short muscle strip inferior to the short extensor muscle of the thumb and posterior to the radius near its distal end.

Adductor Muscle of the Thumb (Adductor Pollicis Muscle). An oblong (superior-to-inferior) muscle slice within the palmar region of the hand, anterior to the third metacarpal bone.

Appendix to Part VII: Clinical Correlation for the Upper Limb Region

CASE ONE (Fig. A VII–1)

History. A bone radiograph and a bone scan show abnormality of the base of the second right metacarpal bone in this patient. The patient experienced pain in this location.

Findings. A mass lesion 1.5 cm in diameter and approximately 18 mm in vertical length is seen. It involves the proximal diaphysis of the second right metacarpal bone. A portion of the epiphysis of the second metacarpal is not involved in this mass lesion, and in fact there is a quite well demarcated, low signal intensity separating the epiphysis from the mass. A small amount of periosteal reaction is noted around the mass. Its distal end shows a clear-cut border merging into the bone marrow of the midportion of the metacarpal. In the T1-weighted images, the signal intensity of the lesion is slightly lower than that of adjacent bone marrow; in the T2-weighted images, its signal intensity becomes similar to that of the bone marrow of adjacent bones. The aforementioned findings, therefore, tend to favor the diagnosis of an enchondroma rather than any other entity. Although chondrosarcoma cannot be totally excluded, the characteristics of an aggressive tumor are not present in the current MRI study to suggest that diagnosis. No other significant findings are noted.

Impression. A 1.5 cm × 1.5 cm × 1.8 cm probable enchondroma involving the proximal shaft of the right second metacarpal bone.

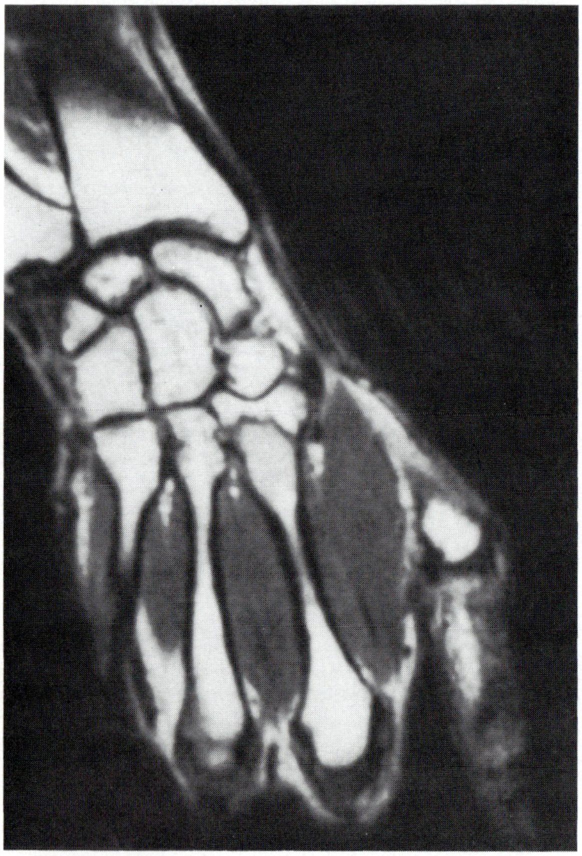

Figure AVII–1
Magnetic resonance image, showing probable enchondroma involving the proximal shaft of the right second metacarpal bone.

PART VIII

THE LOWER LIMB

INTRODUCTION

There are 31 bones in each lower limb: (1) os coxae (hip or pelvic bone), (2) thigh (femur), (3) patella (kneecap), (4) the leg, which includes the tibia and fibula, and (5) the foot, which includes 26 bones (7 tarsal, 5 metatarsal, and 14 phalanges of the digits). The joints of the lower limb are all members of the synovial joint class except the syndesmosis joint between the tibia and fibula.

The muscles of the lower limb are innervated by nerves from both the lumbar and sacral plexuses, as can be noted in the tables of muscles included here.

The arterial supply of the lower limb is summarized as follows: Left ventricle → ascending aorta → arch of aorta → descending thoracic aorta that crosses the diaphragm and becomes the descending thoracic aorta → common iliac arteries → external iliac arteries → femoral arteries → popliteal arteries, each of which bifurcates to form the anterior tibial and peroneal arteries.

Table VIII-1 Muscles of Femoral Triangle (Muscles Acting on Hip Joint)

Muscle	Origin	Insertion	Innervation	Vasculature	Action
Long adductor	Pubic symphysis and pubic crest	Inferior 2/3 of linea alba aspera of femur	Anterior obturator (L3, L4)	Obturator artery, medial femoral circumflex	Flexes and rotates thigh
Short adductor	Inferior ramus of pubis	Femur (linea aspera)	Anterior obturator (L3, L4)	Obturator artery, medial femoral circumflex	Flexes and rotates thigh
Great adductor	Anterior belly: pubis and adjacent ischial ramus. Posterior belly: ischial tuberosity	Inferior part of linea aspera of femur, adductor tubercle	Posterior obturator and sciatic nerve	Obturator artery, medial femoral circumflex, branches of profunda femoris, branches of popliteal	Adducts and flexes thigh
Pectineus	Pubis pectineal line and pubic bone immediately inferior	Vertical faint marking between spiral line and gluteal ridge	Femoral (L2–L4)	Obturator, medial femoral circumflex	Adducts and flexes thigh
Gracilis	Inferior half of symphysis pubis and superior half of pubic arch	Medial surface of head of tibia	Anterior obturator	Obturator, medial femoral circumflex, branches of profunda femoris	Flexes and medially rotates leg, adducts thigh
Psoas major	Lumbar vertebrae (bodies), lumbar intervertebral disks, medial ends of transverse processes of lumbar vertebrae	Lesser trochanter	Lumbar nerves 2 and 3	Iliolumbar (lumbar branch)	Flexion of thigh, vertebral column (on pelvis)
Psoas minor (present in 2/3 of population)	Bodies of thoracic vertebra 12 and lumbar vertebra 1	Arcuate line, iliopectineal eminence	Lumbar nerve 1 and sometimes lumbar nerves 2	Iliolumbar (lumbar branch)	Flexes trunk on pelvis
Sartorius	Anterior upper edge of ilium immediately inferior to anterior, superior iliac spine	Superior end of hypodermic surface of tibia	Femoral L2, L3	Femoral	Flexes leg and thigh, rotates thigh laterally
Iliac	Iliac fossa and anterior sacroiliac ligament	Anterior psoas major tendon and small portion of shaft of femur immediately inferior to lesser trochanter	Femoral	Iliolumbar (iliac branch), superior gluteal	Thigh flexion, slants pelvis anteriorly when leg is fixed

Table VIII-2 Thigh Muscles

Muscle	Origin	Insertion	Innervation	Vasculature	Action
Biceps m. of thigh (biceps femoris)	Short head: (1) lateral intermuscular septum, (2) linea aspera of femur, (3) lateral supracondylar line. Long head: medial part of ischial tuberosity	Fascia of leg below the knee joint and head of fibula	Tibial and peroneal branches of sciatic	Branches of profunda femoris and of popliteal	Flexes knee and rotates leg laterally
Semitendinous	Ischial tuberosity on its medial surface	Superior medial part of tibia and leg fascia	Tibial branch of sciatic (L5, S1, S2)	Branches of profunda femoris and of popliteal	Flexes knee and rotates leg medially
Semimembranous	Upper medial part of ischial tuberosity. *Note:* this origin is by an elongated tendon	Posteromedial part of medial condyle to tibia	Tibial branch of sciatic (L5, S1, S2)	Branches of profunda femoris and of popliteal	Flexes knee and rotates leg medially
Quadriceps rectus femoris	Anterior head: anterior inferior iliac spine. Posterior head: dorsal and upper edges of acetabulum	Superior part: base of patella. Inferior part: tibial tuberosity as part of ligamentum patellae	Femoral (L2–L4)	Lateral femoral circumflex	Extends leg, flexes thigh
Quadriceps vastus lateralis	Several areas of proximal femur: (1) inter-trochanteric line, (2) greater trochanter, (3) gluteal tuberosity, (4) linea aspera, (5) lateral intermuscular septum	Quadriceps tendon	Femoral (L2–L4)	Lateral femoral circumflex, lateral superior genicular	Extends leg
Quadriceps vastus medialis	(1) Medial intermuscular septum, (2) intertrochanteric line, (3) spiral line	Quadriceps tendon	Femoral (L2–L4)	Branches of femoral, profunda femoris, and popliteal	Extends leg
Quadriceps vastus intermedius	Superior 2/3 of anterolateral femur and inferolateral intermuscular septum	Quadriceps tendon	Femoral (L2–L4)	Lateral femoral circumflex	Extends leg

Table VIII–3 Leg Muscles

Muscle	Origin	Insertion	Innervation	Vasculature	Action
Gastrocnemius	Lateral head: lateral condyle of femur. Medial head: medial condyle of femur	Calcaneal tendon (Achilles tendon)	Tibial branch (S1, S2)	Branches of popliteal	Flexes knee and ankle, supinates foot
Soleus	Back of fibular head, tibia (soleal line), and intermuscular septum between other origins	Calcaneal tendon (Achilles tendon)	Tibial branch (S1, S2)	Branches of popliteal, peroneal, posterior tibial	Plantar flexion of ankle
Plantar	Femoral condyle superior to lateral head of gastrocnemius and lateral supracondylar line	Medial part of calcaneal tendon or medial aspect of calcaneus	Medial popliteal	Branches of popliteal	Plantar flexion of ankle
Popliteus	Lateral condyle of femur and knee, oblique popliteal ligament	Back surface of tibia superior to soleal line	Tibial branch (L4, L5, S1)	Popliteal (genicular branches)	Flexes and rotates leg medially
Long flexor m. of toe (flexor digitorum longus)	Back part of tibia inferior to soleal line, proximal fibula	Distal phalanges of toes	Tibial branch (L5, S1, S2)	Posterior tibial	Plantar flexion, supination of foot
Long flexor m. of great toe (flexor hallucis longus)	Flexor part of fibula and adjacent aponeurosis of long flexor m. of toe	Base of distal phalanx of big toe	Tibial branch (L5, S1, S2)	Peroneal (muscular branches)	Flexes great toe, flexes and supinates foot
Posterior tibial	Proximal interosseous membrane between tibia and fibula	Metatarsal bones 2–4; tarsal bones: navicular and 3rd cuneiform	Tibial branch (L5, S1)	Peroneal	Plantar flexion, inverts foot
Long peroneal (peroneus longus)	Tibia (lateral condyle), head of fibular and superior 2/3 of lateral fibular shaft	Lateral part of medial cuneiform bone	Branch of superficial peroneal (L4, L5, S1)	Peroneal (muscular branches), anterior tibial (muscular branches)	Plantar flexion, everts foot
Short peroneal (peroneus brevis)	Inferior 2/3 of lateral fibular shaft	Metatarsal bone 5	Branch of superficial peroneal (L4, L5, S1)	Peroneal (muscular branches)	Pronation and plantar flexion of foot

CHAPTER 21

Transverse (Axial) Cadaver Sections and Images of the Lower Limb

The five different cadaver transverse pictures in this chapter are derived from (1) the midthigh, (2) the distal thigh, (3) the knee joint, (4) the midleg, and (5) the ankle joint. In each illustration, the anatomic categories (units) are combined, excluding the absent visceral unit. Two of the views are from plastinated specimens (2.5 to 4 mm thick).

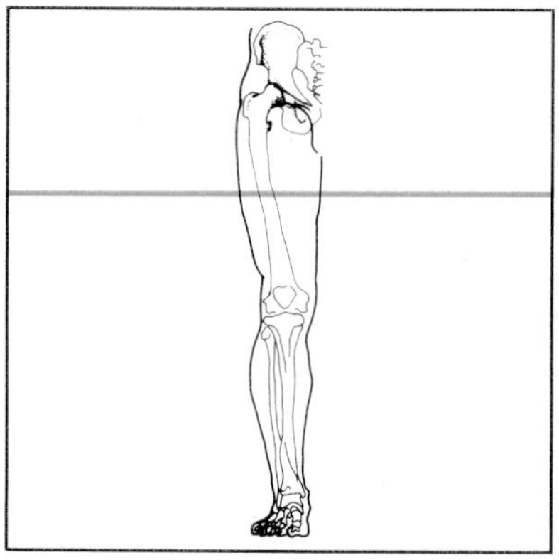

MODULE 1

This is a transverse (axial) section of the midthigh showing quadriceps, hamstrings, sciatic nerve, and greater saphenous vein.

MUSCULOSKELETAL UNIT (Fig. 21–1)

I. Skeletal Portion

Femur Shaft (*E*). A rather large, circular sheet of compact bone that surrounds a cavity (marrow) between the quadriceps muscle and the hamstrings.

Marrow Cavity of Femur (*D*). A circular space surrounded by compact bone of the diaphysis (femur).

II. Muscular Portion

Rectus Femoris Muscle (*A*). An oblong (medial-to-lateral) muscle aggregate anterior to the vastus intermedius and between the vastus lateralis and vastus medialis.

Vastus Medialis Muscle (*H*). An irregularly oval, rather large muscle aggregate medial to the rectus femoris and vastus intermedius muscles. It is anterior to the adductor magnus and lateral to the sartorius muscles.

Vastus Lateralis Muscle (*C*). An oblong (anterior-to-posterior), curved muscle section anterior and lateral to the vastus intermedius muscle. It is lateral to the rectus femoris muscle. The concave surface is posterior to the convex surface of this muscle.

Vastus Intermedius Muscle (*B*). A curved, elongated (medial-to-lateral) muscle section that encloses the anterior lateral part of the femur, as seen in this coronal view. This muscle is deep to the rectus femoris and the vastus lateralis muscles.

Sartorius Muscle (*I*). An irregularly triangular muscle section superficial to the vastus medialis and anterior to the gracilis muscles.

Gracilis Muscle (*M*). An elongated (anterior-to-posterior), rather narrow strip of muscle tissue medial to the adductor magnus and posterior to the sartorius muscles.

Great Adductor (Adductor Magnus) Muscle (*N*). A large, bulky muscle mass medial to the hamstrings and posterior to the vastus medialis muscle. It is medial to the sartorius and gracilis muscles.

Semimembranosus Muscle (*P*). A rounded muscle aggregate medial to the semitendinosus and lateral to the great adductor muscles.

Semitendinosus Muscle (*O*). An oblong (anterior-to-posterior) muscle section between the biceps femoris and the semimembranosus muscles. The anterior part is lateral to the great adductor muscle.

Biceps Femoris Muscle (*G*). An irregularly shaped, oval muscle aggregate posterior to the sciatic nerve and lateral to both the semitendinosus muscle and the anterior part of the great adductor muscle.

NEUROVASCULAR UNIT (see Fig. 21–1)

I. Arteries

Femoral Artery (*L*). A small, partially collapsed vessel lateral to the femoral vein and medial to the sartorius muscle.

II. Veins

Femoral Vein (*J*). A rather small, hollow venous structure lateral to the femoral artery and posterior to the vastus medialis muscle. This vein is the longest in a human body.

Greater Saphenous Vein (*K*). A rather small, oval, superficial venous structure medial to the anterior part of the gracilis muscle.

Origin. It is the medial continuation of the dorsal venous arch.

Termination. Distal part of the femoral vein.

Course. It climbs upward superficially, anterior to the medial malleolus, and takes an oblique medial path across the inferior part of the tibia to reach the medial edge. It continues to climb posteromedial to reach the medial condyle of the tibia. This vein ascends in the subcutaneous tissue on the medial and finally on the anterior medial of the thigh to reach the saphenous opening into the femoral vein.

III. Nerves

Sciatic Nerve (*F*). A rather short, oblong (anterior-to-posterior), mixed nerve tissue between the biceps femoris and the anterior part of the great adductor muscles.

A. Rectus femoris muscle
B. Vastus intermedius muscle
C. Vastus lateralis muscle
D. Marrow cavity of femur
E. Femur shaft
F. Sciatic nerve
G. Biceps femoris muscle
H. Vastus medialis muscle
I. Sartorius muscle
J. Femoral vein
K. Greater saphenous vein
L. Femoral artery
M. Gracilis muscle
N. Great adductor (adductor magnus) muscle
O. Semitendinosus muscle
P. Semimembranosus muscle

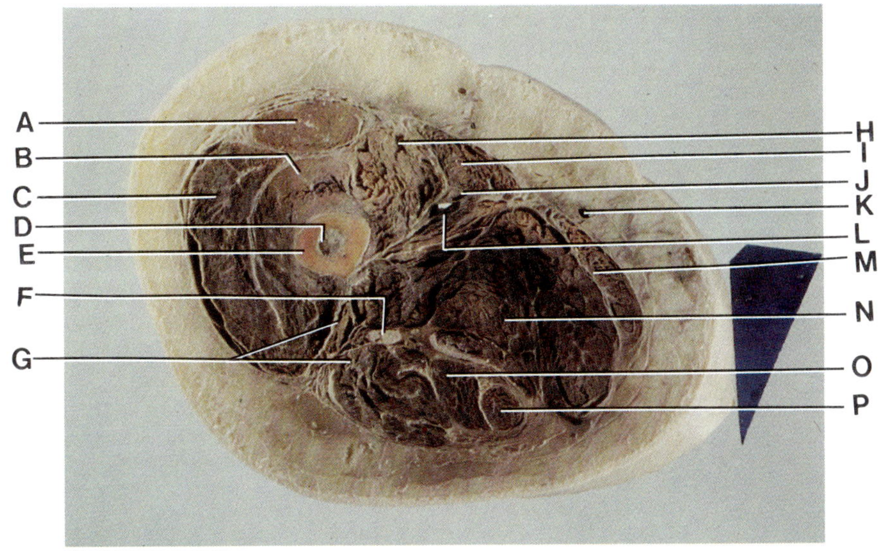

Figure 21-1
Transverse section of a human midthigh, showing musculoskeletal and neurovascular unit structures.

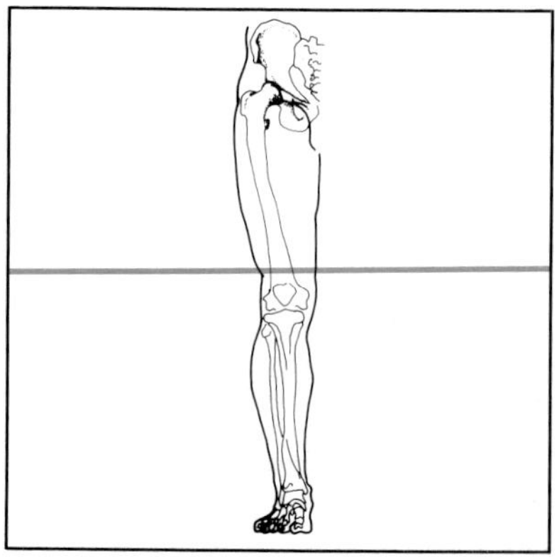

MODULE 2

This is a transverse (axial) section of the lower limb near the distal end of the thigh region, showing the suprapatellar bursa and tendons of the quadriceps and biceps femoris muscles.

MUSCULOSKELETAL UNIT (Fig. 21–2)

I. Skeletal Portion

Femur (*K*). An oblong, oval bone slice posterior to the suprapatellar bursa and tendons of the quadriceps muscle and between the vastus intermedius and vastus medialis muscles. The lateral part of this bone section is wider than the medial part.

II. Muscular Portion

Tendons of the Quadriceps Muscle (*I*). An elongated (medial-to-lateral), sickle-shaped, tendinous slice between the vastus intermedius and vastus medialis muscles. These tendons are anterior to the suprapatellar bursa.

Vastus Medialis Muscle (*L*). An anterior-to-posterior, elongated muscle section medial to the femur, suprapatellar bursa, and the tendons of the quadriceps muscle. It is anterior to the sartorius muscle at this level.

Sartorius Muscle (*N*). An oval muscle aggregate located adjacent to the vastus medialis muscle. The posterior part of this muscle is medial to the gracilis muscle.

Gracilis Muscle (*O*). A rather small, irregular, rounded muscle section between the semimembranosus and the sartorius muscles.

Semimembranosus Muscle (*G*). A rather large, oval muscle slice with a thin posterior projection between the tendon of the semitendinosus and gracilis muscles. The larger portion of this muscle section is more anterior and posterior to the popliteal artery and vein.

Tendon of the Semitendinosus Muscle (*H*). A small, rounded, tendinous tissue posterior to the semimembranosus muscle and medial to the biceps femoris muscle.

Biceps Femoris Muscle (*C*). An oblong (anterior-to-posterior) muscle slice. The posterior part is narrower than the anterior part and lateral to the common peroneal nerve. The anterior part is posterior to the vastus intermedius muscle.

Vastus Intermedius Muscle (*B*). A strip of muscle tissue lateral to the suprapatellar bursa and anterior to the biceps femoris muscle. It is medial to the vastus lateralis muscle.

Vastus Lateralis Muscle (*A*). A rather narrow strip of oblique muscle lateral to the vastus intermedius muscle in this transverse section of the lower thigh. The anterior end is more medial than the posterior end.

NEUROVASCULAR UNIT (see Fig. 21–2)

I. Arteries

Popliteal Artery (*M*). A rounded, hollow vascular structure medial to the popliteal vein and anterior to the semimembranosus muscle.

Origin. Continuation of the femoral artery.

Distribution. The popliteal is the parent of seven arteries: (1) lateral superior genicular, (2) medial superior genicular, (3) lateral inferior genicular, (4) medial inferior genicular, (5) middle genicular, and (6) two sural arteries.

Course. Within the adductor hiatus, the popliteal artery starts as a continuation of the femoral artery and passes down through the popliteal fossa. It ends at the inferior border of the popliteus muscle as it divides into posterior and anterior tibial arteries.

II. Veins

Popliteal Vein (*F*). An oval, hollow vascular structure anterior to the tibial nerve and lateral to the popliteal artery. It is medial to the short head of the biceps.

Origin. It arises from the fusion of the anterior and posterior tibial veins.

Termination. It terminates as it becomes the femoral vein within the adductor hiatus canal.

Course. This vein starts at the inferior border of the popliteus muscle and ascends the popliteal fossa in the following manner: (1) the inferior part of this vein is medial to the popliteal artery, (2) between the heads of the gastrocnemius it is superficial, and (3) the superior part (above the knee) is posterolateral to the popliteal artery.

III. Nerves

Tibial (Medial Popliteal) Nerve (*E*). A rounded section of nerve tissue anterior to the semimembranosus muscle and medial to the common peroneal nerve. It is lateral to the popliteal vein.

Origin. Sciatic nerve.

A. Vastus lateralis muscle
B. Vastus intermedius muscle
C. Biceps femoris muscle (long and short heads)
D. Common peroneal (lateral popliteal) nerve
E. Tibial (medial popliteal) nerve
F. Popliteal vein
G. Semimembranosus muscle
H. Tendon of semitendinosus muscle
I. Tendon of quadriceps muscle
J. Suprapatellar bursa
K. Femur
L. Vastus medialis muscle
M. Popliteal artery
N. Sartorius muscle
O. Gracilis muscle

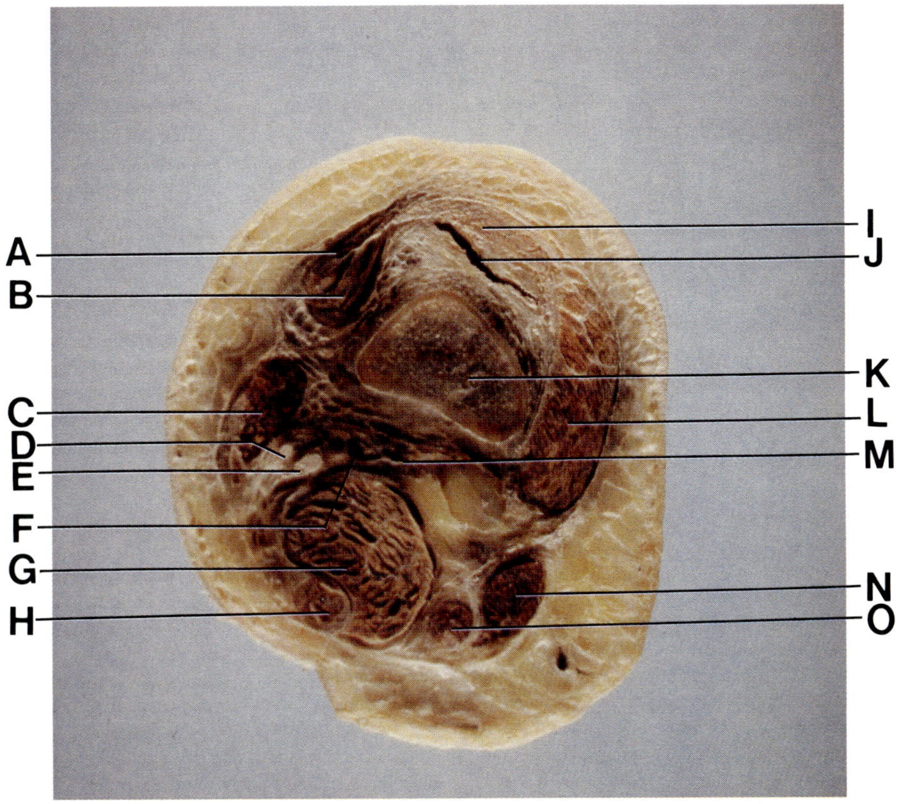

Figure 21-2
Transverse section of the lower limb near the distal end of the thigh region, highlighting the musculoskeletal and neurovascular unit structures.

Distribution. This mixed nerve has a cutaneous branch known as the sural nerve. The muscular branches innervate the following muscles: (1) plantaris, (2) soleus, (3) gastrocnemius, (4) popliteus, (5) long flexor of the fingers, (6) long flexor of the great toe, and (7) posterior tibial. Its genicular branches supply the synovial membrane of the knee joint with a source of innervation.

Course. This nerve courses lateral to the popliteal vessels from its beginning in the superior angle of the popliteal fossa until it crosses posterior to the vessel at the inferior angle of the popliteal fossa.

Common Peroneal (Lateral Popliteal) Nerve (*D*). A rounded section of mixed nerve tissue lateral to the tibial nerve and medial to the short head of the biceps femoris muscle.

Origin. Sciatic nerve.

Distribution. This nerve has the following branches: (1) deep peroneal, (2) superficial peroneal, (3) peroneal communicating, (4) articular, and (5) lateral sural cutaneous.

Course. This nerve arises in the posterior middle thigh region from the sciatic nerve and courses downward to enter the popliteal fossa at its apex. It then continues adjacent to the medial side of the biceps femoris muscle and turns lateral to cross the posterior parts of the gastrocnemius, soleus, and plantaris muscles to reach the long peroneal muscle, which it enters. This vein winds around the fibula toward the front and terminates on the front surface of the leg.

ENCLOSING UNIT (see Fig. 21-2)

Suprapatellar Bursa (*J*). A transversely oblique slit immediately posterior to the tendons of the quadriceps muscle and anterior to the femur at this level. The medial end is more posterior than the lateral end in this transverse section of the body.

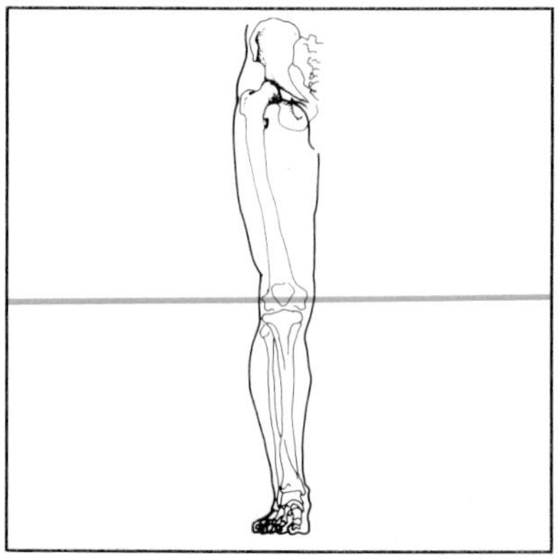

MODULE 3

This is a transverse (axial) section of the distal thigh, showing the patella, proximal portion of the fibula, and medial and lateral condyles of the femur.

MUSCULOSKELETAL UNIT (Fig. 21–3)

I. Skeletal Portion

Patella (*H*). A lens-shaped bone slice anterior to the femur and posterior to the patellar ligament. This structure is seen embedded within the tendon of the quadriceps muscles.

Medial Condyle of the Femur (*L*). A posterior, medial, oval projection of the femur anterior to the sartorius muscle and lateral to the medial head of the gastrocnemius muscle.

Lateral Condyle of the Femur (*B*). A posterior, lateral, oval prominence anterior to the lateral head of the gastrocnemius muscle and medial to the biceps femoris muscle.

Intercondylar Fossa (*K*). An invaginated, curved feature between the medial and lateral condyles of the femur. This feature is anterior to the popliteal artery and the medial head of the gastrocnemius muscle.

Proximal Portion of the Fibula (*C*). An oblong (anterior-to-posterior) bone section lateral to the posterior part of the lateral condyle of the femur. This structure is anterior to the common peroneal nerve.

II. Muscular Portion

Sartorius Muscle (*N*). An oblong (anterior-to-posterior) aggregate of muscle tissue posterior to the medial condyle of the femur and lateral to the tendon of the gracilis muscle.

Tendon of the Gracilis Muscle (*P*). A rather small, irregular tendinous tissue lateral to the sartorius muscle and posterior to the medial condyle of the femur.

Semimembranosus Muscle (*F*). A rounded muscle slice immediately posterior to the medial head of the gastrocnemius muscle and anterior to the tendon of the semitendinosus muscle.

Tendon of the Semitendinosus Muscle (*G*). A rather small, oval tendinous tissue posterior to the semimembranosus muscle. This structure is in the posterior median of this body transverse section.

Medial Head of the Gastrocnemius Muscle (*M*). A rounded muscle slice posterior to the intercondylar fossa and lateral to the medial condyle of the femur. This structure is anterior to the semimembranosus muscle.

Lateral Head of the Gastrocnemius Muscle (*A*). A rather irregularly shaped muscle section anterior to the tibial nerve and lateral to the popliteal vessels.

NEUROVASCULAR UNIT (see Fig. 21–3)

I. Arteries

Popliteal Artery (*J*). An oval, hollow vascular structure posterior to the medial part of the lateral condyle of the femur and anterior to the popliteal vein. This vessel is between the medial and lateral heads of the gastrocnemius muscle.

II. Veins

Popliteal Vein (*I*). An oval, hollow venous structure posterior to the popliteal artery and anterior to the tibial nerve.

Long (Greater) Saphenous Vein (*O*). A rounded, hollow, superficial venous structure lateral to the posterior part of the sartorius muscle.

III. Nerves

Tibial (Medial Popliteal) Nerve (*E*). A rounded, mixed nerve tissue slice posterior to the popliteal vein and lateral to the semimembranosus muscle.

Common Peroneal Nerve (*D*). A rounded aggregate of nerve tissue posterior to the fibula. This nerve is lateral to the lateral head of the gastrocnemius muscle.

A. Lateral head of gastrocnemius muscle
B. Lateral condyle of femur
C. Proximal portion of fibula
D. Common peroneal nerve
E. Tibial (medial popliteal) nerve
F. Semimembranosus muscle
G. Tendon of semitendinosus muscle
H. Patella
I. Popliteal vein
J. Popliteal artery
K. Intercondylar fossa
L. Medial condyle of femur
M. Medial head of gastrocnemius muscle
N. Sartorius muscle
O. Long (greater) saphenous vein
P. Tendon of gracilis muscle

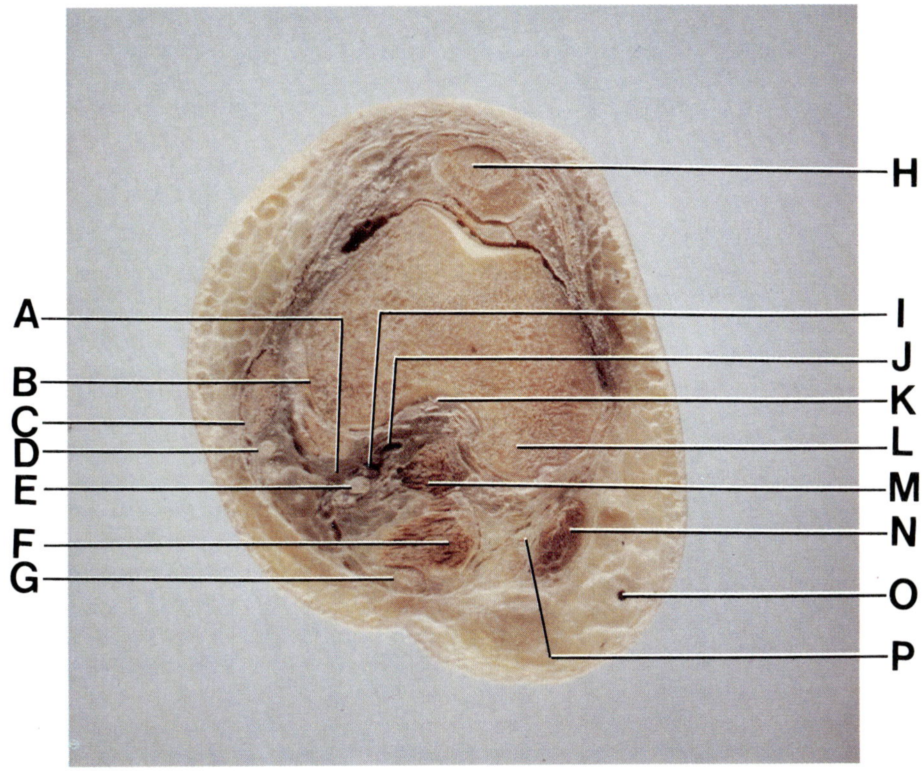

Figure 21-3
Transverse section of the distal thigh, highlighting musculoskeletal and neurovascular unit structures.

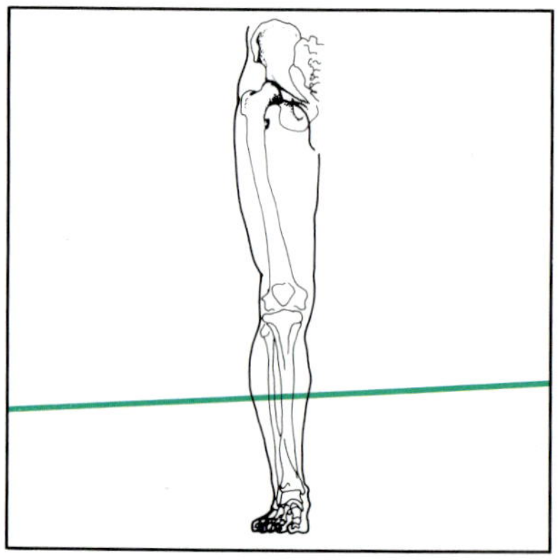

MODULE 4

This is a sheet plastinated (2.5 mm thick) transverse (axial) section of the leg (that portion of the lower limb between the knee and the ankle joint), showing the gastrocnemius muscle, peroneal artery, posterior tibial artery, and posterior tibial and long peroneal muscles.

MUSCULOSKELETAL UNIT (Fig. 21–4)

I. Skeletal Portion

Tibia (*L*). An oblong bone slice medial to both the anterior and posterior tibial muscles. It is anterior to both the popliteal and long flexor muscles of the toes.

Fibula (*C*). A rather small, irregularly oval bone section seen lateral to the posterior tibial muscle and posterior to the long extensor muscle of the toes. It is medial to the long peroneal muscle and anterior to the lateral part of the soleus muscle.

II. Muscular Portion

Anterior Tibial Muscle (*I*). A rather large, irregularly rectangular muscle aggregate anterior to the lateral part of the tibia and anterior to the posterior tibial muscle. It is medial to the long extensor muscle of the toes.

Posterior Tibial Muscle (*K*). A fan-shaped slice of muscle tissue between the tibia and fibula. The wider portion of this muscle is adjacent to the lateral surface of the tibia. This muscle is posterior to the anterior tibial muscle and anterior to the soleus muscle.

Long Extensor Muscle of the Toes (Extensor Digitorum Longus Muscle) (*A*). A muscle aggregate shaped like a trapezoid. The anterior part of this muscle section is broader than its posterior part, and its posterior part is adjacent to the anterior surface of the fibula. This muscle is lateral to the anterior tibial muscle and medial to the lateral portion of the soleus muscle.

Soleus Muscle (*G*). A large, sickle-shaped muscle slice that extends from the posterior part of the tibia to the posterior and lateral part of the fibula. This muscle is anterior to the gastrocnemius muscle, and its anterior surface shows a more continuous edge than its posterior surface.

Lateral Head of the Gastrocnemius Muscle (*H*). An elongated (anterior-to-posterior), rather thick muscle aggregate. Its anterior end is adjacent to the lateral end of the posterior part of the soleus muscle and is narrower than its posterior end.

Medial Head of the Gastrocnemius Muscle (*O*). An elongated (medial-to-lateral), rather thick muscle aggregate. Its medial end is adjacent to the posterior part of the medial end of the soleus muscle and is narrower than its lateral end.

Lateral Part of the Popliteal Muscle (*N*). A rather small, sickle-shaped muscle section immediately posterior to the tibia and medial to the tendon of the long flexor muscle of the toes, as seen in this plastinated leg transverse section.

Tendon of the Long Flexor Muscle of the Toe (*M*). An oval aggregate of tendinous tissue posterior to the lateral part of the tibia and anterior to the medial part of the soleus muscle.

NEUROVASCULAR UNIT (see Fig. 21–4)

I. Arteries

Anterior Tibial Artery (*J*). This is one of the terminal branches of the popliteal artery. It is seen at this level of the leg as a partially collapsed vascular structure medial to the anterior tibial vein and posterior to the lateral part of the anterior tibial muscle.

Origin. Popliteal artery.

Distribution. This artery disperses blood to seven branches: (1) the dorsal artery of the foot (dorsalis pedis), (2) the posterior tibial recurrent artery, (3) the anterior tibial recurrent artery, (4) many muscular branches, (5) the arteries around the ankle joint, (6) the anterior medial malleolar, and (7) the anterior lateral malleolar.

Course. This artery originates as a continuation of the popliteal artery at the inferior edge of the popliteal muscle in the posterior leg. It takes the following courses: (1) between the two heads of the posterior tibial muscle, (2) the superior part of the interosseous membrane to reach the anterior part of the leg, (3) downward on the anterior interosseous membrane to reach the inferior part of the tibia, and (4) midway between the medial and lateral malleoli on the anterior part of the ankle.

Posterior Tibial Artery (*F*). An oval, hollow arterial structure posterior to the medial part of the posterior tibial muscle and medial to the posterior tibial vein.

Origin. Inferior end of the popliteal artery.

Distribution. This artery disperses blood to eight branches: (1) communicating, (2) lateral posterior malleolar, (3) medial posterior malleolar, (4) fibular nutrient, (5) tibial nutrient, (6) peroneal, (7) calcaneal, and (8) plantar.

Course. It starts at the inferior edge of the popliteal

A. Long extensor muscle of the toes (Extensor digitorum longus muscle)
B. Anterior tibial vein
C. Fibula
D. Posterior tibial vein
E. Peroneal artery
F. Posterior tibial artery
G. Soleus muscle
H. Lateral head of gastrocnemius muscle
I. Anterior tibial muscle
J. Anterior tibial artery
K. Posterior tibial muscle
L. Tibia
M. Tendon of long flexor muscle of toe (flexor digitorum longus)
N. Lateral part of popliteal muscle
O. Medial head of gastrocnemius muscle

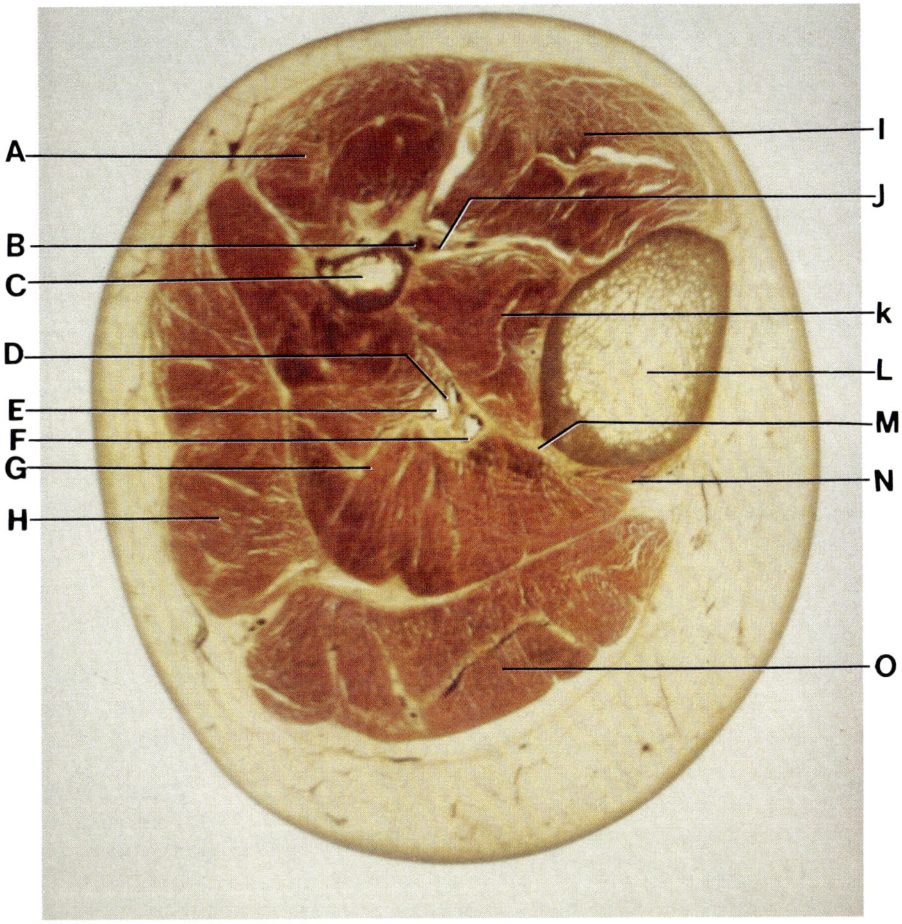

Figure 21-4
Plastinated transverse section of the leg between the knee and the ankle joint highlighting labels of musculoskeletal and neurovascular unit structures.

muscle between the fibula and the tibia and courses inferiorly as well as medially on the posterior leg. As it passes the ankle, it is midway between the medial tubercle of the calcaneus and the medial malleolus. Then it divides and gives rise to the medial and lateral plantar arteries.

Peroneal Artery (*E*). A rounded, hollow vascular structure posterior to the posterior tibial vein and lateral to the posterior tibial artery.

Origin. Posterior tibial artery.

Distribution. This artery shares blood with five branches: (1) communicating, (2) nutrient, (3) muscular, (4) lateral calcaneal, and (5) perforating.

Course. It begins as one of the terminal branches of the posterior tibial artery and courses deep to the long flexor muscle of the great toe to reach the calcaneal artery on the lateral part of the foot.

II. Veins

Anterior Tibial Vein (*B*). A rounded, hollow venous structure lateral to the anterior tibial artery and medial to the anterior part of the fibula. It is anterior to the lateral part of the posterior tibial muscle.

Origin. Dorsal artery of the foot as a superior continuation.

Termination. Popliteal vein as it joins with the posterior tibial vein.

Course. It passes from the anterior part of the inferior part of the leg between the tibia and the fibula. It courses through the superior part of the interosseous membrane to reach the popliteal vein at the inferior border of the popliteal muscle.

Posterior Tibial Vein (*D*). An oval, hollow venous structure lateral to the posterior tibial artery and posterior to the posterior tibial muscle. It is anterior to the peroneal artery.

Origin. A combination of the medial and lateral plantar veins.

Termination. The popliteal vein at the distal edge of the popliteal muscle.

Course. It runs lateral but adjacent to the posterior tibial artery to reach the popliteal vein.

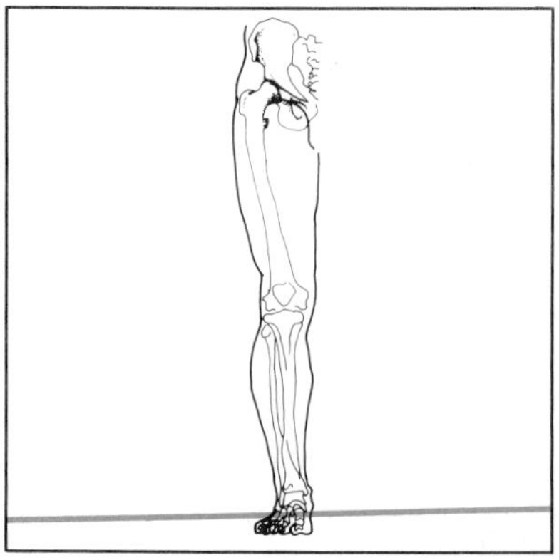

MODULE 5

This is a sheet plastinated (2.5 mm thick) transverse section of the foot near the distal end of the metatarsal bone, showing tendons of the long and short flexor muscles of the toes, the short flexor muscle of the great toe, the abductor muscle of the great toe, and the short flexor muscle of the little toe.

MUSCULOSKELETAL UNIT (Figure 21–5)

I. Skeletal Portion

First Metatarsal Bone (*D*). An oval, rather large bone slice anterior to the short flexor muscle of the great toe and lateral to the posterior part of the second metatarsal bone.

Fifth Metatarsal Bone (*L*). A rounded bone section anterior to the abductor muscle as well as the short flexor muscle of the little toe. This is the most lateral metatarsal bone. The second, third, and fourth metatarsal bones appear medial to lateral, in that order, between the first and fifth metatarsal bones.

II. Muscular Portion

Short Flexor Muscle of the Great Toe (Flexor Hallucis Brevis Muscle) (*F*). A transversely positioned muscle strip. The medial end of this muscle is immediately posterior to the first metatarsal bone. The remainder of this muscle extends laterally to reach the plantar aponeurosis.

Abductor Muscle of Great Toe (Abductor Hallucis Muscle) (*H*). A transversely placed muscle aggregate with a downward bulge between its lateral and medial ends. This muscle is superficial (inferior) to the short flexor muscle of the great toe. The lateral end of this muscle is adjacent to the plantar aponeurosis, and the medial end is adjacent to the inferior lateral part of the short muscle of the great toe.

Long Flexor Tendon of the Great Toe (Flexor Hallucis Longus Tendon) (*E*). An oval, tendinous structure lateral to the first metatarsal bone and medial to the fibrous sheath of the flexor tendons.

Abductor Muscle of the Little Toe (Abductor Digiti Minimi) (*O*). A rather large, three-quarter-moon–shaped muscle aggregate lateral and slightly inferior to the fifth metatarsal bone. It is lateral to the short flexor muscle of the little toe.

Short Flexor Muscle of the Little Toe (Flexor Digiti Minimi Brevis Muscle) (*P*). A muscle sheet medial to the abductor muscle of the little toe and inferior to the fifth metatarsal bone.

Dorsal Interosseous Muscles (*C*). Muscle strips between the upper parts of the metatarsal bones. There are four of these muscles.

Plantar Interosseous Muscles (*M*). Muscle sections adjacent to the inferior part of metatarsal bones 3, 4, and 5. There are three of these muscles.

Extensor Tendons of the Toes (Extensor Digitorum Tendons) (*I*). Rather short, transversely placed tendinous structures superior to the metatarsal bones.

Long Flexor Tendons of the Toes (Flexor Digitorum Longus Tendons) (*K*). Oval structures seen in the upper part of the fibrous sheath of the flexor tendons. These tendons are superior to the short flexor tendons of the toes.

Short Flexor Tendons of the Toes (Flexor Digitorum Brevis Tendons) (*G*). Oval structures in the lower part of the fibrous sheath of the flexor tendons. These tendons are immediately superior to the plantar aponeurosis of the foot and inferior to the long flexor tendons of the toes.

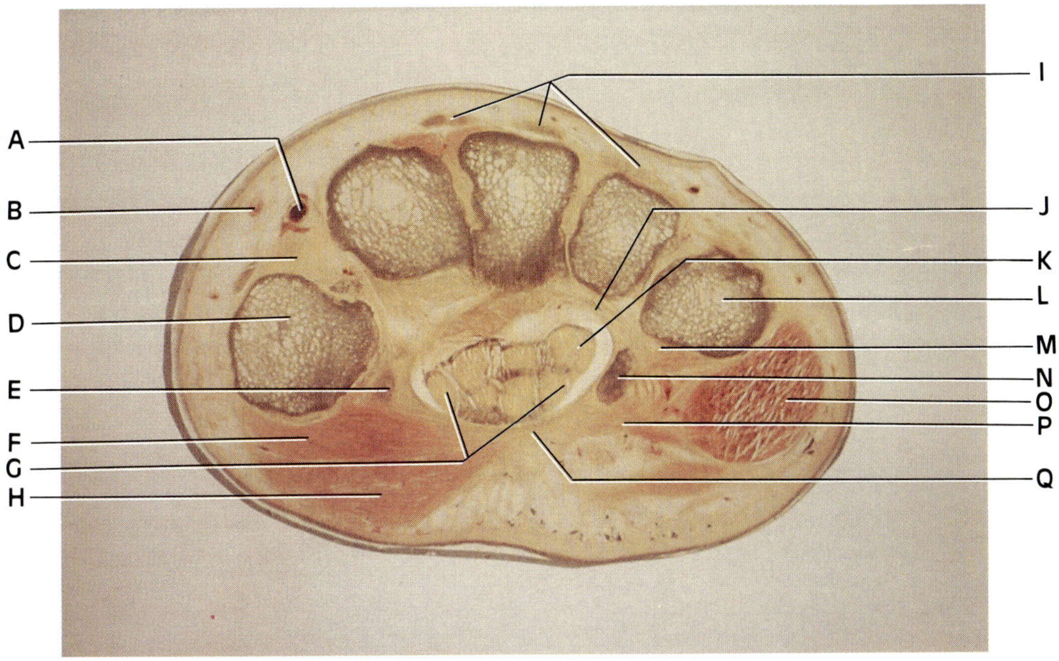

A. First dorsal metatarsal artery
B. First dorsal metatarsal vein
C. Dorsal interosseous muscles
D. First metatarsal bone
E. Long flexor tendon of great toe (flexor hallucis longus tendon)
F. Short flexor muscle of great toe (flexor hallucis brevis muscle)
G. Short flexor tendons of toes (flexor digitorum brevis tendons)
H. Abductor muscle of great toe (abductor hallucis muscle)
I. Extensor tendons of the toes (extensor digitorum tendons)
J. Fibrous sheath of flexor tendons
K. Long flexor tendons of toes (flexor digitorum longus tendons)
L. Fifth metatarsal bone
M. Plantar interosseous muscles
O. Abductor muscle of little toe (abductor digiti minimi)
P. Short flexor muscle of little toe (flexor digiti minimi brevis muscle)
Q. Plantar aponeurosis

Figure 21-5
Plastinated transverse section of the foot near the distal end of the metatarsal bone, showing musculoskeletal and neurovascular unit structures.

NEUROVASCULAR UNIT (see Fig. 21-5)

I. Arteries

First Dorsal Metatarsal Artery (*A*). A rounded, hollow vascular structure between the upper parts of the first and second metatarsal bones. It is inferior and lateral to the first dorsal metatarsal vein.

Origin. Dorsalis pedis artery.

Distribution. It shares blood with the digital branches for the adjacent sides of the first two toes.

Course. It courses from its origin between the superior parts of the first and second metatarsal bones.

II. Veins

First Dorsal Metatarsal Vein (*B*). A venous structure within the superficial fascia, superior and medial to the first dorsal metatarsal artery.

Origin. Dorsal digital veins.

Termination. Dorsal venous arch.

Course. It courses in the superficial fascia above the digit and terminates in the dorsal venous arch.

ENCLOSING UNIT (see Fig. 21-5)

Fibrous Sheath of the Flexor Tendons (*J*). A circular aponeurotic canal within the center of this transverse section of the foot. The wall of this canal surrounds the flexor tendons of the foot.

Plantar Aponeurosis (*Q*). A rather thick sheet of connective tissue medial to the abductor muscle of the little toe and lateral to short flexor muscle of the great toe. It is posterior to the flexor tendons of the foot.

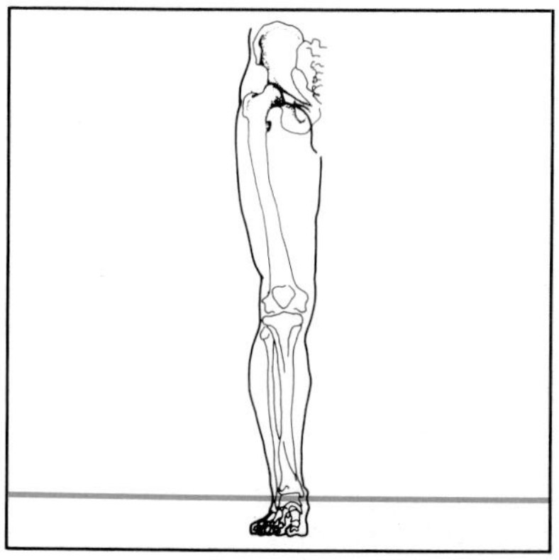

Figure 21–6 shows a magnetic resonance transverse image of the foot.

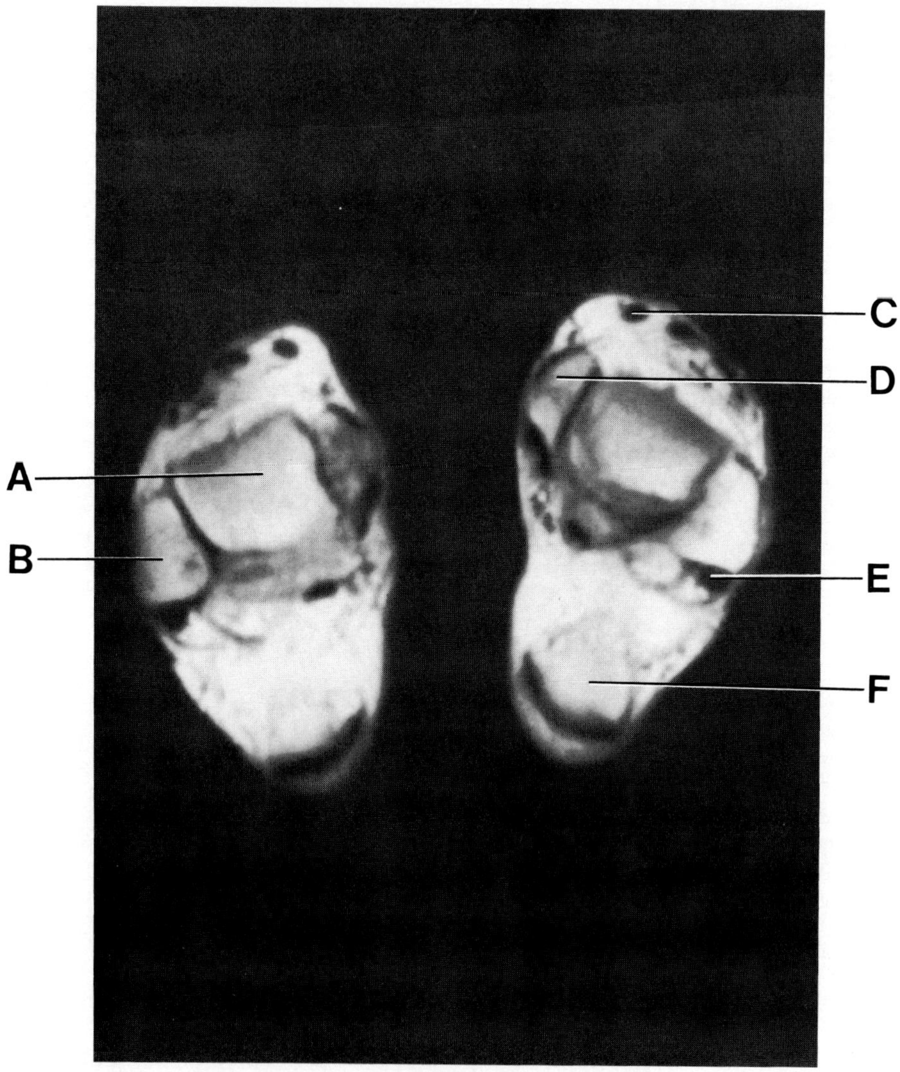

A. Talus bone
B. Lateral malleolus
C. Anterior tibial tendon
D. Medial malleolus
E. Long peroneal tendon
F. Calcaneus bone

Figure 21–6
Magnetic resonance transverse images of the foot, showing some of its structures.

CHAPTER 22

Coronal (Frontal) Cadaver Section and Image of the Lower Limb

This coronal cadaver section shows proximal portions of the lower limb and structures adjacent to both. The structures in this chapter vary from the same structures of other views. Superior and inferior as well as medial and lateral relationships are compared. For further information on other views, see the appropriate chapters.

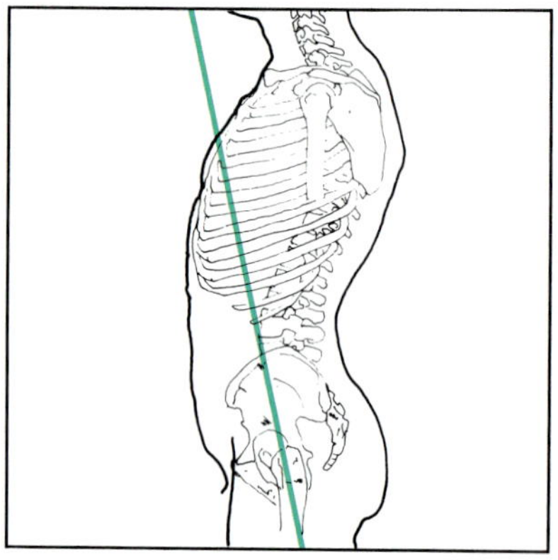

MODULE 1

This is a coronal (frontal) section of the pelvis and the upper part of the lower limbs. Lower limb structures are discussed in this module. Some of the structures seen in this plane are the head of the femur, pectineus muscle, iliopsoas muscle, external iliac artery, and femoral nerve.

MUSCULOSKELETAL UNIT (Fig. 22–1)

I. Skeletal Portion

Head of the Femur (*I*). A rounded, proximal bone part of the femur seen within the acetabulum. This connection of the head of the femur and the acetabulum forms the hip joint.

Epiphyseal Line (*K*). A transversely placed, rather thin bone slice near the superior part of the head of the femur and superior to the neck of the femur. This feature represents a synostosis joint.

Neck of the Femur (*L*). A slightly constricted portion of the proximal epiphysis of the femur, superior and slightly medial to the greater trochanter.

Greater Trochanter (*M*). A large, lateral projection of the femur inferior and lateral to its neck.

Superior Part of the Diaphysis (Shaft) of the Femur (*P*). A double slice of compact bone vertically placed with a medullary (marrow) cavity between these bone slices.

Medullary (Marrow) Cavity (*Q*). A vertical space between two perpendicular slices of compact bone tissue.

Acetabulum (*J*). An inward, oval, cup-shaped pelvic bone feature adjacent to the head of the femur. This structure is referred to as a socket and ball and socket-type of synovial joint of the hip.

II. Muscular Portion

Adductor Magnus Muscle (*N*). An oval aggregate of muscle tissue inferior to the obturator internus and medial to the pectineus muscles. It is superior to the adductor brevis muscle.

Pectineus Muscle (*O*). A vertical, oblong, bilateral muscle aggregate medial to the iliopsoas muscle and lateral to the great adductor muscle. It is superior to the lateral part of the short adductor muscle.

Iliopsoas Muscle (*C*). A vertically elongated, bilateral muscle slice lateral to the pectineus muscle and adjacent to the medial part of the proximal femur opposite the greater trochanter.

Adductor Brevis Muscle (*N*). A rather large, oblong (medial-to-lateral), bilateral muscle aggregate inferior to the great adductor muscle, as seen in this coronal view.

Gracilis Muscle (*D*). A vertical strip of bilateral muscle tissue medial to both the great adductor and short adductor muscles.

Vastus Lateralis Muscle (*E*). A vertical, rather wide, bilateral muscle slice inferior to the greater trochanter and lateral to the vastus intermedius muscle.

Vastus Intermedius Muscle (*G*). A rather narrow, triangular, bilateral muscle aggregate seen immediately medial to the vastus lateralis muscle.

Vastus Medialis Muscle (*H*). A rather narrow, triangular, bilateral muscle slice immediately medial to the medial part of the femur and inferior to the iliopsoas muscle.

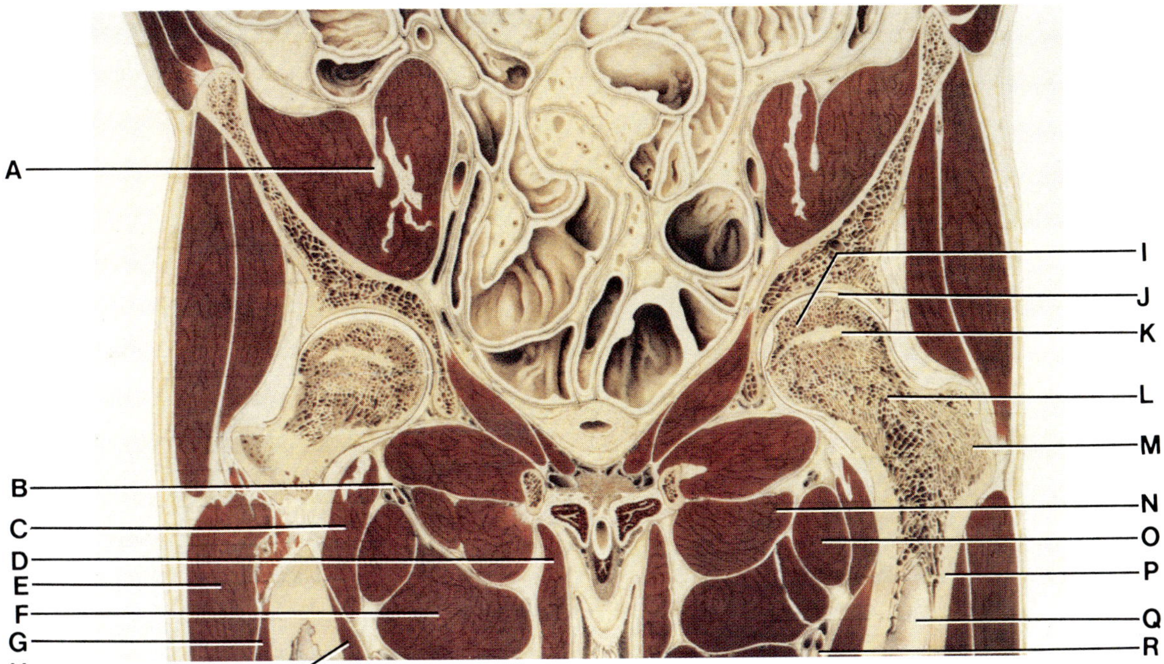

A. Femoral nerve
B. Medial femoral circumflex artery
C. Iliopsoas muscle
D. Gracilis muscle
E. Vastus lateralis muscle
F. Adductor brevis muscle
G. Vastus intermedius muscle
H. Vastus medialis muscle
I. Head of femur
J. Acetabulum
K. Epiphyseal line
L. Neck of femur
M. Greater trochanter
N. Adductor magnus muscle
O. Pectineus muscle
P. Superior part of diaphysis (shaft) of femur
Q. Medullary (marrow) cavity
R. Profunda femoris artery

Figure 22–1
Approximately midcoronal section of a male, highlighting some of the lower limb structures.

NEUROVASCULAR UNIT (see Fig. 22–1)

I. Arteries

Medial Femoral Circumflex Artery (*B*). An oval, hollow, bilateral vascular structure between the great adductor and the pectineus muscles. This vessel is medial to the iliopsoas muscle.

Origin. Profunda femoris artery.

Distribution. This artery disperses blood to the following branches: (1) acetabular, (2) transverse, (3) ascending, and (4) twigs (lateral epiphyseal arteries).

Course. It passes posteriorly between the pectineus and psoas muscles to reach the acetabulum.

Profunda Femoris Artery (*R*). An oval, oblique, hollow, bilateral vascular structure medial to the vastus intermedius and lateral to the short adductor muscles.

Origin. It arises from the proximal part of the femoral artery.

Distribution. This artery distributes blood to the following branches: (1) muscular, (2) perforating, (3) medial femoral circumflex, and (4) lateral femoral circumflex.

Course. It passes medial from the posterior part of the femoral artery and becomes adjacent to the medial side of the femur next to the short and great adductor muscles. It pierces the great adductor where it terminates.

II. Nerves

Femoral Nerve (*A*). A short, vertical, bilateral mixed nerve structure between the iliac and psoas major muscles. The largest branch of the lumbar plexus is the femoral nerve.

Origin. Lumbar plexus (mainly the anterior primary ramus of lumbar nerve 4).

Distribution. Iliac, sartorius, quadriceps, and pectineus muscles and anterior and medial skin of the thigh.

Course. It travels down a groove between the iliac and psoas major muscles and exits from the pelvis posterior and deep to the middle part of the inguinal ligament to reach the thigh. It enters the femoral triangle lateral to the femoral vessels and begins to form many branches.

CHAPTER 23

Sagittal Cadaver Sections and Images of the Lower Limb

The two sagittal illustrations of this chapter have combined anatomic units (musculoskeletal, neurovascular, and enclosing). One picture shows structures at the knee, distal thigh, and proximal leg. The other picture shows structures at the ankle joint, foot, and distal leg.

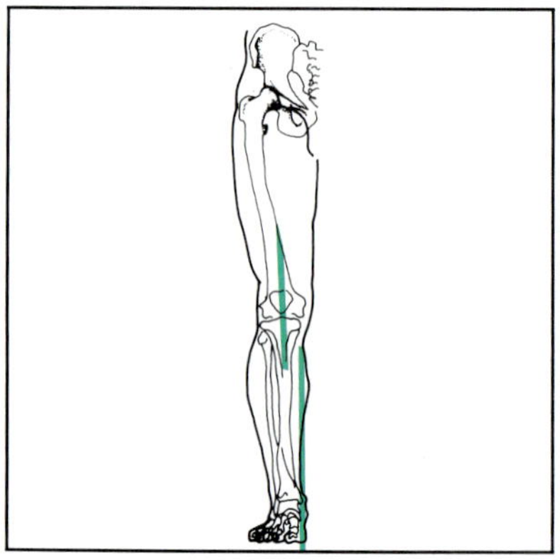

MODULE 1

This module includes two sagittal sections. One sagittal section shows structures of the knee, such as the subcutaneous prepatellar bursa, anterior cruciate ligament, patellar ligament, vastus lateralis muscle, peroneal nerve, and popliteal artery. The other sagittal section is of the foot, showing the ankle joint, calcanean (Achilles) tendon, posterior tibial artery, talus, navicular bone, and lumbrical muscle.

MUSCULOSKELETAL UNIT (Fig. 23–1)

I. Skeletal Portion

Femur (*E*). A vertical, obliquely positioned bone between the vastus intermedius and vastus medialis muscles. The inferior end is anterior to the superior end. Its distal end is adjacent to the knee joint.

Patella (*H*). An oblong bone slice anterior to the knee joint and superior to the patellar ligament. This is a sesamoid bone, and it is embedded within the tendon of the quadriceps muscle.

Tibia (*P, h*). A vertical bone slice inferior to the knee joint and anterior to the plantar muscle. It is anterior to the soleus muscle.

Distal End of Tibia. A bone section adjacent to the talus and anterior to the calcanean tendon.

Ankle (Talocrural) Joint (*f*). A transverse junction between the distal end of the tibia and the proximal part of the talus.

Talus Bone (*d*). An irregularly rectangular bone section superior to the calcaneus and posterior to the navicular bone.

Calcaneus Bone (*m*). A cone-shaped, obliquely placed bone slice inferior to the talus and posterior to the abductor muscle of the little toe, as seen in this sagittal view. The inferior end is posterior to the superior end and forms the heel.

Navicular Bone (*b*). A short, rectangular bone slice anterior to the talus and posterior to the medial cuneiform bone.

Medial Cuneiform Bone (*Z*). A quadrilateral bone section anterior to the navicular bone and posterior to the metacarpal bone.

Metatarsal Bone (*V*). A rather long foot bone section anterior to the medial cuneiform bone and posterior to the proximal phalanx. The distal end is slightly smaller than the proximal end.

Proximal Phalanx (*T*). A rather thin bone slice anterior to the metacarpal bone. The distal end is narrower than the proximal end.

II. Muscular Portion

Vastus Lateralis Muscle (*B*). A vertical, rather wide, superficial anterior muscle section located anterior to the vastus intermedius muscle and superior to the patella. The superior part is wider than the inferior part.

Vastus Intermedius Muscle (*C*). A vertical and obliquely positioned muscle strip posterior to the vastus lateralis. It is anterior to the femur. Its inferior end is anterior to the superior end.

Vastus Medialis Muscle (*A*). A vertical, obliquely placed, thin muscle strip posterior to the tibia and anterior to the short head of the biceps femoris muscle. The inferior end is more anterior than the superior end.

Short Head of the Biceps Femoris Muscle (*D*). A vertical and obliquely positioned, moderately wide muscle strip posterior to the vastus medialis muscle, as seen in this sagittal view, and anterior to the long head of the biceps femoris. The inferior end is wider and more anterior than the superior end.

Long Head of the Biceps Femoris Muscle (*F*). A vertical and obliquely placed, rather wide, elongated muscle section. Its inferior end is narrower and more anterior than the superior end.

Posterior Tibial Muscle (*S*). A vertical strip of muscle tissue immediately posterior to the shaft of the tibia, as seen in this sagittal section.

Soleus Muscle (*Q*). A vertical, rather large muscle slice posterior to the posterior tibial muscle and anterior to the gastrocnemius muscle.

Gastrocnemius Muscle (*R*). A vertical, elongated, superficial, posterior muscle slice posterior to the soleus and the plantar muscle.

Plantar Muscle (*N*). An oblong (superior-to-inferior) muscle slice anterior to the superior part of the gastrocnemius muscle and posterior to the knee joint.

Tendon of the Long Flexor Muscle of the Great Toe (Flexor Hallucis Longus) (*U*). An elongated, tendinous structure perpendicular to the foot bones. This structure extends from the calcaneus to the proximal phalanx. As it courses in the plantar part of the foot, it runs superior to the flexor muscle of the toes, medial plantar nerve, tendon of the long flexor muscle of the toes, quadrate muscle of the sole (quadratus plantae), adductor muscle of the little toe, and the lumbrical muscle. It runs inferior to the short flexor muscle of the great toe.

Short Flexor Muscle of the Great Toe (Flexor Hallucis Brevis) (*W*). A rather short (anterior-to-posterior) strip of muscle tissue superior to the tendon of the long flexor muscle of the great toe and inferior to the metacarpal bone.

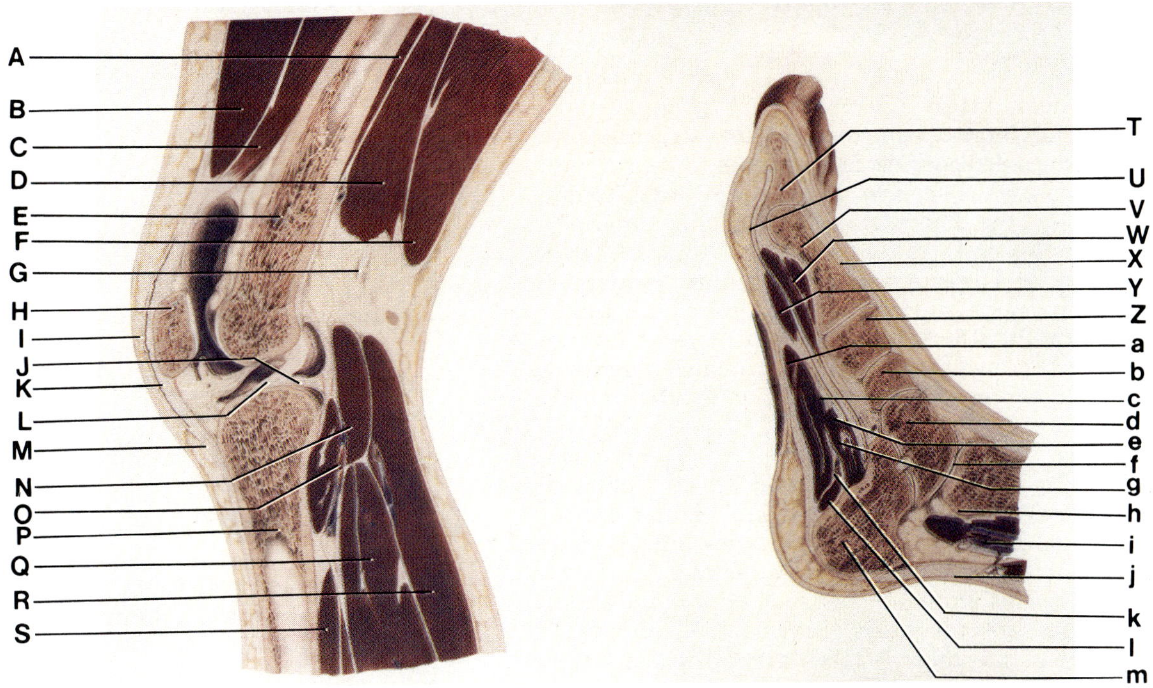

A. Vastus medialis muscle
B. Vastus lateralis muscle
C. Vastus intermedius muscle
D. Short head of biceps femoris muscle
E. Femur
F. Long head of biceps femoris muscle
G. Common peroneal (lateral popliteal) nerve
H. Patella
I. Subcutaneous prepatellar bursa
J. Lateral meniscus
K. Patellar ligament (ligamentum patellae) (upper part)
L. Anterior cruciate ligament
M. Patellar ligament (ligamentum patellae) (lower part)
N. Plantar muscle
O. Popliteal artery
P. Tibia
Q. Soleus muscle
R. Gastrocnemius muscle
S. Posterior tibial muscle
T. Proximal phalanx
U. Tendon of long flexor muscle of great toe (flexor hallucis longus)
V. Metatarsal bone
W. Short flexor muscle of great toe (flexor hallucis brevis)
X. Tendon of long extensor muscle of great toe (extensor hallucis longus)
Y. Lumbrical muscle
Z. Medial cuneiform bone
a. Short flexor muscle of the toes (flexor digitorum brevis muscle)
b. Navicular bone
c. Medial plantar nerve
d. Talus bone
e. Tendon of long flexor muscle of toes (flexor digitorum longus)
f. Ankle (talocrural) joint
g. Quadratus muscle of sole (plantae)
h. Tibia
i. Posterior tibial artery
j. Calcanean (Achilles) tendon
k. Lateral plantar artery
l. Abductor muscle of little toe (digiti minimi)
m. Calcaneus bone

Figure 23–1
Sagittal sections at the level of the knee and foot, highlighting some of their structures.

Lumbrical Muscle (*Y*). A rather short, oblong muscle slice with anterior and posterior tapering ends, located inferior to the tendon of the long flexor muscle of the great toe.

Short Flexor Muscle of the Toes (Flexor Digitorum Brevis) (*a*). A moderately long (anterior-to-posterior) muscle strip inferior to the quadratus muscle of the sole and anterior to the abductor muscle of the little toe.

Tendon of the Long Flexor Muscle of the Toes (Flexor Digitorum Longus) (*e*). A rather short (anterior-to-posterior) strip of tendinous tissue inferior to that part of the tendon of the long flexor muscle of the great toe near its posterior end.

Abductor Muscle of the Little Toe (Digiti Minimi) (*l*). An oblong, obliquely placed muscle aggregate anterior to the calcaneus and posterior to the lateral plantar artery. The inferior end is more posterior than the superior end.

Tendon of the Anterior Tibial Muscle. A vertical, obliquely placed tendinous tissue seen in this foot sagittal section, crossing anterior to the distal end of the tibia, ankle joint, and tarsal bones. The inferior end is more anterior than the superior end.

Tendon of the Long Extensor Muscle of the Great Toe (Extensor Hallucis Longus) (*X*). A strip of tendinous tissue that starts in this sagittal section of the foot, superior to the tarsal bones, and extends to the phalanx of the great toe.

Calcanean (Achilles) Tendon (*j*). A vertical strip of tendinous tissue posterior to the ankle joint, which attaches to the posterior part of the calcaneus. This tendon represents tissue as an extension of the gastrocnemius, soleus, and plantar muscles.

NEUROVASCULAR UNIT (see Fig. 23–1)

I. Arteries

Popliteal Artery (*O*). A vertical, obliquely positioned, moderately elongated vascular structure anterior to the soleus muscle and superior to the posterior tibial muscle, as seen in this slightly flexed, sagittal knee joint area. The inferior end is slightly anterior to the superior end.

Posterior Tibial Artery (*i*). A vertical, moderately elongated vessel anterior to the calcanean tendon.

Lateral Plantar Artery (*k*). A moderately elongated (anterior-to-posterior) vascular structure anterior to the abductor muscle of the little toe and inferior to the quadratus muscle of the sole.

Origin. Posterior tibial artery.

Distribution. It disperses blood to the following branches: (1) cutaneous to the lateral plantar skin, (2) medial calcanean, (3) muscular arteries, and (4) terminal branches for the deep plantar arch.

Course. This artery advances superior to the abductor muscle of the great toe and the short flexor muscles of the toes. It takes a course in a groove between the foregoing muscles and the abductor muscle of the little toe. It appears perpendicular with the lateral plantar nerve and terminates in the plantar arch.

II. Nerves

Common Peroneal (Lateral Popliteal) Nerve (*G*). A vertical strip of nerve tissue posterior to the distal end of the femur and anterior to the tendon of the long head of the biceps femoris muscle. It is seen in this sagittal view of the knee joint area inferior to the short head of the biceps femoris muscle.

Medial Plantar Nerve (*c*). A rather small, slightly elongated (anterior-to-posterior) mixed nerve tissue superior to the short flexor muscle of the toes and anterior to the tendon of the long flexor muscle of the toes.

Origin. Tibial (medial popliteal) nerve.

Distribution. It distributes blood to the following branches: (1) cutaneous branches to the plantar skin, (2) muscular branches, and (3) digital branches.

Course. It courses superior to the abductor muscle of the great toe of the sole of the foot and comes to lie in a groove between the short flexor muscle of the toes and the abductor muscle of the great toe. It runs perpendicular to the medial plantar artery before dividing into digital branches adjacent to the heel.

ENCLOSING UNIT (see Fig. 23–1)

Subcutaneous Prepatellar Bursa (*I*). A vertical, curved, narrow space anterior to the tendon of the quadriceps, patella, and patellar tendon. In this slightly flexed sagittal section of the knee joint area, the concave part of this feature is posterior.

Patellar Ligament (Ligamentum Patellae) (*K, M*). A vertical, oblique strip of tendinous tissue that extends from the patella to the tibial tuberosity. The inferior end in this slightly flexed sagittal section of the knee joint area is posterior to the superior end.

Anterior Cruciate Ligament (*L*). A transverse and obliquely positioned ligamentous tissue seen between the distal end of the femur and the proximal end of the tibia. The posterior end is slightly superior to the anterior end.

Figure 23–2 is a magnetic resonance sagittal image of the lower limb. This image passes through the ankle joint.

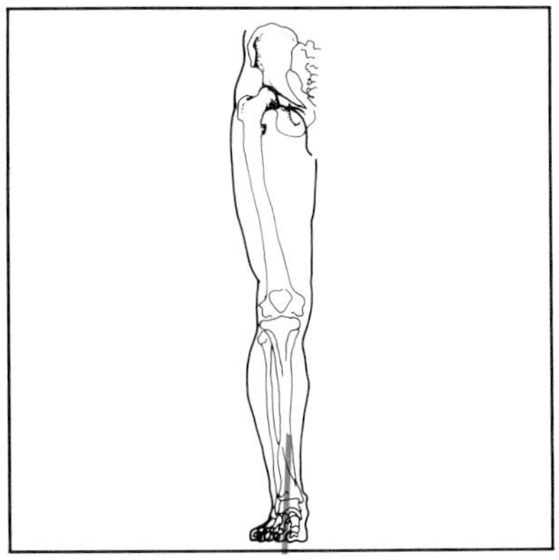

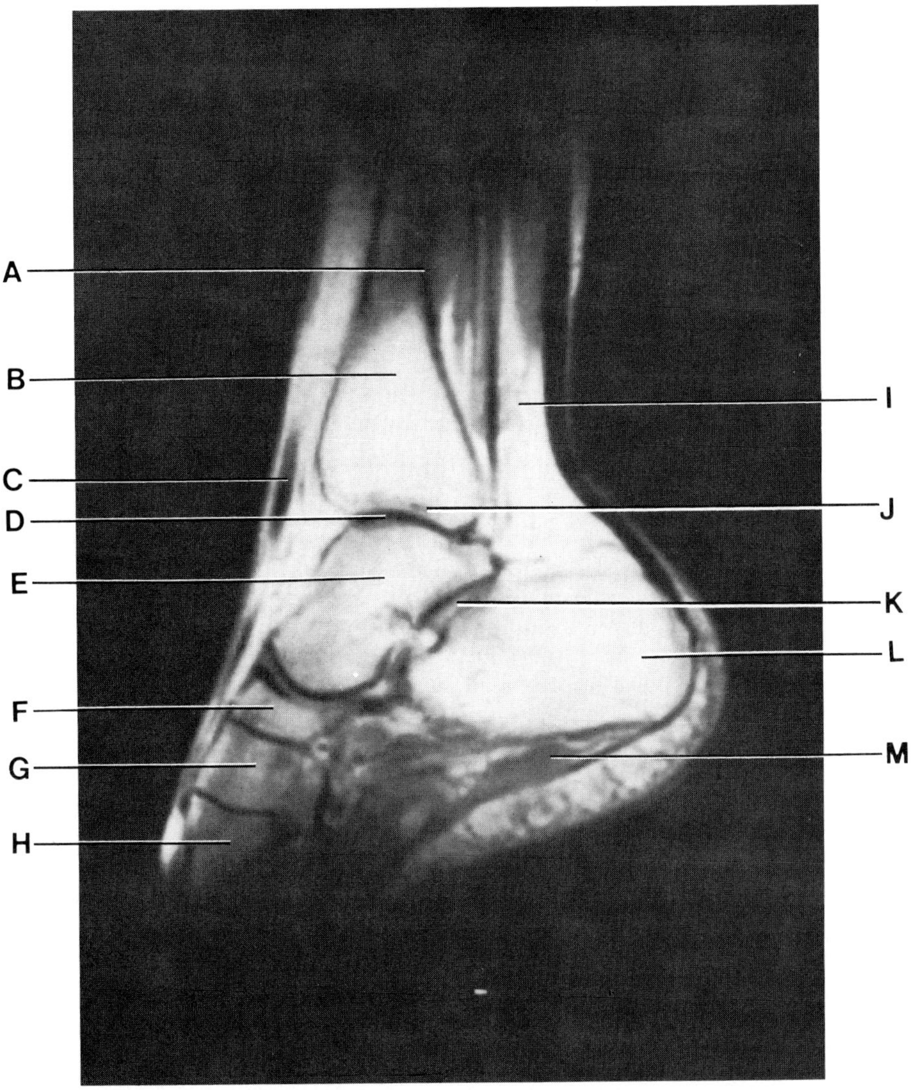

A. Posterior tibial muscle
B. Tibia
C. Long extensor tendon of toes (extensor digitorum longus tendon)
D. Tibiotalar joint
E. Talus bone
F. Navicular bone
G. Lateral cuneiform bone
H. Second metatarsal bone
I. Achilles tendon
J. Distal end of tibia
K. Posterior talocalcaneal joint
L. Calcaneus bone
M. Quadratus muscle of sole (plantae)

Figure 23–2
Magnetic resonance sagittal image of the lower limb, showing some of the associated structures.

Appendix to Part VIII: Clinical Correlations for the Lower Limb Region

CASE ONE (Fig. AVIII–1)

History. The patient was injured in the right knee. Now there is persistent swelling and pain in the knee. The patient had no previous surgery.

Findings. The anterior cruciate ligament is completely torn from its femoral attachment.

A minimal degree of fluid is seen within the knee joint. No abnormality is seen in the suprapatellar tendon, patellar position, patellar cartilage thickness, infrapatellar tendon, and both collateral ligaments of the knee, as well as in the lateral meniscus.

A minimal high signal in the posterior horn of the lateral meniscus may suggest borderline degenerative change. A linear high signal is visualized in the posterior horn of the medial meniscus that does not seem to reach its surface. This is compatible with a degenerative lesion rather than a tear.

Impression. Torn anterior cruciate ligament with minimal fluid within the knee joint. In addition, the free margin of the medial meniscus is slightly blunted. Minimal signal in this posterior horn of the medial meniscus also suggests early degenerative change.

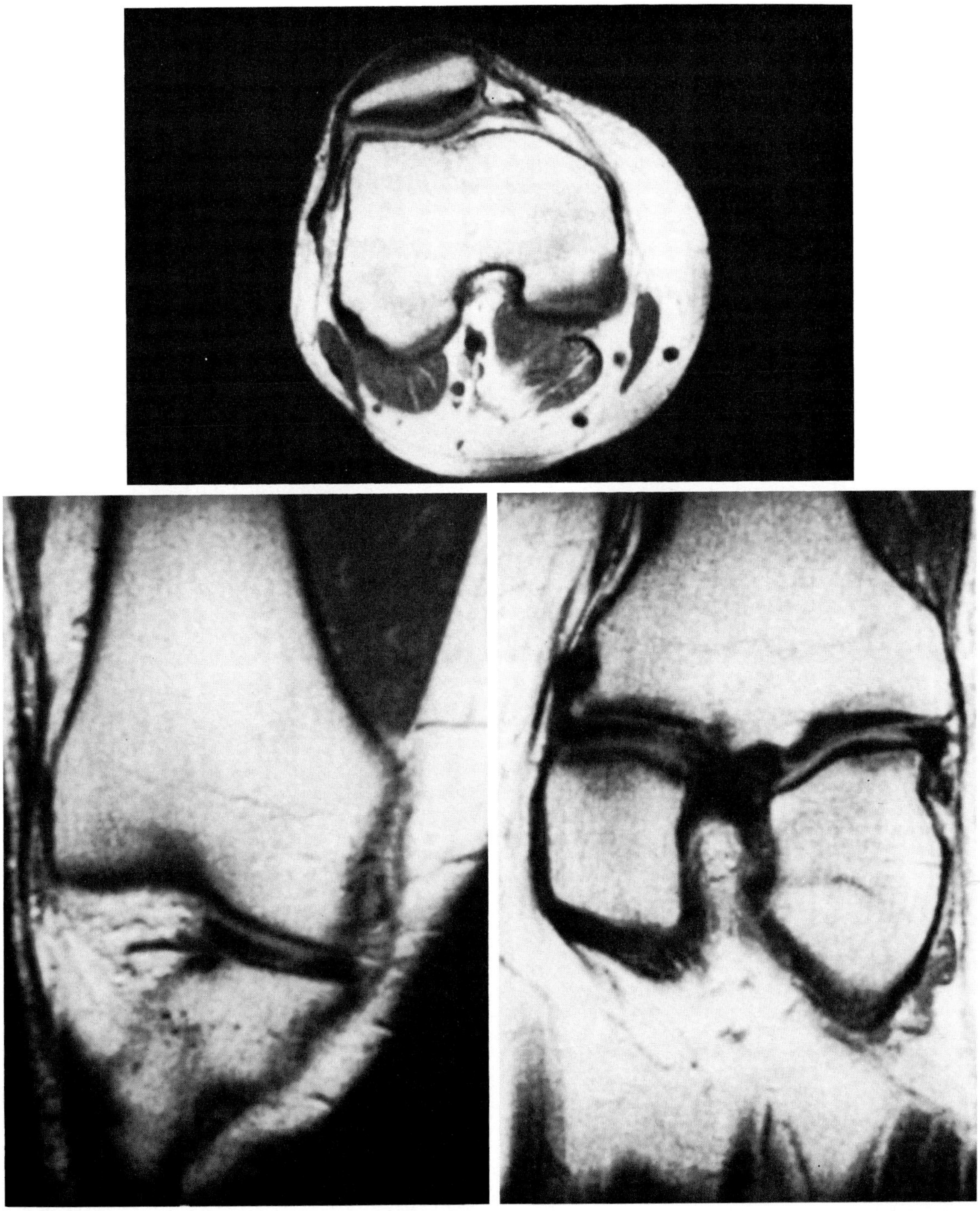

Figure AVIII–1
Magnetic resonance images, showing that the anterior cruciate ligament is completely torn from its femoral attachment. *A*, Transverse view; *B* and *C*, coronal views.

CASE TWO (Fig. AVIII–2)

History. The patient was examined regarding the status after surgery on the left knee several years ago, in which a piece of cartilage was removed. Now the left knee intermittently gives out and swells up and sometimes there is inability to straighten the left knee. The patient reports pain in the medial aspect of the patella. The patient does not recall any recent trauma to the knee.

Findings. A large, irregular high signal passing through the posterior portion of the medial meniscus, reaching its undersurface axial and coronal images.

Moderate fluid is seen within the knee joint.

The lateral meniscus shows no definite abnormality; however, a tiny low signal in the medial anterior aspect of the lateral meniscus suggests a loose cartilaginous or bone fragment. This may be a migrated cartilaginous piece from the medial meniscal tear. Clinical correlation is suggested. Anterior and posterior cruciate ligaments, both collateral ligaments of the knee, the suprapatellar tendon, the infrapatellar tendon, patellar cartilage thickness, and patellar position show no definite abnormality.

Impression. Large tear involving the posterior medial meniscus, with moderate fluid within the knee joint. A tiny loose body within the lateral compartment of the knee. A local linear scar is visualized in the medial aspect of the infrapatellar tendon, compatible with previous surgery.

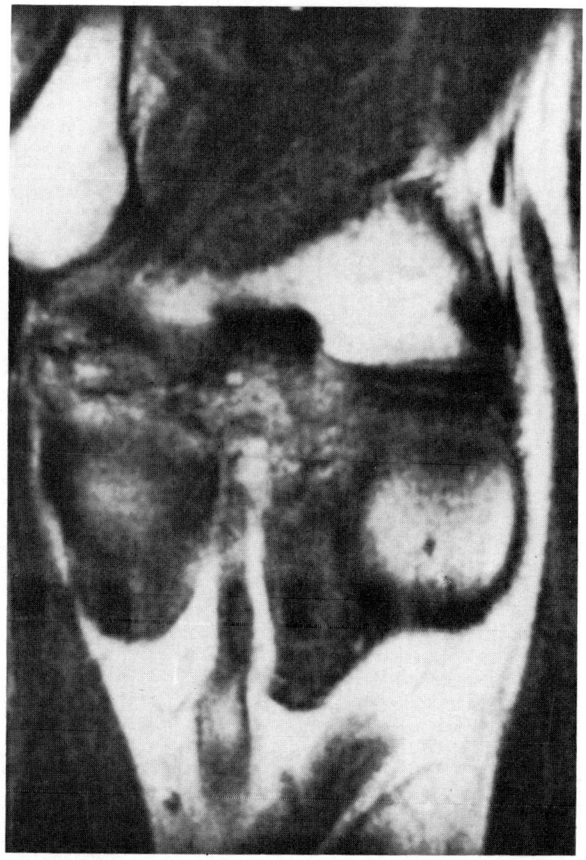

Figure AVIII–2
Magnetic resonance coronal image, showing a large tear involving the posterior medial meniscus, with moderate fluid within the knee joint.

Bibliography

Alfidi, Ralph J., et al. *Computed Tomography of the Human Body: An Atlas of Normal Anatomy.* St. Louis: C. V. Mosby Company, 1977.

Burk, D. Lawrence, et al. Acetabular fractures: Three-dimensional computed tomographic imaging and interactive surgical planning. *Journal of Computed Tomography* 10: 1–10, 1986.

Cholankeril, V. John, et al. Computed tomography in urothelial tumors of the renal pelvis and related filling defects. *Journal of Computed Tomography* 10: 263–272, 1986.

Christoforidis, A. John. *Atlas of Axial, Sagittal, and Coronal Anatomy with CT and MRI.* Philadelphia: W.B. Saunders Company, 1988.

Han, Man-Chang, and C.-W. Kim. *Sectional Human Anatomy, Correlated with CT and MRI.* Seoul, Korea: Ilchokak, 1985.

McLoud, C. Theresa, Jack Wittenberg, and Joseph T. Ferrucci. Computed tomography of the thorax and standard radiographic evaluation of the chest: A comparative study. *Journal of Computer Assisted Tomography* 3: 170–180, 1979.

Meizner, Israel. Sonographic observation of in utero fetus. *Journal of Ultrasound Medicine* 6: 111, 1987.

Mendelson, S. David, et al. Computed tomography findings of tuberculous mediastinitis. *Journal of Computed Tomography* 11: 123–127, 1987.

Nakata, Hajime, et al. Computed tomography for preoperative evaluation of lung cancer. *Journal of Computed Tomography* 10: 147–151, 1986.

Peterson, Roy R. *A Cross-Sectional Approach to Anatomy.* Chicago: Year Book Medical Publishers, 1980.

Rapoport, L. Judith. The biology of obsessions and compulsions. *Scientific American* Mar. 1989, pp. 83–89.

Rennels, D. Glenn, and Edward H. Shortliffe. Advanced computing for medicine. *Scientific American* Oct. 1987, pp. 154–161.

Roy, P. *Ultrasound of the Abdomen.* (Translated by Marie Wackenhem.) Berlin, Springer-Verlag, 1988.

Suslick, S. Kenneth. The chemical effects of ultrasound. *Scientific American* Feb. 1989, pp. 80–86.

Truex, C. Raymond, and Malcolm B. Carpenter. *Human Neuroanatomy.* Baltimore: Williams and Wilkins Company, 1969.

Wagner, Marvin, and L. Lawson Thomas. *Segmental Anatomy; Application to Clinical Medicine.* New York: Macmillan Publishing Company, 1982.

Wragg, Laurence E. *Neck. Part I: Orientation, Fascia, Superficial Structures.* Evanston, Illinois: Medical Teaching Films: Cine Projector Series, 1968.

Glossary

abduction—a process of drawing or moving a body part away from median.

abscess—a collection of pus and fluid in a local area in the form of a cavity derived from disintegrated tissue.

adduction—a process of drawing or moving a body part toward the median.

afferent—a carrying or conveying structure toward another structure.

amphiarthrodial—pertaining to a double junction or joint, such as joints with a fibrocartilage disk between the bone ends.

anastomose—the joining of, or to communicate with, two structures, such as one artery with another artery.

aneurysm—an abnormal widening of a vessel such as an artery.

angiography—a record or picture of a vessel as revealed by an imaging modality.

anterior—toward the front.

aortic hiatus—an opening in the diaphragm for the passage of the aorta, thoracic duct, and azygos vein.

aponeurosis—a broad sheet of tendon for the attachment of a skeletal muscle.

aqueduct of Sylvius—an elongated vertical tubular structure within the midbrain.

arachnoid—a meninge with a spider web–like appearance deep to the dura mater.

arcuate line—a curved feature of the internal anterior abdominal wall below the level of the umbilicus.

areolar—a tissue structure with small spaces.

atlanto-occipital membrane—a sheet of connective tissue that extends from the atlas to the occipital bone.

axial—pertaining to a feature or section perpendicular to the median.

axillary—pertaining to the armpit.

belly—the bulge of a skeletal muscle between its origin and insertion.

brachial—pertaining to the arm.

bursa—a sac-like synovial fluid–filled structure associated with joints and that provides a means for tendons to slide without bone contact.

cancellous—spongy or soft bone tissue.

capitular—pertaining to a head-like or head structure.

carotid triangle—one of the smaller anterior cervical triangles.

cartilaginous joint—a junction having a cartilage disk between bone ends.

cavernous sinus—a vascular structure formed by a fold of the dura mater.

central sulcus—the same as fissure of Rolando, which is a deep groove between the frontal and parietal lobes.

cerebellum—a part of the brain smaller than the cerebrum and posterior to the pons and medulla oblongata.

cerebrum—the largest part of the brain, with numerous folds (gyri) and grooves (sulci) on its surface.

cervical—pertaining to the neck, such as cervical vertebrae.

chordae tendineae—a feature of the heart that resembles tendons attached to both a papillary muscle and an atrioventricular valve cusp.

circle of Willis—a cerebral arterial circle at the base of the brain formed by the internal carotid and basilar arteries.

circumduction—a process of moving a part of the body in a circular fashion.

cleft—an elongated opening or fissure usually found in the fetus or embryo.

clinical image—a section or slice of the body generated by one of the imaging modalities.

commissure—a structure that unites two structures,

such as the anterior cerebral commissure connects the right and left cerebral structures.

compact—a class of hard bone tissue.

computed tomography—a computed record or section picture of the body generated by a computed tomography device.

concha—a shell-like bone feature of the ethmoid (superior and middle nasal conchae) on the lateral wall of the nasal cavity. The inferior nasal conchae are separate bones.

condyle—a knuckle-like articulating process usually found at the distal end of long bones.

coronal—a plane that divides the body or an organ into anterior and posterior portions.

corpus callosum—a major connecting tract of the cerebral hemispheres composed of white matter. This structure has three major parts: (1) body, (2) genu, and (3) splenium.

cortex—the outer layer of an organ, such as the cerebral cortex.

course—a route or path taken by a structure in the body to reach its destination, such as an artery, a vein, or a nerve.

cranial—pertaining to the direction toward the head.

crest—a ridge on a bone or some other structure.

demifacet—a slight depression that has the form of a half depression, that is, an imcomplete circular depression.

diaphysis—a part of a long or short bone between the proximal and distal epiphyses.

dimensional morphology—the study of the shape of a structure in a multiplane fashion.

diploë—a layer of spongy or cancellous bone sandwiched between the inner and outer tables of compact bone.

distal—pertaining to a direction toward the far end of a structure.

dorsal—pertaining to a direction toward the back.

dura mater—the tough, rather thick, outermost meninge that lines the inner wall of both the cranial and vertebral cavities.

dural sinus—a vascular structure formed by a fold of dura mater, such as the superior sagittal sinus.

edema—an abnormally large swelling with fluid, especially in subcutaneous tissue.

efferent—a carrying or conveying structure away from or out of another structure.

enclosing unit—a category of tissues and structure within a section, such as membranes, fossae, and spaces.

epigastrium—a region of the abdomen above the stomach.

falx cerebri—a fold of the meningeal portion of the dura mater that is found within the longitudinal cerebral fissure.

fascia—a band or sheet of tissue, usually fibrous, such as the hypodermis or subcutaneous tissue.

fibrous joint—the same as a synarthrodial joint, with many fibers between the bone ends. This class of joint shows no movement.

fissure—a slit-like feature of a bone or another structure.

fissure of Sylvius—the same as the lateral cerebral fissure found on the side of the brain between the temporal lobe below and the parietal and frontal lobes above.

foramen—a hole.

foramina—many holes (plural of foramen).

fossa—a depression, such as temporal fossa.

frontal—the same as coronal, that is, a plane that divides the body or an organ into anterior and posterior positions.

fundus—an anatomic term that refers to the base or bottom of organs. Sometimes it refers to the part of a hollow organ farthest from its entrance.

glomerulus—a ball-shaped cluster of capillaries within the kidney.

gutter—a vertical space within the abdomen, such as the gutter to the right of the mesentery.

gyrus—a fold of the cerebral cortex.

hematoma—a tumor with effused blood as its major component.

hemidiaphragm—one-half of a diaphragm.

hilum—the entrance of an organ for blood vessels and nerves.

hyaline—a cartilaginous, glassy, and transparent connective tissue.

hypochondriac—referring to a feature under or below the rib cartilage. Generally this is referred to as an abdominal region below the cartilage of the ribs, such as the right hypochondriac region.

inferior—referring to a direction toward the lower parts of the body.

infracolic—pertaining to a portion of the internal abdomen below the transverse colon.

infundibulum—a passage or connecting structure, usually funnel shaped, such as the infundibulum of the uterine tubes.

inguinal—referring to the groin area.

innervation—nerve supply to an organ.

inner table—an internal compact layer of flat bones.

insertion—the distal attachment of a skeletal muscle.

interictal—referring to something that occurs between attacks or paroxysms.

in vivo—in the living.

ischemic—referring to a suppression of blood flow.

lambdoid suture—same as parieto-occipital suture, that is, a seam-like feature between the parietal and occipital bones.

lateral—pertaining to a direction toward the side of the body.

lateral line—a vertical imaginary line midway between the median and the lateral surface of the body.

linea alba—a white line formed by connective tissue and found in the median of the abdomen between the right and left rectus abdominis muscles.

lumbar—referring to bones of the back and also an abdominal region.

lumen—a hollow, elongated cavity within hollow organs.

lymphadenopathy—disease of lymph nodes that causes a glandular-like swelling.

magnetic resonance imaging—a computed body scanning modality that generates views of body slices using a magnetic field and radio waves.

meatus—a canal in a bone.

median—the middle.

mediastinum—middle region of the thorax.

meningeal—referring to a meningeal feature or structure associated with membranes (meninges) of the cranial and vertebral cavities.

meniscus—a crescent-shaped disk of fibrocartilage found on the superior articular surface of the tibia.

mesentery—a fold of membrane that attaches the small intestine to the posterior abdominal wall.

mesocolon—a peritoneal membranous structure that attaches colon segments to the dorsal abdominal wall. Each mesocolon is named based on the segment of colon it attaches to the abdominal wall. For example, sigmoid mesocolon attaches the sigmoid colon to the posterior wall, and transverse mesocolon attaches the transverse colon to the posterior abdominal wall.

metastasis—the transfer of a disease agent from a primary site to a secondary site within the body.

midaxillary line—a midcoronal plane that divides the body into equal anterior and posterior portions.

midclavicular line—a vertical imaginary line that divides one half of the body into two equal portions.

midcoronal—a plane showing equal anterior and posterior portions.

midshaft—the midway point between the proximal and distal end of a long or short bone.

Morrison's pouch—same as the right subhepatic space, that is, the space posterior to the right lobe of the liver.

multiplane sections—slices in at least three different planes.

musculoskeletal unit—a category of structures that includes muscle and bone structures.

necrosis—the death of tissue or cells in a localized area of the body.

needle biopsy—a small sample of living tissue or cells obtained by a needle for study.

neoplasm—new abnormal tissue growth, usually indicating a tumor.

neurovascular unit—a group of structures including nerves, arteries, veins, and lymphatic vessels.

origin—the beginning.

outer table—outer layer of compact bone as a part of flat bones.

paracolic—pertaining to structures and features, such as gutters, associated with the colon.

parasympathetic—a division of the autonomic nervous system, which counterbalances the action of the other division—the sympathetic division.

parenchyma—the functional elements located within an organ.

parietal—pertaining to the inner wall of a body cavity.

pectineal line—a faint marking on the superior pubis, which connects the arcuate line of the ilium with the pubic crest to form most of the pelvic inlet.

pericardium—a rather thick membrane that surrounds the heart.

pericranium—a connective tissue sheath that surrounds the outer skull.

periosteum—a membrane that surrounds bone and takes part in appositional growth of bone.

pia mater—a tender, delicate connective tissue membrane adjacent to the brain and spinal cord surfaces.

plastination—a process of preserving tissue, including sections, by four major steps: (1) fixation, (2) dehydration—defatting, (3) impregnation, and (4) curing.

pleura—a membrane associated with the chest wall (parietal pleura) and lungs (visceral pleura).

plexus—a network of nerves or vessels.

positron—a subatomic particle released by a radioactive substance.

positron emission tomography—a computer graphic picture or record of organs following exposure to a radioactive substance such as Carbon[14].

posterior—referring to a direction toward the back.

pouch of Douglas—a deep vertical space between the urinary bladder and rectum in the male.

protuberance—any marked projection.

proximal—referring to a direction where a structure or organ begins.

radiograph (radiogram)—a graphic film or record that results from the action of actinic rays on a surface that has been sensitized.

ramus—referring to a branch or smaller structure that extends from a larger structure.

rectouterine pouch—a deep vertical space between the uterus and rectum in the female.

Reid's baseline—a horizontal or transverse landmark of the head indicated by features such as the infraorbital rim, external auditory meatus, and occipital bone.

retropubic—referring to a feature posterior to the pubic bone.

roentgen—an international unit of x-ray or gamma radiation.

roentgenogram—a graphic film or record of organs or structures that results from x-ray or gamma radiation.

sagittal—a longitudinal plane that divides the body or one of its parts into right and left portions.

sectional anatomy—a study of the body using slices in a multiplane fashion.

septum—a partition.

sesamoid—referring to a bone or features that resembles a sesame seed.

sheath—a membranous enclosure for a structure or group of structures, such the carotid sheath encloses the neurovascular unit of the neck.

sheet plastinated specimen—a plastinated structure, organ, or slice that includes impregnation with epoxy-resin.

sialography—a graphic picture or record of the salivary gland ducts using x-ray or some other imaging modality.

sonography—a graphic picture or record of a structure, including a body section, which results from an ultrasound device.

subdural space—a narrow area immediately deep to the dura mater.

sulcus—a shallow groove.

superior—referring to a direction toward the upper parts of the body.

supination—the same as lateral rotation of the forearm.

supracolic—referring to an area of the abdominal cavity above the transverse colon.

suture—a seam-like joint, such as the sutures of the skull.

symphysis—a term that indicates a joint with a fibrocartilage disk between bone ends.

syndesmosis—a joint type, characteristically with a band or ligament between bone ends.

synergist—that which works together with another muscle, structure, or process.

synovial joint—a class of the most movable joints in the body.

tendon—a connective tissue structure that attaches a muscle to bone.

tentorium cerebelli—a fold of the meningeal portion of the dura mater located between the cerebellum and posterior cerebrum.

thoracic—referring to the chest area.

transtubercular plane—a horizontal landmark that marks the end of the abdomen and the most superior part of the pelvis.

transverse—referring to a direction across the body or an organ.

trochanter—a very large process, such as the greater trochanter of the femur.

tubercle—a small swelling or hump of a bone or some other organ.

tumor—a swelling or bump of neoplastic cells that continues and grows independently of its surrounding tissue.

ultrasound—a device for graphic pictures or records of body slices, organs, and structures based on sound waves.

ventral—referring to a direction to the anterior or belly part of the body.

ventricle—a small cavity within a hollow organ such as the heart or a space inside the brain.

ventriculography—a graphic picture or record of cavities inside the brain known as ventricles.

vermis—a worm-like median structure of the cerebellum between its right and left hemispheres.

visceral unit—a group of internal organs of a slice as used in this text.

viscus—any internal organ, especially of the abdominal cavity.

Wharton's duct—same as the submandibular duct, that is, a tube that conducts mucus and fluid from the submandibular salivary gland.

Illustration Credits

The authors wish to acknowledge the following sources for allowing us to use or adapt their figures:

Figures 4–1 to 4–4; 4–7 to 4–13; 4–16; 4–17; 6–22 to 6–25; 7–1; and 7–2 of the coronal (frontal) sections of the head region were redrawn by Patricia Borowiak of Triton College from Kampmeier OT, Cooper AR, Jones TS: A Frontal Section Anatomy of the Head and Neck. Urbana, Illinois, University of Illinois Press, 1957.

Figures 5–1; 5–2; 8–1; 8–2; 10–1; 10–2; 11–1; 11–2; 14–1; 16–1; 17–1; 19–1; 20–1; 22–1; and 23–1 of midsagittal cadaver sections were adapted from Braune W: Topographisch-Anatomischer Atlas: Nach Durchschnitten an gefrornen Cadavern, published in 1872.

Figures 3–14; 3–15; 3–17; 3–18; 3–21; 3–22; 3–25; 6–7 to 6–10; 9–1 to 9–4; 9–15 to 9–18; 12–18; 12–20; 12–22; 12–24; 15–8 to 15–10; 18–1; 18–3; 21–4; and 21–5 were shot from specimens obtained from the Carolina Biological Supply Company, Burlington, North Carolina.

Figures 3–1 to 3–5; 3–8; 3–9; 3–11; 3–27 to 3–31; 3–33 to 3–35; 6–1 to 6–4; 6–12 to 6–15; 6–17 to 6–20; 9–6 to 9–9; 9–11 to 9–14; 9–20 to 9–25; 9–27; 9–28; 9–30 to 9–33; 12–1 to 12–4; 12–6 to 12–9; 12–11; 12–12; 12–14; 12–15; 15–1; 15–2; 15–5 to 15–7; 18–2; and 21–1 to 21–3 of transverse cadaver sections are from sections of cadavers from the University of Illinois at Chicago.

Figures 9–26; 9–29; 12–13; 12–16; 12–19; 12–21; 12–23; 12–25; 15–3; 15–4; and 15–13 of CT images were obtained from the Department of Radiology (Imaging), Westlake Community Hospital, Melrose Park, Illinois.

Figures 2–1; 3–7; 3–13; 3–20; 3–37; 4–5; 4–6; 4–14; 4–15; Appendix II–1 to Appendix II–5; 6–5; 6–6; 6–11; 6–16; 6–21; 8–3 to 8–5; Appendix III–1, Appendix III–2; 9–5; 9–10; 9–19; 10–3; 11–3; 11–4; Appendix IV–1; Appendix IV–2; 12–5; 12–10; 12–17; Appendix V–1; Appendix V–2; 15–11; 15–12; 16–2; 16–3; Appendix VI–1; 19–2; Appendix VII–1; 21–6; 23–2; Appendix VIII–1; and Appendix VIII–2 are from MRI Chicago, Chicago, Illinois.

Figures 3–3; 3–6; 3–10; 3–12; 3–16; 3–19; 3–23; 3–24; 3–26; 3–32; 3–36; and 5–4 to 5–13 were obtained from the Department of Radiology (Imaging), Gottlieb Memorial Hospital, Melrose Park, Illinois.

Figure 2–3, the PET scan, is from Montgomery G: The mind in motion. DISCOVER magazine, Vol. 10, no. 3, p. 59, March 1989. Copyright © 1989 Discover Publications. Photograph courtesy of Mark Mintun, Mallinckrodt Institute of Radiology.

Figures 2–2; Appendix II–1; and 5–3 are borrowed from Debbie Lim.

Figures Appendix IV–1 to IV–3 are borrowed from Peg Garrity.

Some of the ultrasound images (Figs. 4–18 to 4–20; 5–14 to 5–16; and 15–14) were obtained from Ultrasound, Triton College, River Grove, Illinois.

Coronal and sagittal views (Figs. 13–1 to 13–4; and 17–2 to 17–5) are from Gower Medical Publishing, London and New York.

INDEX

Page numbers in *italics* refer to illustrations; page numbers followed by t indicate tables.